Martian Geomorphology

Geological Society books refereeing procedures

The Society makes every effort to ensure that the scientific and production quality of its books matches that of its journals. Since 1997, all book proposals have been refereed by specialist reviewers as well as by the Society's Books Editorial Committee. If the referees identify weaknesses in the proposal, these must be addressed before the proposal is accepted.

Once the book is accepted, the Society Book Editors ensure that the volume editors follow strict guidelines on refereeing and quality control. We insist that individual papers can only be accepted after satisfactory review by two independent referees. The questions on the review forms are similar to those for *Journal of the Geological Society*. The referees' forms and comments must be available to the Society's Book Editors on request.

Although many of the books result from meetings, the editors are expected to commission papers that were not presented at the meeting to ensure that the book provides a balanced coverage of the subject. Being accepted for presentation at the meeting does not guarantee inclusion in the book.

More information about submitting a proposal and producing a book for the Society can be found on its web site: www.geolsoc.org.uk.

It is recommended that reference to all or part of this book should be made in one of the following ways:

BALME, M. R., BARGERY, A. S., GALLAGHER, C. J. & GUPTA, S. (eds) 2011. *Martian Geomorphology*. Geological Society, London, Special Publications, **356**.

ASTON, A. H., CONWAY, S. J. & BALME, M. R. 2011. Identifying Martian gully evolution. *In*: BALME, M. R., BARGERY, A. S., GALLAGHER, C. J. & GUPTA, S. (eds) *Martian Geomorphology*. Geological Society, London, Special Publications, **356**, 151–169.

GEOLOGICAL SOCIETY SPECIAL PUBLICATION NO. 356

Martian Geomorphology

EDITED BY

M. R. BALME
Open University, UK

A. S. BARGERY
Lancaster University, UK

C. J. GALLAGHER
University College Dublin, Ireland

and

S. GUPTA
Imperial College London, UK

2011
Published by
The Geological Society
London

THE GEOLOGICAL SOCIETY

The Geological Society of London (GSL) was founded in 1807. It is the oldest national geological society in the world and the largest in Europe. It was incorporated under Royal Charter in 1825 and is Registered Charity 210161.

The Society is the UK national learned and professional society for geology with a worldwide Fellowship (FGS) of over 10 000. The Society has the power to confer Chartered status on suitably qualified Fellows, and about 2000 of the Fellowship carry the title (CGeol). Chartered Geologists may also obtain the equivalent European title, European Geologist (EurGeol). One fifth of the Society's fellowship resides outside the UK. To find out more about the Society, log on to www.geolsoc.org.uk.

The Geological Society Publishing House (Bath, UK) produces the Society's international journals and books, and acts as European distributor for selected publications of the American Association of Petroleum Geologists (AAPG), the Indonesian Petroleum Association (IPA), the Geological Society of America (GSA), the Society for Sedimentary Geology (SEPM) and the Geologists' Association (GA). Joint marketing agreements ensure that GSL Fellows may purchase these societies' publications at a discount. The Society's online bookshop (accessible from www.geolsoc.org.uk) offers secure book purchasing with your credit or debit card.

To find out about joining the Society and benefiting from substantial discounts on publications of GSL and other societies worldwide, consult www.geolsoc.org.uk, or contact the Fellowship Department at: The Geological Society, Burlington House, Piccadilly, London W1J 0BG: Tel. +44 (0)20 7434 9944; Fax +44 (0)20 7439 8975; E-mail: enquiries@geolsoc.org.uk.

For information about the Society's meetings, consult *Events* on www.geolsoc.org.uk. To find out more about the Society's Corporate Affiliates Scheme, write to enquiries@geolsoc.org.uk.

Published by The Geological Society from:
The Geological Society Publishing House, Unit 7, Brassmill Enterprise Centre, Brassmill Lane, Bath BA1 3JN, UK

(*Orders*: Tel. +44 (0)1225 445046, Fax +44 (0)1225 442836)
Online bookshop: www.geolsoc.org.uk/bookshop

British Library Cataloguing in Publication Data

A catalogue record for this book is available from the British Library.
ISBN 978-1-86239-330-1

Distributors
For details of international agents and distributors see:
www.geolsoc.org.uk/agentsdistributors

Typeset by Techset Composition Ltd, Salisbury, UK
Printed by MPG Books Ltd, Bodmin, UK

Contents

Preface

The past few decades have seen extraordinary advances in our understanding of the planet Mars. In particular, our knowledge of its surface topography, composition, morphology and climate history has dramatically improved. This stems largely from the veritable armada of spacecraft, both orbiters and landers, sent to Mars during this time. These spacecraft have collected spectacular high-resolution remotely sensed data, together with remarkable *in situ* observations and measurements by the six missions that have successfully landed on the surface. This treasure trove of data has been further enhanced by high-precision geochemical studies performed on Earth of meteorites thought to have come from Mars. Arguably, some of the most significant findings about the surface evolution of Mars have come from the interpretation of high-resolution image and topographical data acquired from orbit. This has led to a renaissance in the study of Martian geomorphology and surface processes.

The collection of papers that compose this Special Publication was inspired by contributions to the planetary geomorphology sessions at the European Geophysical Union's annual General Assembly between 2007 and 2010. The aim of these sessions has been to bring together scientists specializing in remote sensing of planetary surfaces with terrestrial geomorphologists who have in-depth knowledge of specific landforms and processes. The selection of topics covered here, therefore, represents a snapshot of what was most significant at the interface between these two communities at that time.

We hope that readers with little experience of Mars geomorphology will find this book inspiring, and that seasoned planetary scientists will appreciate the new data and analysis presented here. We would like to thank the reviewers listed on the following page who gave up their time and all of the staff at the Geological Society of London – particularly Angharad Hills – who helped us throughout the process.

MATTHEW R. BALME, ALISTAIR S. BARGERY,
COLMAN J. GALLAGHER & SANJEEV GUPTA

Acknowledgements

The volume editors would like to acknowledge the following colleagues who kindly helped with reviewing the papers submitted for this volume:

Vic Baker, Alexander Basilevsky, Daniel Berman, Susan Conway, Frank Chuang, David Ferrell, Corey Fortezzo, Frank Fueten, Stephan van Gasselt, Peter Grindrod, Ross Irwin, Joseph Levy, Nicolas Mangold, Daniel Mege, Grant Meyer, Gareth Morgan, Julian Murton, Cliff Ollier, Geoffrey Pearce, Angelo Pio Rossi, Louise Prockter, Dennis Reiss, Richard Soare, Nick Warner, and four other reviewers who preferred to remain anonymous.

The following institutions are credited for producing images used in this volume, the acronyms are provided in the captions:

ASI	Agenzia Spaziale Italiana
ASU	Arizona State University
CIW	Carnegie Institution of Washington
DLR	Deutschen Zentrums für Luft und Raumfahrt
ERSDAC	Earth Remote Sensing Data Analysis Center of Japan
ESA	European Space Agency
FUB	Freie Universität Berlin
GSFC	Goddard Space Flight Centre
JAROS	Japan Resources Observation System and Space Utilization Organization
JHUAPL	Johns Hopkins University Applied Physics Laboratory
JPL	Jet Propulsion laboratory
METI	Ministry of Economy, Trade, and Industry of Japan
MSSS	Malin Space Science Systems
NASA	National Aeronautics and Space Administration
UofA	University of Arizona
UMD	University of Maryland

Martian Geomorphology: introduction

M. R. BALME[1]*, A. S. BARGERY[2], C. J. GALLAGHER[3] & S. GUPTA[4]

[1]*Department of Earth Science, Open University, Walton Hall, Milton Keynes MK7 6AA, UK*

[2]*Lancaster Environment Centre, Lancaster University, Lancaster LA1 4YQ, UK*

[3]*UCD School of Geography, Planning and Environmental Policy, Newman Building, University College Dublin, Belfield, Dublin 4, Ireland*

[4]*Department of Earth Science and Engineering, Imperial College, Prince Consort Road, London SW7 2PB, UK*

**Corresponding author (e-mail: m.r.balme@open.ac.uk)*

This book concerns the Martian landscape; that collection of volcanoes, valleys, impact craters and ice caps that recent images reveal both to be strikingly familiar but also strangely alien to the surface of our own planet. The primary aim of studying planetary landscapes is to understand the process(es) by which they formed, with the larger goal of unravelling key questions about the origin, evolution and potential habitability of our solar system.

Compared with Earth, Mars' surface erosion rates are extremely low (Golombek & Bridges 2000), so Martian landscapes ranging in age from the very ancient to the recent still remain preserved and amenable to observation. Because so much of the planet's geological history remains visible, Martian geomorphology has the potential to provide even deeper insights into the early evolution of the planet than is the case for terrestrial geomorphology. Furthermore, the lack of precipitation (at least for much of Martian geological history: Craddock & Howard 2002), vegetation or human influence have preserved landforms on the surface of Mars that on Earth are obscured, degraded or buried, and only recognizable from interpretation of the sedimentary rock record. These observations, together with the fact that virtually all of the geological processes seen on Earth are believed to have also occurred on Mars, make it a powerful laboratory for comparative studies of geomorphological processes.

Like any dominantly remote-sensing approach, studies of the Martian surface must account for *in situ* data, but outcrop and hand-sample examination is a luxury afforded to only a few locations on Mars and then only through robotic missions. Similarly, the meteorite samples from Mars are few in number (Meyer 2009) and also lack information on their source location. Targeted sample return, for the examination of thin sections, analysis of geochemistry and age determination (among others), awaits future missions, funding and new technology.

This lack of *in situ* data, combined with issues of *equifinality* (or convergence of form wherein similar landforms are created by dissimilar processes), presents a challenge to Martian geomorphological interpretations. Thus, we must be circumspect when linking form to process, and must highlight where and when more than one working hypothesis exists. These challenges are not insurmountable, and we suggest that the number of viable hypotheses decreases as the breadth of data types increases, and as their spatial resolution improves. For example, recent and ongoing orbiting missions, including *Mars Global Surveyor*, *Mars Odyssey*, *Mars Express* and *Mars Reconnaissance Orbiter*, are generating a vast quantity of visible-light, near-infrared and thermal spectral data that allow researchers to characterize the surface texture and composition of Mars in evermore spectacular detail. With the 30 cm per pixel imaging data from the HiRISE (High Resolution Imaging Science Experiment) camera (McEwen *et al.* 2007) located on board the *Mars Reconnaissance Orbiter*, we are now able to subject competing hypotheses to closer and closer scrutiny until the weight of consilient evidence for one hypothesis brings it to the fore.

On Mars, geomorphological analysis also lays the groundwork for future targeted studies. Areas of Mars that the planetary community identifies as being of particularly high interest have the potential to eventually become the destinations for *in situ* missions. A good example of this is the Mars Exploration Rover mission *Opportunity* (Squyres *et al.* 2004) that was sent to the Terra Meridiani region largely on the strength of orbital spectroscopy observations of enhanced concentrations of the mineral hematite and its association with

From: BALME, M. R., BARGERY, A. S., GALLAGHER, C. J. & GUPTA, S. (eds) *Martian Geomorphology*.
Geological Society, London, Special Publications, **356**, 1–3.
DOI: 10.1144/SP356.1 0305-8719/11/$15.00

specific surface morphologies (Christensen *et al.* 2000). The mission found evidence of an ancient groundwater table within aeolian sandstones – providing an explanation for the remotely sensed interpretation that the hematite formed in the presence of water (Squyres *et al.* 2009). While field trips such as this take a little more money and a little more time than most such expeditions on Earth, they are the natural end result of the process that began with remotely sensed geomorphological observations and analysis, and the development and testing of multiple working hypotheses.

The chapters of this Special Publication include examples both of the analysis of new datasets and the application of methodologies new to Mars science. Chapter 2, by **Bargery *et al.***, provides context for readers new to Mars by presenting some background material on Martian geology, climate and exploration. In Chapter 3 **Murray & Illiff**'s updated mapping of Mars' larger moon, Phobos, sheds new light on an ongoing debate: the work uses new images from the High Resolution Stereo Camera (HRSC) on the European Space Agency's (ESA's) *Mars Express* spacecraft to constrain the origin of Phobos' enigmatic grooves. Chapters 4–12 of this Special Publication cover various aspects of the influence of water in the Martian near-surface. Ice and water are most certainly a 'hot topic' in Mars science, and one naturally reflected by the number of papers on that theme in this volume. Of particular interest is the question of whether the Martian climate has generally been too cold to allow thaw or whether melting of near-surface ice has been a geomorphologically important process; in other words, what has the balance been between landscapes dominated by sublimation and landscapes dominated by thaw? In Chapter 4 **van Gasselt *et al.*** discuss the evolution of lobate debris aprons in the northern mid-latitude Tempe Terra region. These landforms are thought to have formed by creep of rock–ice mixtures. In Chapter 5 **Rossi *et al.*** find evidence for a suite of glacial and periglacial landforms in the southern mid-latitude Thaumasia Highlands. In both of these chapters evidence is presented that these landforms have been evolving over at least hundreds of millions of years, and that they might still be active today. This is, perhaps, a reflection of periodic climate change driven by the extreme variations in axial tilt that Mars undergoes (Laskar *et al.* 2004).

In Chapter 6 **Gallagher *et al.*** present very-high-resolution imaging data of high-latitude northern impact craters, and describe geologically young patterned grounds and lobate hillslope features that point to a thaw origin. In Chapter 7 **Hauber *et al.*** present a synthesis of terrestrial observations made in Svalbard that can serve as an analogue for Martian periglacial domains. Hauber *et al.* note that, although the two climates are different, the landforms assemblages are closely matched. They conclude that the Martian mid-latitudes are evolving along the same lines as Svalbard, although much more slowly. Chapter 8 presents a 'drier' take on ice in the Martian near-surface, as **Mangold** reviews landforms on Mars thought to have formed by sublimation of ice, rather than of thaw.

Chapters 9 and 10 discuss Martian 'gullies', fluvial-like chutes and debris aprons first discovered in 2000 (Malin & Edgett 2000), and which heralded new interest in the concept of geologically recent liquid water flowing on the Martian surface. In Chapter 9 **Aston *et al.*** use a morphological classification of gullies to demonstrate that two or more generations of gully formation occurred. In Chapter 10 **Conway *et al.*** present a methodology that until now has not been applied to Mars by using a combination of slope-area and cumulative-area distribution analyses of very-high-resolution digital elevation models (DEMs). This is the type of work that has only become possible with the advent of approximately 30 cm per pixel stereo imaging data provided by NASA's HiRISE instrument. Larger and older flows are discussed in Chapters 11 and 12. In Chapter 11 **Balme *et al.*** discuss catastrophic flood channel evolution in the Elysium Planitia region of Mars, while in Chapter 12 **Towner *et al.*** consider whether sinuous channels associated with volcanic landscapes in the Sucii Gordi region of Mars were carved by water or lava.

The final two chapters of the book look at even more ancient Martian landscapes. In Chapter 13 **Kneissl *et al.*** investigate the origin and erosion rate and style of the Tenuis Rupes – a distinctive morphological unit that underlies the north polar cap. They use an exhaustive array of data, combining observations of morphology, topography from the Mars Orbiter Laser Altimeter (MOLA) instrument and shallow ground-penetrating RADAR. Finally, in Chapter 14, **Sowe *et al.*** compare the geomorphology and mineralogy of Interior Layer Deposits, multi-kilometre-scale stacks of strata that occur in settings such as canyons, jumbled 'chaos' terrain and larger impact craters. They conclude that layered mounds in chaos terrains and within the Vallis Marineris canyon system have similar origins and underwent similar post-depositional weathering processes.

The production of this Special Publication was made possible by the support afforded to the editors by the following agencies: M.R. Balme was supported by an 'Aurora' Research Fellowship awarded by the UK Science and Technologies Facilities Council (STFC): and S. Gupta was supported by a UK STFC Astronomy Standard Grant and a UK Royal Society/Leverhulme Trust Senior Research fellowship.

References

CHRISTENSEN, P., BANDFIELD, J. L. *ET AL.* 2000. Detection of crystalline hematite mineralization on Mars by the Thermal Emission Spectrometer: evidence for near-surface water. *Journal of Geophysical Research – Planets*, **105**, 9623–9642.

CRADDOCK, R. A. & HOWARD, A. D. 2002. The case for rainfall on a warm, wet early Mars. *Journal of Geophysical Research (Planets)*, **107**, 5111, doi: 10.1029/2001JE001505.

GOLOMBEK, M. R. & BRIDGES, N. T. 2000. Erosion rates on Mars and implications for climate change: constraints from the Pathfinder landing site. *Journal of Geophysical Research – Planets*, **105**, 1841–1853, doi: 10.1029/1999JE001043.

LASKAR, J., CORREIA, A. C. M., GASTINEAU, M., JOUTEL, F., LEVRARD, B. & ROBUTEL, P. 2004. Long term evolution and chaotic diffusion of the insolation quantities of Mars. *Icarus*, **170**, 343–364.

MALIN, M. C. & EDGETT, K. S. 2000. Evidence for recent groundwater seepage and surface runoff on Mars. *Science*, **288**, 2330–2335.

MCEWEN, A. S., ELIASON, E. M. *ET AL.* 2007. Mars reconnaissance Orbiter's High Resolution Imaging Science Experiment (HiRISE). *Journal of Geophysical Research (Planets)*, **112**, E05S02, doi: 10.1029/2005JE002605.

MEYER, C. 2009. *The Mars Meteorite Compendium*. Astromaterials Research & Exploration Science (ARES), JSC #27672 Revision C. Lyndon B. Johnson Space Center, Houston, Texas. World Wide Web Address: http://curator.jsc.nasa.gov/antmet/mmc/ accessed December 2010.

SQUYRES, S., ARVIDSON, R. E. *ET AL.* 2004. The opportunity Rover's Athena Science investigation at Meridiani Planum, Mars. *Science*, **306**, 1698–1703.

SQUYRES, S. W., KNOLL, A. H. *ET AL.* 2009. Exploration of victoria crater by the Mars Rover opportunity. *Science*, **324**, 1058–1061.

A background to Mars exploration and research

ALISTAIR S. BARGERY[1]*, MATTHEW R. BALME[2], NICHOLAS WARNER[3], COLMAN J. GALLAGHER[4] & SANJEEV GUPTA[3]

[1]*Lancaster Environment Centre, Lancaster University, Lancaster LA1 4YQ, UK*

[2]*Department of Earth Science, Open University, Walton Hall, Milton Keynes MK7 6AA, UK*

[3]*Department of Earth Science and Engineering, Imperial College, Prince Consort Road, London SW7 2PB, UK*

[4]*Geography, Planning & Environmental Policy, Newman Building, University College Dublin, Belfield, Dublin 4, Ireland*

**Corresponding author (e-mail: a.bargery@lancaster.ac.uk)*

Abstract: Mars is the fourth planet in our Solar System and orbits roughly 230 $\times$ 10^6 km from the Sun. It has an orbital period of 687 Earth days and a solar day that is approximately 40 min longer than an Earth day. Mars is less dense and has half the radius of the Earth, and so has about one-tenth the mass; hence, the surface gravity of Mars is about four-tenths that of the Earth. Mars has no oceans and its surface area is therefore almost as large as that of Earth's continents. In this chapter, we present a summary of the Martian environment, global geography and geology, and provide some background on the missions and instruments that have played a role in developing our current understanding. Our aim is to provide a broad overview for those unfamiliar with Mars, rather than providing an exhaustive summary of every aspect of the planet's evolution.

Mars exploration

Pre-Space-Age telescopic observations of Mars showed large features with different albedo, transient clouds, seasonal and perennial polar caps, and dust storms. Initially, the surface of Mars was classified into two types of regions on the basis of albedo (Fig. 1). The lighter-toned, high-albedo plains were once thought of as Martian 'continents' and given names like Arabia Terra or Amazonis Planitia. The darker-toned, low-albedo features were thought to be 'seas', hence their names Mare Erythraeum, Mare Sirenum and Aurorae Sinus. From this, astronomers inferred Mars to be a very Earth-like planet, and even went so far as to interpret albedo features as canals constructed by intelligent Martians (Fig. 1).

Since the Soviet Union's failed attempt to launch *Marsnik 1* in 1960, the space-faring nations of the world have sent 39 missions to Mars to study the planet's surface and climate. Fewer than half of these missions have been successful. The impression left by the first fly-bys (*Mariner 4*, *6* and *7*) was that the surface of Mars is similar to that of the Moon (Snyder & Moroz 1992), characterized by a rocky surface with numerous ancient impact craters (Binder 1966; Opik 1966). However, the observed images were low resolution, and were limited to the older southern hemisphere of the planet. The first successful object to land on the surface was the Soviet probe *Mars 3*, but it lost contact within seconds of landing (see Snyder & Moroz 1992 for a review of early missions to Mars).

The NASA orbiter mission *Mariner 9*, launched in 1971, defined our modern view of the surface characteristics of Mars as the higher image resolution and global coverage of the mission revealed a complex geomorphology (McCauley *et al.* 1972; Hartmann & Raper 1974). The northern hemisphere of Mars was observed to be low in elevation, flat and had relatively few craters. This was in contrast to the earlier observations of the southern hemisphere, obtained from the *Mariner 4*, *6* and *7* missions, which showed it to be rugged and dominated by impact craters, suggesting a planet-wide dichotomy in surface age. Large shield volcanoes were identified, primarily in the northern hemisphere, that suggested a vigorous volcanic history. Highly degraded impact craters were observed that required periods of enhanced surface erosion relative to the Moon. Most significantly, relict dendritic channel networks and catastrophic flood channels were

From: BALME, M. R., BARGERY, A. S., GALLAGHER, C. J. & GUPTA, S. (eds) *Martian Geomorphology*. Geological Society, London, Special Publications, **356**, 5–20.
DOI: 10.1144/SP356.2 0305-8719/11/$15.00

Fig. 1. (Left) Mars as seen by the Hubble Space Telescope. NASA Photojournal image PIA03154. Image credit NASA/Hubble Heritage Team; see prelim viii for acronym definitions. (Right) A map of Mars made by Lowell in 1905 showing the locations of 'canals' (after Lowell 1908).

discovered on the southern highlands and at the equator, providing the first conclusive evidence that liquid once flowed across the surface of the planet.

Measurements from the two *Viking Orbiter* missions launched in 1975 confirmed that the modern Martian atmosphere was thin and very dry (Snyder 1979). We know today, from these observations, that liquid water is not stable under the thin atmosphere (Carr 1983); it either freezes to ice or boils to form vapour (Bargery *et al.* 2010). However, significant advances in image resolution, quality and coverage with *Viking* missions revealed stunning new evidence for dense valley networks in the southern hemisphere (Carr 1987), and giant flood valleys and branching networks of channels and tributaries (Baker & Kochel 1979; Baker 1982). While these observations support the concept that Mars might once have been a warmer, wetter planet (Malin & Edgett 2000*a*), new questions about the fate of the water, the cause of the climate change and the likelihood that Mars supported life arose from the *Viking* missions. These same questions guide the majority of Martian geomorphology research in the modern era.

In the last decade, Mars exploration has been a high priority for planetary exploration, due partly to the will to discover whether a warmer, wetter Mars was once an abode for life. Recent missions have seen a wealth of new instruments sent both into orbit and onto the surface of Mars, and the breadth and quality of data now being returned have eclipsed even the achievements of the *Viking* missions. The recent datasets that are most commonly referred to in this Special Publication are from orbiter missions, and are summarized in Table 1.

The global geography and topography of Mars

The most obvious aspect of Mars' topography is the hemispheric dichotomy (Fig. 2). The flat northern plains contrast with the pitted and cratered ancient southern highlands, and are several kilometres lower in elevation. Mars has ice-caps several kilometres thick at both poles and a bulging equatorial igneous province called 'Tharsis' that straddles the dichotomy boundary. Tharsis is approximately 4000 km across and has an average elevation of 10 km above the surrounding plains. The northern plains contain several other huge volcanoes such as Elysium Mons and the shield volcano Olympus Mons: at 26 km in height, it is the tallest known volcano in the Solar System and more than three times the height of Mount Everest (Fig. 3a). Olympus Mons covers an area approximately equal to the land area of Italy. The largest confirmed impact crater on Mars, the Hellas impact basin in the southern highlands, is over 2000 km in diameter and at least 7 km deep. The largest canyon, Valles Marineris, splits the eastern side of Tharsis, and has a length of 4000 km – equivalent to the distance between New York and Los Angeles – and a depth of 2–7 km (Fig. 3b); by comparison, the Grand

Table 1. *Recent Mars missions and the key instruments and datasets most relevant to Mars geomorphology*

Date	Name	Agency	Relevant instruments*
1996–2006	*Mars Global Surveyor*	NASA	Wide-angle and high-resolution imaging (MOC – up to 1.5 m per pixel); infrared spectrometer (TES); laser altimeter (MOLA)
2001 and ongoing	*Mars Odyssey*	NASA	Visible and infra-red imaging spectrometer (THEMIS – up to 18 m per pixel in visible, up to 100 m per pixel in infrared); Gamma Ray Spectrometer (GRS)
2003 and ongoing	*Mars Express*	ESA	High-Resolution Stereo Camera (HRSC – up to 12 m per pixel); visible and infrared imaging spectrometer (OMEGA – up to *c*. 100 m per pixel); ground-penetrating RADAR (MARSIS)
2005 and ongoing	*Mars Reconnaissance Orbiter*	NASA	Very-high-resolution imaging camera (HiRISE – up to *c*. 30 cm per pixel); Context Imaging Camera (CTX – up to 6 m per pixel); visible and infrared imaging spectrometer (CRISM – up to *c*. 20 m per pixel)

*CRISM, Compact Reconnaissance Imaging Spectrometer for Mars; CTX, Context Camera; ESA, European Space Agency; HiRISE, High Resolution Imaging Science Experiment; MARSIS, Mars Advanced Radar for Subsurface and Ionospheric Sounding; MOC, Mars Orbiter Camera; MOLA, Mars Orbiter Laser Altimeter; NASA, National Aeronautics and Space Administration; OMEGA, Observatoire pour la Minéralogie, l'Eau, les Glaces et l'Activité; TES, Thermal Emission Spectrometer; THEMIS, Thermal Emission Imaging System.

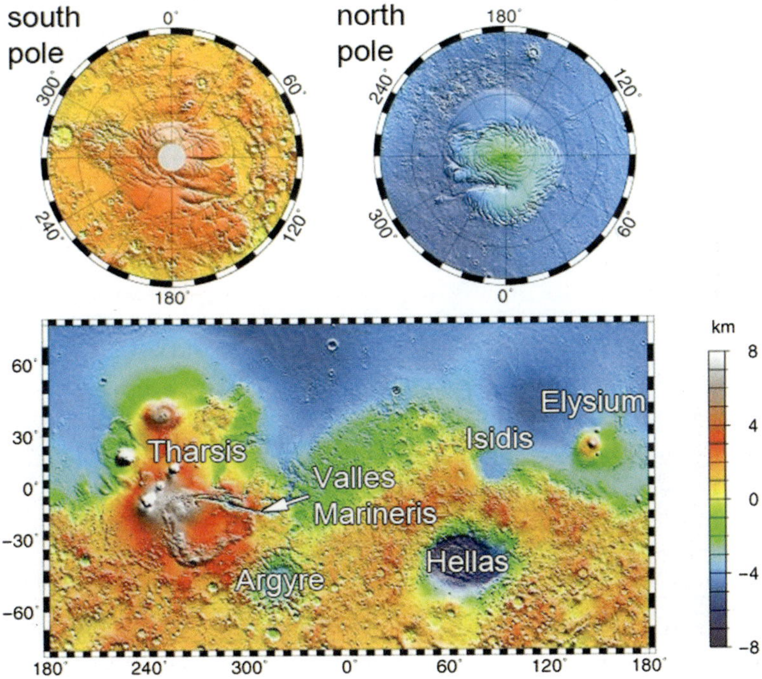

Fig. 2. Maps of Mars' global topography from MOLA data. (Bottom) Mercator projection of Mars to 70° latitude. (Top) Stereographic projections at the South Pole (left) and North Pole (right). Note the elevation difference between the northern and southern hemispheres. The Tharsis volcanotectonic province is centred near the equator in the longitude range 220–300°E and contains the east-west-trending Valles Marineris canyon system and several major volcanic shields. Major impact basins include Hellas (45°S, 70°E), Argyre (50°S, 320°E) and Isidis (12°N, 88°E). Image credit NASA/JPL-Caltech.

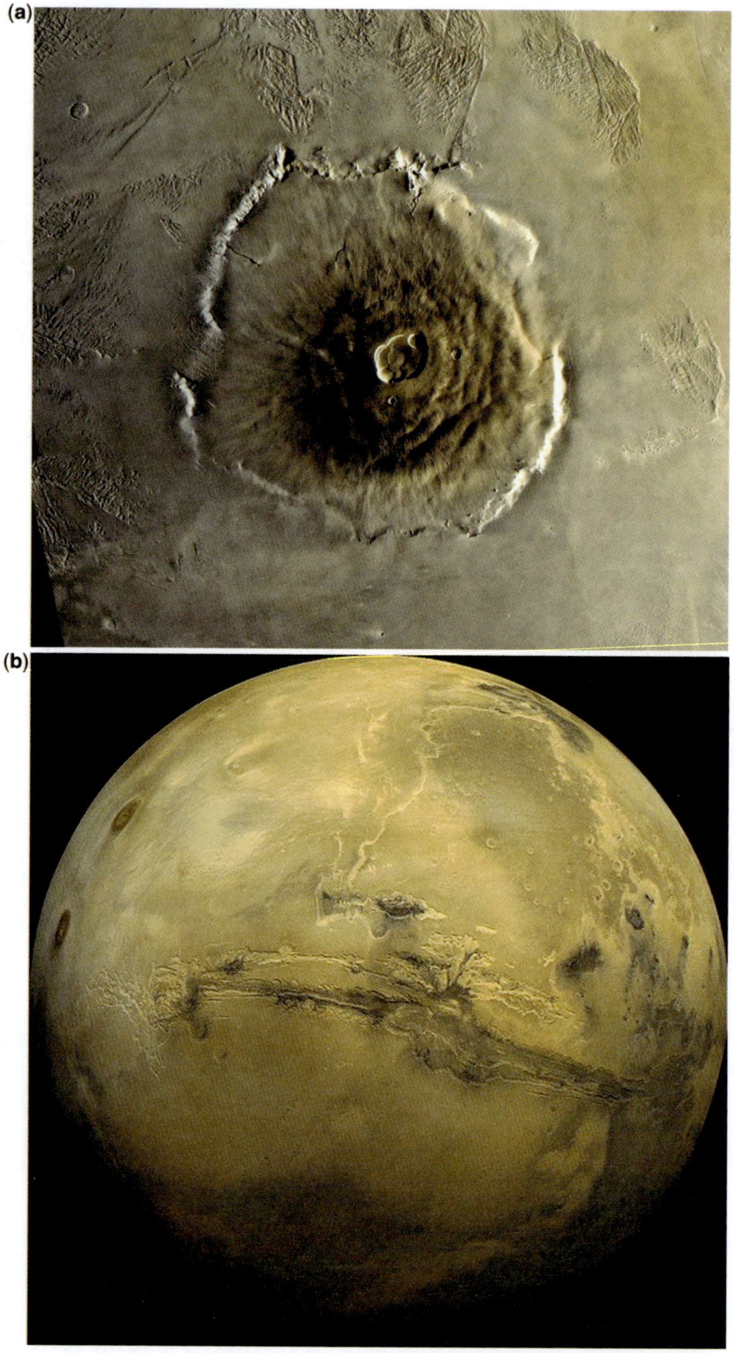

Fig. 3. (**a**) Olympus Mons, the largest volcano in the Solar System. The volcano is about 600 km across and over 25 km high. This image was taken by the *Viking 1 Orbiter*. NASA Planetary Photojournal Image PIA02982. Image credit NASA/JPL. (**b**) Valles Marineris, visible at the centre of this mosaic of 102 *Viking 1 Orbiter* images (MG07S078-334SP) of Mars taken during orbit 1334. Valles Marineris is over 3000 km long and up to 8 km deep. Note the channels running up (north) from the central and eastern portions of Valles Marineris to the dark area, Acidalia Planitia, in the upper right. On the left are the three Tharsis volcanoes and to the south is ancient, heavily impacted terrain. Image credit NASA/NSSDC.

Canyon is 446 km long and has a maximum depth of 1.83 km.

Mars geological time

Mars, as well as the other terrestrial planets and asteroids, is thought to have formed from the primordial planetary nebular about 4.5×10^9 (billion) years ago. The oldest geological surface on Mars is the oldest surface that can be dated using impact crater statistics (Carr & Head 2009), and is the boundary between Pre-Noachian and Noachian. The geological boundaries between the other three Martian epochs (Noachian, Hesperian and Amazonian) (Table 2) (Hartmann & Neukum 2001; Carr & Head 2009) are also defined by impact crater statistics, but the base of the Pre-Noachian cannot be defined by crater counts. The epochs are named after type localities on Mars that were emplaced during those periods. Unlike the Earth, where time divisions are precisely defined by age measurements of rock samples and by fossil records, the age boundaries of Martian eras are uncertain owing to competing models describing the rate of meteoroid impact on Mars and how this rate is converted to absolute time. The Noachian–Hesperian boundary has an uncertainty of 0.19 Ga (3.74–3.55 Ga) and the Hesperian–Amazonian boundary varies between models by almost 1 Ga (3.54–2.70 Ga) (Hartmann & Neukum 2001; Ivanov 2001). For a summary of the Hartmann & Neukum impact crater chronologies, refer to Fassett & Head (2008*a*).

Based on measurements made using OMEGA (Observatoire pour la Minéralogie, l'Eau, les Glaces et l'Activité), a multi-spectral visible light–near-infrared spectrometer on board the European Space Agency (ESA) *Mars Express* mission, an alternative timeline has been proposed from data showing a correlation between the mineralogy and stratigraphy of the planet (Bibring *et al.* 2006). However, this is not a widely excepted epoch/time division as mineralogical constraints based on spectral observations are not practical methods for age determination, as opposed to isotopic chronology, for example.

Volcanism and the Mars interior

Like the Earth, Mars is a differentiated body with a crust, mantle and core. Current models of the planet's interior infer a core region approximately

Table 2. *Commonly used Martian epochs based on impact cratering statistics*

Epoch	*Key events*
Pre-Noachian 4.5–*c.* 4.1 Ga	Planetary differentiation, large impacts and formation of the planetary dichotomy, presence of a magnetic field
Noachian *c.* 4.1–3.7 Ga	Formation of the oldest extant surfaces on Mars (southern highlands), including many large impact craters (e.g. Hellas). The bulk of the Tharsis region formed (Raitala 1988) and extensive flood lavas were emplaced (Edwards *et al.* 2008). Fluvial valley networks developed (Fassett & Head 2008*a*) and catastrophic flooding began late in the epoch (Irwin & Grant 2009; Warner *et al.* 2009). Open and closed lacustrine environments (e.g. crater lakes) were common (Fassett & Head 2008*b*). Erosion rates were relatively high, as demonstrated by highly degraded impact craters in the southern highlands (Craddock *et al.* 1997)
Hesperian *c.* 3.7–2.9 Ga	Formation of extensive lava plains in the northern lowlands, Tharsis and Hesperia Planum (among others). Surface water flow was dominated by megafloods originating from chaos terrains (Coleman & Baker 2009). Early in the Hesperian period, it has been postulated that an ocean may have existed in the northern lowlands (Carr & Head 2003). Localized valley networks and small, individual, fluvial channels formed, and were associated with crater lakes, permafrost melt features (Fig. 4a) and volcanic edifices (Ansan & Mangold, 2006; Fassett & Head, 2006; Di Achille *et al.* 2007; Warner *et al.* 2010*a*, *b*). By the end of this period, most of Mars' water is thought to have been locked away as ice in the regolith, forming an extensive cryosphere buried beneath dryer material (Carr 2000)
Amazonian *c.* 3.0 Ga–present	Volcanism limited to isolated regions of Mars including Tharsis and Elysium (Greeley & Spudis 1981). Catastrophic floods occurred from fissures in Elysium Planitia, forming the Athabasca Valles (Burr *et al.* 2009). Water on Mars was dominantly locked up in ice in the polar regions and in the subsurface. Climate cycles driven by orbital mechanics drove the formation of mid-latitude glaciation, snow packs and tropical mountain glaciers (Head *et al.* 2003*a*, *b*). Gullies formed on the interiors of impact craters and other steep slopes (Malin & Edgett 2000*b*)

1480 km in radius, consisting primarily of iron with about 15–17% sulphur (Kavner *et al.* 2001). This iron sulphide core is partially fluid, with twice the concentration of light elements that exists at the Earth's core, and is surrounded by a silicate mantle (e.g. Fuller & Head 2002).

Unlike the Earth, Mars is not thought to have experienced significant plate tectonics (Zuber 2001); hence, it has not undergone significant crustal recycling and much of the surface is ancient. With some notable exceptions (e.g. dacite lavas identified in Christensen *et al.* 2005), both spectral data and meteorites from Mars indicate that the crust and surface materials are predominantly basaltic or have been derived from basaltic materials (McSween *et al.* 2009). The lack of plate tectonics may explain the general absence of spectral or geomorphic evidence (lava domes, stratovolcanoes and short, thick lava flows: Davidson & De Silva 2000) of high silicic volcanism across the planet. Instead, Mars exhibits hotspot-style volcanism that is similar to inter-plate basaltic systems on Earth (e.g. Hawaii) (Greeley & Spudis 1981; Hodges & Moore 1994). This style of low-viscosity, effusive volcanism dominates Mars, and has resulted in the development of massive shield volcanoes (e.g. Olympus and Elysium Mons) and flood lava provinces which contain individual flows that are over 1000 km long (Mouginis-Mark & Yoshioka 1998). From the *Viking* missions era data, it was generally accepted that volcanism on modern Mars is dormant (Greeley & Spudis 1981). However, recent studies have identified small lava flows with a crater retention age of 2.0–100 Ma (e.g. Neukum *et al.* 2004), suggesting geologically recent, but volumetrically limited, activity.

Atmosphere, climate and polar caps

The Martian atmosphere consists of 95% carbon dioxide, 2.7% nitrogen, 1.6% argon, and traces of oxygen, carbon monoxide and water (Grinspoon 1997). Although Mars' atmospheric surface pressure is approximately 600 Pa at datum, it can be as high as 1160 Pa in the deepest part of Hellas Basin, and as low as 30 Pa at the top of Olympus Mons (Carr & Head 2009). The scale height of the atmosphere is about 11 km, larger than Earth's 6 km (Gierasch & Goody 1968).

Mars' seasons are currently Earth-like, a result of the similar inclinations of the two planets' obliquities (i.e. the angle that the planet's rotational axis makes to the plane of its orbit; currently Mars' obliquity is 25.2° and Earth's is 23.4°). Martian surface temperatures vary from approximately 130K during the polar winters to 293K in tropical summers (Haberle *et al.* 2001), the wide range being a result of the thin atmosphere. Atmospheric temperature sounding and the presence of equatorial glacial deposits on volcano flanks dating from a few million years suggest that Mars is subject to short-term regional climate changes (Taylor *et al.* 2006), and this is supported by models showing periodic changes of up to tens of degrees in Mars' obliquity on timescales of the order of 100 000 years (Laskar *et al.* 2004). The approximately 9% eccentricity of the Martian orbit also has a significant effect on the planet's seasons. At present, Mars is near perihelion when it is summer in the southern hemisphere and winter in the north. Hence, the southern summer is currently shorter and hotter than the northern summer.

Mars possesses polar caps at both poles, which consist mainly of water ice with thin, seasonal layers of CO_2 (Smith *et al.* 1999). The northern polar cap has a diameter of about 1000 km during the northern Mars summer (Fig. 4b), and contains about 1.6 million km^3 of ice, giving it a mean thickness of 2 km (Carr & Head 2003), compared with a volume of 2.85 million km^3 for the Greenland Ice Sheet (Bingham & Siegert 2007). The southern polar cap has a diameter of 350 km and a mean thickness of 3 km (Byrne & Ingersoll 2003). Both polar caps show spiral troughs, which are believed to form as a result of differential solar heating coupled with the sublimation of ice and the condensation of water vapour (Byrne & Ingersoll 2003; Pelletier 2004).

In the winter months, when the poles are in continuous darkness, the surface becomes so cold that as much as 25–30% of the entire atmosphere condenses out as CO_2 ice (Jakosky *et al.* 2003). This accumulates on the northern ice cap in the northern winter only, while the southern ice cap has a permanent CO_2 ice cover. When the poles are exposed to sunlight, the CO_2 ice sublimes, causing winds that sweep off the poles at speeds of up to 400 km h^{-1} (Hess *et al.* 1979).

Despite its thin atmosphere, aeolian activity is widespread on Mars. Large dune fields are seen around both poles (Fig. 4c) and in some larger craters (Hayward *et al.* 2007); also, smaller ripple-like bedforms are ubiquitous at lower latitudes (Wilson & Zimbelman 2004; Balme *et al.* 2008). The Martian atmosphere is persistently dusty, and every few years Mars undergoes global dust storms that shroud the surface of nearly the entire planet from view (e.g. Martin & Zurek 1993). During these storms, the temperature of the atmosphere can rise by several degrees as insolation is absorbed and re-radiated by atmospheric dust; however, surface temperatures can fall markedly during night-time because of the thin atmosphere (Gurwell *et al.* 2005). It is not yet known what triggers these global storms, nor what process maintains

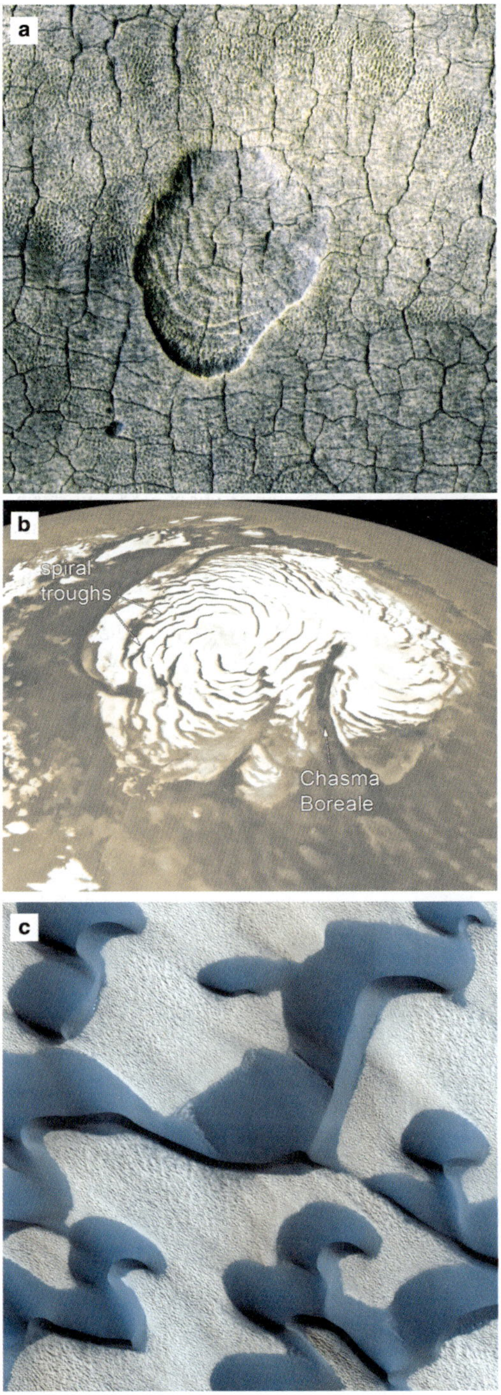

Fig. 4. (**a**) Permafrost/periglacial terrain in the Martian northern mid-latitudes. The polygonal surface patterns probably reflect thermal contraction cracking in an ice-rich regolith (Mellon 1997; Malin & Edgett 2001; Mangold *et al.* 2003). The central pit/depression in this image was probably formed by sublimation of ice (e.g. Morganstern *et al.* 2007), although others have suggested these pits might be thermokarst (i.e. formed as a result of the melting of ice) (Costard & Kargel 1985; Soare *et al.* 2008). The pit is about 350 m across. Image is false colour from HiRISE (High Resolution Imaging Science Experiment) image PSP_010034_2250. Image credit NASA/JPL/Univ. of Arizona. (**b**) Mars' north polar cap, viewed from approximately 270°E. The spiral troughs are clearly visible. Image created by combining MOLA (Mars Orbiter Laser Altimeter) and MOC data from *Mars Global Surveyor*. Image credit NASA/JPL-Caltech/MSSS. NASA Planetary Photojournal PIA13163. (**c**) Small, dark barchan dunes in the sand sea that surrounds the northern polar cap. Dark dunes like these are very common on Mars (Hayward *et al.* 2007), and generally are of basaltic composition (Fenton *et al.* 2003). The field of view of this image is about 1 km across. False colour HiRISE image PSP_009324_2650. Image credit NASA/JPL/Univ. of Arizona.

the persistent atmospheric haze of dust between the storms. Dust devils, small thermal vortices very common on Mars, may help to maintain the haze, but are not thought to trigger global dust storms (Balme & Greeley 2006).

Water on Mars

Two types of large-scale fluvial systems are apparent on Mars: valley systems (Fig. 5a) and outflow channels (Fig. 5b). Valley networks are commonly found on older surfaces and consist either of well-developed dendritic drainage systems that exhibit multiple orders of branching channels or poorly developed, single-branch sinuous channels that initiate from their source as a fully developed, wide channel (Carr 1996*a*, *b*). Generally, their similarity to terrestrial bedrock river valleys suggests formation by fluvial erosion, with the most likely agent of erosion being water. Most Martian valley networks date from the Late Noachian (Irwin *et al.* 2005), when both impact cratering and erosion rates were higher than at present (Carr 1996*a*; Craddock *et al.* 1997*a*, *b*; Hartmann & Neukum 2001). This indicates that the valley networks are probably correlated with processes that are associated with warmer and more humid climate conditions on ancient Mars. The water that supplied the valley networks may have come from either groundwater (Pieri 1980; Gulick 2001) or precipitation (Craddock & Howard 2002; Mangold *et al.* 2004*b*), but both sources probably contributed.

Outflow channels are large canyon systems, some exceeding 1000 km in length, that exhibit few tributaries, and contain streamlined islands, longitudinal grooves and cataracts that indicate

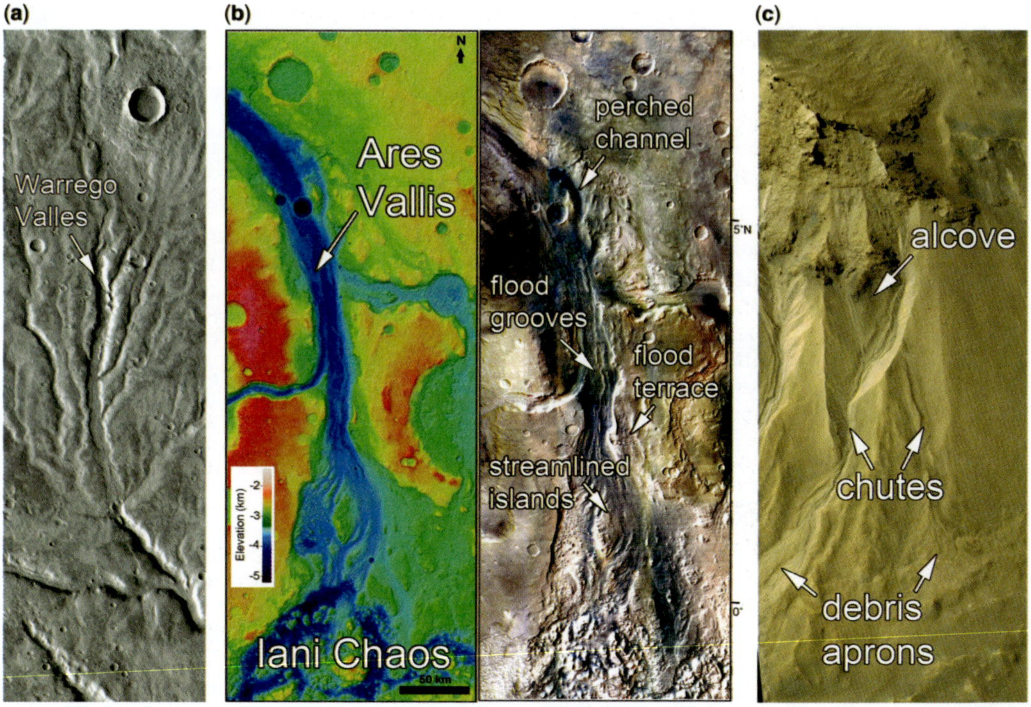

Fig. 5. Examples of fluvial landforms observed on Mars. (**a**) THEMIS image showing an area in the Warrego Valles region, an example of a valley network on Mars, at 42.3133°S, 267.511°E. Image width is 1024 pixels (17 km), image height is 3648 pixels (62 km). Vertical resolution is 0.017061 km per pixel and the horizontal resolution is 0.017185 km per pixel. Image credit NASA/JPL/ASU. (**b**) Proximal Ares Vallis outflow channel at 336°E, 9°N. (Left) Coloured and shaded relief digital terrain model (50 m grid spacing) derived from HRSC (High-Resolution Stereo Camera) stereo images. (Right) HRSC ortho-image (Warner *et al.* 2009). The source of flood waters carving the main canyon is a chaos region (Iani Chaos) at the southern extremity of the image. (**c**) Martian 'gullies'. Martian gullies are mass-wasting landforms that have a 'fluvial-like' form. The classic triangular alcove, sinuous channel and lobate debris deposit are clearly visible in this image. This image shows evidence for episodic activity, as a relict channel and debris apron are seen at the centre of the image. The field of view is about 1 km across. Part of HiRISE image ESP_014153_1430. Image credit NASA/JPL/Univ. of Arizona.

significant bedrock erosion by high-discharge megafloods (Baker 2001, 2009). Some relatively young outflow channels, including Athabasca Valles (Burr *et al.* 2002*a*, *b*; Rice *et al.* 2002) and Mangala Valles (Tanaka & Chapman 1990; Zimbelman *et al.* 1992), originate from fracture systems, suggesting that they were carved by large volumes of groundwater that were released by fissure-style crustal extension (Carr 1979; Baker 1982; Burr *et al.* 2009). Other outflow channels originate in 'chaos' regions that are defined by individual collapse depressions and extensionally fractured bedrock that indicate subsurface volume loss, probably by groundwater release. Collapse of the crust and water release may have occurred as a result of: a thickened cryosphere that overpressurized a regional aquifer (Carr 1979), volcanic melting of subsurface or near-surface ice (Head & Wilson 2002; Bargery & Wilson 2010), impact events (Wang *et al.* 2005) or de-watering of hydrated materials (Montgomery & Gillespie 2005).

An important ongoing debate is whether an ocean was once present in the northern lowlands of Mars (Pechmann 1980; McGill 1985; Phillips *et al.* 2001; Craddock & Howard 2002). Putative delta deposits, wave-cut terraces and shorelines have been identified that are potential signatures of an extensive standing body of water within the northern lowlands (Di Achille *et al.* 2007; Baker 2009). Furthermore, a recent study described high drainage densities on the highland regions that are proximal to the proposed shorelines of the northern ocean (Di Achille & Hynek 2010). The northern ocean may have resulted from catastrophic flooding (Malin & Edgett 2000*a*); however, age estimates, hydrological models and the geomorphology of the outflow channels suggest that many of the outflow channels were carved by multiple, lower volume floods over an extended period of Mars' history (Andrews-Hanna & Philips 2007; Harrison & Grimm 2008; Warner *et al.* 2009; Bargery & Wilson 2011). This period of flooding may be in excess of the interpreted residence time of a standing body of water (Kreslavsky & Head 2002), thus implying that a large northern ocean would have been unstable (Parker *et al.* 1993) otherwise it could have been sustained by flood discharges.

Smaller and geologically younger fluvial features known simply as 'gullies' are also common on Mars (Fig. 5c), first identified in high-resolution Mars Orbiter Camera (MOC) data (Malin & Edgett 2000*b*). They were originally interpreted to be groundwater seepage features, but the consensus has now moved towards them being a result of the melting of ice (e.g. Costard *et al.* 2002; Christensen 2003; Dickson & Head 2009; Levy *et al.* 2009). Gullies formed in the very late Amazonian, and may even be active today (Malin *et al.* 2006).

Finally, recent cycles in Mars' obliquity (Laskar *et al.* 2004) are thought to have allowed water ice to be transported from the polar regions to the mid-latitudes, forming regional 'mantles' of ice-rich dusty material that drape the topography (e.g. Mustard *et al.* 2001; Kreslavsky & Head 2002) and, possibly, controlling gully formation (Costard *et al.* 2002). The apparent link between the timescales of the Martian obliquity cycle and the young ages of pits, polygonally patterned ground (Fig. 4a), possible glacial morphologies and the degraded areas of this mantle itself suggest that Mars has recently undergone a series of 'ice ages' (Head *et al.* 2003*a*; Mangold *et al.* 2004*a*; Kreslavsky *et al.* 2008). Obliquity cycles have also been linked to the deposition of layered deposits at the poles (Laskar *et al.* 2002).

In situ observations: landers and rovers

Although rough or high-relief regions such as chaos terrains have been consistently ruled out as potential landing sites by mission managers because of safety reasons, a range of Martian landscapes have been investigated *in situ* (Table 3). The six missions that successfully landed on the Martian surface were each able to characterize their local environments using a combination of surface imaging and a variety of analytical instruments.

Viking 1 and *Mars Pathfinder* were sent to the Chryse Planitia region, where it was hoped that they would be able to sample a variety of different materials that had been deposited by catastrophic floods (Golombek *et al.* 1999). Both found a surface covered by a jumble of rocks infilled by fine-grained material interpreted to be aeolian mantles. The rocks at both sites were found to be an assortment of angular, cobble- to boulder-sized blocks of mostly basaltic composition that were inferred to be of volcanic and impact breccia origin (Golombek *et al.* 1999). In some cases, the rocks had been sculpted by wind-born sediment into ventifacts (Greeley & Iversen 1985; Bridges *et al.* 1999). *Pathfinder* found imbricated boulders (Smith & Mars Pathfinder Team 1997), interpreted to be left-stacked against one another as a result of waning catastrophic floods during the Hesperian.

Viking 2 and *Phoenix* were sent to the northern plains at latitudes where ice might be expected in the shallow subsurface. *Viking 2* landed on a relatively featureless, boulder-strewn plain where there were only hints of patterned ground visible from the surface. The site was visually quite similar to the *Viking 1* site, with the higher-than-expected abundance of rocks inferred to be the result of

Table 3. *Successful Mars lander missions*

Mission	Date and duration	Type	Landing site and approximate location
Viking 1 Lander	July 1976–November 1982	Lander	Outflow channel terminal deposits: 23°N, 312°E
Viking 2 Lander	September 1976–April 1980	Lander	Mid-latitude northern plains: 48°N, 134°E
Mars Pathfinder and *Sojourner Rover*	July 1997–September 1997	Lander and short-range (tens of metres) rover	Outflow channel terminal deposits: 19°N, 327°E
MER *Spirit*	January 2004–March 2010	Medium-range (kilometres) rover	Floor of an equatorial impact crater thought to contain a palaeolake: 15°S, 175°E
MER *Opportunity*	January 2004 and ongoing	Medium-range (kilometres) Rover	Low relief equatorial layered terrain with considerable aeolian cover: 2°S, 354°E
Mars Phoenix	May 2008–November 2008	Lander	Polygonally patterned ground in the high-latitude northern plains: 68°N, 234°E

debris from a nearby 100 km-diameter impact crater (Mutch *et al.* 1977).

Phoenix landed further north than *Viking 2*, away from large craters, and found a gently undulating landscape of polygonally patterned ground with fewer rocks (Fig. 6a). From orbit, the terrain was interpreted to represent fine-grained material with a significant ice content; the polygonal patterning being attributed to thermal contraction fracturing (Mellon *et al.* 2008*a*). Using a trenching tool, *Phoenix* confirmed this interpretation, finding water ice at a mean depth of 4.6 cm beneath the surface (Mellon *et al.* 2009) – a depth close to that predicted from pre-landing models (Mellon *et al.* 2008*b*). Recent work (Byrne *et al.* 2009) suggests that *Viking 2* would also have found ice if it had trenched only about 10 cm deeper.

The twin Mars Exploration Rovers (MER) were tasked with investigating the history of Martian water and the potential habitability of their landing sites (Squyres *et al.* 2003). They were the first rovers that could travel significant distances to investigate nearby terrains. MER *Spirit* landed in Gusev Crater, an impact crater of Noachian age and approximately 170 km in diameter. Gusev is breached to the south by a fluvial channel and contains what might be an ancient delta of Hesperian age; hence, Gusev is interpreted to have once contained a lake (Cabrol *et al.* 1996). The second MER, *Opportunity*, was sent to low-relief, layered

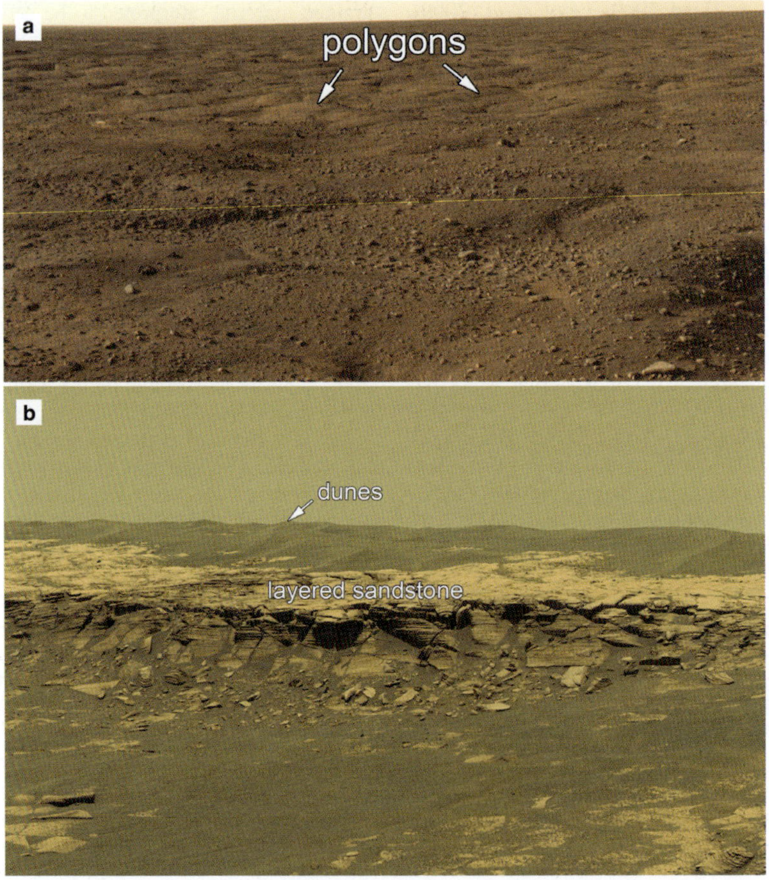

Fig. 6. (**a**) View from the *Phoenix Lander* on 8 June 2008. This is a portion of a larger panorama acquired by the *Phoenix*' Surface Imager instrument. Note the undulating polygonal terrain and the relative paucity of rocks. The polygonal mounds visible in the near field are 2–3 m across and the rocks are generally cobble sized or smaller. Image credit NASA/JPL/Caltech/Univ. of Arizona. (**b**) Sedimentary rocks in the Meridiani Planum as observed by the *Opportunity* MER. This scene is part of a larger panorama acquired on 26 February 2006 and shows layered sandstone deposits within Erebus Crater. Also visible in the top of the image are aeolian bedforms ubiquitous on the plains around the crater. The outcrop in the near-field is about 1 m high. A full description of the sedimentology of this outcrop is given by Arvidson *et al.* (2011). Image credit NASA/JPL-Caltech/USGS/Cornell.

terrain in the Terra Meridiani region (Fig. 6b) that is inferred to be of late Noachian–early Hesperian age (Hynek *et al.* 2002; Squyres & Knoll 2005). This target was identified from orbit as a unique site in that it contains high surface concentrations of coarse-grained, crystalline hematite (Christensen *et al.* 2000).

Spirit made many important discoveries, including spectacular observations of dust devils (Cantor *et al.* 2006; Greeley *et al.* 2006) and the discovery of possible ancient hydrothermal activity (e.g. Yen *et al.* 2008). However, any traces of fluvial sediments associated with a palaeolake were found to be inaccessible (Squyres *et al.* 2004*a*) as they were buried beneath impact debris and later infilling aeolian and volcanic deposits.

Over the course of its ongoing 6-year mission, *Opportunity* has travelled over 25 km and investigated layers of stratified rock many metres thick that are exposed within several small (100 m–1 km) impact craters. *Opportunity* also scrutinized the composition and morphology of loose sediments and aeolian bedforms found on the inter-crater plains. The hematite identified from orbit was found to be concentrated in gravel-sized concretions present both within the layered rocks and strewn across the plains (Squyres *et al.* 2004*b*). The sedimentary rocks, ubiquitous across the *Opportunity* traverse, were found to be siliciclastic, of evaporite or altered basaltic origin (Squyres & Knoll 2005) – probably as a result of the influence of acidic aqueous conditions – and to contain high proportions of sulphate minerals (Squyres *et al.* 2004*c*).

The sedimentology of the exposed strata includes large-scale cross-bedding and ripple lamination suggestive of an aeolian dune and sand-sheet origin (Grotzinger *et al.* 2005). There is also evidence for erosional disconformities within the stacks of sediments. The occurrence of ripple trough cross-lamination has been interpreted as a consequence of local subaqueous reworking of aeolian sediments within inter-dune ephemeral lakes (Squyres *et al.* 2004*c*; Grotzinger *et al.* 2005). Exploration of outcrops in the walls of Erebus Crater revealed ripple patterns, providing compelling evidence of water transport of sulphate-rich sands, which were later cemented to form sandstones (Metz *et al.* 2009). Interpretations of *Opportunity*'s observations suggest an arid environment in which groundwater and, perhaps, subaerial water have played a minor, but important, role (Fig. 6b). The action of acidic groundwater, in particular, is thought to be responsible for the formation of the hematite concretions and the dissolution of primary minerals to form vugs (Fig. 7). *Opportunity* is now sitting in a period of solar conjunction, with Endeavour Crater its next destination. Endeavour Crater is a much larger and deeper impact crater that allows an even more extensive stratigraphy to be observed (Arvidson *et al.* 2011).

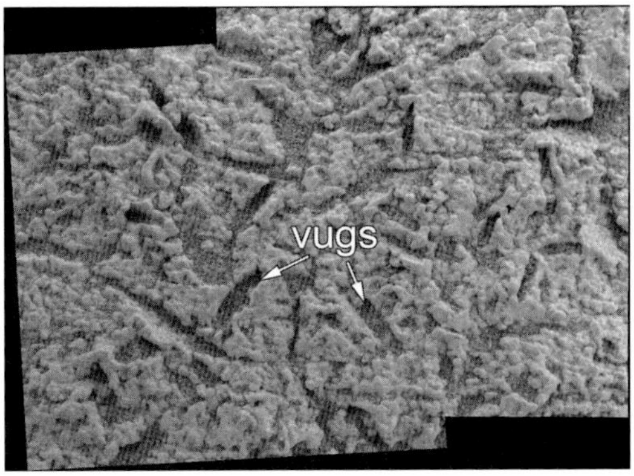

Fig. 7. MER *Opportunity* microscopic image of vugs, or small cavities, located on the region dubbed 'El Capitan', part of the rock outcrop at Meridiani Planum, Mars. The image provides insight into the nature of the rock matrix surrounding the vugs. Several vugs have disc-like shapes with wide mid-points and tapered ends. This is consistent with sulphate minerals that crystallize within a rock matrix, either pushing the matrix grains aside or replacing them. These crystals are either dissolved in water or eroded by wind activity to produce vugs. The rock matrix here exhibits a granular texture, delicately enhanced through wind abrasion. The primary sediment particles in the granular layer are relatively uniform in size, ranging up to 1 mm. Some of these grains are well rounded, which could result from the transport of rock fragments in air or water, or from the precipitation of mineral grains in water. Image credit NASA/JPL/Cornell/USGS.

Summary

Mars has followed an evolutionary path in which substantial early geological activity was followed by declining levels of activity and long periods of quiescence. Very early intense impact cratering was followed by significant volcanism and tectonism that ended 2–3 billion years ago. The accumulation of regional fluvial and oceanic sediments was likely to have been confined to these, or earlier, times. Episodic volcanic, fluvial and tectonic events have occurred over the last 1–2 billion years.

Aeolian erosion and deposition have probably been ongoing throughout the planet's history and continue today. The late Amazonian period appears to have been dominated by climate-driven processes involving wind, ice and water, many of which are probably driven by orbital cycles.

While the age relationships between the various elements of the geomorphology of Mars are difficult to determine, the process record is clearer, at increasingly higher resolution, both visually and spectroscopically. Consequently, while Mars is sufficiently Earth-like for planetary scientists to be confident that many of its surface processes and environments can be understood using analogues from Earth, new data regularly reveal morphologies that require new interpretations and remind us that exploration of the planet is still in its youth.

The authors thank D. Rothery and an anonymous reviewer for useful comments and discussions that helped to improve the manuscript.

References

ANDREWS-HANNA, J. C. & PHILLIPS, R. J. 2007. Hydrological modeling of outflow channels and chaos regions on Mars. *Journal of Geophysical Research*, **112**, E08001, doi: 10.1029/2006JE002881.

ANSAN, V. & MANGOLD, N. 2006. New observations of Warrego Valles, Mars: evidence for precipitation and surface runoff. *Planetary and Space Science*, **54**, 219–242.

ARVIDSON, R. E., ASHLEY, J. W. *ET AL*. 2011. Opportunity Mars rover mission: overview and selected results from purgatory ripple to traverses to endeavour crater. *Journal of Geophysical Research*, **116**, E00F15, doi: 10.1029/2010JE003746.

BAKER, V. R. 1982. *The Channels of Mars*. University of Texas, Austin, TX.

BAKER, V. R. 2001. Water and the Martian landscape. *Nature*, **412**, 228–236.

BAKER, V. R. 2009. Overview of megaflooding: Earth and Mars. *In*: BURR, D. M., CARLING, P. & BAKER, V. (eds) *Megaflooding on Earth and Mars*. Cambridge University Press, Cambridge, 1–12.

BAKER, V. R. & KOCHEL, R. C. 1979. Martian channel morphology – Maja and Kasei Valles. *Journal of Geophysical Research*, **84**, 7961–7983.

BALME, M. & GREELEY, R. 2006. Dust devils on Earth and Mars. *Review of Geophysics*, **44**, RG3003, doi: 10.1029/2005RG000188.

BALME, M. R., BERMAN, D. C., BOURKE, M. C. & ZIMBELMAN, J. 2008. Transverse Aeolian Ridges (TARs) on Mars. *Geomorphology*, **101**, 703–720.

BARGERY, A. S. & WILSON, L. 2010. Dynamics of the ascent and eruption of water containing dissolved CO_2 on Mars. *Journal of Geophysical Research*, **115**, E05008, doi: 10.1029/2009JE003403.

BARGERY, A. S. & WILSON, L. 2011. Erosive flood events on the surface of Mars: application to Mangala and Athabasca Valles. *Icarus*, **212**, 520–540, doi: 10.1016/j.icarus.2011.01.001.

BARGERY, A. S., LANE, S. J., BARRETT, A., WILSON, L. & GILBERT, J. S. 2010. The initial responses of hot liquid water released under low atmospheric pressures: experimental insights. *Icarus*, **210**, 488–506, doi: 10.1016/j.icarus.2010.06.019.

BIBRING, J.-P., LANGEVIN, Y. *ET AL*. & THE OMEGA TEAM 2006. Global mineralogical and aqueous mars history derived from Omega/Mars express data. *Science*, **312**, 400–404, doi: 10.1126/science.1122659.

BINDER, A. B. 1966. Mariner IV: analysis of preliminary photographs. *Science*, **152**, 1053–1055.

BINGHAM, R. G. & SIEGERT, M. J. 2007. Radio-Echo sounding over polar ice masses. *Journal of Environmental & Engineering Geophysics*, **12**, 47–62, doi: 10.2113/JEEG12.1.47.

BRIDGES, N. T., GREELEY, R., HALDEMANN, A. F. C., HERKENHOFF, K. E., KRAFT, M., PARKER, T. J. & WARD, A. W. 1999. Ventifacts at the pathfinder landing site. *Journal of Geophysical Research*, **104**, 8595–8615.

BURR, D. M., GRIER, J. A., MCEWEN, A. S. & KESZTHELYI, L. P. 2002*a*. Repeated aqueous flooding from the cerberus fossae: evidence for very recently extant, deep groundwater on Mars. *Icarus*, **155**, 53–73.

BURR, D. M., MCEWEN, A. S. & SAKIMOTO, S. E. H. 2002*b*. Recent aqueous floods from the Cerberus Fossae. *Geophysical Research Letters*, **29**, 1013, doi: 10.1029/2001GL013345.

BURR, D. M., WILSON, L. & BARGERY, A. S. 2009. Amazonian-aged, fissure-headed, megaflood channels on Mars: a review of morphology and triggering mechanisms. *In*: BURR, D. M., CARLING, P. & BAKER, V. (eds) *Megaflooding on Earth and Mars*. Cambridge University Press, Cambridge, 194–208.

BYRNE, S. & INGERSOLL, A. P. 2003. A sublimation model for Martian South polar ice features. *Science*, **299**, 1051–1053, doi: 10.1126/science.1080148.

BYRNE, S., DUNDAS, C. M. *ET AL*. 2009. Distribution of mid-latitude ground ice on mars from new impact craters. *Science*, **325**, 1674–1676.

CABROL, N. A., GRIN, E. A. & DAWIDOWICZ, G. 1996. Ma'adim vallis revisited through new topographic data: evidence for an Ancient Intravalley Lake. *Icarus*, **123**, 269–283.

CANTOR, B. A., KANAK, K. M. & EDGETT, K. S. 2006. Mars orbiter camera observations of Martian dust devils and their tracks (September 1997 to January 2006) and evaluation of theoretical vortex models. *Journal of Geophysical Research*, **111**, E12002, doi: 10.1029/2006JE002700.

CARR, M. H. 1979. Formation of Martian flood features by release of water from confined aquifers. *Journal of Geophysical Research*, **84**, 2995–3007.

CARR, M. H. 1983. Stability of streams and lakes on Mars. *Icarus*, **56**, 476–495, doi: 10.1016/0019-1035(83)90168-9.

CARR, M. H. 1987. Water on Mars. *Nature*, **326**, 30–35.

CARR, M. H. 1996*a*. *Water on Mars*. Oxford University Press, New York.

CARR, M. H. 1996*b*. Channels and valleys on Mars: cold climate features formed as a result of a thickening cryosphere. *Planetary and Space Science*, **44**, 1411–1423.

CARR, M. H. 2000. Martian oceans, valleys and climate. *Astronomy & Geophysics*, **41**, 3.20–3.26, doi: 10.1046/j.1468-4004.2000.00320.x.

CARR, M. H. & HEAD, J., III. 2003. Oceans on Mars: an assessment of the observational evidence and possible fate. *Journal of Geophysical Research*, **108**, 5042, doi: 10.1029/2002JE001963, 24.

CARR, M. H. & HEAD, J., III 2009. Geological history of Mars. *Earth and Planetary Science Letters*, **294**, 185–203, doi: 10.1016/j.epsl.2009.06.042.

CHRISTENSEN, P. 2003. Formation of recent Martian gullies through melting of extensive water-rich snow deposits. *Nature*, **422**, 45–48.

CHRISTENSEN, P., BANDFIELD, J. L. ET AL. 2000. Detection of crystalline hematite mineralization on Mars by the thermal emission spectrometer: evidence for near-surface water. *Journal of Geophysical Research*, **105**, 9623–9642.

CHRISTENSEN, P. R., MCSWEEN, H. Y. ET AL. 2005. Evidence for magmatic evolution and diversity on Mars from infra-red observations. *Nature*, **436**, 504–509.

COLEMAN, N. M. & BAKER, V. R. 2009. Surface morphology and origin of outflow channels in the Valles Marineris region. *In*: BURR, D. M., CARLING, P. & BAKER, V. (eds) *Megaflooding on Earth and Mars*. Cambridge University Press, Cambridge, 172–191.

COSTARD, F. M. & KARGEL, J. S. 1995. Outwash plains and thermokarst on Mars. *Icarus*, **114**, 93–112.

COSTARD, F., FORGET, F., MANGOLD, N. & PEULVAST, J. P. 2002. Formation of recent Martian debris flows by melting of near-surface ground ice at high obliquity. *Science*, **295**, 110–113.

CRADDOCK, R. A. & HOWARD, A. D. 2002. The case for rainfall on a warm, wet early Mars. *Journal of Geophysical Research*, **107**, 5111, doi: 10.1029/2001JE001505.

CRADDOCK, R. A., CRUMPLER, L. S., AUBELE, J. C. & ZIMBELMAN, J. R. 1997*a*. Geology of central Chryse Planitia and the viking 1 landing site: implications for the mars pathfinder mission. *Journal of Geophysical Research*, **102**, 4161–4183.

CRADDOCK, R. A., MAXWELL, T. A. & HOWARD, A. D. 1997*b*. Crater morphometry and modification in the sinus sabaeus and margaritifer sinus regions of Mars. *Journal of Geophysical Research*, **102**, 13 321–13 340.

DAVIDSON, J. & DE SILVA, S. 2000. Composite volcanoes. *In*: DAVIDSON, J., DE SILVA, S., SIGURDSSON, H. & HUGHTON, B. F. (eds) *Encyclopedia of Volcanoes*. Academic Press, San Francisco, CA.

DI ACHILLE, G. & HYNEK, B. M. 2010. Ancient ocean on Mars supported by global distribution of deltas and valleys. *Nature Geoscience*, **3**, 459–463, doi: 10.1038/ngeo891.

DI ACHILLE, G., ORI, G. G. & REISS, D. 2007. Evidence for late Hesperian lacustrine activity in Shalbatana Vallis, Mars. *Journal of Geophysical Research*, **112**, E07007, doi: 10.1029/2006JE002858.

DICKSON, J. L. & HEAD, J. W. 2009. The formation and evolution of youthful gullies on Mars: gullies as the late-stage phase of Mars' most recent ice age. *Icarus*, **204**, 63–86.

EDWARDS, C. S., CHRISTENSEN, P. R. & HAMILTON, V. E. 2008. Evidence for extensive olivine-rich basalt bedrock outcrops in Ganges and Eos chasmas, Mars. *Journal of Geophysical Research*, **113**, E11003, doi: 10.1029/2008JE003091.

FASSETT, C. I. & HEAD, J. W. 2006. Valleys on Hecates Tholus, Mars: origin by basal melting of summit snowpack. *Planetar and Space Science*, **54**, 370–378.

FASSETT, C. I. & HEAD, J. W. 2008*a*. The timing of Martian valley network activity: constraints from buffered crater counting. *Icarus*, **195**, 61–89.

FASSETT, C. I. & HEAD, J. W. 2008*b*. Valley network-fed, open-basin lakes on Mars: distribution and implications for Noachian surface and subsurface hydrology. *Icarus*, **198**, 37–56.

FENTON, L. K., BANDFIELD, J. L. & WARD, A. W. 2003. Aeolian processes in proctor crater on Mars: sedimentary history as analyzed from multiple data sets. *Journal Geophysical Research*, **108**, 5129, doi: 10.1029/JE002015.

FULLER, E. R. & HEAD, J. W. 2002. Amazonis Planitia: the role of geologically recent volcanism and sedimentation in the formation of the smoothest plains on Mars. *Journal of Geophysical Research*, **107**, 5081, doi: 10.1029/2002JE001842.

GIERASCH, P. & GOODY, R. 1968. A study of the thermal and dynamical structure of the Martian lower atmosphere. *Planetary and Space Science*, **16**, 615–646, doi: 10.1016/0032-0633(68)90102-5.

GOLOMBEK, M. P. & MARS PATHFINDER TEAM 1999. Overview of the Mars pathfinder mission: launch through landing surface operations, data sets and science results, special Mars pathfinder issue. *Journal of Geophysical Research*, **104**, 8523–8554.

GREELEY, R. & IVERSEN, J. 1985. *Wind as a Geologic Process on Earth, Mars, Venus and Titan*. Cambridge University Press, Cambridge.

GREELEY, R. & SPUDIS, P. D. 1981. Volcanism on Mars. *Review of Geophysics*, **19**, 13–41.

GREELEY, R., WHELLEY, P. L. ET AL. 2006. Active dust devils in Gusev crater, Mars: observations from the Mars exploration rover spirit. *Journal of Geophysical Research*, **111**, E12S09, doi: 10.1029/2006JE002743.

GRINSPOON, D. H. 1997. Earth: atmosphere *In*: SHIRLEY, J. H. & FAIRBRIDGE, R. W. (eds) *Encyclopedia of Planetary Sciences*. Springer, Berlin, 198.

GROTZINGER, J. P., ARVIDSON, R. E. ET AL. 2005. Stratigraphy and sedimentology of a dry to wet aeolian depositional system, burns formation, Meridiani Planum, Mars. *Earth and Planetary Science Letters*, **240**, 11–72.

GULICK, V. C. 2001. Origin of the valley networks on Mars: a hydrological perspective. *Geomorphology*, **37**, 241–268.

GURWELL, M. A., BERGIN, E. A., MELNICK, G. J. & TOLLS, V. 2005. Mars surface and atmospheric temperature during the 2001 global dust storm. *Icarus*, **175**, 23–31, doi: 10.1016/j.icarus.2004.10.009.

HABERLE, R. M., MCKAY, C. P., SCHAEFFER, J., CABROL, N. A., GRIN, E. A., ZENT, A. P. & QUINN, R. 2001. On the possibility of liquid water on present-day Mars. *Journal of Geophysical Research*, **106**, 23 317–23 326, doi: 10.1029/2000JE001360.

HARRISON, K. P. & GRIMM, R. E. 2008. Multiple flooding events in Martian outflow channels. *Journal of Geophysical Research*, **113**, E02002, doi: 10.1029/2007JE002951.

HARTMANN, W. K. & NEUKUM, G. 2001. Cratering chronology and the evolution of Mars. *Space Science Reviews*, **96**, 1–4, doi: 10.1023/A:1011945222010.

HARTMANN, W. K. & RAPER, O. 1974. *The New Mars – The Discoveries of Mariner 9*. NASA Special Publications, **337**.

HAYWARD, R. K., MULLINS, K. ET AL. 2007. Mars global digital dune database and initial science results. *Journal of Geophysical Research*, **112**, E11007, doi: 10.1029/2007JE002943.

HEAD, J. W. & WILSON, L. 2002. Mars: a review and synthesis of general environments and geological settings of magma–H_2O interactions. *In*: SMELLIE, J. L. & CHAPMAN, M. G. (eds) *Volcano–Ice Interaction on Earth and Mars*. Geological Society, London, Special Publications, **202**, 27–58.

HEAD, J. W., MUSTARD, J. F., KRESLAVSKY, M. A., MILLIKEN, R. E. & MARCHANT, D. R. 2003*a*. Recent ice ages on Mars. *Nature*, **426**, 797–802.

HEAD, J. W., WILSON, L. & MITCHELL, K. L. 2003*b*. Generation of recent massive water floods at Cerberus Fossae, Mars by dike emplacement, cryospheric cracking, and confined aquifer groundwater release. *Geophysical Research Letters*, **30**, 1577, doi: 10.1029/2003 GL017135.

HESS, S. L., HENRY, R. M. & TILLMAN, J. E. 1979. The seasonal variation of atmospheric pressure on Mars as affected by the south polar cap. *Journal of Geophysical Research*, **84**, 2923–2927, doi: 10.1029/JB084iB06p02923.

HODGES, C. A. & MOORE, H. J. 1994. *Atlas of Volcanic Landforms on Mars*. Geological Survey, Map Division, Denver, CO.

HYNEK, B. M., ARVIDSON, R. & PHILLIPS, R. J. 2002. Geologic setting and origin of Terra Meridiani hematite deposits on Mars. *Journal of Geophysical Research*, **107**, 5088, doi: 10.1029/2002JE001891.

IRWIN, R. P., III & GRANT, J. A. 2009. Large basing overflow floods on Mars. *In*: BURR, D. M., CARLING, P. & BAKER, V. (eds) *Megaflooding on Earth and Mars*. Cambridge University Press, Cambridge, 209–221.

IRWIN, R. P., III, CRADDOCK, R. A. & HOWARD, A. D. 2005. Interior channels in Martian valley networks: discharge and runoff production. *Geology*, **33**, 489–492, doi: 10.1130/G21333.1.

IVANOV, B. 2001. Mars/Moon cratering ratio estimates. *Space Science Reviews*, **96**, 87–104.

JAKOSKY, B. M., HENDERSON, B. G. & MELLON, M. T. 1993. The Mars water cycle at other epochs: recent history of the polar caps and layered terrain. *Icarus*, **102**, 286–297.

KAVNER, A., DUFFY, T. S. & SHEN, G. 2001. Phase stability and density of FeS at high pressures and temperatures: implications for the interior structure of Mars. *Earth and Planetary Science Letters*, **185**, 25–33, doi: 10.1016/S0012-821X(00)00356-3.

KRESLAVSKY, M. A. & HEAD, J. W. 2002. Mars: nature and evolution of young latitude-dependent water-ice-rich mantle. *Journal of Geophysical Research*, **29**, 1719, doi: 10.1029/2002GL015392.

KRESLAVSKY, M. A., HEAD, J. W. & MARCHANT, D. R. 2008. Periods of active permafrost layer formation during the geological history of Mars: implications for circum-polar and mid-latitude surface processes. *Planetary and Space Science*, **56**, 289–302.

LASKAR, J., CORREIA, A. C. M., GASTINEAU, M., JOUTEL, F., LEVRARD, B. & ROBUTEL, P. 2004. Long term evolution and chaotic diffusion of the insolation quantities of Mars. *Icarus*, **170**, 343–364, doi: 10.1016/j.icarus.2004.04.005.

LASKAR, J., LEVRARD, B. & MUSTARD, J. F. 2002. Orbital forcing of the Martian polar layered deposits. *Nature*, **419**, 375–377, doi: 10.1038/nature01066.

LEVY, J. S., HEAD, J. W., MARCHANT, D. R., DICKSON, J. L. & MORGAN, G. A. 2009. Geologically recent gully–polygon relationships on Mars: insights from the Antarctic dry valleys on the roles of permafrost, microclimates, and water sources for surface flow. *Icarus*, **209**, 113–126.

LOWELL, P. 1908. *Mars as the Abode of Life*. Macmillan, New York.

MALIN, M. C. & EDGETT, K. S. 2000*a*. Sedimentary rocks of early Mars. *Science*, **290**, 1927–1937.

MALIN, M. C. & EDGETT, K. S. 2000*b*. Evidence for recent groundwater seepage and surface runoff on Mars. *Science*, **288**, 2330–2335.

MALIN, M. C. & EDGETT, K. S. 2001. Mars global surveyor mars orbiter camera: interplanetary cruise through primary mission. *Journal of Geophysical Research*, **106**, 23 429–23 570.

MALIN, M. C., EDGETT, K. S., POSIOLOVA, L. V., MCCOLLEY, S. M. & NOE DOBREA, E. Z. 2006. Present day impact cratering rate and contemporary gully activity on Mars. *Science*, **314**, 1573–1577.

MANGOLD, N., MAURICE, S., FELDMAN, W., COSTARD, F. & FORGET, F. 2003. Geographical relationships between small scale polygons and ground ice distribution from neutron spectrometer on Mars. *In*: *Conference Proceedings of Third International Conference on Mars Polar Science and Exploration. Alberta, Canada, 13–17 October 2003*. Abstract no. 8043.

MANGOLD, N., MAURICE, S., FELDMAN, W. C., COSTARD, F. & FORGET, F. 2004*a*. Spatial relationships between patterned ground and ground ice detected by the Neutron Spectrometer on Mars. *Journal of Geophysical Research*, **109**, E08001, doi: 10.1029/2004JE002235.

MANGOLD, N., QUANTIN, C., ANSAN, V., DELACOURT, C. & ALLEMAND, P. 2004*b*. Evidence for precipitation on Mars from dendritic valleys in the valles Marineris area. *Science*, **305**, 78–81.

MARTIN, L. J. & ZUREK, R. W. 1993. An analysis of the history of dust storm activity on Mars. *Journal of Geophysical Research*, **98**, 3221–3246.

MCCAULEY, J. F., CARR, M. H. *ET AL*. 1972. Preliminary Mariner 9 report on the geology of Mars. *Icarus*, **17**, 289–327.

MCGILL, G. E. 1985. Age and origin of large Martian polygons. *Lunar and Planetary Science*, **16**, 534–535.

MCSWEEN, H. Y., TAYLOR, G. J. & WYATT, M. B. 2009. Elemental composition of the Martian crust. *Science*, **324**, 736–739.

MELLON, M. T. 1997. Small-scale polygonal features on Mars: seasonal thermal contraction cracks in permafrost. *Journal of Geophysical Research*, **102**, 25 617–25 628.

MELLON, M. T., ARVIDSON, R. E. *ET AL*. 2009. Ground ice at the Phoenix landing site: stability state and origin. *Journal of Geophysical Research*, **114**, E00E07, doi: 10.1029/2009JE003417.

MELLON, M. T., ARVIDSON, R. E., MARLOW, J. J., PHILLIPS, R. J. & ASPHAUG, E. 2008*a*. Periglacial landforms at the Phoenix landing site and the northern plains of Mars. *Journal of Geophysical Research*, **113**, E00A23, doi: 10.1029/2007JE003039.

MELLON, M. T., BOYNTON, W. V. *ET AL*. 2008*b*. A prelanding assessment of the ice table depth and ground ice characteristics in Martian permafrost at the Phoenix landing site. *Journal of Geophysical Research*, **113**, E00A25, doi: 10.1029/2007JE003067.

METZ, J. M., GROTZINGER, J. P., RUBIN, D. D., LEWIS, K. W., SQUYRES, S. W. & BELL, J. F. 2009. Sulfate-rich eolian and wet interdune deposits, Erebus crater, Meridiani Planum, Mars. *Journal of Sedimentary Research*, **79**, 247–264.

MONTGOMERY, D. R. & GILLESPIE, A. 2005. Formation of Martian outflow channels by catastrophic dewatering of evaporite deposits. *Geology*, **33**, 625–628, doi: 10.1130/G21270AR.1.

MORGENSTERN, A., HAUBER, E., REISS, D., VAN GASSELT, S., GROSSE, G. & SCHIRRMEISTER, L. 2007. Deposition and degradation of a volatile-rich layer in Utopia Planitia, and implications for climate history on Mars. *Journal of Geophysical Research*, **112**, E06010, doi: 10.1029/2006JE002869.

MOUGINIS-MARK, P. & YOSHIOKA, M. T. 1998. The long lava flows on Elysium Planitia, Mars. *Journal of Geophysical Research*, **103**, 19 389–19 400.

MUSTARD, J. F., COOPER, C. D. & RIFKIN, M. K. 2001. Evidence for recent climate change from the identification of youthful, near-surface ice. *Nature*, **412**, 411–414.

MUSTARD, J. F., MURCHIE, S. L. *ET AL.* 2008. Hydrated silicate minerals on Mars observed by the Mars Reconnaissance Orbiter CRISM instrument. *Nature*, **454**, 305–309.

MUTCH, T. A., ARVIDSON, R. E., BINDER, A. B., GUINESS, E. A. & MORRIS, E. C. 1977. The geology of the Viking 2 Landing site. *Journal of Geophysical Research*, **82**, 4452–4467.

NEUKUM, G., BASILEVSKY, A. T. *ET AL.* 2010. The geologic evolution of Mars: episodicity of resurfacing events and ages from cratering analysis of image data and correlation with radiometric ages of Martian meteorites. *Earth and Planetary Science Letters*, **294**, 204–222.

NEUKUM, G., JAUMANN, R. *ET AL.* 2004. Recent and episodic volcanic activity and glacial activity on Mars revealed by the High Resolution Stereo Camera. *Nature*, **432**, 971–979.

OPIK, E. J. 1966. The Martian surface. *Science*, **153**, 255–265.

PARKER, T. J., GORSLINE, D. S., SAUNDERS, R. S., PIERI, D. C. & SCHNEEBERGER, D. M. 1993. Coastal geomorphology of the Martian northern plains. *Journal of Geophysical Research*, **98**, 11 061–11 078.

PECHMANN, J. C. 1980. The origin of polygonal troughs on the Northern Plains of Mars. *Icarus*, **42**, 185–210.

PELLETIER, J. D. 2004. How do spiral troughs form on Mars? *Geology*, **32**, 365–367.

PHILLIPS, R. J., ZUBER, M. T. *ET AL.* 2001. Ancient geodynamics and global-scale hydrology on Mars. *Science*, **291**, 2587–2591, doi: 10.1126/science.1058701.

PIERI, D. C. 1980. Martian valleys: morphology, distribution, age, and origin. *Science*, **210**, 895–897.

PLESCIA, J. B. 1990. Recent flood lavas in the Elysium region on Mars. *Icarus*, **88**, 465–490.

RAITALA, J. T. 1988. Tharsis block tectonics on Mars. *Earth, Moon, and Planets*, **41**, 201–216, doi: 10.1007/BF00113983.

RICE, J. W., PARKER, T. J., RUSSEL, A. J. & KNUDSEN, O. 2002. Morphology of fresh outflow channel deposits on Mars. *In*: *Proceedings of the 33rd Annual Lunar and Planetary Science Conference, March 11–15, 2002, Houston, Texas*. Lunar and Planetary Institute, Houston, TX, Abstract 2026.

SAGAN, C., TOON, O. B. & GIERASCH, P. J. 1973. Climatic change on Mars. *Science*, **181**, 1045–1049.

SMITH, D. E., ZUBER, M. T. *ET AL.* 1999. The global topography of mars and implications for surface evolution. *Science*, **284**, 1495–1503, doi: 10.1126/science.284.5419.1495.

SMITH, P. H. & MARS PATHFINDER TEAM. 1997. Results from the Mars pathfinder camera. *Science*, **278**, 1758–1765.

SNYDER, C. W. 1979. The planet mars as seen at the end of the viking mission. *Journal of Geophysical Research*, **84**, 8487–8519.

SNYDER, C. W. & MOROZ, V. I. 1992. Spacecraft Exploration of Mars. *In*: KIEFFER, H. H., JAKOSKY, B. M., SNYDER, C. W. & MATTHEWS, M. S. (eds) *Mars*. University of Arizona Press, Tucson, AZ.

SOARE, R. J., OSINSKI, G. R. & ROEHM, C. L. 2008. Thermokarst lakes and ponds on Mars in the very recent (late Amazonian) past. *Earth and Planetary Science Letters*, **272**, 382–393.

SQUYRES, S. W. & KNOLL, A. H. 2005. Sedimentary rocks at Meridiani Planum: origin, diagenesis, and implications for life on Mars. *Earth and Planetary Science Letters*, **240**, 1–10, doi: 10.1016/j.epsl.2005.09.038.

SQUYRES, S. W., ARVIDSON, R. E. *ET AL.* 2003. Athena mars rover science investigation. *Journal of Geophysical Research*, **108**, 8062, doi: 10.1029/2003JE002121.

SQUYRES, S. W., ARVIDSON, R. E. *ET AL.* 2004*a*. The spirit rover's athena science investigation at Gusev Crater, Mars. *Science*, **305**, 794–799.

SQUYRES, S. W., ARVIDSON, R. E. *ET AL.* 2004*b*. The Opportunity rover's Athena science investigation at Meridiani Planum, Mars. *Science*, **306**, 1698–1703.

SQUYRES, S. W., GROTZINGER, J. P. *ET AL.* 2004*c*. In situ evidence for an ancient aqueous environment at Meridiani Planum, Mars. *Science*, **306**, 1709–1714.

TANAKA, K. L. & CHAPMAN, M. G. 1990. The relation of catastrophic flooding of Mangala Valles, Mars, to faulting of Memnonia Fossae and Tharsis volcanism. *Journal of Geophysical Research*, **95**, 14 315–14 323.

TAYLOR, F. W., CALCUTT, S. B., READ, P. L., LEWIS, S. R., MCCLEESE, D. J., SCHOFIELD, J. T. & ZUREK, R. W. 2006. Atmospheric temperature sounding on Mars, and the climate sounder on the 2005 reconnaissance orbiter. *Advances in Space Research*, **38**, 713–717.

WANG, C., MANGA, M. & WONG, A. 2005. Floods on Mars released from groundwater by impact. *Icarus*, **175**, 551–555, doi: 10.1016/j.icarus.2004.12.003.

WARNER, N., GUPTA, S., KIM, J., LIN, S. & MULLER, J. P. 2010*a*. Hesperian equatorial thermokarst lakes in Ares Vallis as evidence for transient warm conditions on Mars. *Geology*, **38**, 71–74.

WARNER, N., GUPTA, S., LIN, S., KIM, J., MULLER, J. P. & MORLEY, J. 2010*b*. Late Noachian to Hesperian climate change on Mars: evidence of episodic warming from transient crater lakes near Ares Vallis. *Journal of Geophysical Research*, **115**, doi: 10.1029/2009JE003522.

WARNER, N., GUPTA, S., MULLER, J. P., KIM, J. & LIN, S. 2009. A refined chronology of catastrophic outflow events in Ares Vallis, Mars. *Earth and Planetary Science Letters*, **288**, 58–69.

Wilson, S. A. & Zimbelman, J. R. 2004. Latitude-dependent nature and physical characteristics of transverse aeolian ridges on Mars. *Journal of Geophysical Research*, **109**, E10003, doi: 10.1029/2004JE002247.

Yen, A. S., Morris, R. V. *et al.* 2008. Hydrothermal processes at Gusev Crater: an evaluation of paso robles class soils. *Journal of Geophysical Research*, **113**, E06S10, doi: 10.1029/2007JE002978.

Zimbelman, J. R., Craddock, R. A., Greeley, R. & Kuzmin, R. O. 1992. Volatile history of Mangala Valles, Mars. *Journal of Geophysical Research*, **97**, 18 309–18 317.

Zuber, M. T. 2001. The crust and mantle of Mars. *Nature*, **412**, 220–227, doi: 10.1038/35084163.

Morphological and geographical evidence for the origin of Phobos' grooves from HRSC *Mars Express* images

JOHN B. MURRAY[1]* & JONATHAN C. ILIFFE[2]

[1]*Department of Earth & Environmental Sciences, The Open University, Milton Keynes MK7 6AA, UK*

[2]*Department of Civil, Environmental and Geomatic Engineering, University College London, Gower Street, London WC1E 6BT, UK*

**Corresponding author (e-mail: j.b.murray@open.ac.uk)*

Abstract: The surface of Phobos, the 27 × 22 × 18 km inner moon of Mars, is dominated by several families of parallel grooves. At least seven different groups of hypotheses have been advanced to explain their origin, but studies have always been limited by the fact that, until recently, much of Phobos was imaged at a resolution too low to show grooves. Now, however, the High Resolution Stereo Camera (HRSC) on board the European *Mars Express* mission has made 134 imaging fly-bys past Phobos. The pictures of the previously poorly imaged regions and much of the rest of the satellite have been returned with resolutions down to a few metres, facilitating the construction of a more complete map of the grooves. Each of the seven hypotheses was tested against the new data on groove morphology, positions and orientations, and it was found that six of the previous hypotheses could be discarded. The only hypothesis to pass all tests was that they are chains of secondary impact craters from primary impacts on Mars. An implication of these results is that previous estimates of an unusually thick Phobos regolith of 100–200 m depth are no longer necessary, and our conclusions place no constraints on the interior of Phobos, so recent evidence that Phobos is a 'rubble pile' is consistent with our work. The preferred hypothesis also sheds light on the origin of crater chains on Eros, and on impact processes in the early stages of crater excavation.

Phobos is the largest of Mars' two satellites, both of which are tiny asteroid-sized bodies orbiting close to the planet in near-circular, equatorial orbits. Its rotation is synchronous with its orbital period, which is only 7 h 39 min, faster than Mars' rotation period of 24 h 37 min, so that from Mars' surface it appears to travel in the opposite direction to other celestial objects, rising and setting twice each Martian day. Its orbital radius averages 9377 km, but it orbits at a mean distance of only 5988 km from Mars' surface, by far the closest of any satellite to its primary – the next nearest being Charon, which orbits at 16 380 km above Pluto's surface. Only Mars- or Earth-crossing asteroids, such as Eros, can approach closer to a planetary surface. Phobos is often likened to a potato in shape, with dimensions approximately 27 × 22 × 18 km. Its density is extremely low, 1.876 ± 20 kg m^{-3} (Andert *et al.* 2010), and recent spectral measurements (e.g. Murchie & Erard 1996; Rivkin *et al.* 2002; Murchie *et al.* 2008) indicate that it is closer to a D-type asteroid than a C-type, although Phobos' spectrum appears quite different from both (Bibring 2010; Giuranna *et al.* 2010; Palomba *et al.* 2010).

The most striking features of Phobos are the prominent grooves and crater chains that run across its surface for distances of more than 25 km in almost artificial-looking parallel lines covering the leading half of the satellite in its orbit. Since these grooves were first discovered from *Viking* images in 1977 (Duxbury & Veverka 1977), many hypotheses have been advanced to explain them. Few of the early papers considered their morphology, orientation or geographical distribution in any detail, with the notable exception of Thomas *et al.* (1979), yet these characteristics are a primary key to their origin and to eliminating many of the hypotheses. The history of the investigation of Phobos' grooves is a nice illustration of the vital role that geomorphology plays in determining the origin of planetary surfaces and the processes that formed them. The different hypothetical origins of the grooves that have been proposed can be summarized as follows.

Chains of secondary impact craters from Stickney

The resemblance of some grooves to secondary impact craters was noted very early on (Veverka & Duxbury 1977) and led to the idea that the

From: Balme, M. R., Bargery, A. S., Gallagher, C. J. & Gupta, S. (eds) *Martian Geomorphology*. Geological Society, London, Special Publications, **356**, 21–41.
DOI: 10.1144/SP356.3 0305-8719/11/$15.00

grooves were secondary impact craters from Stickney Crater, which at 10.5 km in diameter is the largest crater on Phobos (Head & Cintala 1979; Davis *et al.* 1980). This idea was supported by *Viking* images that seemed to show that many of the grooves were prominent near Stickney and appeared to emanate from it. More supporting evidence was that the grooves fade out near the opposite side of Phobos.

Rolling boulders from the Stickney impact

Similar to the previous idea, this hypothesis envisages boulders ejected from Stickney at low velocity in the final stages of crater excavation re-impacting gently on Phobos and rolling across its surface, causing the grooves (Wilson & Head 1989).

Fractures caused by the Stickney impact

The hypothesis that the grooves were radial fractures resulting from the Stickney impact was proposed by Thomas *et al.* (1978), based on the fact that this impact was almost large enough to have broken the satellite apart altogether, so a Phobos-wide family of fractures was considered a likely consequence.

Fractures caused by tidal forces

A fourth idea was that tidal forces might have induced a global fracture pattern (Soter & Harris 1977; Weidenschilling 1979). Dobrovolskis (1982) used a semi-analytical treatment to produce maps of the stress field within Phobos at past, present and future dates.

Fractures caused by drag forces during capture

Following the theory that Phobos is a captured asteroid, it was proposed that during capture Phobos might have experienced drag forces sufficiently different from one side of the satellite to the other to have fractured the body throughout, thus causing the grooves (Pollack & Burns 1977).

Reopening of drag force fractures by the Stickney impact

A combination of ideas from the subsections on 'Fractures caused by the Stickney impact' and 'Fractures caused by drag forces during capture', above, was also suggested: that the Stickney impact opened pre-existing Phobos-wide fractures caused by drag forces during capture (Thomas *et al.* 1979). This probably remains the most popular theory to this day (e.g. Veverka & Burns 1980; Fujiwara 1991; Spohn *et al.* 1998; Buczkowski *et al.* 2008).

Secondary impact crater chains from primary impacts on Mars

Finally, it was proposed that the grooves on Phobos were caused by the secondary impact of ejecta from primary craters on Mars (Murray *et al.* 1994, 2006). In this model, melted and fluidized material ejected in the earliest stages of crater excavation from large impact events on Mars forms itself into strings of ejecta that impact Phobos, probably as partially melted clots. Each impact event forms one family of parallel grooves.

A new groove map from *Mars Express* HRSC images

Mars Express has imaged Phobos more than any previous spacecraft, by a very large margin. At present (January 2010) the number of passes stands at 134 (Table 1).

The HRSC camera has imaged areas that had not been covered by previous missions, and has repeatedly covered both known and unknown areas from different viewing angles and under different lighting conditions, and sometimes at greatly improved resolutions, down to a few metres. A map of coverage is included in Willner *et al.* (2010). The only areas of Phobos that remain to be more extensively covered by HRSC surround the trailing apex, where *Viking* coverage is still the best.

From this large amount of new data, it has been possible to construct a new map of the grooves of Phobos, which shows about three times the number of grooves than previously documented (Fig. 1). Viewing direction, sun angle and direction of illumination can all be critical to the detection of grooves, as a feature can be quite invisible when these are unfavourable (Fig. 2). The great variety of illumination and viewing angle now available has greatly aided the discovery of new grooves, and only those images where the viewing angle and lighting is sufficient to see clearly defined shadows were used to construct the map. All elongated depressions or crater chains were mapped as grooves: ridges or albedo features were not included. This map provides new information vital to the understanding of how these grooves were formed, and allows us to discriminate between the above-mentioned seven hypotheses.

The new groove map illustrates a number of characteristics of the groove distribution, which are summarized in the list below. By far the most thorough of the early papers on Phobos was that of

Table 1. *List of HRSC Phobos imaging passes up to orbit 7742*

Orbit	Date	*D* (km)	*R* (m)	Orbit	Date	*D* (km)	*R* (m)
413	2004-05-18	1881	17	2673	2006-02-10	2077	19
649	2004-07-23	1837	17	2682	2006-02-12	5190	47
682	2004-08-01	1466	13	2706	2006-02-19	2127	20
715	2004-08-11	1216	11	2739	2006-02-28	1767	16
748	2004-08-20	1245	11	2747	2006-03-02	931	36
756	2004-08-22	149	6	2756	2006-03-05	4153	38
1064	2004-11-16	4677	43	2780	2006-03-12	613	6
1163	2004-12-14	3816	35	2805	2006-03-19	1962	18
1212	2004-12-28	1969	79	2813	2006-03-21	828	8
1558	2005-04-03	3588	33	2846	2006-03-30	1322	12
1574	2005-04-08	3798	35	2854	2006-04-01	887	9
1607	2005-04-17	3986	36	2912	2006-04-18	2385	22
1769	2005-06-02	1303	111	2979	2006-05-06	5075	46
1901	2005-07-09	3107	28	3005	2006-05-13	5307	48
2151	2005-09-16	3826	35	3245	2006-07-20	4929	45
2192	2005-09-28	2863	238	3310	2006-08-07	567	6
2233	2005-10-09	2113	19	3761	2006-12-11	2050	19
2381	2005-11-20	3288	31	3769	2006-12-14	744	7
2397	2005-11-24	1784	17	3802	2006-12-23	877	8
2405	2005-11-27	1423	13	3835	2007-01-01	1235	11
2446	2005-12-08	2026	19	3843	2007-01-03	650	6
2463	2005-12-13	4253	39	3868	2007-01-10	1656	15
2479	2005-12-17	2964	28	3876	2007-01-13	1223	11
2487	2005-12-20	2581	24	3909	2007-01-22	1906	18
2501	2005-12-24	5279	49	3942	2007-01-31	2436	22
2583	2006-01-15	4346	177	3999	2007-02-16	3816	35
2601	2006-01-20	5254	48	4000	2007-02-16	3585	143
2643	2006-02-01	4669	44	4030	2007-02-25	11 268	103
4233	2007-04-22	2005	19	4880	2007-10-21	1288	12
4274	2007-05-04	1084	10	4888	2007-10-23	1083	10
4307	2007-05-13	580	5	4913	2007-10-30	1850	17
4332	2007-05-20	1828	17	4946	2007-11-08	2364	22
4340	2007-05-22	682	6	5163	2008-01-08	4090	37
4233	2007-04-22	2005	19	5277	2008-02-10	3141	29
4274	2007-05-04	1084	10	5305	2008-02-18	1349	12
4307	2007-05-13	580	5	5343	2008-02-29	1040	10
4332	2007-05-20	1828	17	5362	2008-03-05	1053	10
4340	2007-05-22	682	6	5381	2008-03-11	1268	12
4348	2007-05-25	473	4	5409	2008-03-19	666	6
4373	2007-06-01	1268	12	5428	2008-03-24	796	8
4381	2007-06-03	785	8	5447	2008-03-29	1055	10
4414	2007-06-12	1415	13	5504	2008-04-15	2242	21
4447	2007-06-21	2026	18	5552	2008-04-28	5081	47
4529	2007-07-14	2742	25	5604	2008-05-13	4565	42
4554	2007-07-21	3442	32	5699	2008-06-09	3709	35
4568	2007-07-25	4826	44	5766	2008-06-28	2443	23
4603	2007-08-04	3548	33	5850	2008-07-22	4448	41
4636	2007-08-13	3916	36	5851	2008-07-23	91	1
4683	2007-08-27	3774	34	5861	2008-07-26	2317	21
4698	2007-08-31	4786	44	5870	2008-07-28	352	3
4765	2007-09-18	2067	19	5889	2008-08-03	655	6
4773	2007-09-21	1015	9	5908	2008-08-08	982	9
4806	2007-09-30	1229	11	5984	2008-08-30	2367	22
4814	2007-10-02	130	1	6042	2008-09-15	3713	34
4847	2007-10-11	657	6	6128	2008-10-10	4358	40
4855	2007-10-14	771	7	6217	2008-11-04	4818	44
6551	2009-02-08	5493	50	7038	2009-06-27	2742	25
6637	2009-03-04	5281	48	7048	2009-06-30	2826	26

(*Continued*)

Table 1. *Continued*

Orbit	Date	D (km)	R (m)	Orbit	Date	D (km)	R (m)
6745	2009-04-04	5293	48	7088	2009-07-12	4005	37
6748	2009-04-05	5307	48	7109	2009-07-18	3863	35
6757	2009-04-08	4419	41	7225	2009-08-20	5094	46
6896	2009-05-18	971	9	7407	2009-10-11	937	10
6906	2009-05-21	529	5	7478	2009-11-01	479	19
6916	2009-05-23	518	5	7488	2009-11-04	586	5
6926	2009-05-26	957	9	7492	2009-11-05	11 695	107
6987	2009-06-13	1734	16	7719	2010-01-09	5355	49
7017	2009-06-21	2830	26	7742	2010-01-16	4405	40

D (km), distance in kilometres; R (m), approximate highest spacial resolution in metres.

Thomas *et al.* (1979), who not only mapped the groove distribution for the first time, but also listed the curiously organized characteristics of the groove distribution, orientation and shape. They first noted many of the following characteristics.

(a) All grooves lie at the intersections of planes with the surface, that is, they form small circles (Veverka & Duxbury 1977; Thomas *et al.* 1978, 1979) or would do so if Phobos were spherical. This latter point is illustrated in Figure 3, which shows the same groove from two different viewpoints: on the top the groove is viewed from near the plane of the groove and appears straight; whilst at the bottom it is viewed from well outside the plane, and appears wavy in response to Phobos' non-spherical topography.
(b) There are several distinct families of grooves; the groove planes of members of the same family being parallel to each other (Thomas *et al.* 1979).
(c) For each family, all grooves are parallel to a line joining the leading and trailing apex of Phobos. This means that grooves from all families become parallel to each other along the sub-Mars (Fig. 4) and anti-Mars meridians (Thomas *et al.* 1979).
(d) Each groove family extends over no more than one hemisphere of the satellite.

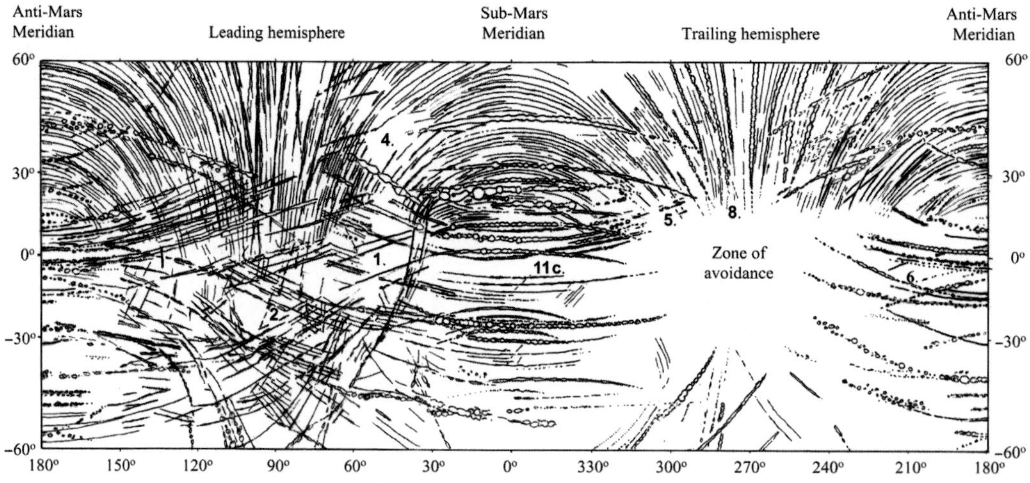

Fig. 1. Sketch map of grooves on the surface of Phobos derived from HRSC *Mars Express*, *Viking* and MGS (*Mars Global Surveyor*) images. A Mercator projection between −60° and +60° is used, and locations and orientations of features were assembled using crater positions from an existing control network (Duxbury & Callahan 1989). (1) Marks the centre of Stickney crater, the largest on Phobos; other numbers (2, 4, 5, 8 and 11c) refer to the approximate centres of figures in the text. Note that grooves become parallel along the sub-Mars and anti-Mars meridians (0° and 180°), and that there is a 'zone of avoidance' around which all grooves fade out and disappear surrounding the trailing apex (0° latitude, 270° longitude). The leading apex (0° latitude, 90° longitude) is characterized by groove families crossing each other at all orientations.

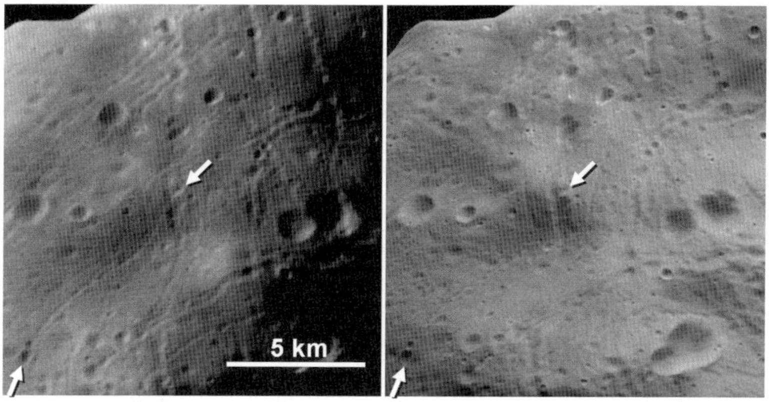

Fig. 2. The effect of lighting angle on perception of planetary features. The image on the right (h4307_nd2), despite having better resolution, fails to show the arrowed sinuous groove shown in the left image (h2854_nd). The difference in direction of illumination is about 90°. Image credit: ESA/DLR/FUB.

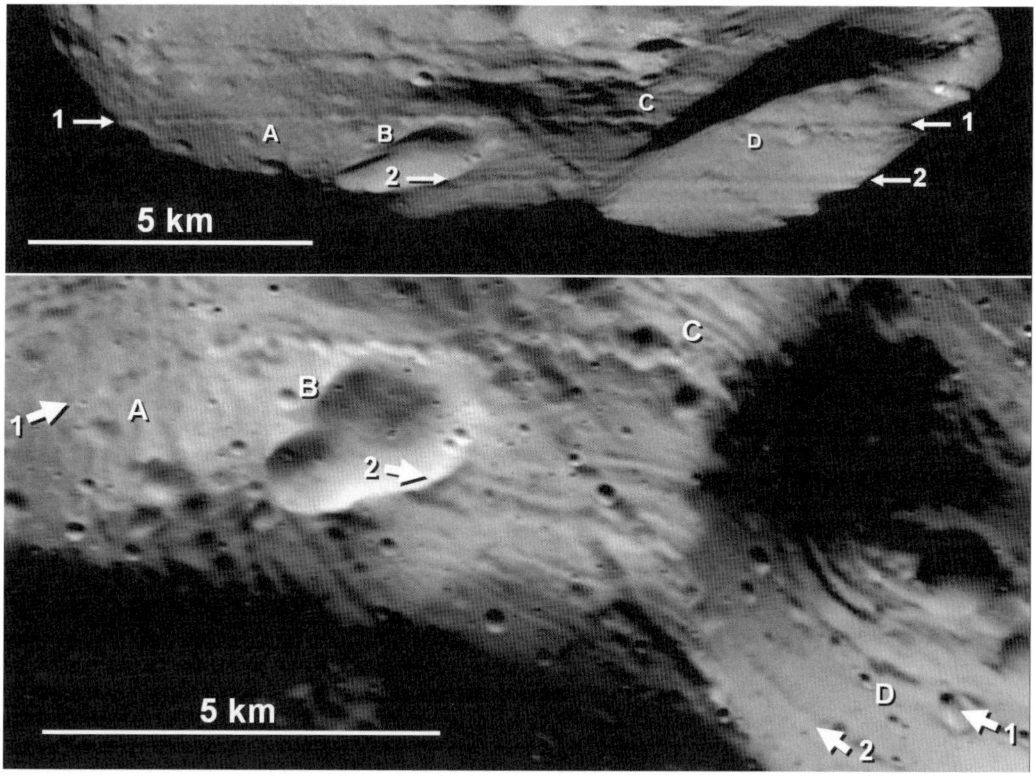

Fig. 3. The same groove viewed from different viewing points. Grooves 1 and 2 (arrowed) appear as straight and regular troughs exactly parallel to each other in the top image (h7478_0000.s22.03), where the viewing angle to the local surface is about 30° from the horizontal. In the lower image (h4307_nd2) the surface is seen from a viewing direction closer to the vertical. Despite the surface illumination being similar, the grooves appear as wavy, disjointed segments of roughly aligned groups of craters of different diameters, as at C and D, and above A and B. Craters A, B, C and D are the same in each image, labelled to aid identification of the groove traces in the lower image. Note the clear raised rim to the crater chain between C and B in the lower image. Image credit: ESA/DLR/FUB; see prelim viii for acronym definitions.

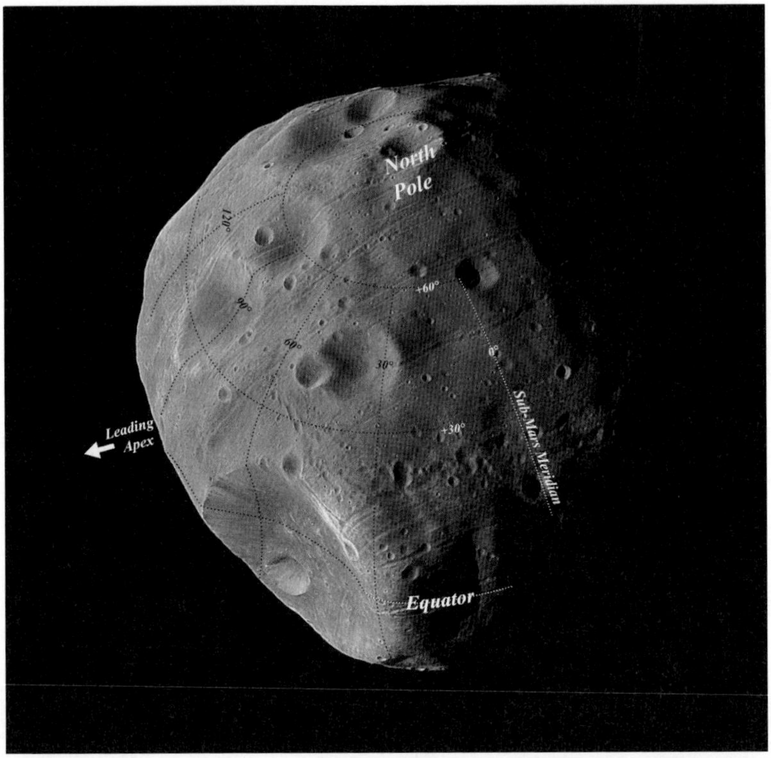

Fig. 4. *Mars Express* HRSC image h756_nd of the hitherto poorly imaged leading hemisphere of Phobos at a resolution of about 7 m per pixel, with approximate coordinate grids superimposed, and the leading apex and direction of Phobos' motion marked by an arrow. Note how all grooves and crater chains become parallel to each other along the sub-Mars meridian. Grooves inside craters enable pre- and post-groove impact craters to be distinguished. The large crater on the lower left limb is Stickney. Image credit: ESA/DLR/FUB.

(e) There is evidence that topographical highs, such as ridges and impact crater rims, act as a barrier to groove families close to the edge of their hemispheres (Fig. 5).
(f) The groove families were formed at different times; grooves from one family consistently cutting across those of another where relationships are clear (Thomas *et al.* 1979). This observation has been confirmed by the *Mars Express* images (Fig. 6) and has important consequences in constraining the origin of the grooves.
(g) Nowhere is the pattern of grooves radial. The apparently radial pattern of grooves in Figure 1 emanating from the equator at 90° and 270° is an artefact of the map projection: Figure 7 (left and right) is an orthographical projection that shows the same grooves to be parallel to each other.
(h) All grooves in its vicinity are younger than Stickney. They overlie its floor, rim and any rudimentary ejecta blanket that might exist (Veverka & Duxbury 1977). The grooves also overlie at least three impact craters that lie inside Stickney (Murray *et al.* 1994) and on its rim, suggesting that the ages of these grooves are substantially younger than Stickney.
(i) There is a 'zone of avoidance' surrounding the trailing apex of Phobos in its orbit (i.e. 270° longitude, 0° latitude) where there are no grooves found (Thomas *et al.* 1979). Grooves from each family fade out about 20°–30° from the trailing apex (Fig. 8), so this barren area forms a circle about 12 km wide.
(j) Grooves diminish in width towards the zone of avoidance near the trailing apex (Thomas *et al.* 1979).

Groove morphology and relative age

At high resolution, grooves can be seen to consist of contiguous pits, in places separating into distinct craters (Veverka & Duxbury 1977; Head & Cintala 1979). Under low lighting, they can be seen to have raised rims (Figs 3 & 9). Many grooves resemble chains of secondary impact craters on airless

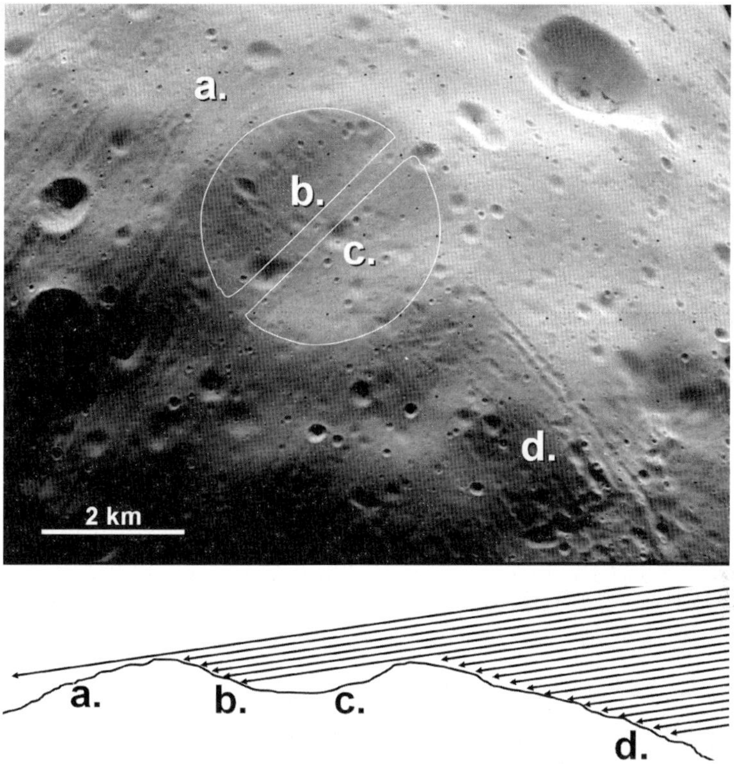

Fig. 5. (Top) Image of crater and superimposed grooves at longitude 300°, latitude +15°, close to the 'zone of avoidance' in the top right-hand part of the image where all grooves fade out (*Viking* image 248a01). North is to the bottom left and illumination is from the top of the picture. Note that the family of parallel grooves at d disappear on the east-facing slope (c) despite the favourable illumination angle (*c.* 20°), but reappears on the west-facing slope (b), only to disappear again on the outer east-facing slopes (a). (Bottom) Schematic section through a, b, c and d; east is to the left. The parallel arrows represent directions of impacting ejecta from a large impact on Mars (see the diagrams illustrating the formation of Phobos' grooves by the impact of ejecta from large impacts on Mars later in this chapter). The uncratered inner slope (c) is protected from ejecta impact by the crater rim to its west, and the outer slope (a) is similarly protected by the crater rim between a and b. The ejectum represented by the top arrow misses Phobos altogether. Image credit: NASA.

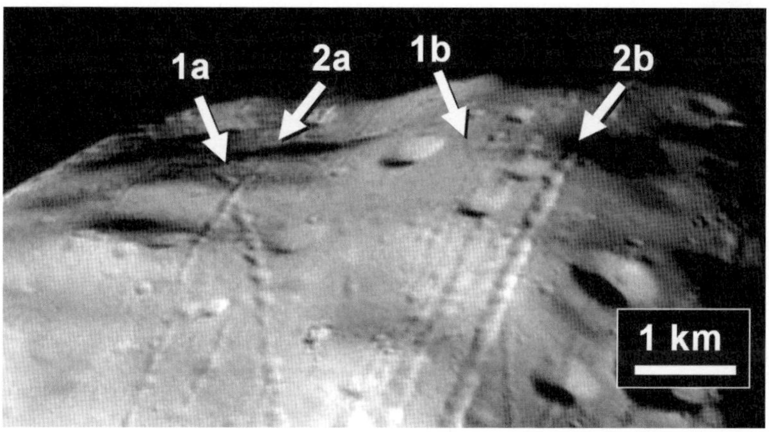

Fig. 6. The two parallel grooves 1a and 1b are cut by grooves 2a and 2b, respectively, indicating that the groove family represented by 2a and 2b is younger. From image h6906_0000.nd2. Image credit: ESA/DLR/FUB.

Fig. 7. (Top) Orthographic projections of groove positions on Phobos centred on (left) the leading apex (0° latitude, 90° longitude), (centre) the sub-Mars point and (right) the trailing apex. Note that grooves appear straight and in groups of parallel families when viewed from the leading and trailing apex (left and right). Four prominent families were named by Thomas *et al.* (1979); family A (the most prominent and numerous), B and D are marked. Also, note that all grooves become parallel along the sub-Mars meridian.

bodies, such as the Moon or Mercury, in that they comprise both linear troughs and elongate craters in roughly linear irregular groups (Fig. 10). It should be noted that the appearance of both grooves and secondary impact crater chains is strongly dependent upon lighting angle and the direction of illumination. The same feature can appear as either a linear trough or a roughly aligned chain of craters depending upon image resolution, viewing direction and illumination direction (Figs 3 & 9).

Fig. 8. View of Phobos near the trailing apex, indicated at the top right, from Viking image 246A08. The densely packed parallel grooves in the bottom half of the image become narrower and break up into irregular pits as they approach the dashed line, where they fade out altogether. Above this line, the terrain is smoother at the 100 m scale and entirely without any grooves. The image is about 7 km wide and north is towards the bottom left. Image credit: NASA.

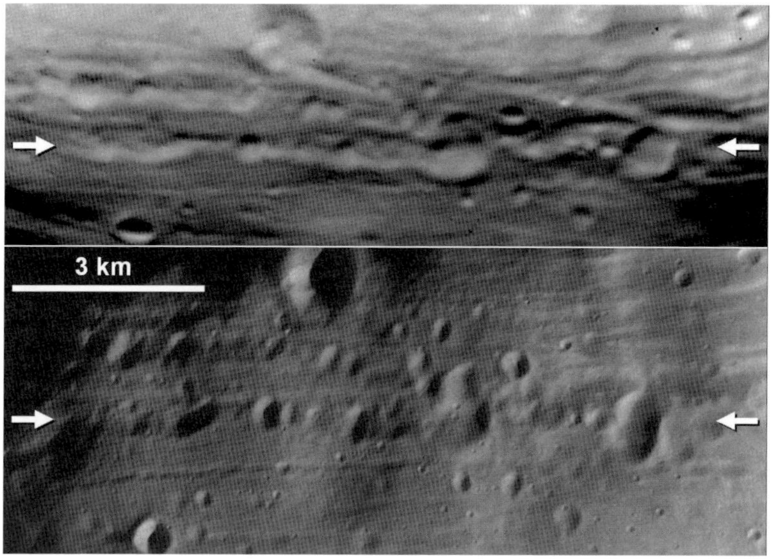

Fig. 9. Two views of the same groove (arrowed) under different illumination. The top image, with illumination from the top, shows a groove with a raised rim, composed of craters of different sizes. In the lower image with lighting from the right, the groove appears as a roughly aligned string of irregular pits and craters. Details from HRSC images h4447_sr2_0005 (top) and h756_nd2 (bottom); the centre of the images is 0° longitude, +23° latitude. Image credit: ESA/DLR/FUB.

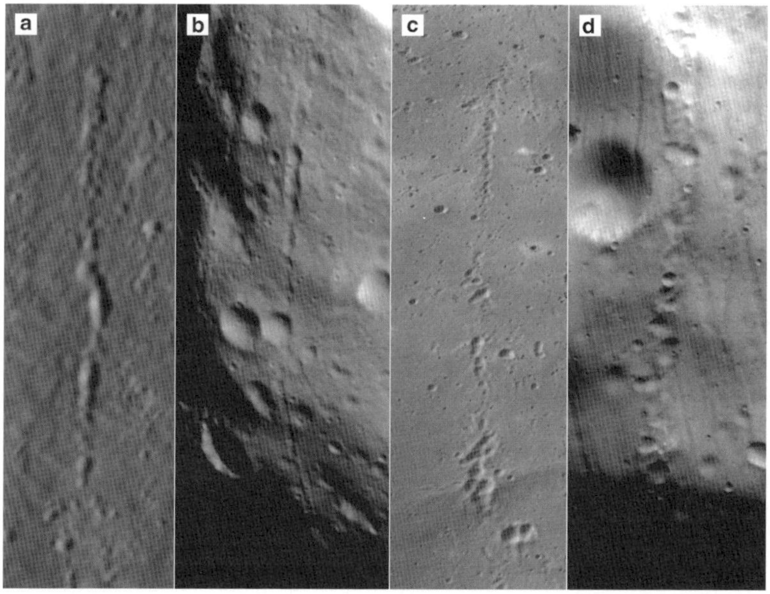

Fig. 10. (**a**) A linear chain of secondary impact craters radial to the crater Eminescu on Mercury (image width 38 km, from *Messenger* image PIA10610). (**b**) A groove on Phobos of similar appearance (image width 6 km, HRSC h4340_s22). (**c**) A more loosely grouped chain of irregular secondary impact craters emanating from the crater Copernicus on the Moon (image width about 18 km, *Apollo* image AS17-M-2441). (**d**) A similar chain of craters on Phobos (approximate width of image 4 km). Despite up to an order of magnitude difference in scale, note the morphological similarities between (a) and (b), and between (c) and (d). Image credit: ESA/DLR/FUB, NASA and NASA/JHUAPL/CIW.

The relative ages of the grooves from superposition relations show them to be younger than all craters larger than about 1 km in diameter. More importantly, in addition to being younger than Stickney, grooves are also superimposed on (and are therefore younger than) at least two craters on its northern lip that are 0.7 and 1 km in diameter. One groove also runs up the outer rim (i.e. the overturned flap) of Limtoe, a 2 km-diameter crater on the floor of Stickney. The fact that the grooves formed later than these three primary craters, which randomly impacted into Phobos subsequent to the Stickney impact, suggests that the grooves are likely to have formed millions of years later.

Discussion

We now examine each of the seven hypotheses described in the Introduction, and then use both first principles and the groove characteristics described in new groove map from the *Mars Express* HRSC images to test each of these hypotheses in turn.

Stickney secondary crater chains

Of the seven hypotheses advanced to explain the origin of Phobos' grooves listed in the Introduction, the first idea (see the subsection on 'Chains of secondary impact craters from Stickney'), that they were secondary impacts from Stickney crater, has some fundamental objections that cannot be overcome. The principal of these is the escape velocity of Phobos, between 3 and 11 m s^{-1} (Dobrovolskis & Burns 1980), which is so low that any ejecta re-impacting the surface will have much too low a velocity to form craters. A second objection is that the grooves are not radial to Stickney, many being tangential, and most forming small circles far from it. Another problem was noted by Dobrovolskis & Burns (1980): because of the interaction with Mars' gravity field, material ejected on its eastern side, where there are plenty of prominent grooves (Figs 9 & 11c), has an escape velocity of less than 3.4 m s^{-1}. This means that very little ejecta would land east of Stickney and any that did would form loops, as in Figure 12a (Thomas 1998). Ejecta ejected westwards at speeds of between 6 and 10 m s^{-1} would all re-impact the surface outside the crater, yet there are very few grooves emanating from Stickney in this direction, and those that do are not radial to the centre of Stickney (see Figs 1 & 13a). A fourth objection is the age difference: the grooves were formed after both Stickney and the three later impact craters superimposed on it. There are other objections, but for these four reasons alone we believe that this hypothesis is untenable.

Rolling boulder tracks from Stickney crater

The second hypothesis given earlier (see the subsection on 'Rolling boulders from the Stickney impact' in the Introduction), that the grooves were created by rolling boulders following the Stickney impact, also comes up against strong objections. Primary amongst these is the fact that the rolling boulders should be visible at the ends of the grooves thus created, but nowhere does this occur. Conversely, the few large boulders (>10 m) found on Phobos have no groove traces associated with them. Secondly, many grooves trace uphill sections for many kilometres. Thirdly, at least in their terminal stages, rolling boulders should keep close to the bottoms of valleys and roll normal to contour lines. Again, this never happens (Fig. 12b).

Fracture hypotheses

The ideas listed in the subsections 'Fractures caused by the Stickney impact' through to 'Reopening of drag force fractures by the Stickney impact' in the Introduction all postulate that the grooves and crater chains of Phobos are fractures formed by one mechanism or another. However, there are strong objections to the idea that the grooves can be fractures at all. In the first place, Phobos' grooves are quite unlike fractures in both appearance and behaviour. Those fracture fields on the Earth, the Moon and Mars of similar size and number-density to the grooves of Phobos exhibit concomitant faulting, whereas on Phobos there is no case of the ground being downfaulted on one side of a groove, nor any grabens; neither do crater rims cut by the grooves show any horizontal shifts indicating either strike-slip movement or lateral opening (Thomas *et al.* 1979). For such a large, complex, cross-cutting fracture field, this would be unprecedented. Fracture fields on the Earth, Mars and the Moon do not comprise uniformly long, straight lines such as those visible in Figure 14, but normally exhibit small changes of direction, are habitually segmented and en echelon sections are common.

Another difficulty for any fracture hypothesis is the angle of dip of the fractures. Figure 7 (top left) shows the large family of grooves that covers almost half of the satellite's surface (labelled A) from the plane of the grooves. The angle of dip becomes progressively shallower away from the centre of the grooves family, until those near the edge have dip angles of around 30° below the horizontal, again unprecedented for tensional faults.

Faults and fractures do not normally consist of the coalesced pits seen on Phobos, but this morphological difference has been explained by Thomas *et al.* (1979) as being the result of fracturing in solid rock beneath a thick regolith. They postulate

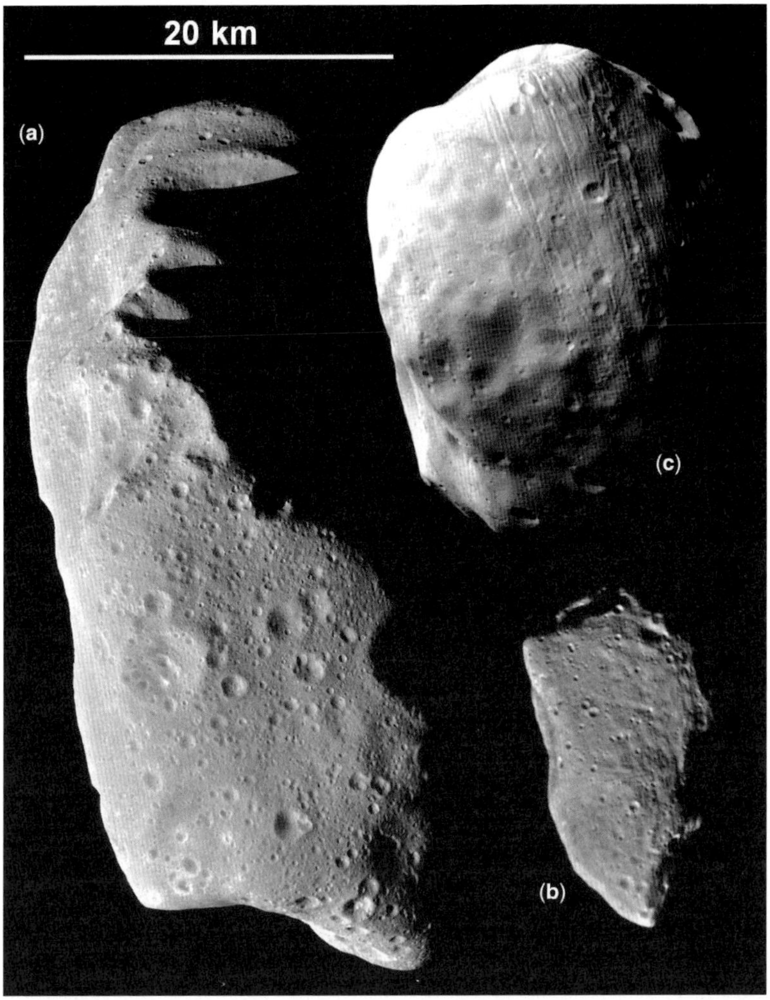

Fig. 11. Images of asteroids Ida (**a**, left), Gaspra (**b**, below right) and Phobos (**c**, top right), all shown at the same scale and similar illuminations. Neither Ida nor Gaspra have any well-marked parallel sets of grooves like those of Phobos. Models of Eros' unstable orbit have shown that it could have made repeated close approaches to both Mars and the Earth–Moon system over prolonged periods in the past, and thus have been in range of impact ejecta from any of these three bodies in a similar manner to Phobos. Ida and Gaspra images from NASA Planetary Photojournal Image PIA00332. Image credit: ESA/DLR/FUB and NASA/JPL.

that the observed pits are drainage (of dry, granular material) pits of Phobos' loose regolith into open fractures, as can occur under similar circumstances on Earth and Mars (Ferrill *et al.* 2004). A critical part of this model is that the regolith must already be fully formed at the surface when the fractures open. Such pit craters need a substantial subsurface void to accommodate collapse (Wyrick *et al.* 2004); however, the postulated extensional fractures must be very narrow, 20 m wide at the most, otherwise they would have caused visible distortion of pre-existing impact crater rims, which they do not (Thomas *et al.* 1979). Yet, the grooves are typically 100–200 m wide. Thomas *et al.* (1979) and subsequent authors have therefore suggested the presence of a regolith about 100–200 m thick that drains into the fractures, thereby producing grooves much wider than the fractures which caused them. Without this thick regolith in place at the time of fracture formation, any fracturing hypothesis cannot be maintained. More recent models have put the regolith depth at 20 m (Kuzmin *et al.* 2003), closer to that found on the Moon (Oberbeck & Quaide 1968).

Unlike Phobos' grooves, such pit crater chains are rimless (Wyrick *et al.* 2004). Nonetheless, Veverka

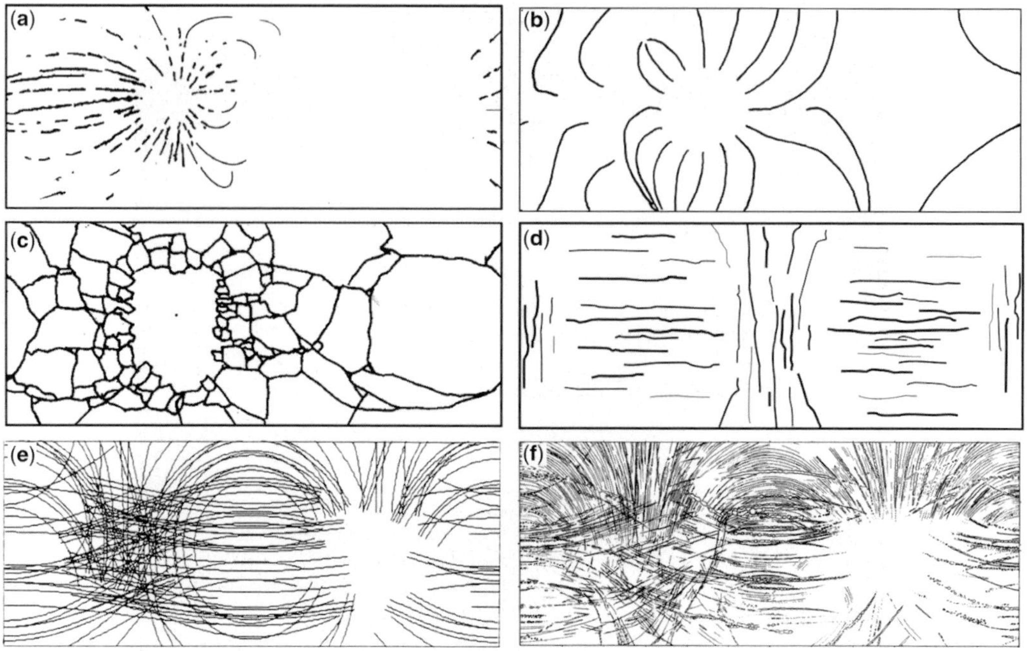

Fig. 12. Mercator projected models of groove orientation expected from four of the seven different theories of groove formation discussed. The theory involving fracturing by drag forces during capture, and the theory involving drag force fractures reopened by Stickney impact, are not shown, as no numerical or analogue model of either scenario has been published. (**a**) (Top left) Groove orientations expected if ejecta from Stickney could form secondary craters (redrawn based on Thomas 1998). Note that very little ejecta could be ejected in the prograde direction (right of the impact) and will form looped patterns due to Phobos' proximity to the Roche limit. (**b**) (Top right) Approximate groove orientations expected if they were formed by rolling boulders ejected from Stickney. In this presentation, the influence of Mars' gravity on the downhill direction is not included; orientations will be significantly altered for distances close to Mars. (**c**) (Middle left) Groove orientations expected if they were fractures formed by the impact of Stickney crater. These are fracture patterns from laboratory experiments of impacts into a Phobos ellipsoid, redrawn after Fujiwara & Asada (1983) from their model 44, which most resembles the Phobos groove distribution. Note that no patterns of straight parallel fractures occur. Instead, the overall pattern is one of polygonal fracturing, quite unlike the closely-spaced planar grooves of Phobos, although the authors point out that some fractures follow radial and concentric directions with reference to Stickney. (**d**) (Middle right) Map of grooves formed by tidal stress. No part of Phobos is in tension, so only thrust faults could occur (Dobrovolskis 1982), but this map shows the orientations of lines of minimal compression, which are normal to thrust fault directions. (**e**) (Lower left) Predicted secondary crater chain orientations from impacts at 12 different latitudes on Mars, chosen to match those seen on Phobos. Note the resemblance between this model and the map of grooves at the bottom. The model is simplified as a spherical Phobos, so does not fit the real situation as well as if it were modelled as a triaxial ellipsoid; nevertheless, the resemblance between theory and model illustrated in (e) is strong. (**f**) (Bottom right) Map of Phobos' grooves from HRSC images.

& Duxbury (1977) and Thomas *et al.* (1979) suggest that the Stickney impact may have generated enough heat to release volatiles from within Phobos along the groove fractures, which would then have entrained material from within the satellite and the regolith to form the raised rims. However, Phobos' tiny gravity field causes problems with these explanations. Even assuming a high mean ejection angle of 84° from the horizontal (Reidel *et al.* 2003), such material would have to have been ejected at a velocity of less than 2 m s^{-1} to form the raised rims adjacent to the grooves, which are typically 100–200 m wide. For comparison, Ma *et al.* (2002) found ejection velocities of material from comets one–two orders of magnitude greater, between 28 and 120 m s^{-1}, and Thomas *et al.* (1979) suggested several metres per second velocity for particles ejected from the grooves. This suggests that entrained material is likely to be either ejected at greater than escape velocity or widely distributed across the surface: this appears to be the case for comets, which show no signs of build-up of debris cones around volatile ejection sites (Fig. 15).

A further objection to fracturing is that all of Phobos is under compression and no part of the surface is in tension (Dobrovolskis 1982). This

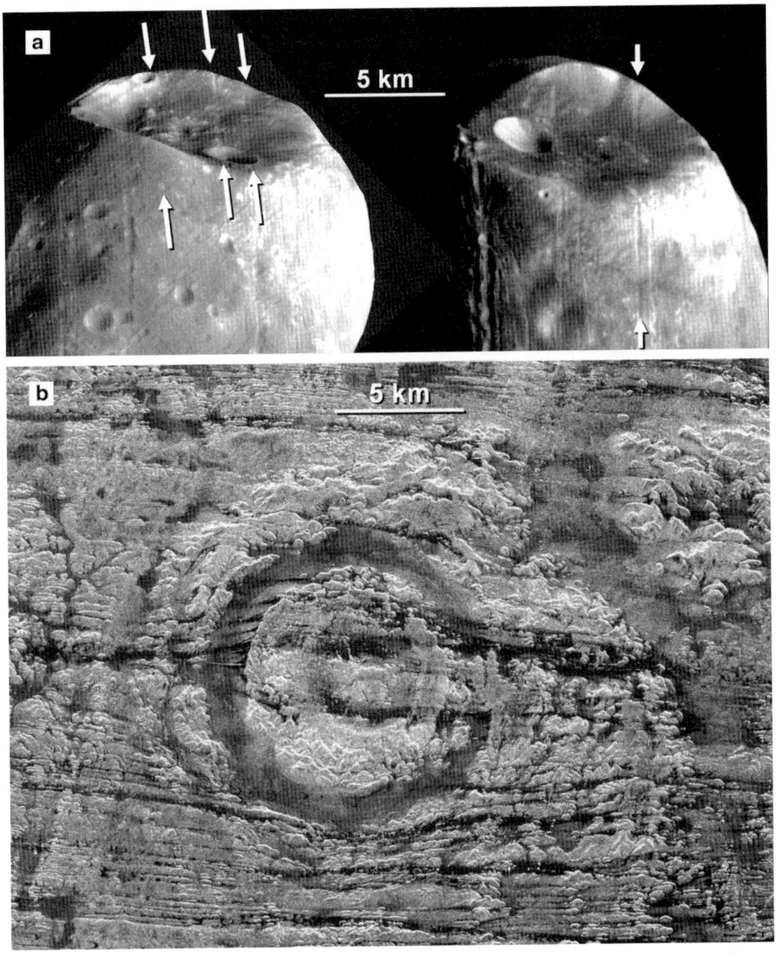

Fig. 13. (**a**) Two oblique views of the crater Stickney (diameter about 10.5 km) seen from the plane of two different groove families (arrowed), whose planes intersect each other at an angle of about 45°. There is no measurable deviation in direction as they cross the crater. (**b**) The deep structure of the 12.6 km-diameter Aorounga impact crater, Chad (centre), visible as circular moats in Devonian sandstones. Note the outward displacement of lineaments in the sandstones caused by radial compression during the impact and the resulting deformation. The impact crater is between 3500 and 12 000 years old (Becq-Giraudon *et al.* 1992). The SAR (synthetic aperature radar) image is from NASA Space Shuttle *Endeavour*, 1994. Image credit: ESA/DLR/FUB and NASA.

means that surface readjustments consequent upon the continuous impact cratering would tend to close fractures rather than open them. The possibility that grooves formed at a time when Phobos was part of a larger body is unlikely because their morphometry and orientation argue against it.

There are also fundamental objections to the cross-cutting grooves on Phobos being fractures of any kind. On Phobos the parallel planes of each younger family cut across older grooves without ever causing any deviation of the older set, at dip angles that differ from each other by 60°–80° (Fig. 7). The fact that each groove family has been formed at a different time (Thomas *et al.* 1979) means that it would not be possible for later parallel fracture sets to propagate through the voids created by earlier fracturing events (Fig. 16). For these reasons we believe that the grooves of Phobos cannot be the result of faults or fractures of any origin.

Fractures caused by the impact of Stickney crater

The principal argument for the third idea, that they are fractures consequent upon the Stickney

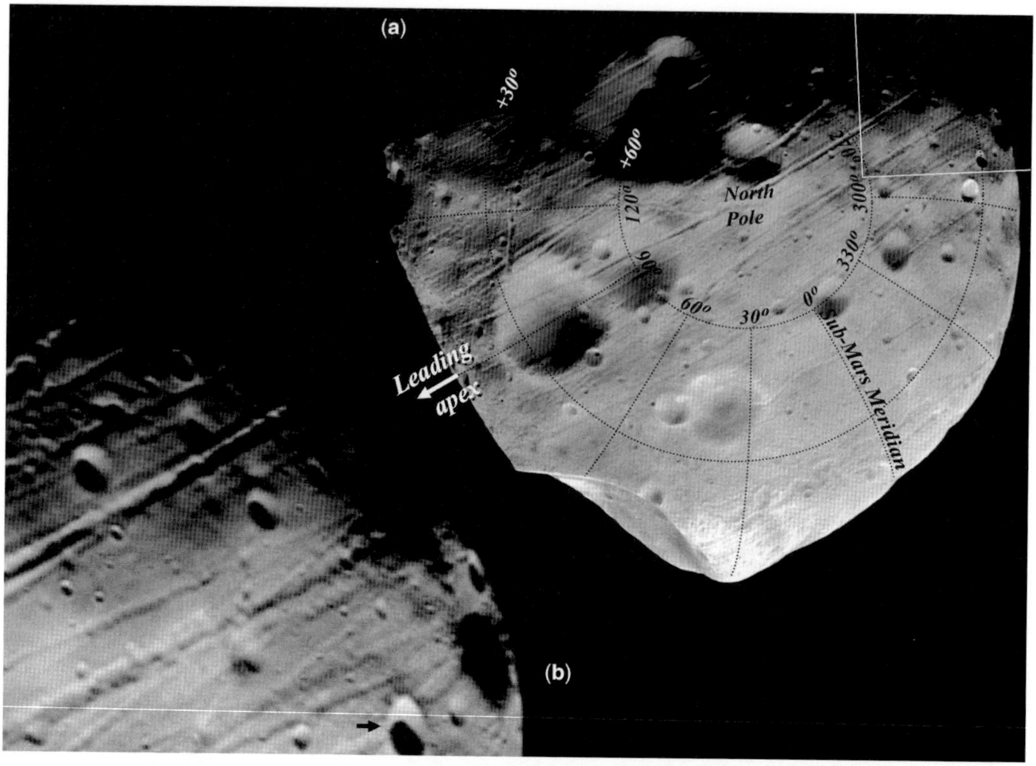

Fig. 14. (**a**) (Top right) Image h3310_0000.nd2 centred close to the north pole, dominated by the subparallel striations and crater chains of family A, the most complete groove family on Phobos that covers most of the northern hemisphere. Note that some individual grooves run unbroken for nearly 180° of latitude, and that the central groove of this family passes close to both the north pole and the leading apex. The tightly spaced grooves appear very straight and linear, but the super-resolution image (**b**, lower left) of the area in the box at top right shows that they comprise contiguous pits with raised rims (image h3310_sr2_0006). The top right-hand image is about 23 km left to right. Image credit: ESA/DLR/FUB.

impact, has been that the grooves appeared to emanate from Stickney crater and are most prominent near it. However, the HRSC mapping shows the association with Stickney crater to be an artefact of the previous coverage: groove positions and orientations form a moon-wide pattern that is quite independent of Stickney and bears no relation to it (see Fig. 1). Furthermore, work by Fujiwara & Asada (1983) has shown that laboratory experiments in which Phobos ellipsoid clay models were impacted by high-velocity projectiles do not produce the closely spaced straight lines that are observed on Phobos (Fig. 12). Instead, the model surface is broken by a polygonal fracture pattern, with size of polygons increasing with distance from the impact. Although the authors point out that some lines in this pattern are radial and concentric to the impact, on Phobos no fractures concentric to Stickney occur, and the overall model pattern is very different from Phobos. In addition, the large difference in age between the Stickney impact and the groove formation preclude such an explanation. Further problems are caused by the characteristics (d), (e) and (h) mentioned earlier, none of which is explained by this idea.

Fractures caused by tidal forces

The fourth idea mentioned in the Introduction, that the grooves result from fracturing caused by tidal forces, has received comparatively little attention. Dobrovolskis (1982), treating Phobos as a triaxial ellipsoid, has calculated the internal stress field taking into account tidal, rotational and self-gravitational stresses at different epochs in the past and future. He found that the minute values of the stresses and strains inside Phobos were too small to initiate widespread failure unless it were essentially a rubble pile. However, if Phobos were a rubble

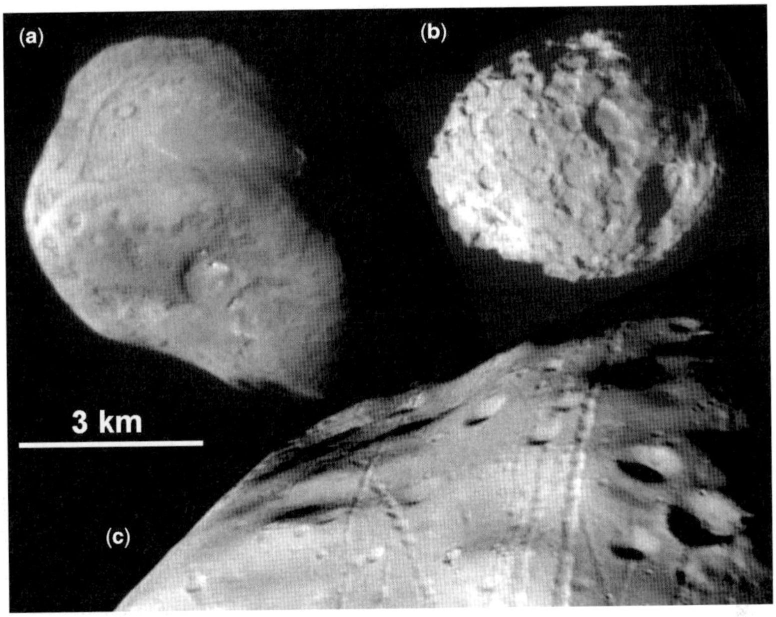

Fig. 15. Images of comet Tempel I (**a**) and comet Wild 2 (**b**) compared with Phobos at the same scale (**c**). Phobos' grooves and crater chains might be expected to resemble those of comets if outgassing from its interior had ever taken place. However, apart from a few impact craters, on neither comet is there any sign of chains of cones or craters with raised outer walls at the sites of volatile emission, and the appearance of both comets is distinctly different from Phobos. Tempel I image from NASA Planetary Photojournal image PIA02142. Image credit: NASA/JPL-Caltech/UMD, NASA/JPL and ESA/DLR/FUB.

pile, as now seems likely (Andert *et al.* 2010), then linear fractures could not have propagated through it because of the multiple discontinuities. Furthermore, the analysis by Dobrovolskis (1982) demonstrated that only thrust faults could occur on the surface because no part of the surface is actually in tension, despite Phobos' position within the Roche limit. (The Roche limit is defined as the distance from Mars (or any other body) within which a satellite held together only by its own gravity will disintegrate due to Mars' tidal forces exceeding the satellite's gravitational self-attraction.) Even if the stresses were sufficient to initiate fracture, these (tensional) fractures should run normal to the thrust faults, that is, north–south along the sub-Mars and anti-Mars meridian, but east–west at the leading and trailing apex, which does not match what is observed (Fig. 12). Therefore, for all these reasons we believe this idea to be contrary to the evidence.

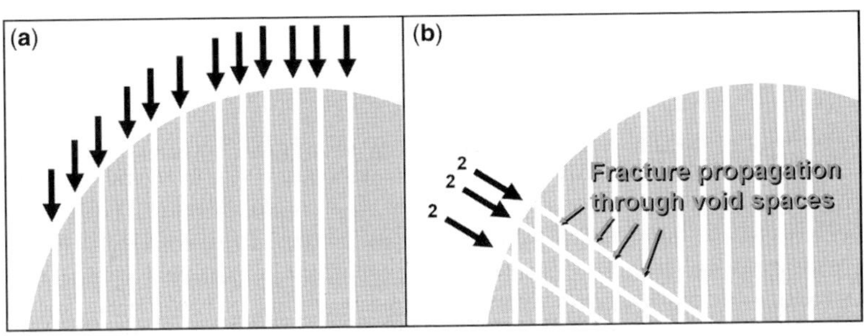

Fig. 16. (**a**) Schematic diagram of a set of hypothetical parallel fractures (arrowed; width greatly exaggerated) opened on Phobos, seen from the leading edge. (**b**) Shows a second family of cross-cutting parallel fractures formed at a later time and at a different orientation (labelled 2). The second fractures would have to propagate through the voids created by the first set of fractures, which is not possible.

Fractures caused by drag forces during capture

The fifth hypothesis, that the grooves are the result of drag forces during capture, cannot be sustained because Phobos' gravitational field is so tiny that drag forces sufficient to fracture Phobos would also have initiated enough acceleration to eject the regolith altogether (Thomas *et al.* 1979), and, as previously noted, a thick regolith at the time of fracture formation is essential to all fracture models. The fact that Phobos' spectral measurements are in some respects similar to a D-type body (Murchie & Erard 1996; Rivkin *et al.* 2002; Murchie *et al.* 2008) support the idea that it could be a captured asteroid; however, Phobos appears different from both C-type and D-type bodies in some important respects (Bibring 2010; Giuranna *et al.* 2010; Palomba *et al.* 2010), and, furthermore, there are important dynamical objections to the idea that Phobos was captured at all. Dynamicists maintain that it could not have been captured, but must have formed in orbit around Mars (see the discussion in Burns 1992). Nevertheless, the tidal capture necessary to produce fracturing cannot have taken place. This leaves aerodynamic drag in a huge protoatmosphere as the only means of capture (Pollack *et al.* 1979), with all of its problems of fine tuning to arrive at its present position without rapid orbital decay or escaping Mars' gravitational field. These results are confirmed by the latest *Mars Express* results, which are inconsistent with the proposition that Phobos is a captured asteroid (Andert *et al.* 2010).

Reopening of drag force fractures by the Stickney impact

The sixth hypothesis, involving the reopening of capture drag force fractures by the Stickney impact, is a combination of the Stickney impact fracturing idea and the drag forces model above. Thomas *et al.* (1979) proposed this model because drag forces during capture would have been sufficient to eject all of the regolith. They therefore proposed that a regolith 100–200 m thick had re-formed between the time of Phobos' capture and the Stickney impact. The impact itself reopened the fractures formed during capture, and the regolith drained into the fractures causing the lines of circular pits that form the grooves. However, this theory cannot be sustained because the impact that caused Stickney would also have produced very large surface tremors, causing all surface regolith particles to be launched from the surface at greater than Phobos' tiny escape velocity, the entire regolith thus being lost into space for a second time (Horstman & Melosh 1989). As noted earlier, any fracturing hypothesis requires 100–200 m of regolith to be in place at the time of formation of the grooves, and to remain there until the present. A subsequent regolith that forms after fracture opening does not result in collapse pits. This idea is also subject to the same objections as the fourth hypothesis regarding the questions over Phobos' capture.

Another test of this hypothesis is to look at the grooves that cross Stickney crater (Fig. 13a). During impact crater excavation, the transient cavity is created by downward and outward compression of the target rock, causing radial deformation of pre-existing features. If the grooves in Figure 13a are fractures that formed prior to Stickney, they would have been subject to radial outward movements that deformed the groove traces away from the centre of Stickney, as occurs in impact craters on Earth (e.g. the Aorounga impact structure: Becq-Giraudon *et al.* 1992) (Fig. 13b). This is clearly not the case: grooves cross Stickney without any deviation in their planar nature (Fig. 13a).

Secondary impact chains from primary craters on Mars

The final hypothesis, that the grooves of Phobos are secondary impact crater chains from impacts on Mars, is explained in more detail in Figure 17. Unlike all of the other ideas, the pattern of grooves on Phobos almost exactly matches that predicted by theory (Fig. 12). On this hypothesis, each groove family originates from a large impact on Mars and is composed of radial (effectively parallel at the distance of Phobos) coalesced crater chains. These would, therefore, create the parallel plane intersections observed, each family having a different orientation, but the motion of Phobos would ensure that the plane passing through the leading apex of Mars would also pass through the centre of Phobos. This idea is also the only one that explains why each groove family covers only one hemisphere of Phobos, and also why the groove families are of different ages. The 'zone of avoidance' at the trailing apex of Phobos ties in exactly with what this hypothesis predicts: this is the only location that ejecta from Mars cannot reach because Phobos' forward motion in its orbit exceeds the ejecta velocity. The same is true of the barrier effect of the crater seen in Figure 5: the groove family concerned is close to the edge of the hemispheres it occupies, so the crater's rim would act as a topographical barrier (Fig. 5). Finally, oblique impacts result in smaller craters (Gault & Wedekind 1978); those impacting at less than 5° to the horizontal excavating an order of

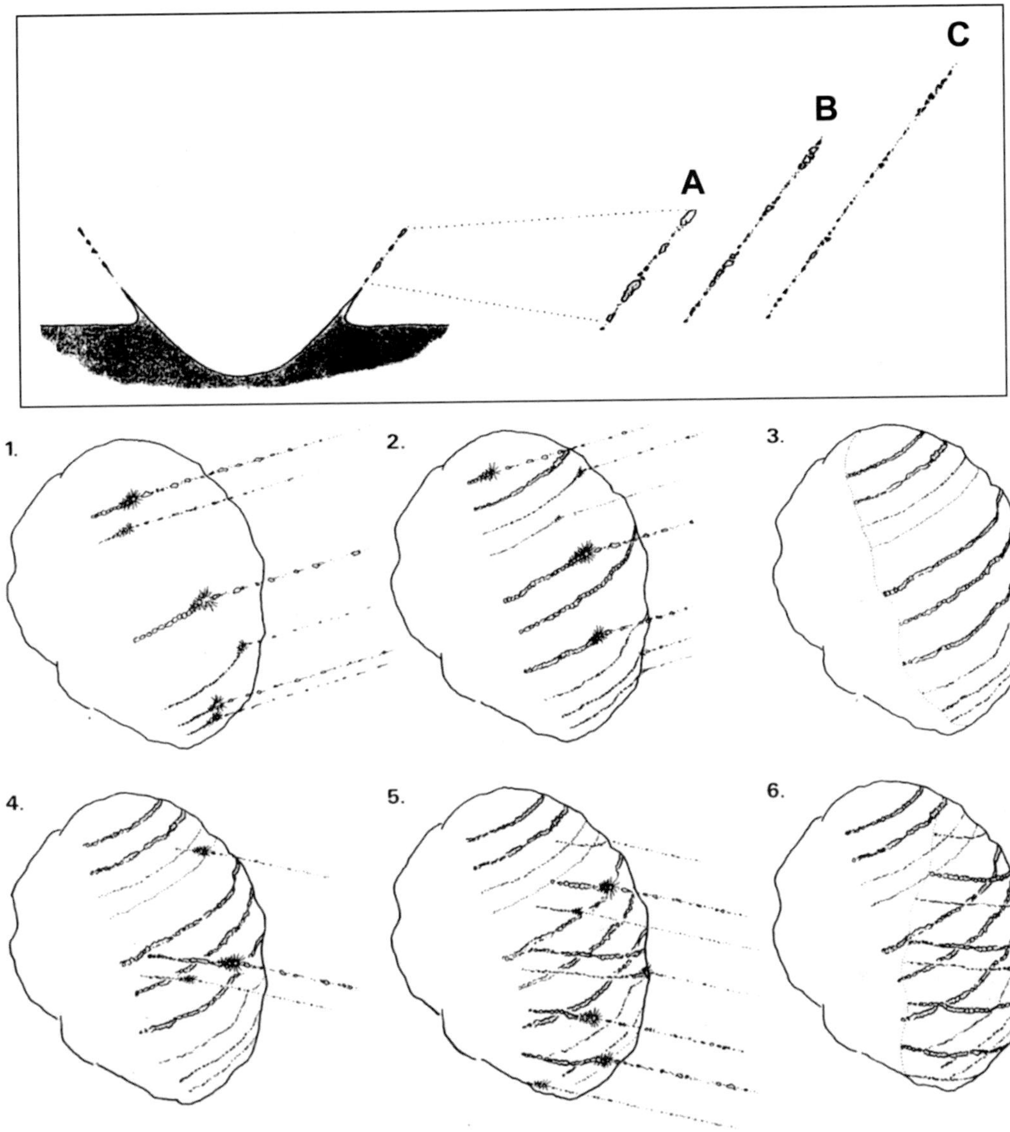

Fig. 17. Diagrams illustrating the formation of Phobos' grooves by the impact of ejecta from large impacts on Mars, from Murray *et al.* (1994). The top diagram shows a section through a large impact event on Mars, early in crater excavation. Highly shocked and melted material from both the impactor and the impacted surface is ejected at velocities of several km s^{-1} in a cone whose apex is the crater centre. The velocity of ejecta rapidly decays with time, so the leading part of each individual ejected jet of melt will steadily draw away from the following end and cause it to be stretched out into a progressively longer string of ejecta, as at A, B and C. The lower series of diagrams (1–6) shows the situation at Phobos. (1) and (2) Show radial (effectively parallel at the distance of Phobos) ejecta strings from the same impact on Mars arriving at Phobos, where Phobos' motion in its orbit (towards the lower right of the observer) causes long chains of secondary impact craters to form on its surface. (3) Shows the situation after the shower of debris has passed: nearly one half of Phobos is crossed with parallel grooves composed of contiguous secondary impact craters with raised rims. (4) and (5) Show the situation at a later time, when a different large impact on Mars results in the arrival of a second shower of ejecta, causing new parallel grooves to form at a different orientation from the earlier set. (6) Shows the final situation, with two families of parallel grooves crossing each other, each family covering only one half of Phobos.

magnitude less mass than vertical impacts. This ties in with the observation that Phobos' grooves become narrower and shallower as they approach the trailing apex (Fig. 8).

Objections to this hypothesis might be that, in the case of groove family A, the pits that form the grooves are often so close together as to give the appearance of very straight troughs of uniform width (Fig. 14a), much more so than are usual for secondary craters. However, these are not ordinary secondary craters. The dynamics of ejection and arrival at Phobos indicate that the ejecta that formed them impacted at velocities of the order of 4 km s^{-1} in every case (Murray *et al.* 1994, 2006), that is, very early in the crater excavation stage of the impact, and an order of magnitude higher in launch velocity than ejecta that forms secondary craters on the planet's surface. Ejecta at this early stage of ejection is more highly compressed, shocked and fractured, and much is melted material (Stöffler *et al.* 1991). By contrast, normal secondary impact craters on the Moon, Mercury and other airless bodies are caused by fragments ejected late in crater excavation that have velocities approaching an order of magnitude lower (0.1–1.0 km s^{-1} for the Moon and Mercury) and fall close to the crater where they are subject to non-gravitational effects (Guest & Murray 1971).

As a second point, these grooves appear as unnaturally straight troughs only at low resolution (Fig. 14a). Higher resolution picks out the individual craters and raised rims quite clearly (Fig. 14b). Despite the effects of lighting, there is no doubt that there is a wide variation in groove morphology, from straight grooves in which the individual pits are contiguous and of fairly uniform width, to individual craters that are more loosely aligned. But this variation is exactly what is found in secondary craters around lunar and planetary impact craters. In addition, all types of 'groove' morphology conform to the same strict geographical rules – for example, those that resemble secondary craters lie along planes exactly parallel to other grooves in the same family, and all disappear at the following apex – so it is inconceivable that the two types would have different origins. Furthermore there is a clear continuum, probably related to age, with those that are similar to young secondary craters at one end and those that are smoothed over by meteoritic gardening and appear less crateriform at the other.

The question as to why there are no parallel grooves on Deimos, or on satellites of other planets, is mainly a result of their distance from the primary (Deimos orbits at 20 070 km and Phobos at 5988 km above Mars' surface), and also due to the very small chances of a satellite passing through the narrow and sparse cone of high-velocity ejecta that is ejected in the earliest stages of impact. Such events have reached Phobos only a few times in its entire history, and the chances of reaching Deimos would be very much smaller. Deimos also lies outside the synchronous orbit distance and so could be a captured body, an event that might have occurred significantly later than the formation of Phobos.

Another point that might be raised against this hypothesis (although it does not affect the validity of the evidence presented here) is more interesting: why do grooves appear on asteroids that are not orbiting close to a large planet? In the first place, it should be noted that grooves reported on Gaspra and Ida (Sullivan *et al.* 1996) are very subdued, and are different from those on Phobos in several critical respects. They are not straight or planar, and do not form parallel families widespread

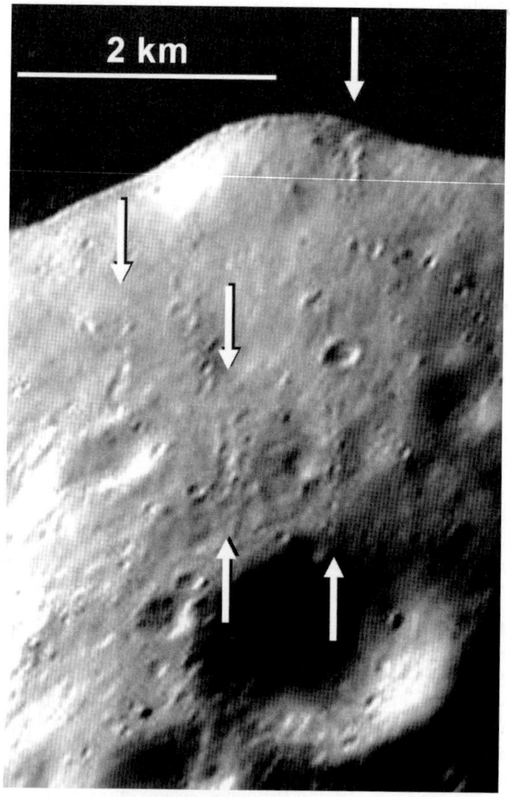

Fig. 18. Image NEAR 127389178 of part of the asteroid Eros, showing the crater Narcissus (bottom) and a series of straight parallel chains of irregular craters with raised rims (arrowed), similar to the 'grooves' of Phobos in width (80–130 m), morphology and disposition. They also resemble secondary impact craters in their alignment, irregular plan shape, and flank to diameter ratios (see Fig. 10). Image credit: NASA/JPL/JHUAPL.

across one hemisphere. They are also very much shorter, most being less than 1 km in length (Veverka *et al.* 1994), yet their width is similar to those on Phobos, so that 'valleys' might be a better description than grooves (see Fig. 11). The grooves on Eros, however, are another matter. They are clearly visible and very similar to those on Phobos (Fig. 18) in that the best-defined ones are composed of coalesced craters with raised rims, and some are parallel to each other (Prockter *et al.* 2002). There are two points to make with reference to Eros: Buczkowski *et al.* (2008) suggested that it may have been part of a larger parent body broken apart by a large impact, in which case it could have received secondary ejecta at close range. Secondly, Michel *et al.* (1998) showed that Eros has a Mars-crossing, unstable orbit, and can also become an Earth-crossing asteroid for extended periods and then return to being a Mars-crosser. A small inclination while in the Earth-crossing state leads to very frequent close approaches to the Earth–Moon system. We suggest that it is possible that this was the case, and that those parallel crater chains with raised rims seen on its surface in Figure 18 are not tectonic, as proposed by Prockter *et al.* (2002), but secondary craters from primary impacts on the Moon, the Earth or Mars that occurred during multiple close approaches to one or more of these bodies.

Conclusions and implications

In summary, the new data from the European Space Agency's (ESA's) *Mars Express* has provided evidence of hitherto unconfirmed characteristics of Phobos' grooves, and has confirmed most of the earlier observations of Thomas *et al.* (1979). Our data show that each groove family extends over no more than one hemisphere of Phobos, that there are topographical barriers to some groove families and that all grooves are substantially younger than Stickney crater, overlying its rim and floor, and also parts of at least three subsequent impact craters within Stickney, but grooves are not radially distributed around Stickney. We have confirmed the observations of Thomas *et al.* (1979) that all grooves lie at intersections of planes with Phobos' surface, and that for each groove family these planes are parallel, and that the groove planes are parallel to a line joining Phobos' leading and trailing apex. We have also confirmed their observation that each groove family is of a different age, and that there are no grooves found within an area 12 km wide centred on the trailing apex of Phobos.

We find that these characteristics, together with structural and other considerations, rule out the possibility that the grooves of Phobos are fractures of any kind, and we also conclude that they cannot be chains of secondary craters from Stickney or any other crater on Phobos. However, all of these characteristics point to an origin from successive bombardment by secondary ejecta from primary impacts on Mars.

Our conclusions have some input into other current debates. First, our model indicates that it is no longer necessary to invoke an excessively thick Phobos regolith of 100–200 m as proposed by previous authors because this would only be required if the grooves were collapse pits over fractures, which we have demonstrated is not the case. Secondly, our interpretation does not require Phobos to have substantial internal strength, nor an extremely homogeneous solid composition throughout, which if the grooves were fractures would be implied by their unprecedentedly straight planar nature. Thirdly, our results indicate that there could be large void spaces within Phobos, which again would not be possible if the grooves were planar fractures because they could not propagate from one side of the satellite to the other through void spaces. Our work is therefore quite consistent with the hypothesis that Phobos is essentially a 'rubble pile', as now seems likely (Andert *et al.* 2010).

Further implications lie within the mechanism of impact cratering and ejection of impact debris. It has long been known that ejecta can form into strings, which results in crater chains within a few hundred kilometres of the impact site (e.g. Guest & Murray 1969). However, the fact that these ejecta strings can, in rare special cases, reach heights of nearly 6000 km above the surface in sufficient numbers to create contiguous crater chains on orbiting bodies is not widely appreciated, although again the range and volume of impact ejecta has long been well documented (e.g. Gault *et al.* 1963; Gladman *et al.* 1996). The width of the Phobos grooves suggests that the individual impactors within the strings had diameters of the order of 10 m and spacings of around 100 m, and the range would be in thousands of kilometrs from the impact site. The special cases where impact debris can attain these heights and distances in high concentrations presumably relate either to the very large size of the event or to the serendipitous orientation of an oblique impact, in which a far greater proportion of ejecta at higher velocity can be concentrated in the downrange direction (Anderson *et al.* 2003).

An interesting observation is that the Phobos grooves seem to be usually quite uniform in width and appearance. In our model this suggests that the ejecta jets that formed them contained a high proportion of melt initially, enabling them to stretch out into strings and break up into relatively uniform clots, in the manner of Pélé's hair and volcanic glass beads, although on a larger scale. A 100 km-diameter impact crater will create of the order of 10 000 km^3 of melted rock (Pierazzo *et al.* 1997), that is, a greater volume than any known lava flow on Earth, so there is plenty of melt available in large impact events.

An interesting by-product of our results is their application to asteroids, such as Eros, that have chaotic orbits which could have been both Earth-crossing and Mars-crossing for long periods in the past. Because Eros is the only asteroid yet imaged close up with long, straight chains of craters on the surface similar to those on Phobos, we suggest that it has, indeed, made multiple close approaches to the Earth–Moon system or Mars, and has encountered ejecta strings from impacts on one or more of these bodies.

We thank G.Neukum and S.Werner for improvements and corrections to an earlier version of this manuscript, and K. Willner for providing the data for Table 1. Louise Prockter and David Ferrill are thanked for their thorough reviews that greatly improved the final manuscript.

References

Anderson, J. L. B., Schultz, P. H. & Heineck, J. T. 2003. Asymmetry of ejecta flow during oblique impacts using three-dimensional particle image velocimetry. *Journal of Geophysical Research–Planets*, **108**, 5094–5103, doi: 10.1029/2003JE002075.

Andert, T. P., Rosenblatt, P., Pätzold, M., Häusler, B., Dehant, V., Tyler, G. L. & Marty, J. C. 2010. Precise mass determination and the nature of Phobos. *Geophysical Research Letters*, **37**, L09202, doi: 10.1029/2009GL041829.

Becq-Giraudon, J.-F., Rouzeau, O., Goachet, E. & Solages, S. 1992. Impact hyperveloce d'une météorite géante à l'origine de la dépression circulaire d'Aorounga au Tchad (Afrique). *Comptes rendus de l'Académie des sciences, Paris*, **315**, serie II, 83–88.

Bibring, J.-P. 2010. Phobos origin: a reappraisal. *In*: *The First Moscow Solar System Symposium, Space Research Institute, Moscow, Russia 11–15 October 2010*. EPSC Abstracts, Volume 5, EPSC2010-554. European Planetary Science Congress, 36.

Buczkowski, D. L., Barnouin-Jha, O. S. & Prockter, L. M. 2008. 433 Eros lineaments: global mapping and analysis. *Icarus*, **193**, 39–52.

Burns, J. A. 1992. Contradictory clues as to the origin of the Martian moons. *In*: Kieffer, H. H., Jakowsky, B. M., Snyder, C. & Matthews, M. (eds) *Mars*. University of Arizona Press, Tucson, AZ, 1283–1301.

Davis, D. R., Weidenschilling, S. J., Chapman, C. R. & Greenberg, R. 1980. Dynamical studies of Phobos and Deimos: groove origin and ejecta dynamics. *Reports of the Planetary Geology Program, 1979–1980*, 14–15.

Dobrovolskis, A. R. 1982. Internal stresses in Phobos and other triaxial bodies. *Icarus*, **52**, 136–148.

Dobrovolskis, A. R. & Burns, J. A. 1980. Life near the Roche limit: behavior of ejecta from satellites close to planets. *Icarus*, **42**, 422–441.

Duxbury, T. C. & Callahan, J. D. 1989. Phobos and Deimos control networks. *Icarus*, **77**, 275–286.

Duxbury, T. C. & Veverka, J. 1977. *Viking* imaging of Phobos and Deimos: an overview of the primary mission. *Journal of Geophysical Research*, **82**, 4203–4211.

Ferrill, D. A., Wyrick, D. Y., Morris, A. P., Sims, D. W. & Franklin, N. M. 2004. Dilational fault slip and pit chain formation on Mars. *GSA Today*, **14**, 4–12, doi: 10.1130/1052-5173.

Fujiwara, A. 1991. Stickney-forming impact on Phobos: crater shape and induced stress distribution. *Icarus*, **89**, 384–391.

Fujiwara, A. & Asada, N. 1983. Impact fracture patterns on Phobos ellipsoids. *Icarus*, **56**, 590–602.

Gault, D. E., Shoemaker, E. M. & Moore, H. J. 1963. *Spray Ejected From the Lunar Surface by Meteoroid Impact*. NASA Technical Note, **D-1767**.

Gault, D. E. & Wedekind, J. A. 1978. Experimental studies of oblique impact. *In*: *Proceedings of the ninth Lunar Science Conference, Houston, Texas, March 13–17, 1978, League City, Texas*. Lunar and Planetary Institute, Houston, TX, 3843–3875.

Giuranna, M., Roush, T. L., Duxbury, T., Hogan, R. C., Carli, C., Geminale, A. & Formisano, V. 2010. Compositional interpretation of PFS/MEX and TES/MGS thermal infrared spectra of Phobos. *In*: *The First Moscow Solar System Symposium, Space Research Institute, Moscow, Russia 11–15 October 2010*. EPSC Abstracts, Volume 5, EPSC2010-554. European Planetary Science Congress, 44–45.

Gladman, B. J., Burns, J. A., Duncan, M., Lee, P. & Levison, H. F. 1996. The exchange of impact ejecta between terrestrial planets. *Science*, **271**, 1387–1392.

Guest, J. E. & Murray, J. B. 1969. Nature and origin of Tsiolkovsky crater, Lunar farside. *Planetary and Space Science*, **17**, 121–141.

Guest, J. E. & Murray, J. B. 1971. A large scale pattern associated with the ejecta blanket and rays of Copernicus. *The Moon*, **3**, 326–336.

Head, J. W. & Cintala, M. J. 1979. Grooves on Phobos: evidence for possible secondary cratering origin. *In*: *Reports of the Planetary Geology Program, 1978–1979*. NASA TM-80339, 19–21.

Horstman, K. C. & Melosh, H. J. 1989. Drainage pits in cohesionless materials: implications for the surface of Phobos. *Journal of Geophysical Research*, **94**, 12,433–12,441.

Kuzmin, R. O., Shingavera, T. V. & Zabalueva, E. V. 2003. An engineering model for Phobos surface. *Solar System Research*, **37**, 266–281.

Ma, Y., Williams, I. P. & Cheng, W. 2002. On the ejection velocity of meteoroids from comets. *Monthly Notices of the Royal Astronomical Society*, **337**, 1081–1086.

Michel, P., Farinella, P. & Froeschlé, C. 1998. Dynamics of Eros. *Astronomy Journal*, **116**, 2023–2031.

Murchie, S. & Erard, S. 1996. Spectral properties and heterogeneity of Phobos from measurements by Phobos 2. *Icarus*, **123**, 63–86.

MURCHIE, S., CHOO, T. *ET AL.* & THE CRISM TEAM 2008. MRO/CRISM observations of Phobos and Deimos. *In*: *Proceedings of the 39th Annual Lunar and Planetary Science Conference, March 10–14, 2008, League City, Texas*. Lunar and Planetary Institute, Houston, TX, Abstract 1434.

MURRAY, J. B., ILIFFE, J. C., MULLER, J.-P. A. L., NEUKUM, G., WERNER, S., BALME, M. & THE HRSC CO-INVESTIGATOR TEAM 2006. New evidence on the origins of Phobos' parallel grooves from HRSC *Mars Express*. *In*: *Proceedings of the 37th Annual Lunar and Planetary Science Conference, March 13–17, 2006, League City, Texas*. Lunar and Planetary Institute, Houston, TX, Abstract 2195.

MURRAY, J. B., ROTHERY, D. A., THORNHILL, G. D., MULLER, J.-P. A. L., ILIFFE, J. C., DAY, T. & COOK, A. C. 1994. The origin of Phobos' grooves and crater chains. *Planetary Space Sciences*, **42**, 519–526.

OBERBECK, V. R. & QUAIDE, W. L. 1968. Genetic implications of Lunar regolith thickness variations. *Icarus*, **9**, 446–465.

PALOMBA, E., D'AMORE, M., ZINZI, A., MATURILLI, A., D'AVERSA, E. & HELBERT, J. 2010. Revisiting the thermal infrared spectral observations of Phobos. *In*: *Proceedings of the 41st Annual Lunar and Planetary Science Conference, March 1–5, 2010, The Woodlands, Texas*. Lunar and Planetary Institute, Houston, TX, Abstract 1899.

PIERAZZO, E., VICKERY, A. M. & MELOSH, H. J. 1997. A re-evaluation of impact melt production. *Icarus*, **127**, 408–423.

POLLACK, J. B. & BURNS, J. A. 1977. An origin by capture for the Martian satellites? *Bulletin of the American Astronomical Society*, **9**, 518–519.

POLLACK, J. B., BURNS, J. A. & TAUBER, M. E. 1979. Gas drag in primordial circumplanetary envelopes: a mechanism for satellite capture. *Icarus*, **37**, 587–611.

PROCKTER, L., THOMAS, P. *ET AL.* 2002. Surface expressions of structural features on Eros. *Icarus*, **155**, 75–93.

REIDEL, C., ERNST, G. G. J. & RILEY, M. 2003. Controls on the growth and geometry of pyroclastic constructs. *Journal of Volcanology & Geothermal Research*, **127**, 121–152.

RIVKIN, A. S., BROWN, R. H., TRILLING, D. E., BELL, J. F., III. & PLASSMANN, J. H. 2002. Near-infra-red spectrophotometry of Phobos and Deimos. *Icarus*, **156**, 64–75.

SOTER, S. & HARRIS, A. 1977. Are striations on Phobos evidence for tidal stress? *Nature*, **268**, 421–422.

SPOHN, T., SOHL, F. & BREUER, D. 1998. Mars. *The Astronomical Astrophysics Review*, **8**, 181–236.

STÖFFLER, D., KEIL, K. & SCOTT, E. R. D. 1991. Shock metamorphism of ordinary chondrites. *Geochimica et Cosmochimica Acta*, **55**, 3845–3867.

SULLIVAN, R., GREELEY, R. *ET AL.* 1996. Geology of 243 Ida. *Icarus*, **120**, 119–139.

THOMAS, P. C. 1998. Ejecta emplacement on the Martian satellites. *Icarus*, **131**, 78–106.

THOMAS, P. C., VEVERKA, J. & DUXBURY, T. C. 1978. Origin of the grooves on Phobos. *Nature*, **273**, 282–284.

THOMAS, P. C., VEVERKA, J., BLOOM, A. & DUXBURY, T. C. 1979. Grooves on Phobos: their distribution, morphology and possible origin. *Journal of Geophysical Research*, **84**, 8457–8477.

VEVERKA, J. & BURNS, J. A. 1980. The moons of Mars. *Annual Review of Earth and Planetary Sciences*, **8**, 527–558.

VEVERKA, J. & DUXBURY, T. C. 1977. *Viking* observations of Phobos and Deimos: preliminary results. *Journal of Geophysical Research*, **82**, 4213–4223.

VEVERKA, J., THOMAS, P. *ET AL.* 1994. Discovery of grooves on Gaspra. *Icarus*, **107**, 72–83.

WEIDENSCHILLING, S.J. 1979. A possible origin for the grooves of Phobos. *Nature*, **282**, 697–698.

WILLNER, K., OBERST, J. *ET AL.* 2010. Phobos control point network, rotation, and shape. *Earth and Planetary Science Letters*, **294**, 541–546.

WILSON, L. & HEAD, J. W. 1989. Dynamics of groove formation on Phobos by ejecta from Stickney. *In*: *Proceedings of the 20th Annual Lunar and Planetary Science Conference, March 13–17, 1989, Houston, Texas*. Lunar and Planetary Institute, Houston, TX, Abstracts 1211–1212.

WYRICK, D., FERRILL, D. A., MORRIS, A. P., COLTON, S. L. & SIMS, D. W. 2004. Distribution, morphology, and origins of Martian pit crater chains. *Journal of Geophysical Research–Planets*, **109**, E06005, doi: 10.1029/2004JE002240.

Periglacial geomorphology and landscape evolution of the Tempe Terra region, Mars

S. VAN GASSELT[1]*, E. HAUBER[2], A.-P. ROSSI[3,4], A. DUMKE[1], R. OROSEI[5] & G. NEUKUM[1]

[1]*Institute of Geological Sciences, Planetology and Remote Sensing, Freie Universitaet Berlin, Malteserstrasse 74-100, D-12249 Berlin, Germany*

[2]*Institute of Planetary Research, Department of Planetary Geology, German Aerospace Centre (DLR), Rutherfordstrasse 2, D-12489 Berlin, Germany*

[3]*International Space Science Institute (ISSI), Hallerstrasse 6, CH-3012 Bern, Switzerland*

[4]*Jacobs University Bremen, Campus Ring 1, D-28759 Bremen, Germany*

[5]*Institute of Physics of Interplanetary Space (IFSI), 00133 Rome, Italy*

**Corresponding author (e-mail: Stephan.vanGasselt@fu-berlin.de)*

Abstract: A systematic survey was undertaken and an investigation carried out into the geomorphological characteristics of lobate debris aprons in the Tempe Terra region of Mars. Based on the most recent high-resolution (sub 15 m per pixel) imagery and on new topography data, this study endeavoured to raise and discuss questions regarding their formation (emplacement) and modification (deformation sequence), as well as the role of a mantling deposit found at mid-latitude locations on Mars. Furthermore, a model for the formation of debris aprons in the Tempe Terra–Mareotis Fossae settings is proposed. Image survey, in combination with basic morphometric observations within a geomorphological context, provided additional insights into the source, emplacement and modification of hillslope debris material. Our results imply that lobate debris aprons are not mainly relicts of remnant degradation but are substantially composed of mantling material probably deposited episodically in the course of planetary obliquity changes and over a long timespan, as derived erosion rates suggest. Crater-size frequency statistics and the derivation of absolute ages show ages of sub-recent modification and document earlier resurfacing events.

The fretted terrain at the Martian dichotomy boundary hosts an abundance of landforms related to the creep of mountain debris and ice. These have become known as the so-called lobate debris aprons (LDA), and units of lineated valley fills (LVF) and concentric crater fills (CCF) (e.g. Sharp 1973; Carr & Schaber 1977; Squyres 1978, 1979; Lucchitta 1981, 1984).

These creep-related landforms are generally considered to be indicators for the existence of past and present ice in the near-subsurface of Mars (Sharp 1973; Carr & Schaber 1977; Lucchitta 1981; Rossbacher & Judson 1981; Kargel & Strom 1992; Colaprete & Jakosky 1998; Mangold 2003; Whalley & Azizi 2003; van Gasselt *et al.* 2008; Head *et al.* 2005, 2006*a*, *b*, 2010; Hauber *et al.* 2008). They have been interpreted as being mixtures of rock particles and ice (Squyres 1978, 1979) analogous to terrestrial rock glaciers, that is, debris transport systems comprising a mixture of rock fragments and segregational and/or interstitial ice (e.g. Wahrhaftig 1954; Wahrhaftig & Cox 1959; Haeberli 1985; Barsch 1996).

The analogy between terrestrial rock glaciers and Martian LDA and similar landforms is mainly based on:

- the accumulation of debris at footslopes of escarpments and mountain wall rock (e.g. Squyres 1978; Lucchitta 1981; Martin & Whalley 1987; Vitek & Giardino 1987; Barsch 1996);
- the cross-sectional convex-upwards profile of the LDA indicative of internal deformation akin to glacial ice (e.g. Squyres 1978; Barsch 1996; Paterson 2001; Ikeda & Matsuoka 2002; Turtle *et al.* 2003);
- characteristic sets of surficial ridges and furrows indicating differential compression and extension (e.g. Squyres 1978; White 1987; Barsch 1996; Kääb & Weber 2004);
- the geomorphological relationship to adjacent regions indicative of permafrost-related morphologies (e.g. Vitek & Giardino 1987; Barsch 1996).

Possible Martian rock glacier analogues have been observed primarily along steep escarpments near

From: Balme, M. R., Bargery, A. S., Gallagher, C. J. & Gupta, S. (eds) *Martian Geomorphology*. Geological Society, London, Special Publications, **356**, 43–67.
DOI: 10.1144/SP356.4 0305-8719/11/$15.00

the dichotomy boundary, as well as in the large impact basin of Hellas Planitia (e.g. Squyres 1979; Crown *et al.* 2002), in two latitude bands located between 30° and 60° north and south, and centred at 40° and 45°, respectively.

As on Earth, rock glaciers should be sensitive indicators of the climatic environment during their formation and evolution, and are thought to be potentially large and accessible water reservoirs (e.g. Wahrhaftig & Cox 1959; Johnson 1987; Vitek & Giardino 1987; Whalley & Martin 1992; Barsch 1996; Whalley & Azizi 2003; Kääb & Reichmuth 2005; Kääb & Kneisel 2006).

Although such landforms have been studied and described in detail by different groups, many questions put forward during the early *Viking Orbiter* missions have not yet been successfully addressed and, notwithstanding the general consensus with respect to an apron's composition of icy debris, alternative explanations have been discussed. Ideas about alternative origins include aeolian processes for CCF (Zimbelman *et al.* 1989), pediments, that is, denudational (rather than depositional) hillslope surfaces (Malin & Edgett 2000) and primary glacial deposits *sensu lato* (e.g. Head *et al.* 2006*a*, *b*, 2010; Madeleine *et al.* 2009).

Rock glacier formation and deformation requires periglacial environmental conditions that allow movement of coarse ice-rich wall-rock debris by creep. A glacial origin, however, requires substantial amounts of primary ice precipitated from the atmosphere and stored as a glacial body either as an ice cap or valley glacier (e.g. Head *et al.* 2006*a*, *b*, 2010; Madeleine *et al.* 2009). Consequently, characteristic landforms of glacial environments must be observed not only in the context of lineated valley-fill landforms but also in context of lobate debris aprons.

Carr & Schaber (1977) considered gelifluction and surficial frost-creep processes as primary mechanisms for debris-apron formation, a view that was not supported by Squyres (1978) owing to the inferred thickness of deformed material. Instead, Squyres (1979) and Squyres & Carr (1986) believed that erosion of the escarpment provided the debris and the terrain softening was caused by creep processes acting over the full thickness of aprons. The exact style of debris-apron emplacement is still under discussion, and short-term mass-wasting processes (Squyres 1979; Squyres & Carr 1986; Mangold & Allemand 2001) have been compared to slow creep mechanisms (Squyres & Carr 1986; Lucchitta 1984) for the northern hemisphere. With the advent of topographical data from Mars, provided by the Mars Orbiter Laser Altimeter (MOLA: Zuber *et al.* (1992); Smith *et al.* (2003)) on board the *Mars Global Surveyor* (MGS, Cunningham (1996); Albee *et al.* (1998)), a general consensus has been reached suggesting that these landforms undergo viscous deformation. This was suggested for Mars by Squyres (1978), Mangold & Allemand (2001) and other workers, applying glacial ice-sheet models developed for the Earth (e.g. Paterson 2001).

Recent work on these features on Mars has focused on morphometric characteristics and discussed textural details of debris aprons and valley- and crater-fill units using highest image-resolution data (Mangold 2003; Pierce & Crown 2003; Chuang & Crown 2005; Head *et al.* 2006*a*, *b*; van Gasselt *et al.* 2008). There is general agreement that these landforms have undergone degradation in the recent geological past – a view that was already implicitly supported by Squyres (1979) and Lucchitta (1984) based on the lack of large impact craters. Surface-dating techniques using size–frequency distributions of impact craters showed that apron surfaces are roughly in the range of a few tens to 100 Ma (Mangold 2003; Head *et al.* 2005; van Gasselt *et al.* 2008). For lineated valley units, some surfaces imply even much younger formation ages or resurfacing events (van Gasselt *et al.* 2010).

Hence, one of our aims is to elaborate on aspects of cold-climate geomorphology pertinent to the population of remnant-massif/debris-apron constructs (RACs), that is, LDA circumferential to a central wall-rock unit on the basis of recent image and topography data. An additional aim is to identify how (and what kind of) material was incorporated into debris aprons and how long it took to accumulate slope material at footslopes. We achieve this by comparing observations with morphometric and age information. The derivation of absolute surface ages is, in general, not necessarily indicative of formation ages; however, resurfacing events are usually reflected in the crater statistics. As a third aim we here tried to identify absolute ages or time spans, providing insights into earlier episodes of surface modification. Rather than trying to identify a single age as done previously, higher-resolution data and a more systematic measurement approach enabled us to extract characteristic kinks in the impact crater size–frequency distribution, and to identify and characterize phases of resurfacing more clearly.

Results from these aims are related to each other and lead to the set-up of an idealized model of landscape formation starting at the stage of wall-rock and remnant-massif degradation, and covering the later stages of landscape development.

Methodology and data usage

Using panchromatic-orthoimage data, we systematically investigated high-resolution images with a pixel resolution of better than 15 m from the *Mars Express* (MEX) High Resolution Stereo Camera (HRSC; Table 1) (Neukum *et al.* 2004*a*, *b*; Jaumann *et al.* 2006) and the *Mars Reconnaissance Orbiter* (MRO) Context Camera instrument (CTX; Table 2) (Malin *et al.* 2007). These two sets of images completely and independently cover the area of interest that extends from 275.5°E to 292.5°E and from 45.9°N to 53.3°N. In addition, stereo-photogrammetrically derived products, that is, terrain model data, were included mainly for morphometric characterizations, that is, general settings, slope gradients, volumes, deconvoluted areas and topographical profiles.

Apart from digital terrain model mosaics derived from HRSC data, we used MGS/MOLA altimetry data for higher precision topographical profiles. All data tracks were extracted from the binary archived experiment data records (version L). Image data were processed from raw formats obtained from the Planetary Data System (PDS) and respective instrument websites. Data processing was been carried out using USGS' Integrated Software for Imagers and Spectrometers (ISIS) and JPL's Video Image Communication and Retrieval (VICAR) environment, and were subsequently integrated into a geographical information system (GIS) product (ArcGIS) from the Environmental Systems Research Institute (ESRI) in order to overlay, analyse and visualize image and topographical data.

Subscenes of HRSC and CTX scenes were prepared for statistics on crater-size frequency distributions in order to obtain crater (absolute) retention ages (Shoemaker & Hackman 1962; Baldwin 1964; Hartmann 1966*a*, *b*; Neukum & Dietzel 1971) and to put constraints on possible resurfacing events. This has been achieved by using the chronology-function coefficients by Neukum & Wise (1976) and Ivanov (2001), and the production-function coefficients as defined by Ivanov (2001) and Hartmann & Neukum (2001). Error estimates and a discussion on the issue of secondary impact craters and resurfacing are treated in detail in Neukum *et al.* (2004*a*, *b*), Werner (2005) and Michael & Neukum (2010). Craters and area sizes, as well as crater-size frequency statistics, were computed directly within the ArcGIS environment (Kneissl *et al.* 2011).

A systematic morphometrical GIS-based assessment of 34 RACs was carried out on the basis of digital elevation data and orthoimage data (Fig. 1). This aimed to quantify volumes, areas, slope angles, relief, that is, thicknesses, and length values for LDA, as well as remnant massifs, based on their spatial location in order to assess the influence of, for example, geographical latitude as a climate indicator (see Table 3). This also helped in

Table 1. *Table of MEX HRSC image scenes, pixel resolution and imaging time as used for this study. Illumination conditions as listed in columns 6–8 were derived using algorithms published by Allison (1997) and Allison & McEwen (2000)*

	Image scene	Scale (m per pixel)	Date	Time (h:min:s)	Azimuth, Φ (°)	Solar elevation, ψ (°)	Longitude, L_S (°)
1	3272_0000	13.1	2006-07-27	20:44:49	241.4	58.2	84.99
2	3283_0000	13.1	2006-07-30	22:40:31	240.7	58.6	86.34
3	3294_0000	13.0	2006-08-03	00:36:19	240.1	59.0	87.69
4	3305_0000	13.0	2006-08-06	02:32:12	239.4	59.3	89.05
5	3316_0001	13.0	2006-08-09	04:28:09	238.7	59.6	90.41
6	5286_0000	13.9	2008-02-13	01:37:18	250.5	35.5	31.17
7	5293_0000	14.9	2008-02-15	01:31:02	229.6	48.3	32.08
8	5304_0000*	13.6	2008-02-18	04:46:57	249.9	37.2	33.51
9	5311_0000*	13.5	2008-02-20	04:40:41	228.3	49.9	34.42
10	5329_0000*	13.5	2008-02-25	07:50:19	227.0	51.4	36.75
11	5365_0000	13.6	2008-03-06	14:09:21	224.0	54.4	41.37
12	5383_0000	13.7	2008-03-11	17:18:40	222.3	55.8	43.66
13	5401_0000	13.9	2008-03-16	20:27:53	220.3	57.3	45.95
14	5437_0000	14.1	2008-03-27	02:46:09	215.9	60.0	50.51
15	5473_0000*	14.5	2008-04-06	09:04:20	210.6	62.5	55.03
Mean values		13.6			231.9	53.8	
$\pm\sigma$		0.57			12.1	8.1	

Asterisks refer to HRSC image strips used for the derivation of the 100 m per pixel terrain model. In addition, HRSC scenes in orbits h1180_0000, h1429_0000, h1440_0000, h1462_0000, h1550_0000, h2880_0000, h2913_0000 and h5239_0009 were included for the bundle-block adjustment and terrain model derivation.

Table 2. *Table of MRO CTX image scenes, pixel resolution and imaging time used for this study. Illumination conditions as listed in columns 6–8 were derived using algorithms published by Allison (1997) and Allison & McEwen (2000)*

	Image scene	Scale (m per pixel)	Date	Time (h:min:s)	Solar azimuth, Φ (°)	Solar elevation, Ψ (°)	Solar longitude, L_S (°)
1	P01 001390 2292 XI 49N076W	5.91	2006-11-12	19:05:07	253.4	40.3	134.44
2	P02 001865 2292 XI 49N076W	5.95	2006-12-19	18:46:34	248.1	35.1	153.00
3	P03 002023 2320 XI 52N071W	6.04	2007-01-01	02:16:22	240.8	36.7	159.44
4	P05 002907 2258 XN 45N083W	5.94	2007-03-10	23:25:15	242.3	14.8	198.04
5	P05 003078 2318 XI 51N070W	6.14	2007-03-24	07:14:51	230.6	19.1	206.01
6	P13 006190 2285 XN 48N073W	5.93	2007-11-21	19:04:24	215.9	34.4	350.94
7	P13 006203 2286 XN 48N068W	5.95	2007-11-22	19:23:01	210.5	36.5	351.46
8	P14 006467 2287 XI 48N076W	5.94	2007-12-13	09:08:25	223.4	36.6	1.84
9	P15 006744 2297 XI 49N080W	5.96	2008-01-03	23:13:39	231.8	37.8	12.38
10	P15 006757 2284 XN 48N074W	5.96	2008-01-04	23:31:52	226.3	40.7	12.87
11	P15 006823 2288 XN 48N077W	5.96	2008-01-10	02:58:33	229.9	40.2	15.33
12	P15 007021 2276 XN 47N082W	5.94	2008-01-25	13:16:55	239.7	38.4	22.61
13	P15 007034 2277 XN 47N077W	5.96	2008-01-26	13:36:00	235.1	41.4	23.08
14	P15 007060 2309 XN 50N068W	5.95	2008-01-28	14:14:27	224.0	46.9	24.03
15	P16 007166 2283 XN 48N082W	5.95	2008-02-05	20:28:32	241.6	40.0	27.86
16	P16 007192 2287 XN 48N071W	5.96	2008-02-07	21:06:41	231.3	46.0	28.79
17	P16 007205 2310 XN 51N068W	5.97	2008-02-08	21:26:15	225.8	48.6	29.26
18	P16 007232 2259 XN 45N084W	5.91	2008-02-10	23:55:06	244.9	39.0	30.22
19	P16 007245 2293 XN 49N079W	5.96	2008-02-12	00:15:27	240.5	42.0	30.69
20	P16 007377 2295 XN 49N084W	5.94	2008-02-22	07:09:41	247.2	40.0	35.38
21	P16 007390 2291 XN 49N078W	5.95	2008-02-23	07:28:42	242.5	43.3	35.83
22	P16 007416 2326 XN 52N069W	5.95	2008-02-25	08:07:17	232.2	49.2	36.75
23	P17 007482 2292 XN 49N071W	5.96	2008-03-01	11:33:43	235.7	48.5	39.07
24	P17 007680 2275 XN 47N077W	5.96	2008-03-16	21:54:15	244.1	46.8	45.98
25	P17 007693 2293 XN 49N073W	6.03	2008-03-17	22:14:07	241.6	48.5	46.43
26	P18 007878 2258 XN 45N083W	5.94	2008-04-01	08:14:51	255.4	41.8	52.81
27	P18 007904 2285 XN 48N074W	5.93	2008-04-03	08:53:46	246.5	48.4	53.71
28	P18 008036 2292 XI 49N079W	5.96	2008-04-13	15:48:05	253.0	45.7	58.23
29	P18 008062 2320 XI 52N070W	6.01	2008-04-15	16:26:38	243.2	52.3	59.12
30	P18 008102 2285 XI 48N081W	5.97	2008-04-18	19:14:39	255.6	44.5	60.49
31	P18 008115 2283 XN 48N075W	5.95	2008-04-19	19:33:40	251.1	47.9	60.93
32	P18 008128 2288 XI 48N069W	5.95	2008-04-20	19:52:34	246.2	51.2	61.37
33	P18 008168 2279 XI 47N082W	5.94	2008-04-23	22:41:29	258.3	43.1	62.74
34	P19 008524 2293 XN 49N084W	5.97	2008-05-21	16:37:48	263.0	42.2	74.85
35	P19 008537 2285 XI 48N080W	6.04	2008-05-22	16:56:31	259.0	45.6	75.30
36	P19 008550 2297 XN 49N073W	6.01	2008-05-23	17:16:40	254.9	48.9	75.74
37	P19 008629 2282 XI 48N071W	5.94	2008-05-29	21:02:13	252.9	50.8	78.43
Mean		5.96			241.0	42.0	
$\pm\sigma$		0.04			12.5	7.8	

assessing the relationship between material supply (remnant-massif area) and material depositional area (debris-apron area). This method is valid for the Tempe Terra area as isolated remnant constructs are more or less conical in shape and do not show a flat-topped mesa-like appearance as is characteristic for the Deuteronilus–Protonilus–Nilosyrtis suite of remnant massifs (e.g. Mangold 2003). We decided to measure only conical remnant massifs for the reason that erosion and downwasting directly contributes to the accumulation of material on the debris apron. Flat-topped remnant massifs are often associated with interior depressions and incisions so that a proper delineation of accumulation zones and zones of degradation, that is, source zones, is not unambiguously possible. Although volumes and areas can still be assessed and properly determined, relationships between source and accumulation zones cannot be established as remnant massif degradation is only active at wall rock that is facing towards a debris apron, but not on top of the remnant massif.

Areas of debris aprons have been calculated using digitized polygons covering debris aprons,

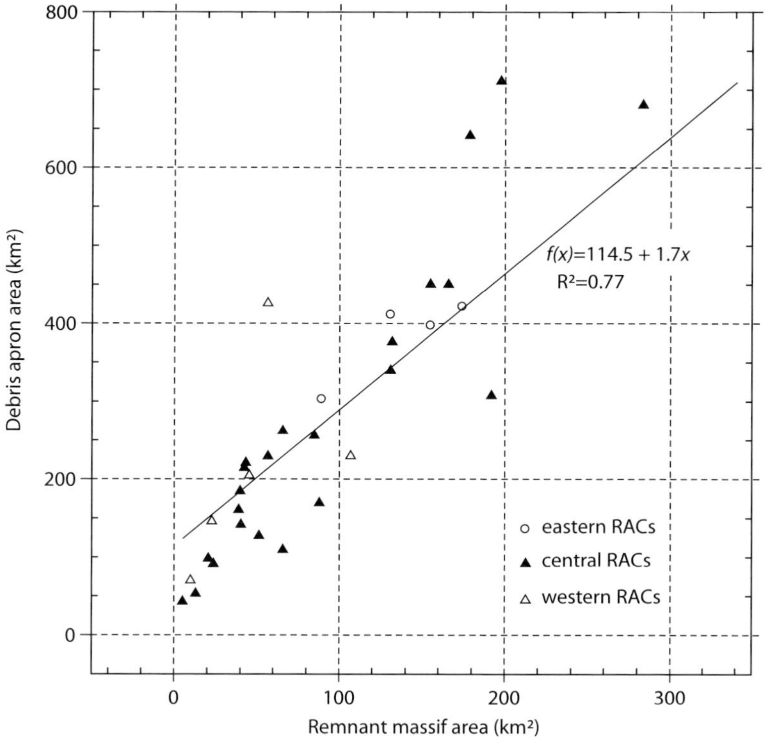

Fig. 1. Surface area of remnant massifs (abscissa) v. surface area of debris aprons (ordinate) as obtained from 34 measurements across the Tempe Terra lobate debris-apron population. On average, remnant-massif/debris-apron area ratios are approximately 1:3.4 ($\sigma_1 = 1.7$), with smaller remnants having smaller ratios owing to prolonged denudation. The linear fit equation ($f(x) = 114.4 + 1.7x$) gives a goodness-of-fit for the data of $R^2 = 0.77$.

that is, extending from the most distal parts of the debris apron to the contact line between debris apron and remnant massif. In this way, area sizes of isolated as well as coalescing aprons can be measured. Areas of remnant massifs were calculated using polygons of areas enclosed by a debris apron's polygon. Volumes of remnant massifs and debris aprons were derived using a minimum-surface plane estimate, that is, by making use of a plane defined through the minimum elevation value of the digitized polygon. This plane was subtracted from the digital terrain model in order to obtain the volume of a debris apron or remnant massif. All area and volume measurements were carried out using an equal-area map projection in order to avoid errors due to cartographic representation. For lengths, a true-scale map projection was used. Both conditions are met using a sinusoidal map projection with the central meridian set to the centre of the measurement area, which means that for each measurement map-projection parameters had to be redefined.

Elevation differences between these two polygons provide a first-order estimate of thicknesses of debris aprons. For all vertices defined through the digitalization process of a polygon feature, a height value was obtained and an average height value could be derived. Length values were derived using manual measurements perpendicular to the debris-apron and the remnant-massif's outlines. In order to obtain reasonable statistics, up to 20 length values were extracted for each RAC.

The derivation of basal shear stresses follows the general approach as outlined and demonstrated in, for example, Paterson (2001). Owing to the sparse knowledge of surface topography covered by the debris apron, we assumed a flat surface ($\alpha = 0°$) and calculated shear stresses using the relation $\tau = (\rho\ gh\sin\ \alpha\ [\text{kg m s}^{-2} = \text{N}]$, with the average density, ρ (= 900 kg m^{-3} for typical rock glaciers: e.g. Barsch 1996), the gravitational acceleration for Mars, $g = 3.72$ m s^{-2} and h [m] as the thickness of the debris apron.

Geological settings and geomorphology

The Tempe Terra–Mareotis Fossae region (Figs 2 & 3) is located between 270°E and 295°E and

Table 3. *Morphometric key values of investigated remnant-apron constructs, where ID denotes the identification number of remnant-apron construct,* L *is the average debris-apron length in km,* ϕ *and* α *are geographical latitude and longitude in degrees,* h *is the average height determined as the difference between apron head and terminus at different locations of each apron,* A_R *and* V_R *are the areas and volumes of remnants in* km^2*, and* t_A*,* A_A *and* V_A *are the respective average thickness, area and volumes of aprons*

ID	L (km)	ϕ (°N)	α (°E)	h (m)	A_R (km^2)	V_R (km^3)	t_A (m)	A_A (km^2)	V_A (km^3)
1	5.3	50.50	278.23	522.9	56.5	27.0	622.0	425.6	90.7
2	3.5	50.42	278.88	285.6	45.4	10.5	290.2	204.6	27.2
3	2.0	50.75	279.50	268.9	9.9	2.5	307.3	69.5	11.0
4	3.6	50.03	280.19	272.9	22.7	5.8	197.0	145.8	14.2
5	3.1	49.35	280.87	260.8	106.7	42.8	243.2	229.5	33.0
6	2.6	48.92	282.43	245.5	20.6	4.3	253.8	98.2	10.4
7	5.5	48.88	283.35	394.5	178.6	53.0	377.0	641.2	187.0
8	4.8	49.41	283.82	288.9	39.8	12.6	229.4	184.4	18.6
9	2.8	48.90	283.91	266.5	51.4	24.6	325.4	127.2	11.7
10	2.1	48.68	283.97	186.7	5.1	0.8	186.9	42.6	4.4
11	4.4	48.52	283.78	419.9	42.3	14.9	367.2	214.2	39.2
12	3.2	48.55	283.47	237.1	87.8	30.0	252.4	169.4	43.0
13	2.8	48.47	284.44	279.1	40.2	10.4	199.6	141.5	26.7
14	5.0	49.05	284.69	437.2	283.2	173.6	594.1	680.3	124.5
15	3.0	49.47	284.76	244.1	56.7	15.1	299.3	229.1	43.4
16	6.3	49.80	285.07	522.7	197.4	139.4	670.4	710.9	158.1
17	4.4	50.12	285.30	438.3	43.2	15.6	399.6	220.7	42.5
18	4.4	49.43	285.47	479.6	84.7	38.8	441.8	256.1	34.4
19	4.0	48.99	285.53	509.2	65.7	33.8	432.4	261.8	57.4
20	2.4	48.70	285.10	266.8	65.8	13.7	279.1	109.4	12.3
21	2.2	49.26	285.94	244.4	13.0	3.8	229.2	52.8	6.9
22	3.9	48.62	285.83	373.1	165.8	112.1	545.5	450.0	145.1
23	3.4	48.88	286.29	349.6	38.8	14.6	381.2	160.5	31.1
24	5.6	49.49	286.38	501.1	154.9	76.6	395.1	450.4	88.5
25	2.1	48.97	286.65	198.8	23.7	4.5	222.7	91.3	23.5
26	4.4	49.91	286.67	277.7	131.7	47.6	398.8	375.9	62.6
27	5.2	51.76	288.73	571.8	154.1	99.2	415.0	397.2	65.0
28	4.2	51.75	289.33	282.4	174.0	102.7	387.8	422.1	59.4
29	5.0	51.23	289.02	483.3	129.9	93.2	504.6	411.9	63.5
30	4.3	52.25	291.52	543.9	89.0	55.9	527.0	303.7	103.6
31	2.1	49.67	277.53	145.8	191.9	117.9	397.2	307.4	191.4
32	5.8	49.78	278.35	441.8	522.6	806.9	628.9	839.8	399.5
33	3.7	49.35	277.65	339.4	130.7	111.1	460.1	339.5	110.7

from 40°N to 55°N, and is characterized by flat and smooth northern lowlands and the southern heavily cratered highland terrain. Both units are separated by a steep escarpment that marks the global dichotomy boundary. The transition between the northern lowlands and the southern highlands was observed in *Mariner 9* image data. This transition is characterized by the so-called fretted terrain (Sharp 1973) in two main locations where highly dissected highland terrain with broad and flat-floored valley incisions are observed: the Tempe Terra–Mareotis Fossae area in the western hemisphere; and the Deuteronilus–Protonilus–Nilosyrtis Mensae area in the eastern hemisphere.

The study area forms a subset of the western-hemispheric region, and extends from 275.5°E to 292.5°E and from 45.9°N to 53.3°N. In contrast to other areas closely related to the dichotomy escarpment, the Tempe Terra–Mareotis Fossae region forms a relatively isolated region that is embayed towards the east and west by the arcuately shaped highland escarpment (Fig. 2).

The highlands have an average elevation level of approximately −1500 m in the east, −500 m in the central parts and up to 1400 m in the west (Fig. 2). The lowlands are smoothly inclined in an eastern direction with elevations of −3600 m in the east and −2400 m in the west (all above aeroid). The escarpment is formed by either an abrupt or gradual topographical step of 1600–2200 m. Valleys incised into the highland units are on an average elevation level of −1700 m to −2500 m. Lowland remnant massifs, that is, relicts of the southern highland, reach elevations of up to −2500 m with a

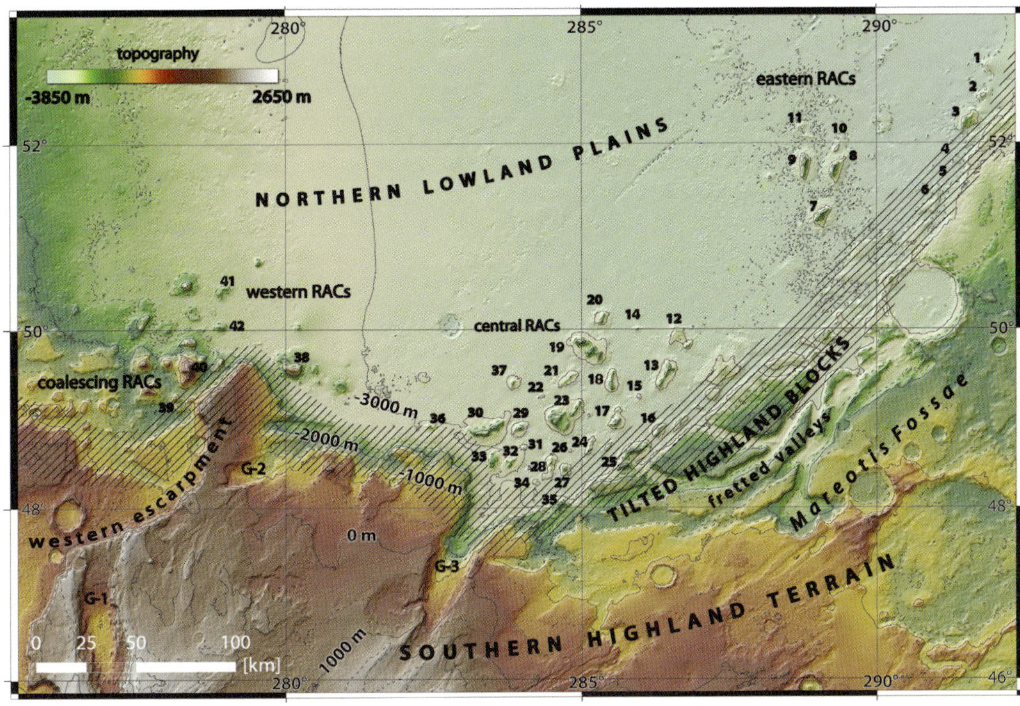

Fig. 2. Topography and general settings of the Mareotis Fossae–Tempe Terra study area; hillshade representation on colour-coded digital terrain model data as represented by a terrain model mosaic derived from bundle-block adjusted HRSC image scenes (Table 1). Labelled remnants/aprons are featured landforms referred to in the main text and used for morphometric studies. Elevations are based on the Mars areoid; the hatched area marks the escarpment transition between the southern highlands and northern lowlands. Isolines have a 500 m spacing (1 km lines are drawn solid), illumination is from upper left, map projection is Mercator. North is up. Image credit: ESA/DLR/FUB; see prelim viii for acronym definitions.

relative local relief of up to 1100 m. The elevation differences are slightly less than that reported from the dichotomy boundary along Deuteronilus, Protonilus and Nilosyrtis Mensae, with values as high as 2–6 km (e.g. Frey *et al.* 1998). The highlands have a generally flat surface, sloping at an angle of less than 0.1° when measured perpendicular to the dichotomy boundary. The surfaces of very large upland segments bounded by a fretted channel have larger slopes towards the lowlands (1°–2°) and might be tilted as blocks (Fig. 2).

The transition between highlands and lowlands is often marked by complex-shaped fault structures that are deeply incised into the highlands either in a perpendicular, that is, north–south to NNE–SSW (Fig. 2, features G1–G3), or in a parallel direction (ENE–WSW) to the dichotomy transition (tilted blocks in Fig. 2). Both sets of structures imply a tectonic control related to the Mareotis Fossae and Tempe Terra rift (e.g. Kochel & Peake 1984; Hauber & Kronberg 2001). Grabens have a depth of up to 3000 m and a width of up to 25 km (Fig. 2). At their terminus and in parallel to their main direction various populations of remnant knobs are visible. While the eastern set of remnant massifs show pronounced debris aprons, the western set is characterized by a basal layer on which some of the aprons are superimposed (Figs 2 & 3). The morphological boundary between highland remnants and terrain and the lowland areas is characterized by two and sometimes three distinct components, as outlined by Carr (2001): (a) a steep upper slope, that is, the wall rock; (b) sometimes an intermediate shallow-sloped unit with downslope-facing striae; and (c) the highly textured apron. This intermediate unit is only rarely observed at the Tempe Terra remnants. At a few sample locations, the cross-sectional angles of intermediate units were measured to be 6°–8° and angles of debris aprons measured at 2°–4° (van Gasselt *et al.* 2008).

Remnant massifs of the highland–lowland boundary are either autochthonous, that is, they represent erosional remnants of highland material as suggested by geological mapping work (Scott & Tanaka 1986; Tanaka *et al.* 2005*a*) and earlier

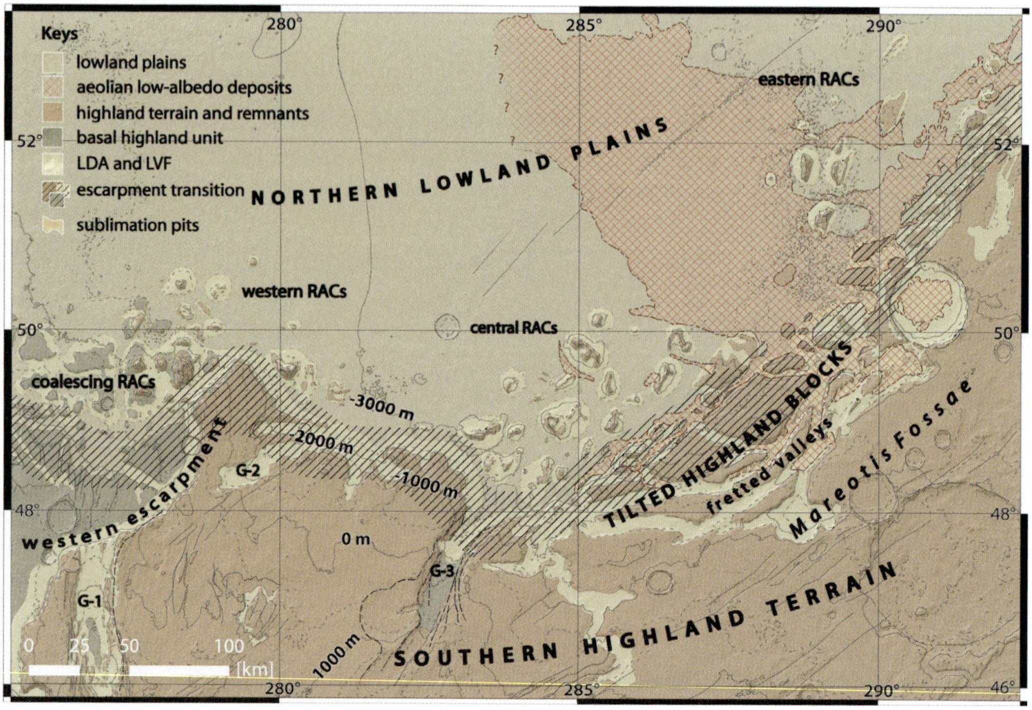

Fig. 3. Geomorphic map of the Tempe Terra region. Isoline spacing is 500 m. Illumination of the hillshade relief map is from the upper left, map projection is Mercator. See the legend for details on mapped units. North is up.

discussions (Sharp 1973; Carr & Schaber 1977; Squyres 1979; Lucchitta 1984; Carr 2001), or they form uplifted crustal material as suggested for the southern hemispheric circum-Hellas and Argyre Planitiae remnants, as mapped by Greeley *et al.* (2006) (cf. question marks in Fig. 4). Alternatively, an allochthonous origin is conceivable although less likely, that is, emplacement by impact processes, similar to alternative explanations for the southern hemispheric remnant-apron features (e.g. Crown *et al.* 1992; Greeley *et al.* 2006).

Stratigraphically, the study area is composed of units that span Martian history from the earliest Noachian to the most recent Amazonian epoch. Remnant massifs are interbedded in Amazonian material of the Arcadia Formation (*Aa1*) on which lobate debris aprons extend. Towards the west this formation borders volcanic material of the Hesperian lower member of the Alba Patera Formation (*Hal*), towards the east it borders a large impact crater. Highland terrain as well as remnant massifs were mapped as Noachian plateau sequence (*Npl1*), composed of highly cratered volcanic material, and Noachian basement (*Nb*) material, respectively. While in the central part of the study area remnant massifs belong stratigraphically to the Noachian basement unit, in the east and west remnants are part of the Noachian plateau sequence (*Nplh*) (basal highland and highland terrain units in Fig. 3). The youngest units are aprons and valley-fill units that were interpreted as surficial deposits related to the creep of ice and debris (e.g. Squyres 1978, 1979; Lucchitta 1984), and mapped as unit *As* by Scott & Tanaka (1986) and Skinner *et al.* (2006) (debris apron units in Fig. 3).

These units, defined in the mid-1980s, have been augmented in recent mapping efforts (Tanaka *et al.* 2005*a*, *b*). The southern highland unit has been remapped as the Noachian Noachis Terra unit (*Nn*). Areas containing isolated remnant massifs and associated debris aprons are located in the Noachian–Hesperian-aged Nepenthes Mensae unit (*HNn*) or the Amazonian-aged Scandia region unit (*ABs*).

Remnant massifs, as well as highland units, from both past and present mapping approaches are thought to be related to Noachian and/or Hesperian units, which are highly fractured and form degraded units as a result of basal-sapping and mass-wasting processes that led to formation of steep aprons (Tanaka *et al.* 2005*a*, *b*). It is considered that the erosion processes leading to the formation of fretted valleys occurred in the Early Hesperian,

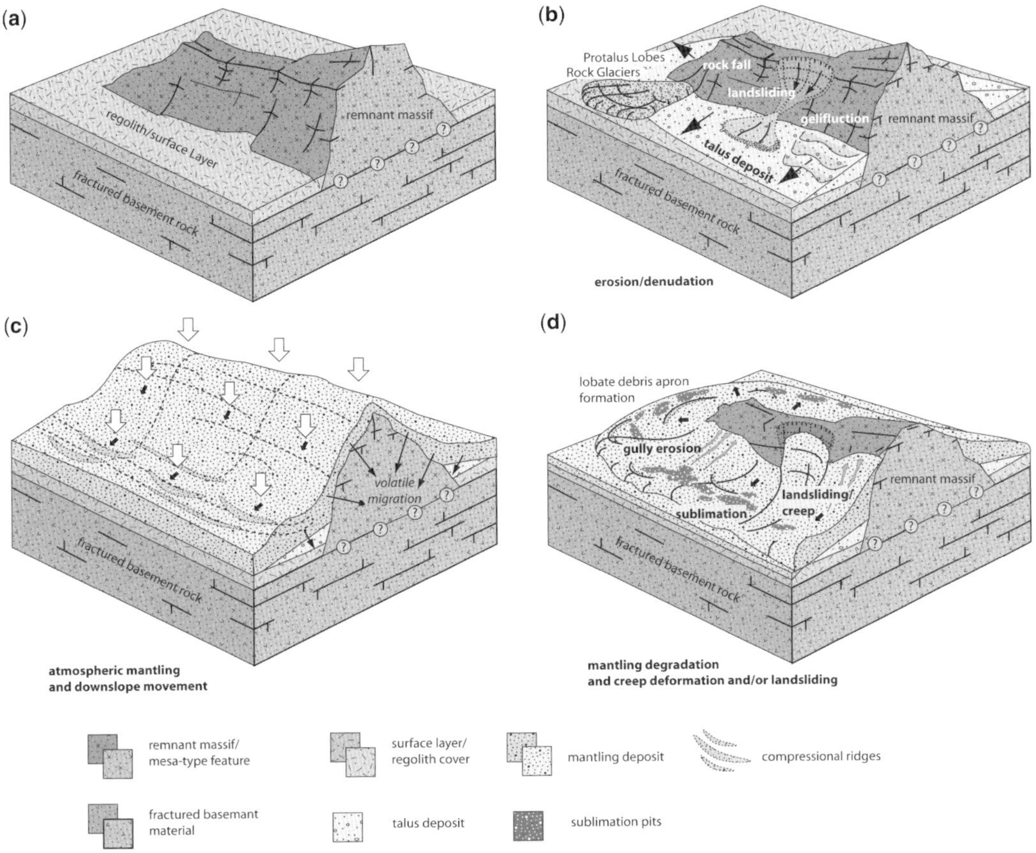

Fig. 4. Reconstruction of landscape evolution in Tempe Terra–Mareotis Fossae (idealized and not to scale). (**a**) Erosional remnant of highland material as the initial landform; (**b**) degradation processes act on the remnant surface, leading to relief smoothening and a variety of landforms associated with landscape denudation; (**c**) homogeneous deposition of an atmospheric/aeolian mantling deposit and the associated downward movement; (**d**) degradation by sublimation and gravitational mass movement and gully erosion. The model applies to the population of Tempe Terra RACs, although the exact formation and denudation (a, b) is different for the western RACs (Fig. 2); for a full explanation see the text and Discussion.

while the formation of valley fill and associated landforms are of Late Hesperian or younger ages (McGill 2000). The timescale for debris-slope development has been estimated to be in the range of 20 Ma–2 Ga (where Ga is 10^9 years) (Perron *et al.* 2003), while initial viscous deformation of material occurs on the timescales of 1–10 ka (Turtle *et al.* 2003).

Observations

Remnant landforms

In contrast to remnant massifs and mesa-like landforms described in, for example, Squyres (1979), Mangold & Allemand (2001) and Mangold (2003) from the Deuteronilus–Protonilus–Nilosyrtis Mensae region, the central Tempe Terra–Mareotis Fossae population of remnant massifs generally shows no flat-topped structures but rather have a conical, convex-upwards or rugged appearance. Flat-topped mesa-like morphologies appear only in the eastern study area as large titled blocks dissected by fretted valleys (Fig. 2). The direction of the long axis of remnant features is usually from east to west, that is, parallel to the escarpment boundary. Isolated features exhibiting a well-pronounced morphology have a length of up to 30 km; however, much of the remnant's surface is covered by apron material so that the true extent is not visible. The latitudinal distribution of remnant knobs and conical hills is restricted to a narrow band of no more than 125 km away from the dichotomy escarpment. The relief of remnants is generally larger near the escarpment but can also occasionally be significant at greater distances away from the boundary (Fig. 2).

A total of 64 remnant massifs with associated debris aprons (remnant-apron constructs; RAC in Figs 2 & 3) are clearly identified. Isolated tilted highland blocks in the eastern study area are categorized as dissected highland material rather than remnants. This study focuses on the central and eastern RACs for which ages and morphometric values have been derived. A number of samples are, however, also taken from the western population. The number of identifiable and, despite high-resolution terrain-model data, morphometrically measurable features is significantly lower (33 features) than the total number (cf. Chuang & Crown 2005). Of these 33 features, 15 are identified as being isolated, that is, aprons extending away from the remnant massif that do not (or appear not to) coalesce with debris aprons of other remnants or those aprons associated with the main escarpment. Most of the isolated RAC features have an elliptical and well-pronounced lobate apron, but at several locations remnant massifs pierce through the apron. Only nine features with a remnant size of at least 15 km exhibit a single well-pronounced, non-coalescing and undisturbed apron, that is, aprons that do not abut onto remnant impact craters or other material. Proper determination of sizes of remnant massifs causes problems with respect to the morphology visible at the surface. All features show a more or less pronounced remnant that is significantly covered by a debris apron so that the true size of such features cannot be determined correctly. Except for the easternmost flat-topped remnants, all features exhibit a sharply defined ridge with wall-rock slopes of 20°–30° according to slope assessment carried out using the HRSC digital elevation model (DEM) mosaic (Fig. 5).

Debris aprons situated in the so-called fretted terrain are confined by valley walls, so that apron material has filled the valley interior and forms the so-called LVF, which is interpreted as material comparable to that of lobate debris aprons but confined to the valley extent. In the study area, fretted valleys have uniform widths of 5–10 km and relatively constant depths of 400–600 m (cf. Table 3). The direction of debris transport remained an open

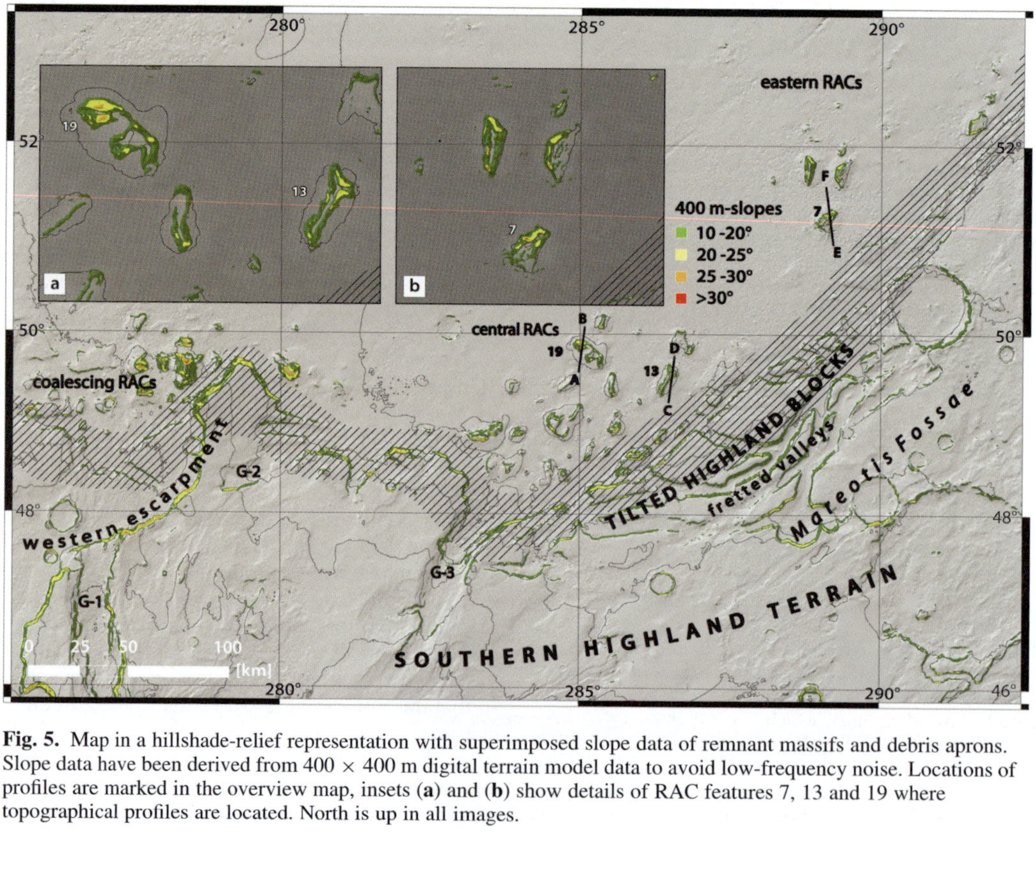

Fig. 5. Map in a hillshade-relief representation with superimposed slope data of remnant massifs and debris aprons. Slope data have been derived from 400 × 400 m digital terrain model data to avoid low-frequency noise. Locations of profiles are marked in the overview map, insets (**a**) and (**b**) show details of RAC features 7, 13 and 19 where topographical profiles are located. North is up in all images.

issue for some time (Squyres 1978, 1979) but there are observations that suggest transport is along the floor, as well as indicated by convex-upwards cross-profiles of LVF at the terminus of valleys (e.g. van Gasselt *et al.* 2008).

Most lowlands remnant massifs that are not completely covered by a surface mantling show a well-pronounced and smooth convex shape, and are often marked by a central irregular and segmented en echelon ridge at the crest (Fig. 6a, c). Over one-third of all investigated remnant massifs (Table 3) show such segmented ridges, which usually extend along the long-axis of remnant massifs. At both sides of the central ridge, remains of a surficial mantling material are observed (Fig. 6a). Such ridges are superposed on the original and highly eroded remnant-massif surface, which is also often covered by sets of subparallel lineations (Figs 6a, b, d & 7). Such lineations usually do not show a preferred direction across individual remnant massifs, suggesting that they are secondary (erosional or depositional) rather than of primary (structural) nature. On uncovered remnant surfaces highly eroded impact craters are often observed (Figs 6d & 7).

Dissected and degraded remnants have an irregular shape, and often show subcircular arcuate and bowl-shaped incisions characteristic of deep-seated landslide scars or ravines reaching down to remnant bedrock units. At various locations gully erosion reaches through the surficial mantling and footslope-apron material and exhibits

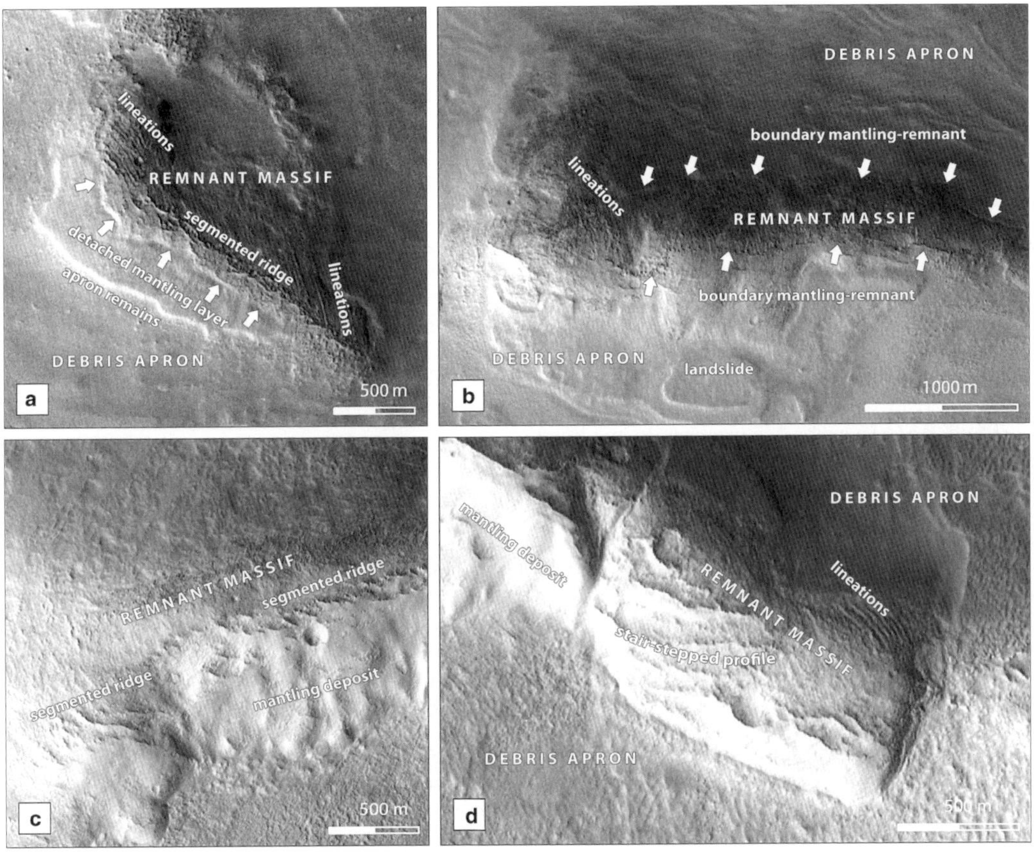

Fig. 6. Remnant massif constructs located in the western study area (278.5°E/49.5°N). (**a**) Remnant massif showing a dissected remnant surface in the detached mantling layer. The relict ridge at the crest indicates the former extent of the mantling material (CTX P19_008537_2285_XI_48N080W, complex feature #41, Fig. 2). (**b**) Remnant massif with a detached mantling layer, which is partly draped over the remnant wall rock (CTX P19_008537_2285_XI_48N080W, complex feature #41, Fig. 2). (**c**) and (**d**) remnant wall rock dissected by deeply incised gully features and stair-stepped profiles indicative of differential downslope mass movement and heavy degradation as suggested by impact-crater obliteration. Traces of remnant mantling layers are indicated at the footslope and remnant ridges on the wall rock (CTX P18_008102_2285_XI_48N081W, feature #39, Fig. 2). North is up in all images. Image credit: NASA/JPL/MSSS.

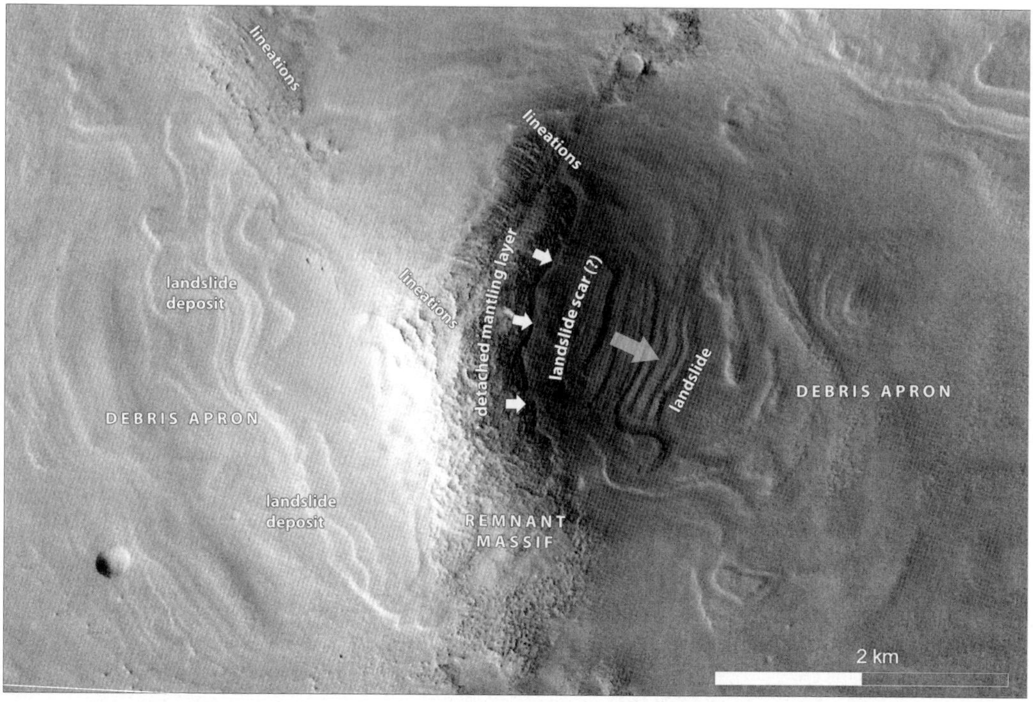

Fig. 7. Complex remnant-apron construct showing several episodes of resurfacing, as indicated by an old landslide scar, obliterated impact craters, parallel surface lineations and several younger phases of deformation of mantling deposits leading to landslides and accumulation on, as well as reworking in, lobate aprons. The mantling deposit is detached from the remnant massif (P19_008537_2285_XI 48N080W, feature #41, Fig. 2). North is up. Image credit: NASA/JPL/MSSS.

deeply incised ravine-like features (Figs 6d, 7, 8a, b, d & 9d) (Christensen 2003). For over 50% of observed RACs we observed a heavily modified remnant unit with characteristic marks of landslide erosion.

Although highest-resolution data have been used, it is not always possible to unambiguously differentiate between a heavily eroded, smooth remnant surface and a thin apron veneer covering a particular feature. The transition between remnant and apron units is mostly diffuse, and a clear identification of colluvial footslope aprons is hindered by the presence of the relatively thick mantling unit that is draped over individual remnants up to the midslope level and beyond (Fig. 6b–d). Topographically, a transitional step can be approximated by the slope gradient, but this does not necessarily mean that this step forms the natural boundary between both units because remnants are often not comparable with regard to their erosional state. When illumination conditions are optimal (cf. Tables 1 & 2) the upper boundary of the extent of the overlying smooth transitional deposit can be determined by a heavily eroded border. This border is exposed at various locations with respect to the overall remnant's relief, that is, it can be situated at the remnant basis or it covers the remnant up to the hilltop (Fig. 6b).

Apron landforms and mantling deposit

Aprons usually show patterns of compressional ridges and, in contrast to other mid-latitude locations and lineated valley-fill units, no extensional patterns (Figs 6a, b, d, 7 & 8a). Compressional ridges occur on the footslope below former landslide scars, indicative of significant post-emplacement mass wasting. Multiple layers and overlapping lobes additionally suggest the continuing or episodically recurring release of hillslope material (Fig. 7). Concentric lineation patterns also occur below niches related to the structural framework of remnant massifs (Fig. 10b, c).

Aprons often exhibit assemblies of irregularly spaced and more or less concentric- to elliptical-shaped flat-floored depressions that partly coalesce and form a pattern similar to those patterns known from the seasonal south polar cap, termed a Swiss-cheese terrain (Figs 8a–c & 9c, d). For the polar cap, such pits are associated with seasonal sublimation of volatiles (Malin & Edgett 2001; Byrne & Ingersoll 2003) and were also discussed

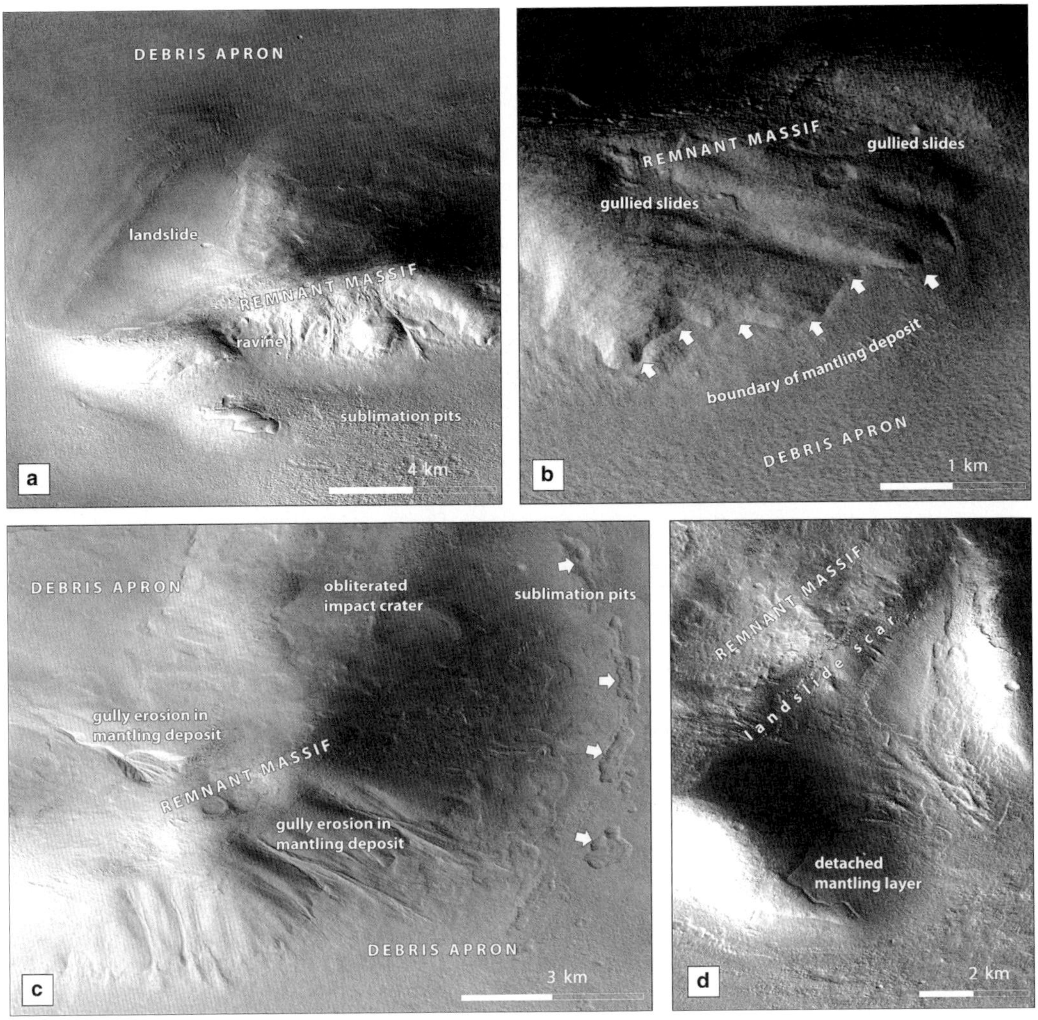

Fig. 8. Remnant massif constructs and mantling deposits. (**a**) Heavily degraded remnant rock unit exhibiting ravines and gullied incisions, and covered by sublimating mantling material; a landslide scar in the west is filled with mantling material that forms compressional ridges at the footslope (P02_001865_2292_XI_49N076W, feature #29, Fig. 2). (**b**) Dissected remnant massif exhibiting deep-seated ravine-like incisions and gullied slide flows; the mantling deposit is draped over remnant in the south (P18_008115_2283_XN_48N075W, feature #29, Fig. 2). (**c**) Smoothly shaped remnant showing obliterated impact-crater structures and gully erosion in overlying mantling deposit; disintegrating mantling surface exhibits characteristic patterns of irregular and coalescing sublimation pits (P18_007904_2285_XN_48N074W, feature #17, Fig. 2). (**d**) Landslide scar filled with mantling material that is partly detached from the remnant surface, morphologically comparable to glacial bergschrund features (P02_001865_2292_XI_49N076W, feature #29, Fig. 2). North is up in all images. Image credit: NASA/JPL/MSSS.

for debris aprons (e.g. Mangold 2003). Sublimation patterns are found in 23 of the studied images (cf. Table 3) and they are more frequently located on south-facing aprons and below gully-erosional landforms (Fig. 9c, d).

Gullies occur in one-third of analysed image data and, similar to the depressions, are mostly located on south-facing hillslopes. Gullies occur in subparallel sets and are deeply incised into a mantling cover located on the steeper mid-slope rather than the apron. Gully heads are often relatively wide (Figs 8c & 9a, b), suggesting that they are cut into a relatively loose substrate. Further uphill, traces of sublimation pits do occur at some places indicating a possible genetic relation between hilltop mantling sublimation, footslope sublimation

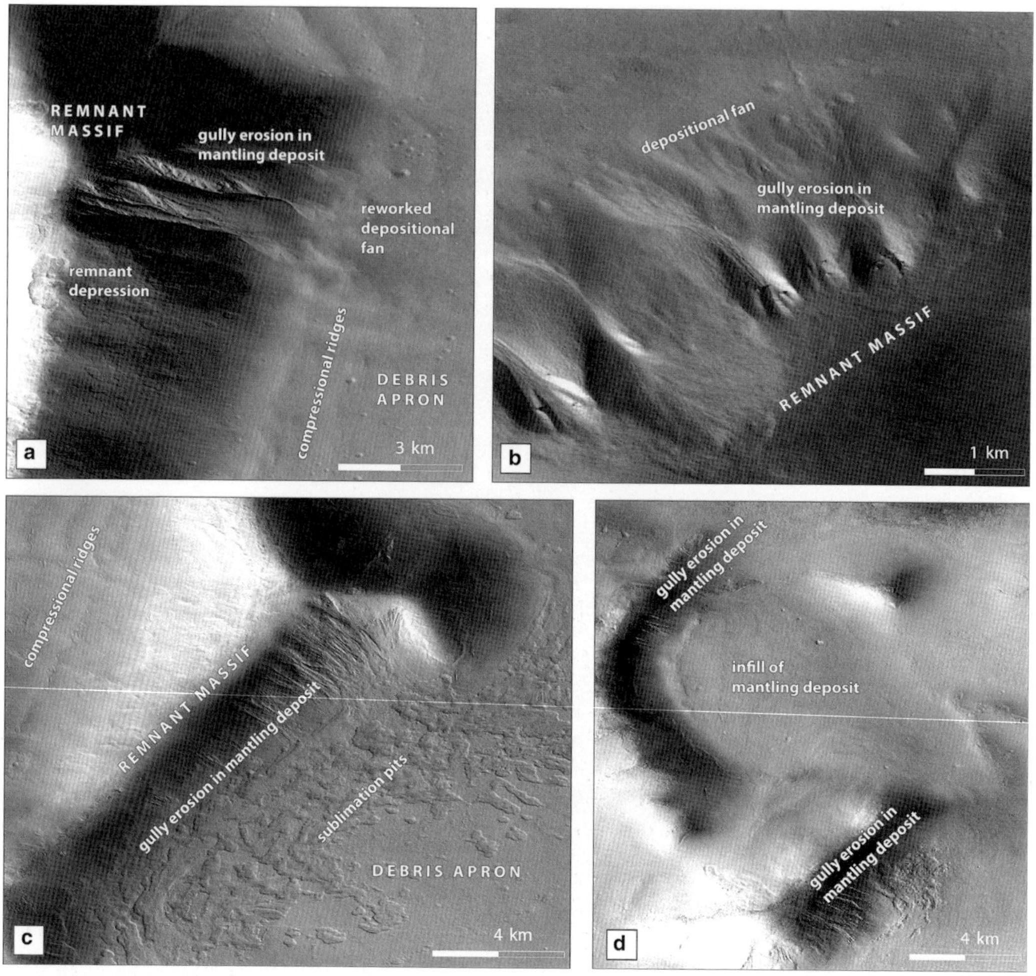

Fig. 9. Gully erosion and sublimation pits in covering mantling cover. (**a**) Gully heads are often associated with surficial depressions found in the mantling deposit covering remnant crests; depositional gully fans are reworked into the footslope apron (P03_002023_2320_XI_52N071W, feature #9, Fig. 2) or (**b**) are superimposed on older hillslope material (P04_002590_2319_XI 51N070W, feature #8, Fig. 2); (**c**) remnant gully erosion and sublimation pattern below gullies (P17_007693_2293_XN_49N073W, feature #13, Fig. 2); (**d**) intra-crater mantling infill and downwasting associated with gully formation (B01_009908_2298_XI_49N073W, feature #12, Fig. 2). North is up in all images. Image credit: NASA/JPL/MSSS.

pits and gully formation (Fig. 9a). Depositional fans formed by gully erosion are partly superimposed on apron footslope material (Figs 8c & 9b) and partly reworked into the debris apron (Fig. 9a, c, d), suggesting multiple phases of hillslope denudation and gully-erosional activity.

The relatively smooth-appearance mantling deposit blankets most parts of the study area and overlies lowland plains and footslope aprons as well as the smoothly convex remnant massifs. It is prominent where landslide scars and small-scaled depressions, such as obliterated impact craters, form local catchment areas (Fig. 9d). At over-steepened walls, mantling material accumulates at the footslope and reaches backwards uphill. Remnant crests marked by a central segmented ridge (Fig. 6a) often exhibit an overlying mantling material that terminates on both sides of the crest. This suggests the retreat of the covering material either by atmospheric loss of volatiles or by downwasting through slumping or creep at steep locations (Figs 6b, c & 7). Some of the displacement of a mantling layer occurs as glide flows, as indicated by polygonal sheets and circumferential ridges (Fig. 6b).

The mantling deposit is (a) superimposed on the remnant massif and (b) often detached from the

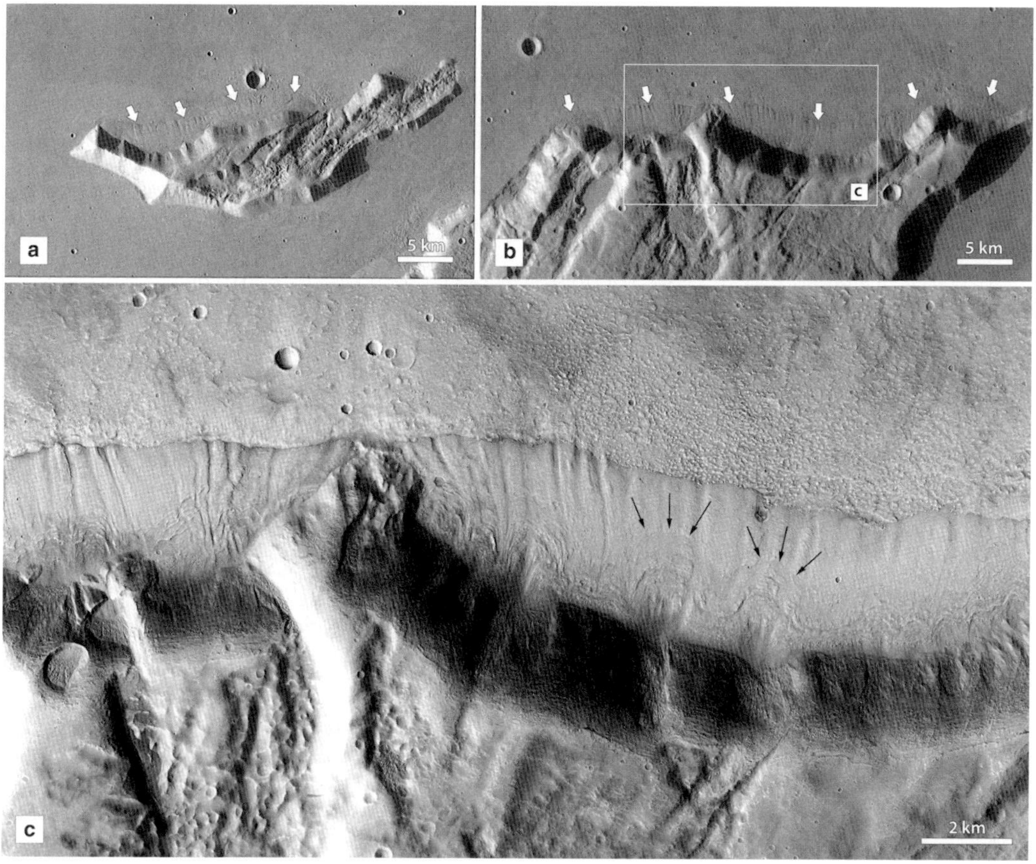

Fig. 10. Lobate debris aprons at isolated remnants in western Tempe Terra. The distribution and dimensions of lobate debris aprons in this more southern part are different from features located in more northern locations of Tempe Terra (Fig. 11). (**a**) Massif at 35.36°N with lobate debris aprons (white arrows) on the northern side, but not on the southern side (detail of HRSC image h5081 0000; centre at 35.36°N and 268.65°E; north is up, illumination is from the west/left). (**b**) Remnant highland massif with marginal lobate debris apron (white arrows). Lineations on lobate debris apron are parallel to the inferred flow direction (detail of HRSC image h5081_0000; centre at 35.0°N and 267.9°E; north is up, illumination is from the west/left). (**c**) Detailed view of lobate debris apron shown in (b); a convoluted or undulating pattern in plan view characterizes the texture of the upper (southern) parts of the apron. The position of lobes (e.g. black arrows) are controlled by indentations of the southern scarp (detail of CTX image P17_007852_2154; north is up, illumination from the SW/lower left). Image credit: ESA/DLR/FUB and NASA/JPL/MSSS.

remnant rock by a narrow trench (Figs 6a, 7 & 8b), which is analogous to a glacier–wall rock detachment, often referred to as randkluft or, less precisely, as bergschrund or rimaye in terrestrial glacial study fields (e.g. Benn & Evans 2003, pp. 213 and 358). These characteristics are not observed for wall-rock debris and are suggestive of a different source for the apron material.

Apart from gully erosion and the formation of sublimation pits that are indicative of a control by insolation (e.g. Morgenstern *et al.* 2007; Soare *et al.* 2008; Lefort *et al.* 2009; Dickson & Head 2009; Morgan *et al.* 2010; Kneissl *et al.* 2010; Lefort *et al.* 2010), apron formation is – at least at lower latitudes – also insolation-controlled, as suggested by observations at massifs and lineated valley-fill units located at 35°N with debris aprons on the northern side, but not on the southern side (Figs 10a–c & 11a, b). The debris apron appears lineated, with stripe orientation parallel to the inferred flow direction. A convoluted or undulating pattern characterizes the texture of the upper parts of the apron with the position of lobes controlled by indentations of remnant scarps. In Figure 11a, b apron units are preserved only at insolation-protected locations, such as the northern margins and interior depressions. At sun-facing sites, their occurrence is limited to graben structures

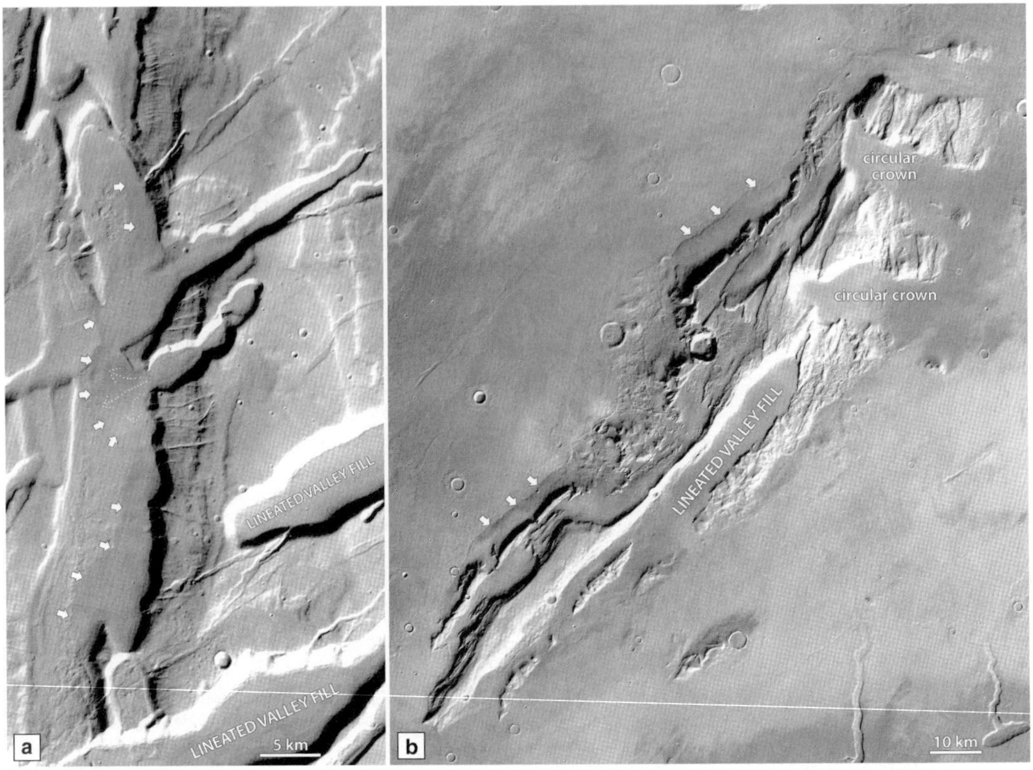

Fig. 11. Lobate debris aprons and lineated valley fill in western Tempe Terra. Lobate debris aprons (white arrows) are seen at the base of scarps associated with an ancient rift. Their dimensions are larger than in more southern locations (Fig. 10). (**a**) A flow lobe (dotted arrows) at the terminus of a valley (consisting of a series of coalesced oval depressions) indicates creep of material out of this valley. Deep depressions are filled with smooth lineated valley fill (detail of HRSC image h1572_0000; centre at 38.0°N and 274.0°E; north is up, illumination from the SE/lower right). (**b**) Isolated, structurally controlled massif in western Tempe Terra. Well-developed lobate debris-aprons frame the massif at its northern margin (white arrows). Interior elongated depressions are filled with lineated valley fill. The southern margin of the massif is devoid of lobate debris aprons, except in protected circular crowns (detail of HRSC image h1594 0000; centre at 39.4°N and 268.9°E; north is up, illumination from the SE/lower right). Image credit: ESA/DLR/FUB.

that are perpendicular to the escarpment transition. Overall, the control of insolation on the LDA population at this latitude seems stronger than in the more northern parts of the study area, but less than in the more southern parts, regardless of wall-rock relief (Fig. 10).

Morphometry

Aprons extend radially from the associated remnant massif, with apron length ranging from 1.6 ± 0.62 km as a minimum up to 12.1 ± 2.4 km at maximum and with a mean length of measured samples of approximately 3.9 km (Table 3). Average lengths of aprons are significantly less than those reported by Mangold & Allemand (2001), 10.8–33 km from the Deuteronilus–Protonilus area, and are also much less than estimates of 15 km given by Colaprete & Jakosky (1998) and Carr (2001).

Apron thicknesses vary between 107 m as a minimum average to up to 750 m as a maximum average, with an overall thickness average of approximately 355 ± 119 m, leading to values that are slightly higher than estimates of approximately 280 m from the Deuteronilus–Protonilus area (Mangold & Allemand 2001) but closely comparable to values of 340 m derived in previous work (Chuang & Crown 2005). Apron-volume estimates at Tempe Terra range from <10 up to 300 km^3, with an average volume of 21 km^3. These values are over one order of magnitude smaller than those derived by Chuang & Crown (2005) and are considered as only approximate values owing to limitations regarding the identification and delineation of extent and the subsurface distribution.

Apron areas are in the range of 40 up to 900 km^2, with remnant sizes of 5–500 km^2. Such area sizes for aprons are much lower than the average values of 550 km^2 given by Chuang & Crown (2005).

The ratio between the area of active potential supply (remnant massif) and area of deposition (debris apron) is roughly 1:3, indicating prolonged denudation if all apron material was derived from remnant massifs. This rough estimate is in general accordance with values given by Barsch (1996) for terrestrial rock glaciers and which are in the range 1:1.36–1:4.4 (Wahrhaftig & Cox 1959; Barsch 1977; Gorbunov 1983).

Calculations of basal shear stress are in the range of 34–108 kPa for the Deuteronilus–Protonilus Mensae areas (Mangold & Allemand 2001) and are slightly lower than photoclinometrically derived values given by Squyres (1978). For Tempe Terra debris aprons, we obtained values of between 6.7 and 82.4 kPa for average apron lengths, with an average of approximately 38 kPa (cf. van Gasselt *et al.* 2008). For terrestrial rock glaciers, values of 100–300 kPa are generally assumed (Whalley 1992). This difference might indicate a higher ice content for Martian analogue landforms, as suggested by other observations (e.g. Li *et al.* 2005; Hauber *et al.* 2008; Plaut *et al.* 2009).

Age constraints

For age determination, crater-size frequency distributions for nine lobate debris aprons were derived from undeformed impact craters. The obtained absolute ages provide proxies for the most recent resurfacing period and give some insight into the early past of apron formation or modification. Ages are in the range of 10–50 Ma, with early traces dating back to 100 Ma and up to 200 Ma ago (Table 4 & Fig. 12). Old ages for #9 are old signatures in the terrain covered by debris aprons; such older impact craters are partially filled by debris-apron material. Segmentation and stair-stepped frequency curves indicate multiple resurfacing events. Shallow branches of frequency curves strongly suggest the continuation of denudation and/or resurfacing and obliteration of older

Table 4. *Statistics for impact-crater size–frequency measurements conducted on nine debris aprons. Numbers refer to the debris aprons labelled in Figure 2; see Figure 12 for size–frequency and isochrone plots*

	Debris apron								
	#7	#8	#9	#13	#18	#19	#20	#24	#29
Crater diameter (km)	Number of impact craters per bin								
0.020	1								
0.030	4	1				3	2		1
0.035	0	1		1		4	2	1	0
0.040	0	2		5		9	3	3	1
0.045	10	4		3	2	5	1	3	3
0.050	22	6		9	6	11	6	8	6
0.060	19	9		13	4	13	11	4	4
0.070	15	18	1	10	1	8	6	2	6
0.080	25	8	1	3	2	5	4	4	3
0.090	11	14	7	6	0	6	3	4	5
0.100	12	8	9	2	1	1	2	0	2
0.110	5	7	5	1	2	1	3	0	1
0.120	4	7	8	2	0	3	0	3	2
0.130	2	5	9	2	1	2	0	0	0
0.140	1	11	6	2	2	0	0	0	2
0.150	0	4	10	2	0	1	0	0	1
0.170	0	6	5	1	1	1	0	0	0
0.200	3	5	4			4	1	1	1
0.250	2	1	3				2	1	
0.300	1	1	0					1	
0.350			1						
0.500			1						
0.600			1						
Sum of craters	137	118	68	62	22	77	46	35	38
Area (km^2)	277.3	470.6	243.6	353.6	212.5	495.9	164.5	132.8	122.0

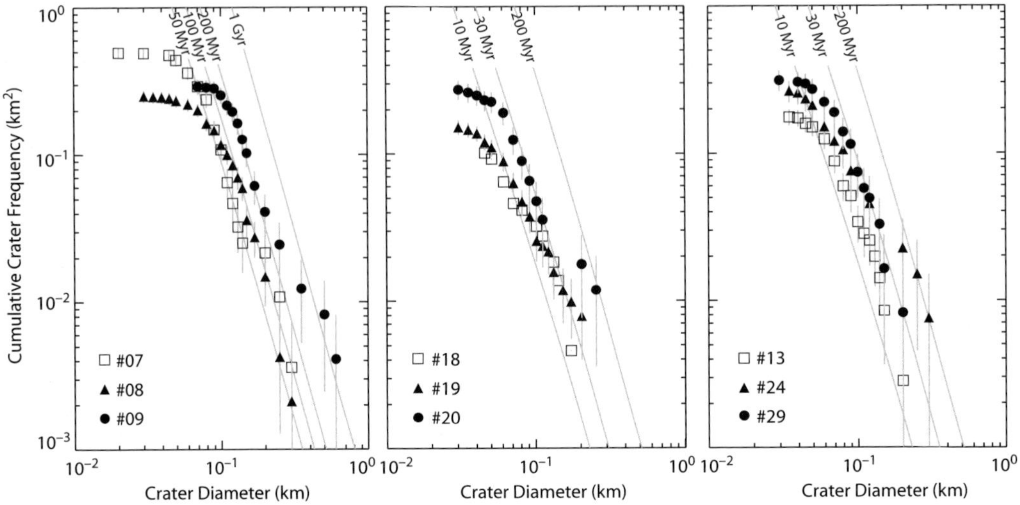

Fig. 12. Isochrones and crater-size frequency measurements for nine debris aprons in Tempe Terra. Numbers in the keys refer to the aprons labelled in Figure 2. Segmented and stair-stepped curves indicate multiple resurfacing events, and shallow segments suggest the continuation of denudation and/or resurfacing. Absolute ages are in the range of few tens of Ma, with early traces dating back to 100 Ma and up to 200 Ma ago. For all aprons, intensive thermokarst degradation with numerous subconcentric pits on the debris apron complicates the derivation of proper ages so that these values are considered as absolute minimums. For the statistics see Table 4.

impact craters. Individual branches indicate a clear phase at 50 Ma ago and some modification process dating back to 10–20 Ma ago. It cannot completely be ruled out that sublimation pits could have been mistaken for impact craters. However, sublimation pits do usually occur in small groups and sample sites exhibiting clustering of circular depressions are excluded from the crater-size statistics in order to avoid secondary craters created during an impact event. On the basis of our observations, we cannot see any clear-cut observational evidence for deformed impact craters. Elliptical-shaped depressions occur frequently on debris-apron surfaces but they are mostly related to clusters of depressions either indicating secondary impact craters or sublimation pits.

Absolute age values of 50–100 Ma are within the error limits generally consistent with estimates from other debris-apron populations (e.g. Squyres 1978; Mangold 2003; Berman *et al.* 2003; Head *et al.* 2005), although the record of impact craters in Tempe Terra might be suggestive of a more youthful population (or more recent modification) when compared to other locations. In addition, resurfacing events have, thus far, not been clearly documented from other debris-apron populations.

Erosion rates are difficult to assess owing to considerable uncertainties regarding the formation age of remnant massifs and the limited knowledge on the proper extent of aprons, and the difficulties regarding the assessment of volumes. Following the geological mapping of Scott & Tanaka (1986), it can be assumed that remnants are considered to be of Noachian–Hesperian age, that is, denudation has been active for 3.5 Ga at variable rates: 1 mm of ground lowering corresponds to the removal of 1000 m^3 km^{-2}; for debris aprons and remnant massifs (source) we obtain average erosion rates of 0.2 mm $\pm$ 0.09 mm $year^{-1}$ for the last 3.5 Ga, corresponding to 200 Bubnoff units (B) (with 1 B corresponding to a surface lowering of 1 m per 10^6 years) (e.g. Selby 1982; Saunders & Young 1983). These rates are considerably faster than values in the range of 10^{-5} B reported from the Mars Exploration Rovers landing site at Gusev or rates of up to 25 nm $year^{-1}$ (0.025 B) for Meridiani Planum (Golombek *et al.* 2006). With regard to the typical Noachian erosion rates of 7.7 μm $year^{-1}$ (7.7 B) as determined by, for example, Hynek & Phillips (2001) and, especially, the average rates of 18 nm $year^{-1}$ (0.018 B) for the Hesperian–Amazonian (Carr 1992), values from the Tempe Terra region seem extraordinarily high. However, in contrast to the Gusev landing site and other locations with low relief energy on Mars, the dichotomy transition provides several kilometres of relief so that such values, despite limitations regarding their derivation, are at least conceivable. High rates, as determined by the accumulation of debris-apron material with respect to the recent extent of remnant massifs, indicate that considerable amounts of material must be of allochthonous origin, which can only be explained by atmospheric deposition.

For comparison, typical terrestrial values for landscapes undergoing comparable processes are in the range of $\leq$2 mm year^{-1} (2000 B) for soil creep, $\leq$20 mm year^{-1} (20 000 B) for solifluction and $\leq$1000 B for polar denudation (Selby 1982, p. 400f).

Discussion on landscape evolution

Based on our observations we propose the following model for landscape evolution in the Tempe Terra region. This model needs further verification in other areas of the dichotomy boundary, especially in the northern hemispheric mid-longitude area of Deuteronilus, Protonilus and Nilosyrtis Mensae, as well as in several other locations, such as the Phlegra Montes and circum Isidis Planitia as outlined by Squyres (1979). However, each setting is unique and constraints on landscape evolution cannot necessarily be made on a global scale because regional environmental conditions as defined by the regional climatic situation as well as the erosional potential, that is, relief, differ significantly.

Remnant massifs might have been emplaced early in Martian history, as demonstrated by mapping (Scott & Tanaka 1986; Tanaka *et al.* 2005*a*), and undergone erosional degradation and denudation by fluvial erosion, or by gravitational processes such as landsliding and mass wasting, respectively (Fig. 4a as supported by Figs 6–8). The rugged and partly conical shape and the generally smooth appearance of exposed remnant material and observations of surficial lineations and ghost impact craters furthermore support the theory of early remnant erosion and deflation since the Noachian.

Release of volatiles under post-Noachian environmental conditions (Carr 1992; Chassefière & Leblanc 2004; Barabash *et al.* 2007) and the formation of typical periglacial surface processes, such as frost creep, gelifluction and rock glaciers, might have occurred on different intensity levels. These denudation processes form the basis for remnant degradation and the lowering of relief. Periglacial acitvity in early Martian history is, however, not a prerequisite and driving factor in the process of formation of the Tempe Terra–Mareotis Fossae lobate debris aprons (Fig. 4b) as other mechanisms could explain the formation of isolated remnant massifs and footslope debris (Büdel 1982). Considering that most of Mars' post-Noachian record lacks a dense atmosphere, however, strongly suggests that a periglacial and hyperarid environment formed the driving boundary conditions for landscape evolution in Tempe Terra.

Our observations in Tempe Terra and research in different regions by other workers have shown that a widespread mantling deposit covers vast areas of the northern and southern mid-latitudes. A general terrain softening and a smoothing of the topography was observed early on the basis of low-resolution data (Soderblom *et al.* 1973; Squyres 1978; Lucchitta 1984). This has since found further support based on observations of topographical and high-resolution image data (Kreslavsky & Head 2000; Mustard *et al.* 2001; Milliken *et al.* 2003; Morgenstern *et al.* 2007). The mantling cover most probably forms a fine-grained material with a considerable volatile content that was either directly derived from the atmosphere or through an aeolian redeposition of polar volatiles in connection with high obliquities (Ward 1974; Toon *et al.* 1980; Laskar & Robutel 1993; Head *et al.* 2003; Laskar *et al.* 2004; Levrard *et al.* 2004; Milkovich *et al.* 2008). This cover has masked and smoothed the old Martian surface with a layer several metres to tens of metres in thickness (Mustard *et al.* 2001; Head *et al.* 2003; Morgenstern *et al.* 2007).

Remnants were covered by an aeolian/atmospheric mantling deposit (Fig. 4c), and buried, at least partially, the underlying topography and traces of any earlier evidence of landscape evolution and wall-rock erosion. Sublimation processes and the formation of debris-apron pits beyond the extent of debris aprons suggest that the mantling deposit covered the Tempe Terra–Mareotis Fossae region homogeneously. As a consequence, estimates of the true relief of remnants are only approximate, as a significant volume is probably masked. Considerable volumes of debris aprons are therefore probably composed of mantling deposits that have been deposited during cyclic climatic variations. As a result of the observed degradation state of individual remnants, mantling deposits either covered the remnant completely and/or it was remobilized gravitationally by downslope movement and revealed the underlying topography. Apron material that subsequently moved downslope and formed lineations aligned perpendicular to the inferred flow directions is interpreted to have been induced by differential flow or creep velocities of the lobate apron's ice–debris mixture and associated shear stresses (Figs 6, 7 & 10). The additional loading from material downwasted along indentations in remnant massifs might have triggered the differential velocities, with higher creep rates at the positions of increased loading. Lineations aligned parallel to the flow directions are considered to have been caused by the sorting of material. Both types of lineations are frequently observed in mass-wasting and/or (peri-)glacial transport

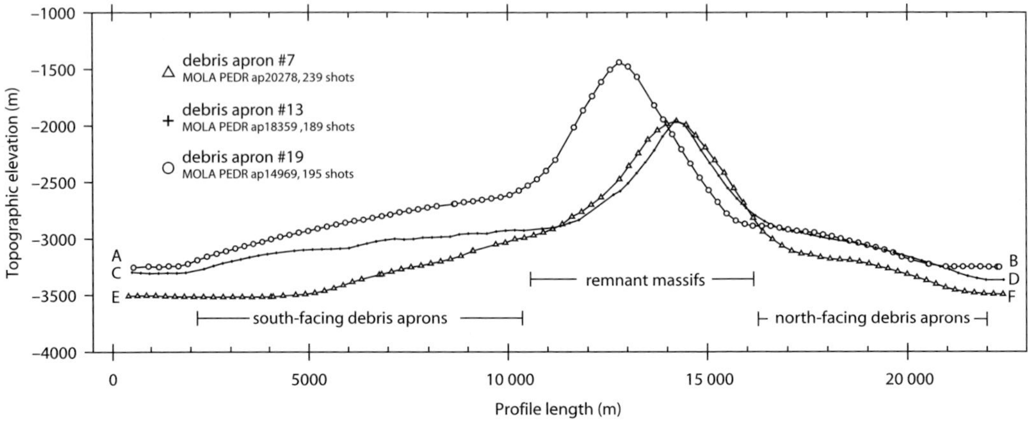

Fig. 13. Topographical profiles from MOLA PEDR data records for central (#13 and #19) and eastern RACs (#7). For display reasons only, each fourth data point of each track segment is plotted.

systems on Earth (e.g. Selby 1982; Barsch 1988; Menzies 2002). Size differences in debris-apron extents at lower latitudes (Figs 10 & 11) might be due to reduced year-integrated insolation on the sun-protected northern side of debris aprons. Such size differences are also observed using high-resolution topographical profiles (Fig. 13).

The process of mantling redeposition and gravitational mass movement, as well as gully formation (Fig. 9), has advanced until geologically recent times. Even episodically occurring events might be conceivable, as indicated by observations of obliterated and filled impact craters and the impact-crater size–frequency data, which has indicators of multiple resurfacing events (Fig. 12). Differential stresses and gravitational mass movement (i.e. creep and gelifluction), perhaps even by reactivation of the underlying periglacial landforms, have ultimately lead to the formation of landforms indicative of characteristic cold-climate phenomena, such as rock glaciers. Subsequent sublimation, perhaps also initiated at cracks and crevasses, has contributed to apron degradation and revealed the underlying surfaces (Mangold 2003; Chuang & Crown 2005). This process is thought to have been active at different levels until at least 10–100 Ma ago, as indicated by crater size–frequency distributions. This activity might also have occurred in recent times. However, seasonal changes have not yet been observed and it is suggested from our observations that the process of apron degradation might be prolonged and slowly paced, as clear indicators for impact-crater deformation on debris-apron slopes have not been observed. There are a variety of factors influencing the shape of impact-crater size–frequency curves and the derivation of absolute surface ages. As recently discussed in, for example, Hartmann & Werner (2010), materials prone to degradation, for example, thermokarst erosion, mask their true age; however, multiple depositional events also cause the obliteration of younger impact craters and only the oldest landscape-forming phases can be observed.

Summary and outlook

Our investigations of the number of Tempe Terra remnant massifs and lobate debris-apron constructs (RACs) indicate that an atmospheric deposition of ice-rich material (icy dust or dusty ice), as suggested by, for example, Mustard *et al.* (2001), covers most of these landforms and the surrounding areas. This plays a crucial role in landscape evolution and the formation of characteristic lowland debris aprons.

Remnant massifs in Tempe Terra have undergone long-term denudation by landslides and rockfall as a coupled process (Figs 7 & 8), with creep deformation of a surficial mantling deposit. Prolonged degradation of remnants led to low-gradient denudation slopes (Figs 5 & 13) on which significant amounts of mantling material could accumulate without being transported downslope, as indicated by a homogeneous coverage and the general shapes and slopes of remnant massifs (Fig. 5). Owing to the low slope gradients, wall-rock supply probably ceased and the contribution of the wall rock to debris-apron formation was therefore limited. This is also confirmed by the mostly undeformed impact craters, which are suggestive of ongoing impact processes, and the lack of the continuation of creep and impact-crater deformation. Crests of remnant massifs do frequently show a detachment of mantling deposits, suggesting that gradients locally are relatively high (Figs 7 & 8).

Relict differential movement of material caused the shaping of crest fractures and compressional, as well as degradational, lineation patterns on debris aprons. The amount of ice and/or debris incorporated into debris aprons is difficult to assess. With respect to area estimates and the comparison to classical terrestrial values of source/catchment ratios, it is likely that slope erosion and denudation contribute a considerable amount to the volume. Spatially overlapping landforms indicate that such events occur either on long-term scales or even episodically. Geologically recent volatile escape might have occurred through vapour diffusion or by sublimation, leading to the formation of characteristic thermokarst pits and debris-apron degradation (Mangold 2003). Average erosion rates for the Tempe Terra features are comparable to terrestrial cold-climate areas but are generally considered too high for Mars, which further supports the idea of a considerable atmospheric component in apron formation. Crater size–frequency analyses for nine debris aprons and crater obliterations indicate that several resurfacing events occurred in the geologically recent past, with early signals starting approximately 200 Ma ago. In the context of ongoing discussions on climatic variations on Mars caused by obliquity changes, it is conceivable that deposition of a mantling deposit occurred periodically and for prolonged periods (Ward 1974; Toon *et al.* 1980; Laskar & Robutel 1993; Head *et al.* 2003; Laskar *et al.* 2004; Levrard *et al.* 2004; Milkovich *et al.* 2008). Atmospheric precipitation of ice-rich material was deposited on top of erosional wall-rock debris and was intermixed subsequently, ultimately leading to thick rock glacier bodies that deformed gravitationally. Cyclic processes over the last few billions (10^9) of years might have led to the suite of landforms observed today, that is, footslope aprons indicative of denudation and the remobilization of a mantling material by creep under periglacial conditions.

For the future, a more detailed assessment of morphometric values could help to assess the amount of eroded material more precisely by using CTX-based topography data of remnant-apron constructs. Significant uncertainties will persist as a mantling material covers most areas of the northern plains and we cannot observe the true lateral extent of each remnant apron body. Such higher-resolution terrain-model data – based also on the imaging of the MRO HiRISE (High Resolution Imaging Science Experiment) camera – could, however, allow the thickness of mantling material exposed at several locations to be properly assessed where gully incisions occur. With the help of additional impact-crater size–frequency analyses on different impact-crater classes (e.g. degradation and deformation state) it will be possible to more clearly distinguish between different surface ages and episodic activity. Initial subsurface radar analysis of high-resolution data has not shown any significant reflector bodies in the subsurface that might indicate the presence of an existing ice table (Holt *et al.* 2008; Plaut *et al.* 2009). A systematic survey will show whether observations of such ice tables are a characteristic feature or an exception, and whether assumptions as to the high ice content of debris aprons are realistic (Hauber *et al.* 2008).

We thank the MRO CTX as well as the *Mars Express* HRSC science and experiment teams for their successful planning and acquisition of data, as well as for making the processed data available to the public. We further wish to acknowledge the MGS teams and Malin Space Science Systems for their successful experimental procedure and for making the data available. This work was funded by the German DLR agency under contract number 500 QM 301 and 500 QM 1001, and was supported by the Helmholtz Alliance 'Planetary Evolution and Life'. This research has made use of NASA's Astrophysics Data System.

References

Albee, A. L., Palluconi, F. D. & Arvidson, R. E. 1998. Mars global surveyor mission: overview and status. *Science*, **279**, 1671, doi: 10.1126/science.279.5357.1671.

Allison, M. 1997. Accurate analytic representations of solar time and seasons on Mars with applications to the Pathfinder/Surveyor missions. *Geophysical Research Letters*, **24**, 1967, doi: 10.1029/97GL01950.

Allison, M. & McEwen, M. 2000. A post-Pathfinder evaluation of areocentric solar coordinates with improved timing recipes for Mars seasonal/diurnal climate studies. *Planetary and Space Science*, **48**, 215–235.

Baldwin, R. B. 1964. Lunar crater counts. *Astronomical Journal*, **69**, 377–391.

Barabash, S., Fedorov, A., Lundin, R. & Sauvaud, J. 2007. Martian atmospheric erosion rates. *Science*, **315**, 501, doi: 10.1126/science.1134358.

Barsch, D. 1977. Nature and importance of mass-wasting by rock glaciers in alpine permafrost environments. *Earth Surface Processes*, **2**, 231–245.

Barsch, D. 1988. Rockglaciers. *In*: Clark, M. (ed.) *Advances in Periglacial Geomorphology*. Wiley, Chichester, 69–90.

Barsch, D. 1996. *Rockglaciers: Indicators for the Present and Former Geoecology on High Mountain Environments*, Springer, Berlin.

Benn, D. I. & Evans, D. J. A. 2003. *Glaciers and Glaciation*, Arnold, London.

Berman, D. C., Hartmann, W. K. & Crown, D. A. 2003. Debris aprons, channels, and volcanoes in the Reull Vallis region of Mars. *In*: *Proceedings of the 34th Annual Lunar and Planetary Science Conference, March 17–21, 2003, League City, Texas*. Lunar and Planetary Institute, Houston, TX, Abstract 1879.

Büdel, J. 1982. *Climatic Geomorphology*, Princeton University Press, Princeton, NJ.

Byrne, S. & Ingersoll, A. P. 2003. A sublimation model for Martian South Polar ice features. *Science*, **299**, 1051–1053, doi: 10.1126/science.1080148.

Carr, M. H. 1992. Post-Noachian erosion rates: implications for Mars climate change. *In*: *Proceedings of the 23rd Annual Lunar and Planetary Science Conference, March 16–20, 1992, League City, Texas*. Lunar and Planetary Institute, Houston, TX, Abstract 205206.

Carr, M. H. 2001. Mars global surveyor observations of Martian fretted terrain. *Journal of Geophysical Research*, **106**, 23,571–23,593, doi: 10.1029/2000JE001316.

Carr, M. H. & Schaber, G. G. 1977. Martian permafrost features. *Journal of Geophysical Research*, **82**, 4039–4054.

Chassefière, E. & Leblanc, F. 2004. Mars atmospheric escape and evolution; interaction with the solar wind. *Planetary and Space Science*, **52**, 1039–1058, doi: 10.1016/j.pss.2004.07.002.

Christensen, P. R. 2003. Formation of recent martian gullies through melting of extensive water-rich snow deposits. *Nature*, **422**, 45–48.

Chuang, F. C. & Crown, D. A. 2005. Surface characteristics and degradational history of debris aprons in the Tempe Terra/Mareotis Fossae region of Mars. *Icarus*, **179**, 24–42, doi: 10.1016/j.icarus.2005.05.014.

Colaprete, A. & Jakosky, B. M. 1998. Ice flow and rock glaciers on Mars. *Journal of Geophysical Research*, **103**, 5897–5909, doi: 10.1029/97JE03371.

Crown, D. A., Price, K. H. & Greeley, R. 1992. Geologic evolution of the east rim of the Hellas basin, Mars. *Icarus*, **100**, 1–25, doi: 10.1016/0019-1035(92)90014-X.

Crown, D. A., Pierce, T. L., McElfresh, S. B. Z. & Mest, S. C. 2002. Debris aprons in the eastern Hellas region of Mars: implications for styles and rates of highland degradation. *In*: *Proceedings of the 33rd Annual Lunar and Planetary Science Conference, March 11–15, 2002, League City, Texas*. Lunar and Planetary Institute, Houston, TX, Abstract 1642.

Cunningham, G. E. 1996. Mars global surveyor mission. *Acta Astronautica*, **38**, 367–375.

Dickson, J. L. & Head, J. W. 2009. The formation and evolution of youthful gullies on Mars: Gullies as the late-stage phase of Mars' most recent ice age. *Icarus*, **204**, 63–86, doi: 10.1016/j.icarus.2009.06.018.

Frey, H., Sakimoto, S. E. & Roark, J. 1998. The MOLA topographic signature at the crustal dichotomy boundary. *Geophysical Research Letters*, **25**, 4409–4412, doi: 10.1029/1998GL900095.

Golombek, M. P., Grant, J. A. *et al.* 2006. Erosion rates at the Mars exploration rover landing sites and long-term climate change on Mars. *Journal of Geophysical Research – Planets*, **111**, 12, doi: 10.1029/2006JE002754.

Gorbunov, A. 1983. Rock glaciers of the mountains of middle Asia. *In*: *Proceedings of the International Conference on Permafrost, Fairbanks, Alaska*, Volume 4. National Academy Press, Washington, DC, 359–362.

Greeley, R., Arvidson, R. E. *et al.* 2006. Gusev crater: Wind-related features and processes observed by the Mars Exploration Rover Spirit. *Journal of Geophysical Research*, **111**, E02S09, doi: 10.1029/2005JE002491.

Haeberli, W. 1985. Creep of mountain permafrost: internal structure and flow of alpine rock glaciers. *Mitteilung der Versuchsanstalt für Wasserbau, Hydrologie und Glaziologie, Zürich*, **77**, 142.

Hartmann, W. K. 1966*a*. Early lunar cratering. *Icarus*, **5**, 406–418, doi: 10.1016/0019-1035(66)90054-6.

Hartmann, W. K. 1966*b*. Martian cratering. *Icarus*, **5**, 565–576, doi: 10.1016/0019-1035(66)90071-6.

Hartmann, W. K. & Neukum, G. 2001. Cratering chronology and the evolution of Mars. *Space Science Reviews*, **96**, 165–194.

Hartmann, W. K. & Werner, S. C. 2010. Martian Cratering 10. Progress in use of crater counts to interpret geological processes: Examples from two debris aprons. *Earth and Planetary Science Letters*, **294**, 230–237, doi: 10.1016/j.epsl.2009.10.001.

Hauber, E. & Kronberg, P. 2001. Tempe Fossae, Mars: a planetary analogon to a terrestrial continental rift? *Journal of Geophysical Research*, **106**, 20,587–20,602, doi: 10.1029/2000JE001346.

Hauber, E., van Gasselt, S., Chapman, M. G. & Neukum, G. 2008. Geomorphic evidence for former lobate debris aprons at low latitudes on Mars: indicators of the Martian paleoclimate. *Journal of Geophysical Research*, **113**, E02007, doi: 10.1029/2007JE002897.

Head, J. W., Mustard, J. F., Kreslavsky, M. A., Milliken, R. E. & Marchant, D. R. 2003. Recent ice ages on Mars. *Nature*, **426**, 797–802.

Head, J. W., Neukum, G. *et al.* 2005. Tropical to mid-latitude snow and ice accumulation, flow and glaciation on Mars. *Nature*, **434**, 346–351.

Head, J. W., Marchant, D. R., Agnew, M. C., Fassett, C. I. & Kreslavsky, M. A. 2006*a*. Extensive valley glacier deposits in the northern mid-latitudes of Mars: evidence for Late Amazonian obliquity-driven climate change. *Earth and Planetary Science Letters*, **241**, 663–671, doi: 10.1016/j.epsl.2005.11.016.

Head, J. W., Nahm, A. L., Marchant, D. R. & Neukum, G. 2006*b*. Modification of the dichotomy boundary on Mars by Amazonian mid-latitude regional glaciation. *Geophysical Research Letters*, **33**, L08S03, doi: 10.1029/2005GL024360.

Head, J. W., Marchant, D. R., Dickson, J. L., Kress, A. M. & Baker, D. M. 2010. Northern mid-latitude glaciation in the Late Amazonian period of Mars: criteria for the recognition of debris-covered glacier and valley glacier landsystem deposits. *Earth and Planetary Science Letters*, **294**, 306–320, doi: 10.1016/j.epsl.2009.06.041.

Holt, J. W., Safaeinili, A. *et al.* 2008. Radar sounding evidence for buried glaciers in the southern mid-latitudes of Mars. *Science*, **322**, 1235–1238, doi: 10.1126/science.1164246.

Hynek, B. M. & Phillips, R. J. 2001. Evidence for extensive denudation of the Martian highlands. *Geology*, **29**, 407–410.

Ikeda, A. & Matsuoka, N. 2002. Degradation of talus-derived rock glaciers in the Upper Engadin, Swiss Alps. *Permafrost and Periglacial Processes*, **13**, 145–161, doi: 10.1002/ppp.413.

Ivanov, B. A. 2001. Mars/Moon cratering rate ratio estimates. *Space Science Reviews*, **96**, 87–104.

Jaumann, R., Neukum, G. *et al.* 2006. The High Resolution Stereo Camera (HRSC) experiment on Mars

Express: instrument aspects and experiment conduct from interplanetary cruise through the nominal mission. *Planetary Space Science*, **55**, 928–952, doi: 10.1016/j.pss.2006.12.003.

JOHNSON, P. 1987. Rock glacier: glacier debris systems or high-magnitude low-frequency flows. *In*: GIARDINO, J., SHRODER, J. J. & VITEK, J. (eds) *Rock Glaciers*. Allen & Unwin, Boston, MA, 175–192.

KÄÄB, A. & KNEISEL, C. 2006. Permafrost creep within a recently deglaciated glacier forefield: Muragl, Swiss Alps. *Permafrost and Periglacial Processes*, **17**, 79–85, doi: 10.1002/ppp.540.

KÄÄB, A. & REICHMUTH, T. 2005. Advance mechanisms of rock glaciers. *Permafrost and Periglacial Processes*, **16**, 187–193, doi: 10.1002/ppp.507.

KÄÄB, A. & WEBER, M. 2004. Development of transverse ridges on rock glaciers: field measurements and laboratory experiments. *Permafrost and Periglacial Processes*, **15**, 379–391, doi: 10.1002/ppp.506.

KARGEL, J. S. & STROM, R. G. 1992. Ancient glaciation on Mars. *Geology*, **20**, 3–7, doi: 10.1130/0091-7613(1992)020(0003:AGOM)2.3.CO;2.

KNEISSL, T., REISS, D., VAN GASSELT, S. & NEUKUM, G. 2010. Distribution and orientation of northern-hemisphere gullies on Mars from the evaluation of HRSC and MOC-NA data. *Earth and Planetary Science Letters*, **294**, 357–367, doi: 10.1016/j.epsl.2009.05.018.

KNEISSL, T., VAN GASSELT, S. & NEUKUM, G. 2011. Map-projection-independent crater size-frequency determination in GIS environments – new software tool for ArcGIS. *Planetary and Space Science*, doi: 10.1016/j.pss.2010.03.015.

KOCHEL, R. C. & PEAKE, R. T. 1984. Quantification of waste morphology in Martian fretted terrain. *Journal of Geophysical Rsearch (Solid Earth)*, **89**, 336, doi: 10.1029/JB089iS01p0C336.

KRESLAVSKY, M. A. & HEAD, J. W. 2000. Kilometer-scale roughness of Mars: results from MOLA data analysis. *Journal of Geophysical Research*, **105**, 26,695–26,712, doi: 10.1029/2000JE001259.

LASKAR, J. & ROBUTEL, P. 1993. The chaotic obliquity of the planets. *Nature*, **361**, 608–612, doi: 10.1038/361608a0.

LASKAR, J., CORREIA, A. C. M., GASTINEAU, M., JOUTEL, F., LEVRARD, B. & ROBUTEL, P. 2004. Long term evolution and chaotic diffusion of the insolation quantities of Mars. *Icarus*, **170**, 343–364, doi: 10.1016/j.icarus.2004.04.005.

LEFORT, A., RUSSELL, P. S., THOMAS, N., MCEWEN, A. S., DUNDAS, C. M. & KIRK, R. L. 2009. Observations of periglacial landforms in Utopia Planitia with the High Resolution Imaging Science Experiment (HiRISE). *Journal of Geophysical Research – Planets*, **114**, 4005–+, doi: 10.1029/2008JE003264.

LEFORT, A., RUSSELL, P. S. & THOMAS, N. 2010. Scalloped terrains in the Peneus and Amphitrites Paterae region of Mars as observed by HiRISE. *Icarus*, **205**, 259–268, doi: 10.1016/j.icarus.2009.06.005.

LEVRARD, B., FORGET, F., MONTMESSIN, F. & LASKAR, J. 2004. Recent ice-rich deposits formed at high latitudes on Mars by sublimation of unstable equatorial ice during low obliquity. *Nature*, **431**, 1072–1075, doi: 10.1038/nature03055.

LI, H., ROBINSON, M. S. & JURDY, D. M. 2005. Origin of Martian northern hemisphere mid-latitude lobate debris aprons. *Icarus*, **176**, 382–394, doi: 10.1016/j.icarus.2005.02.011.

LUCCHITTA, B. K. 1981. Mars and Earth – comparison of cold-climate features. *Icarus*, **45**, 264–303, doi: 10.1016/0019-1035(81)90035-X.

LUCCHITTA, B. K. 1984. Ice and debris in the fretted terrain, Mars. *Journal of Geophysical Research*, **89**(Suppl.), B409–B419.

MADELEINE, J.-B., FORGET, F., HEAD, J. W., LEVRARD, B., MONTMESSIN, F. & MILLOUR, E. 2009. Amazonian northern mid-latitude glaciation on Mars: a proposed climate scenario. *Icarus*, **203**, 390–405, doi: 10.1016/j.icarus.2009.04.037.

MALIN, M. C. & EDGETT, K. S. 2000. Observations of aprons in Martian fretted terrain. *In*: *Proceedings of the 31st Annual Lunar and Planetary Science Conference, March 13–17, 2000, League City, Texas*. Lunar and Planetary Institute, Houston, TX, Abstract 1053.

MALIN, M. C. & EDGETT, K. S. 2001. Mars global surveyor Mars Orbiter Camera: interplanetary cruise through primary mission. *Journal of Geophysical Research*, **106**, 23,429–23,570, doi: 10.1029/2000JE001455.

MALIN, M. C., BELL, J. F. ET AL. 2007. Context camera investigation on board the Mars Reconnaissance Orbiter. *Journal of Geophysical Research*, **112**, E05S04, doi: 10.1029/2006JE002808.

MANGOLD, N. 2003. Geomorphic analysis of lobate debris aprons on Mars at Mars Orbiter Camera scale: evidence for ice sublimation initiated by fractures. *Journal of Geophysical Research*, **108**, 8021, doi: 10.1029/2002JE001885.

MANGOLD, N. & ALLEMAND, P. 2001. Topographic analysis of features related to ice on Mars. *Journal of Geophysical Research Letters*, **28**, 407–410, doi: 10.1029/2000GL008491.

MARTIN, H. & WHALLEY, W. 1987. Rock glaciers, part 1, rock glacier morphology: classification and distribution. *Progress in Physical Geography*, **11**, 260–282.

MCGILL, G. E. 2000. Crustal history of north central Arabia Terra, Mars. *Journal of Geophysical Research – Planets*, **105**, 6945–6960, doi: 10.1029/1999JE001175.

MENZIES, J. 2002. Chapter 3. Gaciers and Ice Sheets. *In*: MENZIES, J. (ed.) *Modern and Past Glacial Environments*. Butterworth-Heinemann, Oxford, 53–78.

MICHAEL, G. G. & NEUKUM, G. 2010. Planetary surface dating from crater size–frequency distribution measurements: partial resurfacing events and statistical age uncertainty. *Earth and Planetary Science Letters*, **294**, 223–229, doi: 10.1016/j.epsl.2009.12.041.

MILKOVICH, S. M., HEAD, J. W. & NEUKUM, G. 2008. Stratigraphic analysis of the northern polar layered deposits of Mars: implications for recent climate history. *Planetary and Space Science*, **56**, 266–288, doi: 10.1016/j.pss.2007.08.004.

MILLIKEN, R. E., MUSTARD, J. F. & GOLDSBY, D. L. 2003. Viscous flow features on the surface of Mars: observations from high-resolution Mars Orbiter Camera (MOC) images. *Journal of Geophysical Research*, **108**, 5057, doi: 10.1029/2002JE002005.

MORGAN, G. A., HEAD, J. W., FORGET, F., MADELEINE, J. & SPIGA, A. 2010. Gully formation on Mars: two recent phases of formation suggested by links between morphology, slope orientation and insolation history. *Icarus*, **208**, 658–666, doi: 10.1016/j.icarus.2010.02.019.

MORGENSTERN, A., HAUBER, E., REISS, D., VAN GASSELT, S., GROSSE, G. & SCHIRRMEISTER, L. 2007. Deposition and degradation of a volatile-rich layer in Utopia Planitia, and implications for climate history on Mars. *Journal of Geophysical Research*, **112**, E06010, doi: 10.1029/2006JE002869.

MUSTARD, J. F., COOPER, C. D. & RIFKIN, M. K. 2001. Evidence for recent climate change on Mars from the identification of youthful near-surface ground ice. *Nature*, **412**, 411–414, doi: 10.1038/35086515.

NEUKUM, G. & DIETZEL, H. 1971. On the development of the crater population on the moon with time under meteoroid and solar wind bombardment. *Earth and Planetary Science Letters*, **12**, 59–66.

NEUKUM, G. & WISE, D. U. 1976. Mars – a standard crater curve and possible new time scale. *Science*, **194**, 1381–1387, doi: 10.1126/science.194.4272.1381.

NEUKUM, G., JAUMANN, R. *ET AL.* & THE HRSC CO-INVESTIGATOR TEAM. 2004*a*. Recent and episodic volcanic and glacial activity on Mars revealed by the High Resolution Stereo Camera. *Nature*, **432**, 971–979, doi: 10.1038/nature03231.

NEUKUM, G., JAUMANN, R. & THE HRSC CO-INVESTIGATOR AND EXPERIMENT TEAM. 2004*b*. HRSC: the High Resolution Stereo Camera of Mars Express. *In*: WILSON, A. (ed.) *Mars Express: The Scientific Payload.* ESA, Noordwijk, The Netherlands, **SP-1240**, 17–35.

PATERSON, W. B. 2001. *The Physics of Glaciers*, 3rd edn. Butterworth-Heinemann, London.

PERRON, J. T., DIETRICH, W. E., HOWARD, A. D., MCKEAN, J. A. & PETTINGA, J. R. 2003. Ice-driven creep on Martian debris slopes. *Geophysical Research Letters*, **30**, 1747, doi: 10.1029/2003GL017603.

PIERCE, T. L. & CROWN, D. A. 2003. Morphologic and topographic analyses of debris aprons in the eastern Hellas region, Mars. *Icarus*, **163**, 46–65, doi: 10.1016/S0019-1035(03)00046-0.

PLAUT, J. J., SAFAEINILI, A. *ET AL.* 2009. Radar evidence for ice in lobate debris aprons in the mid-northern latitudes of Mars. *Geophysical Research Letters*, **36**, doi: 10.1029/2008GL036379.

ROSSBACHER, L. A. & JUDSON, S. 1981. Ground ice on Mars – inventory, distribution, and resulting landforms. *Icarus*, **45**, 39–59, doi: 10.1016/0019-1035(81)90005-1.

SAUNDERS, I. & YOUNG, A. 1983. Rates of surface processes on slopes, slope retreat and denudation. *Earth Surface Processes Landforms*, **8**, 473501.

SCOTT, D. H. & TANAKA, K. L. 1986. *Geologic Map of the Western Equatorial Region of Mars, scale 1:15,000,000.* United States Geological Survey Miscellaneous Investigations Series Map, **I-1802-A**.

SELBY, M. J. 1982. *Hillslope Materials and Processes.* Oxford University Press, Oxford.

SHARP, R. P. 1973. Mars: fretted and chaotic terrains. *Journal of Geophysical Research*, **78**, 4073–4083.

SHOEMAKER, E. M. & HACKMAN, R. J. 1962. Stratigraphic basis for a lunar time scale. *In*: KOPALZ, Z. & MIKHAILOV, Z. K. (eds) *The Moon*. Academic Press, New York, 289–300.

SKINNER, J. A., HARE, T. M. & TANAKA, K. L. 2006. Digital renovation of the atlas of Mars 1:15,000,000-scale global geologic series maps. *In*: *Proceedings of the 37th Annual Lunar and Planetary Science Conference, March 13–17, 2006, League City, Texas.* Lunar and Planetary Institute, Houston, TX, Abstract 2331.

SMITH, D., NEUMANN, G., ARVIDSON, R. E., GUINESS, E. A. & SLAVNEY, S. 2003. *Mars Global Surveyor Laser Altimeter Mission Experiment Gridded Data Record.* NASA Planetary Data System, MGS-M-MOLA-5-MEGDR-L3-V1.0. NASA, Washington, DC.

SOARE, R. J., OSINSKI, G. R. & ROEHM, C. L. 2008. Thermokarst lakes and ponds on Mars in the very recent (late Amazonian) past. *Earth and Planetary Science Letters*, **272**, 382–393, doi: 10.1016/j.epsl.2008.05.010.

SODERBLOM, L. A., KREIDLER, T. J. & MASURSKY, H. 1973. Latitudinal distribution of a debris mantle on the Martian surface. *Journal of Geophysical Research (Solid Earth)*, **78**, 4117–4122, doi: 10.1029/JB078i020p04117.

SQUYRES, S. W. 1978. Martian fretted terrain: flow of erosional debris. *Icarus*, **34**, 600–613, doi: 10.1016/0019-1035(78)90048-9.

SQUYRES, S. W. 1979. The distribution of lobate debris aprons and similar flows on Mars. *Journal of Geophysical Research*, **84**, 8087–8096.

SQUYRES, S. W. & CARR, M. H. 1986. Geomorphic evidence for the distribution of ground ice on Mars. *Science*, **231**, 249–252, doi: 10.1126/science.231.4735.249.

TANAKA, K. L., SKINNER, J. A. & HARE, T. M. 2005*a*. *Geologic Map of the Northern Plains of Mars.* Pamphlet accompanying United States Geological Survey Miscellaneous Investigations Series Map, **2888**.

TANAKA, K. L., SKINNER, J. A. & HARE, T. M. 2005*b*. *Geologic Map of the Northern Plains of Mars, scale 1:15,000,000.* United States Geological Survey Miscellaneous Investigations Series Map, **2888**.

TOON, O. B., POLLACK, J. B., WARD, W., BURNS, J. A. & BILSKI, K. 1980. The astronomical theory of climatic change on Mars. *Icarus*, **44**, 552–607, doi: 10.1016/0019-1035(80)90130-X.

TURTLE, E., PATHARE, A., CROWN, D., CHUANG, F., HARTMANN, W. K., GREENHAM, J. & BUENO, N. 2003. Modeling the deformation of lobate debris aprons on Mars by creep of ice rich permafrost. *In*: *Third International Conference on Mars Polar Science and Exploration, October 13–17, 2003, Alberta, Canada.* Lunar and Planetary Institute, Houston, TX, Abstract 8091.

VAN GASSELT, S., HAUBER, E., ROSSI, A. P. & NEUKUM, G. 2008. Emplacement of lobate rock glacier landforms and landscape modification, Mareotis Fossae, Mars. *In*: *The Ninth International Conference on Permafrost.* Institute of Northern Engineering, University of Alaska Fairbanks, 487–492.

VAN GASSELT, S., HAUBER, E. & NEUKUM, G. 2010. Lineated valley fill at the Martian dichotomy boundary: nature and history of degradation. *Journal of*

Geophysical Research, **115**, E08003, doi: 10.1029/2009JE003336.

Vitek, J. D. & Giardino, J. R. 1987. Rock glaciers: a review of the knowledge base. *In*: Giardino, J. R., Shroder, J. F. & Vitek, J. D. (eds) *Rock Glaciers*. Allen & Unwin, London, 1–26.

Wahrhaftig, A. L. & Cox, A. 1959. Rock glaciers in the Alaska Range. *Bulletin of the Geological Society of America*, **70**, 383–436.

Wahrhaftig, C. 1954. Observations on rock glaciers in the Alaska Range. *Bulletin of the Geological Society of America*, **65**, 1353.

Ward, W. P. 1974. Climate Variations on Mars, 1, astronomical theory of insulation. *Journal of Geophysical Research*, **79**, 3375–3386.

Werner, S. C. 2005. *Major Aspects of the Chronostratigraphy and Geologic Evolutionary History of Mars*. Dissertation thesis, Fachbereich Geowissenschaften, Freie Universität Berlin, Berlin.

Whalley, W. 1992. A rock glacier in South Ellendalen, Lyngen Alps. *Norsk Geografisk Tidsskrift*, **46**, 29–31.

Whalley, W. & Martin, H. 1992. Rock glaciers: II models and mechanisms. *Progress in Physical Geography*, **16**, 127–186.

Whalley, W. B. & Azizi, F. 2003. Rock glaciers and protalus landforms: analogous forms and ice sources on Earth and Mars. *Journal of Geophysical Research*, **108**, 8032, doi: 10.1029/2002JE001864.

White, S. E. 1987. Differential movement across transverse ridges on Arapaho rock glacier, Colorado Front Range, USA. *In*: Giardino, J. R., Shroder, J. F. & Vitek, J. D. (eds) *Rock Glaciers*. Allen & Unwin, London, 145–150.

Zimbelman, J. R., Clifford, S. M. & Williams, S. H. 1989. Concentric crater fill on Mars – an aeolian alternative to ice-rich mass wasting. *In*: *Proceedings of the 20th Annual Lunar and Planetary Science Conference, March 13–17, 1989, League City, Texas*. Lunar and Planetary Institute, Houston, TX, 397–407.

Zuber, M. T., Smith, D. E. *et al.* 1992. The Mars Observer Laser Altimeter Investigation. *Journal of Geophysical Research*, **97**, 7781–7797, doi: 10.1029/92JE00341.

Evolution of periglacial landforms in the ancient mountain range of the Thaumasia Highlands, Mars

ANGELO PIO ROSSI[1,2]*, STEPHAN VAN GASSELT[3], MONICA PONDRELLI[4], JAMES DOHM[5,6], ERNST HAUBER[7], ALEXANDER DUMKE[3], TANJA ZEGERS[8] & GERHARD NEUKUM[3]

[1]*International Space Science Institute (ISSI), Hallerstrasse 6, CH-3012 Bern, Switzerland*

[2]*Present address: Jacobs University Bremen, Campus Ring 1, D-28759 Bremen, Germany*

[3]*Institut für Geologische Wissenschaften, Freie Universität Berlin, Germany*

[4]*IRSPS, Università d'Annunzio, Pescara, Italy*

[5]*Department of Hydrology and Water Resources, University of Arizona, Tucson, AZ 85721, USA*

[6]*The Museum, University of Tokyo, Tokyo 113-0033, Japan*

[7]*DLR Institute for Planetary Research, Berlin, Germany*

[8]*Faculty of Geosciences, Utrecht University, Utrecht, The Netherlands*

**Corresponding author (e-mail: an.rossi@jacobs-university.de)*

Abstract: Possible periglacial and relict glacial landforms in the ancient mountain range of the Thaumasia Highlands, Mars, are described. The landforms include large-scale mantling, lineated crater and valley-fill materials, debris aprons, protalus lobes and ramparts. The most pristine ice-related landforms appear to be small-scale protalus lobes and ramparts with no visible distinct impact craters at both medium (High Resolution Stereo Camera (HRSC)) and high (Mars Orbiter Camera (MOC) narrow angle (NA), Context Camera (CTX)) spatial resolution. These small landforms are possibly active at present and post-date more extensive features such as crater fills, possibly formed during high obliquity climatic periods.

In contrast to the rock glacier-like landforms with distribution preferentially occurring on south-facing slopes, possibly controlled by enhanced exposure to the Sun, older, less pristine lineated fill materials show a less systematic distribution of flow directions, suggesting a more generalized periglacial and possibly glacial environment in the Thaumasia Highlands.

Background

The possible presence of glacial (Kargel & Strom 1992) and periglacial (e.g. Squyres 1978; Lucchitta 1981; Rossbacher & Judson 1981) features on Mars has been proposed and discussed since the *Viking Orbiter* missions era using relatively low-resolution *Viking* images. A system of glacial-like landforms extending from the south polar region into the Hellas impact basin, for example, was interpreted as marking ice-sheet-related activity (Kargel & Strom 1992). In addition to the Hellas glacial system, which was highly controversial, the fretted terrain (e.g. Squyres 1978) and the debris aprons, interpreted as some form of rock glacier (Colaprete & Jakosky 1998), were the main candidates for ice-related landforms during the era.

Post-*Viking* missions data have significantly increased the number of identified ice-related landforms, including candidate rock glaciers, which indicates climatic and environmental conditions vastly different to those observed today. This increase in data is thanks in part to high-resolution *Mars Global Surveyor* (MGS) Mars Orbiter Camera (MOC) imagery (Rossi *et al.* 2000). Most periglacial landforms were simply not easily visible at resolutions lower than the ones achieved, in the first instance, by MGS. In addition, several ice-related landforms appear to have formed during recent geological times through the evaluation of the post-*Viking*-era data (e.g. Head & Marchant 2003).

With this new-found perspective on ice-related modifications to the Martian landscape and the resulting enhanced enquiry by the planetary science community, Whalley & Azizi (2003) pointed to problems in the description of both terrestrial and Martian rock glaciers, including nomenclature and formational mechanisms. Likewise, Mahaney *et al.* (2006) discussed in detail the investigation of rock glaciers on Earth and their Martian counterparts.

From: Balme, M. R., Bargery, A. S., Gallagher, C. J. & Gupta, S. (eds) *Martian Geomorphology*. Geological Society, London, Special Publications, **356**, 69–85.
DOI: 10.1144/SP356.5 0305-8719/11/$15.00

Other relevant post-*Viking*-era observations included in the literature are the presence of ground ice at relatively shallow depths (Boynton *et al.* 2002, 2004; Feldman *et al.* 2002, 2004) and at increasingly lower latitudinal reaches through time (e.g. Head & Marchant 2003; Head *et al.* 2006; Dickson *et al.* 2008; Hauber *et al.* 2008). Similarly, several glacier or rock glacier-like landforms have been discovered in tropical latitudes (e.g. Head & Marchant 2003) which are interpreted as being debris-covered glaciers.

Low-latitude glacial-like morphologies have been documented mostly in the northern hemisphere (Head *et al.* 2006), but recently have also been found in the southern hemisphere (Berman *et al.* 2005, 2009; Dickson *et al.* 2006; Rossi *et al.* 2006, 2008).

With regard to these features newly identified through the analysis of post-*Viking* data, the Thaumasia Highlands region was one of the geological provinces on Mars with very few reported (Dickson *et al.* 2006; Rossi *et al.* 2006, 2008) glacial–periglacial features despite the rich and complex geological history of the region (Dohm & Tanaka 1999; Dohm *et al.* 2001*b*). Therefore, the present work attempts to fill a gap in describing possible periglacial–glacial landforms on Mars, specifically within the high-altitude, ancient mountain range of the Thaumasia Highlands region. Such rugged environments have the potential to yield further clues concerning the palaeoclimatic and palaeoenvironmental conditions of Mars.

In addition to transient endogenic-driven activity and the associated change in climate and environmental conditions that resulted in a landscape modified by water–ice and liquid water (Baker 2001), geologically recent obliquity-driven climatic changes and the associated precipitation have been proposed as the driving forces for the development of glacial–periglacial landforms on Mars (Forget *et al.* 2006). In particular, some model runs also show the growth of ice at high altitudes such as in the Thaumasia Highlands region and its surroundings (Madeleine *et al.* 2007), the region where the present work concentrates.

Thaumasia settings

The Thaumasia Highlands (Fig. 1) is a rugged ancient mountain range on Mars (Dohm *et al.* 2001*a*, *b*), which separates the Tharsis magmatic complex (Dohm *et al.* 2001*a*, *2007*) to the NNW from the Argyre impact-influenced transition zone to the SSE. The highest promontory within the mountain range is Warrego Rise (268°E, 40°S) at an elevation of more than 7.6 km above Martian datum. Therefore, there is a significant difference in elevation between the Thaumasia Highlands region and its surrounding area (e.g. the floor materials of the Argyre Basin, for example, occur below the −2 km Martian datum), and this possibly exerts a significant influence on the regional environmental–climatic conditions.

The distinct Warrego Rise is located near the southernmost margin of the Thaumasia Highlands. Heat flow calculations indicate that the crust beneath the rise may be chemically stratified, with a heat-producing enriched upper layer thinner than the whole crust (Ruiz *et al.* 2009). Stratigraphic and cross-cutting relations, impact-crater statistics, an order of magnitude greater density of tectonic structures in the Noachian mountain-forming materials compared to the Late Hesperian lava plains of the shield complex of Syria Planum, and magnetic signatures indicate that the mountain range formed during an ancient geological period of Mars, prior to the shut down of the magnetosphere (Dohm *et al.* 2001*a*, *b*, 2009).

Faults and folds of diverse orientation resulting from contractional and extensional deformation, complex rift systems, shield volcanoes that occur along rift systems, and hogbacks, cuestas and valley networks such as Warrego Valles (267°E, 42°S), record a complex geological history for the Thaumasia Highlands region (Dohm & Tanaka 1999; Dohm *et al.* 2001*a*, *b*, 2007; Grott *et al.* 2005, 2007; Hauber & Kronberg 2005; Anguita *et al.* 2006), which may include magmatic-driven activity such as igneous plateau uplift (Dohm *et al.* 2001*b*) and possibly some form of plate tectonism (Dohm *et al.* 2002; Anguita *et al.* 2006; Baker *et al.* 2007). Contrary to such a complex history, a gravity-spreading system (mega-slide) related to the geothermal heating and topographical loading of extensive buried deposits of salts and/or mixtures of salts, ice and basaltic debris has been proposed to explain the formation of the Thaumasia Highland mountain range (Montgomery *et al.* 2009).

To date, the parent mountain-forming rock materials of the Thaumasia Highlands are unknown, perhaps largely due to secondary weathering rinds, aoelian mantles, alluvial fans, fluvial deposits and periglacial materials. The latter of these, which is the primary focus of this study, obscures the bedrock materials from an orbital perspective (Dohm *et al.* 2009). Thermal Emission Spectrometer (TES)- and Compact Reconnaissance Imaging Spectrometer for Mars (CRISM)-based analyses indicated spectral signatures distinct from the volcanic lava flows of the shield complex of Syria Planum, which includes phyllosilicates (Dohm *et al.* 2009). Based on the features similar to terrestrial mountain ranges, parent rock materials in addition to basalt and basaltic andesites are anticipated (Scott & Tanaka 1986; Dohm *et al.* 2009).

When a regional mapping investigation of the Thaumasia Highlands region and surrounding area

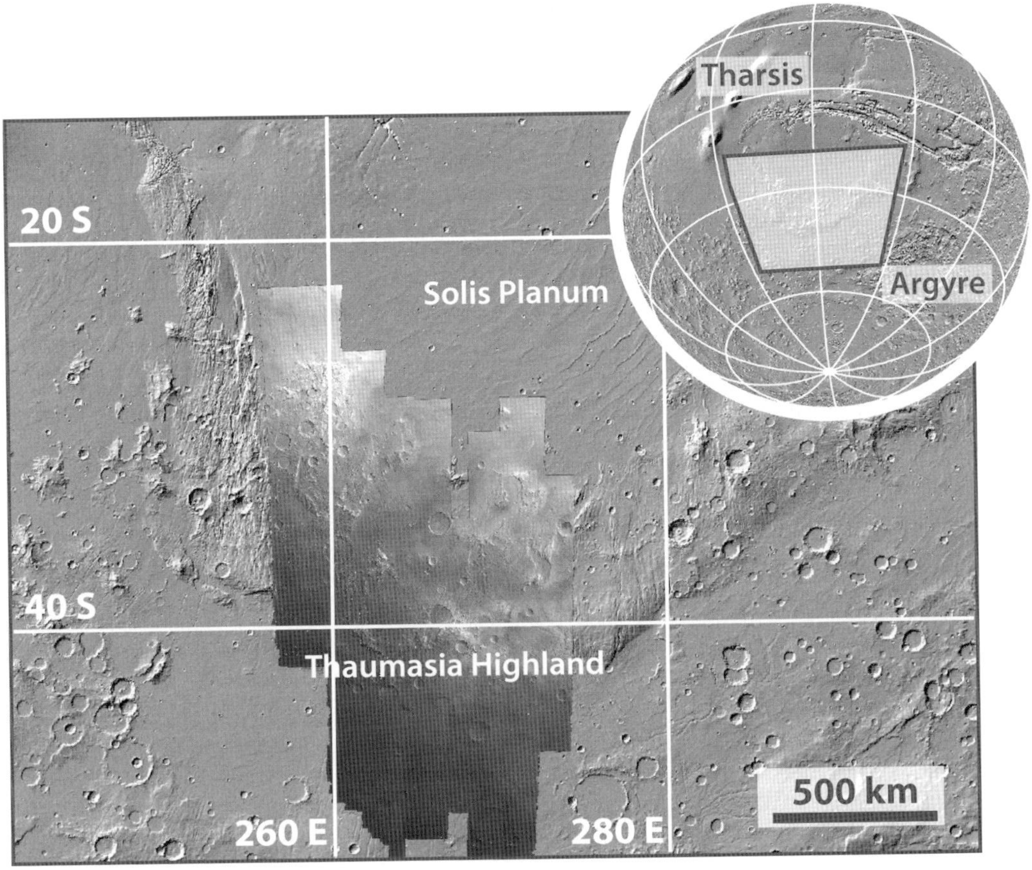

Fig. 1. Location map. HRSC nadir mosaic over MOLA-based shaded relief map. The coverage of the imagery mosaic coincides with the extent of the HRSC stereo-derived 125 m per pixel DEM, obtained from orbits 420, 431, 442, 453, 486, 497, 508 and 530. Image credit: NASA/JPL/MOLA Science Team and ESA/DLR/FUB; see prelim viii for acronym definitions.

was carried out (Dohm *et al.* 2001*b*) the relatively small-scale landforms such as the ones described here were not distinct enough to be resolvable using the *Viking Orbiter* imagery. Our focused survey (Rossi *et al.* 2006, 2008) of possible ice-related landforms in the region has been enhanced with the use of more recent higher-resolution imagery and topographical data.

Data and methods

In this study we use image data from multiple missions, including: MGS Mars Orbiter Laser Altimeter (MOLA) topographical data; both Mission Experiment Gridded Data Records (MEGDR) grids (128 pixel/degree) and Precision Experiment Data Records (PEDR) profiles (*c.* 200 m ground spacing between shots); *Mars Express* (MEX) High Resolution Stereo Camera (HRSC); and 2001 *Mars Odyssey* (MO) Thermal Emission Imaging System (THEMIS). In selected areas we also utilize THEMIS visible (VIS), MGS MOC narrow angle (NA), *Mars Reconnaissance Orbiter* (MRO) Context Camera (CTX) and MRO High Resolution Imaging Science Experiment (HiRISE), where available.

In addition, topographical data derived from HRSC stereo imagery were used. In particular, a custom multi-orbit digital elevation model (DEM) was produced (e.g. Gwinner *et al.* 2005; Dumke *et al.* 2008), based on stereo imagery form MEX HRSC orbits 420, 431, 442, 453, 486, 497, 508 and 530 with a final ground resolution of 125 m per pixel (Fig. 2).

The data were processed with either DLR-VICAR (Video Image Communication and Retrieval, produced by the Jet Propulsion Laboratory and modified by the German Space Agency, DLR) or the United States Geological Survey (USGS) Integrated Software for Imagers and Spectrometers

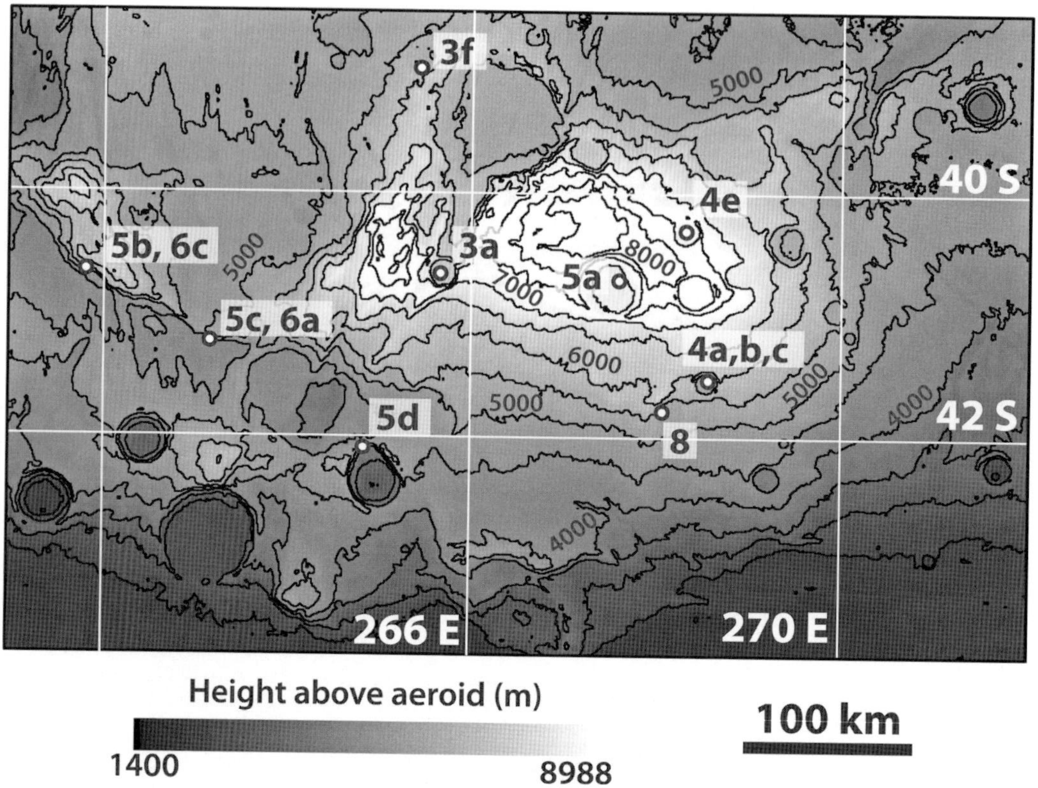

Fig. 2. HRSC digital elevation model subset. Contours vertical spacing: 500 m. Some significant contours labels are displayed. The location of figures in the paper is provided. Figures outside the displayed area: Figure 3b: 265.1°E, 38°S; Figure 3c, d: 270.3°E, 35.8°S; Figure 3e: 265.2°E, 37.7°S; Figure 7: 263.3°E, 36.9°S. Image credit: NASA/JPL/MOLA Science Team and ESA/DLR/FUB.

ISIS3 (Gaddis *et al.* 1997; Torson & Becker 1997; Anderson *et al.* 2004) system, and then integrated and analysed using geographical information systems (GIS) tools. The nomenclature used here is descriptive, trying to avoid genetic terms or implications when possible. The terminology is similar to the one used by Whalley & Azizi (2003).

Landforms described here are usually too small to provide reliable dating with crater counting (Wagner *et al.* 1999). Some of the landforms lack high-resolution coverage. Moreover, where suitable imagery is available, the identification of actual impact craters, often deformed, degraded or modified, among other topographic lows, such as thermokarst features, is difficult, if not unfeasible.

Geologically recent periglacial landforms in the Thaumasia Highlands region

Focusing on the rugged ancient mountain range of Thaumasia Highlands using post-*Viking* data as highlighted earlier, we have newly identified a suite of periglacial landforms that mark changes in climate and environmental conditions in a high-altitude environment of Mars. The features identified using post-*Viking* data includes plateau mantling features, lineated crater/depression-fill materials, debris aprons, protalus lobes and ramparts, as well as their association in space and time as described below.

Plateau mantling

Located outside of the impact craters and other topographical depressions, including grabens (Fig. 3a, b, e), terrain comprised mainly of plateau-forming materials appears to be highly mantled (Mustard *et al.* 2001; Kreslavsky & Head 2003; Milliken *et al.* 2003; van Gasselt *et al.* 2008). The altitude of these plateaus is largely 4000 m above Martian datum (Figs 1 & 2). The plateau surfaces appear to be smooth in places, in contrast to the rough pitted terrains that often occur in a close spatial relationship (Fig. 3f).

The mantles of materials are widespread but probably heterogeneous in thickness and preservation, covering pre-existing topography and relatively small impact craters. Such materials embay the rims of the larger degraded impact craters. Typical lineated fill materials characterize the interiors of these larger impact craters. The widespread mantles of materials may be the result of a combination of geological processes through time, including fluvial, alluvial, colluvial and glacial/ice-sheet deposition.

Lineated crater/depression-fill materials

Lineated crater/depression-fill materials (e.g. Dickson *et al.* 2008), similar to the fretted terrain (e.g. Squyres 1978; Lucchitta 1984) mainly observed in the northern hemisphere in the vicinity of the crustal dichotomy boundary, are widespread in the study area.

Most relatively large impact craters (Figs 2–4) in the highlands contain lineated (either transversal or subconcentric) fill materials; longitudinal textures, however, have not been identified in the study area.

Lineated, textured, pitted crater and depression-fill materials appear to be associated with each other. Within the fill materials, for example, ridges and saddles at lower spatial resolution (e.g. HRSC) appear to be composed of smaller-scale knobs and pits at higher resolution (e.g. CTX). The envelope of these small knobs is arranged slightly concentric to transversal ridges and furrows with respect to the inferred direction of flow(s) geometries (Fig. 3c, d). Unlike the fretted terrains, which occur in elongated depressions (e.g. Squyres 1978), longitudinal ridges are not observed.

At high resolution (decametre/metre scale; e.g. CTX/MOC) the surface texture of lineated fills is very rough. Circular–quasi-circular features, many of which could be small, degraded impact craters (Fig. 3a–d), are visible within the infill materials of various impact craters. These circular features are more widespread and numerous than expected based on analyses using lower-resolution data (Rossi *et al.* 2008). Some are likened to features elsewhere on Mars that have been interpreted to be thermokarst features (Costard & Kargel 1995) in the Chryse and Elysium areas. They include closed irregular depressions. Other lineated (either linear or curvilinear) fill materials on Mars (e.g. Dickson *et al.* 2008) have been interpreted to be the result of ice flow or eolian erosion (Zimbelman *et al.* 1989).

In most cases the topography of lineated fill materials is consistent with flow directions inferred from morphology and texture (Figs 3 & 4). The flow directions of these crater-fill materials correlate with regional slope, but with little or no correlation to local exposure, unlike protalus lobes and ramparts (Figs 5 & 6) (Berman *et al.* 2005, 2009). The slopes of the lineated fill materials are usually less than 1°, as exemplified in two examples (Fig. 4) with average slopes of 0.6° (Fig. 4c, d) and 0.5° (Fig. 4e, f).

The thickness of the fill materials is difficult to determine in the study area because they are enclosed by impact-crater walls or depression-bounding scarps with no observed natural cuts that expose total cross-sections of the materials. However, there are minor fault scarps visible in places within the lineated valley fills with noticeable offsets (Fig. 3f). Although, it is not clear whether these fault scarps post-date or predate the fill materials. In the latter case, the faults may have been covered by material that was subsequently partially sublimated, leaving behind a lag and an exhumed fault.

Apart from local disruption and deformation (Fig. 3a–d), some lineated fill materials show evidence of horizontal deformation/movement. In one case (Fig. 7) the finite deformation of an elliptical feature, interpreted to be a deformed circular impact crater (Fig. 7a, b), can be measured. Linear features in the impact-crater fill materials (Fig. 7a) indicate a direction (indicated with a white arrow in Fig. 7a) consistent with horizontal simple shear (black in Fig. 7c), as deduced from the deformation of an assumed circular impact crater, rather than pure shear (light grey in Fig. 7c). The simple-shear deformation assumes area conservation on the surface of the fill. The assumption of an original circular impact crater in Figure 7c is also consistent with the observed crater features (at available resolution) that tend to rule out an oblique impact event (an alternative explanation for such an oblique structure). A hypothetical projectile with a 5°–15° incidence angle (Gault & Wedekind 1978) necessary to produce such an elliptical crater would probably impact on the outer rim before reaching its interior (Fig. 7d). Also, a thin-rimmed structure, as observed in Figure 7b, which has a major axis length of about 800 m does not display features of a non-impact origin, such as pingos. Moreover, the lineations in the textured floor, although not imaged in very high resolution, seem to be consistent with a direction of movement compatible with simple shear and are marked by a white arrow in Figure 7a. The finite maximum linear deformation (and linear movement within the fill) of the crater is of about 200 m. Both the thickness of the deformed fill materials and the vertical component of the deformation cannot be determined. Such a deformation, if confirmed, would provide an unambiguous determination of actual movement in lineated valley-fill materials. Moreover, although

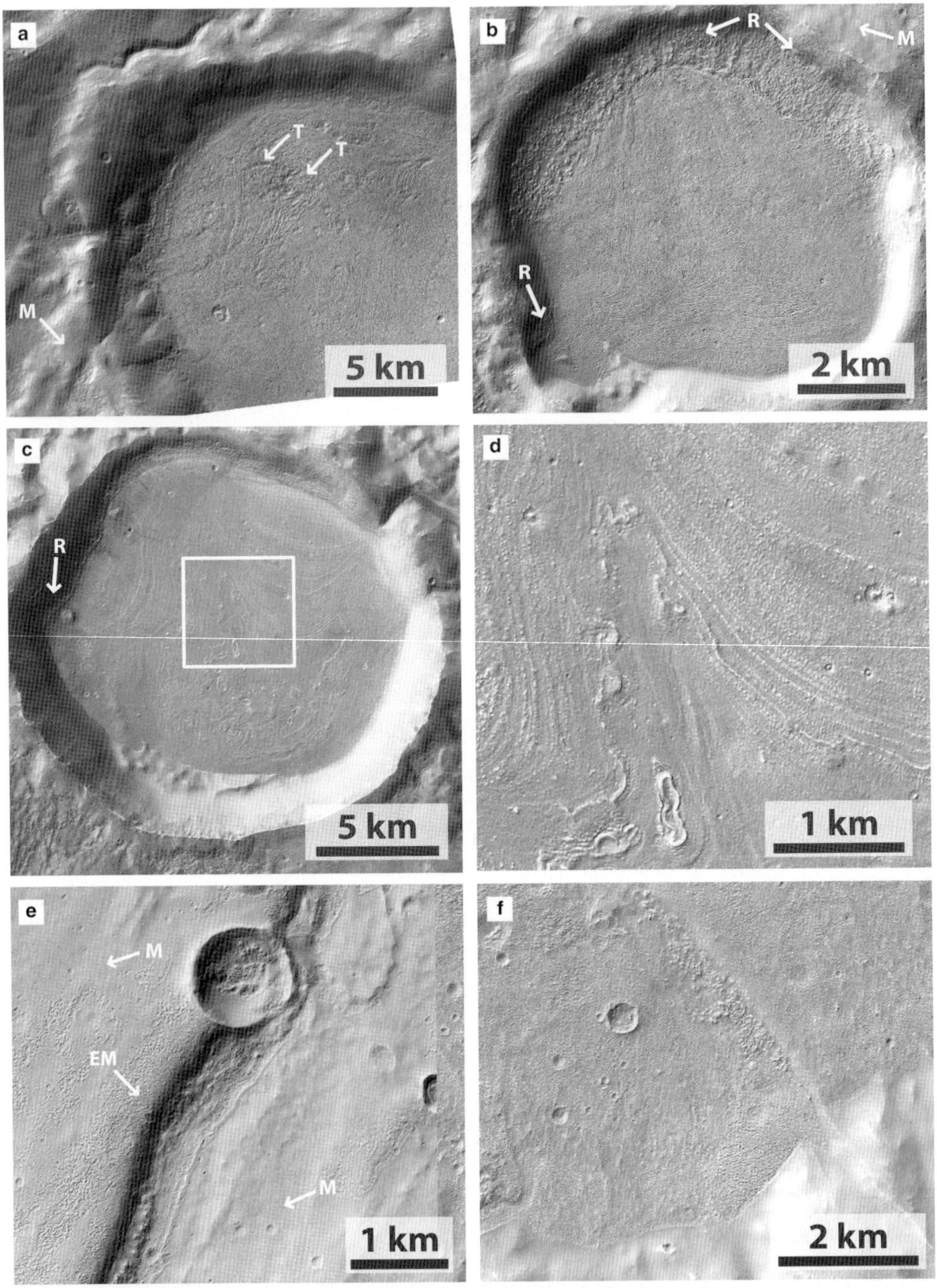

Fig. 3. Lineated valley fill at different scales. North is up for all images. (**a**) Slightly deflated, pitted lineated crater fill (CTX P16_007246_1406_XI_39S094W). 'T' indicates possible thermokarst-like depressions within the crater fill; 'M' indicates mantling. (**b**) Heterogeneous crater fill: lineated very pitted/blocky on the northern part. 'R' indicates ridges, possibly indicated past higher topographic levels of the fill; 'M' indicates mantling. (CTX P16_007246_1406_XI_39S094W). (**c**) Thinly spaced multiple lineations in a crater fill. 'R' indicates ridges. (CTX P13_006191_1456_XN_34S089W); the extent of (d) is outlined in white. (**d**) Detail from (c); multiple lobes of ridged material are visible,

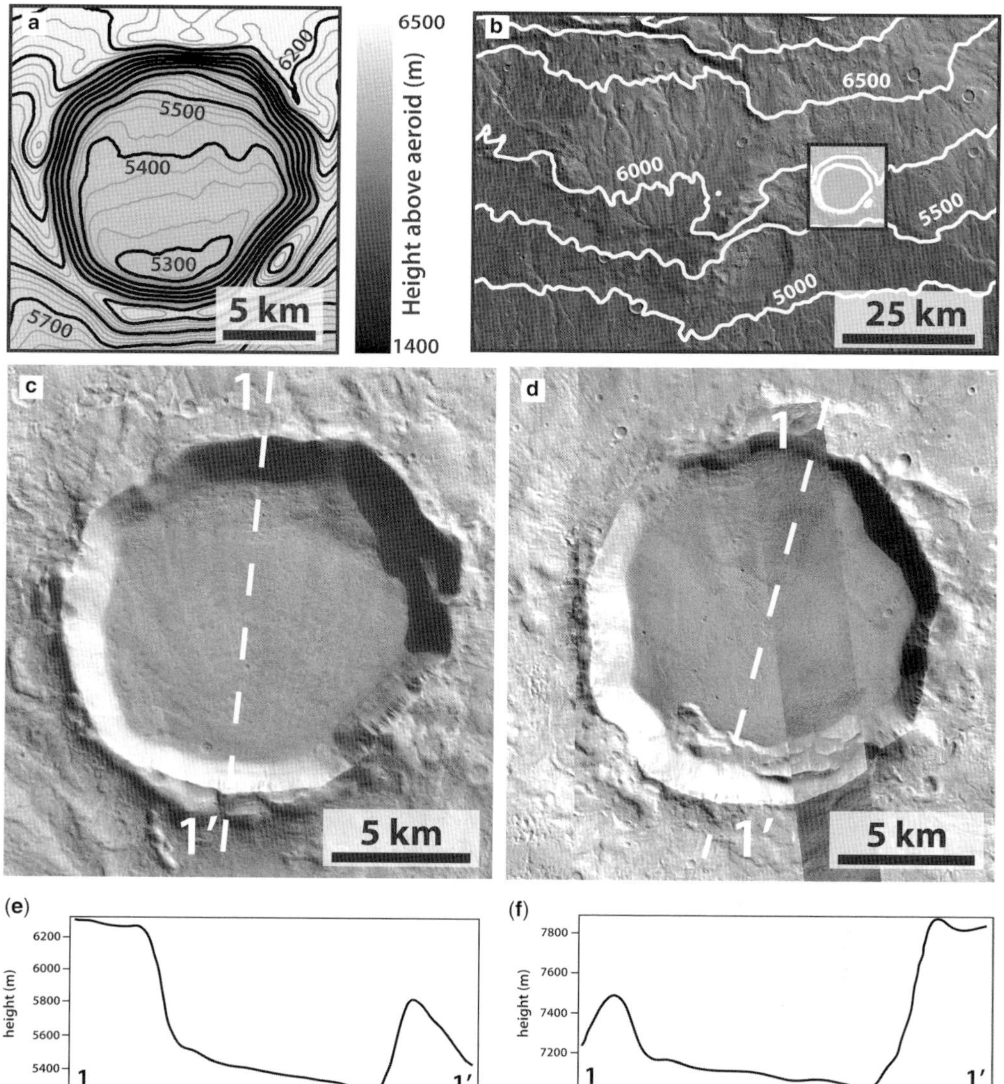

Fig. 4. HRSC DEM topography of lineated crater fills. (**a**) 25-m vertical spacing contour map over a lineated crater (thick lines every 100 m) the same crater is imaged in (c) and the extent of (a) is outlined in (b) (HRSC DEM, 125 m/pixel). (**b**) Local setting of crater in (a) outlined in semi-transparency (HRSC nadir mosaic, overlain by 500 m vertical spacing HRSC DEM contours). (**c**) HRSC nadir image (orbit 497) and the outline of the topographical profile. (**d**) HRSC nadir image (orbit 442) over another lineated crater. (**e**) HRSC DEM topographical profile outlined in (c). (**f**) HRSC DEM topographical profile outlined in (e). Image credit: NASA/JPL/MOLA Science Team and ESA/DLR/FUB.

Fig. 3. (*Continued*) with highly compressed portions of the fill at their junction; several crater-like depressions are present (CTX image as in c). (**e**) Example of mantled plateau and fretted-like terrain in the lowlands, separated by a normal fault scarp. 'M' indicates mantling and 'EM' etched mantling, possible modified by sublimation and/or eolian erosion. Very smooth and highly pitted terrains are co-existing in the mantled units (CTX P16_007246_1406_XI_39S094W). (**f**) Normal fault in a lineated crater fill: it is not clear whether the fault precedes the deposition, movement and likely sublimation of the deposits or not (CTX P16_007246_1406_XI_39S094W). Image credit: NASA/JPL/MSSS.

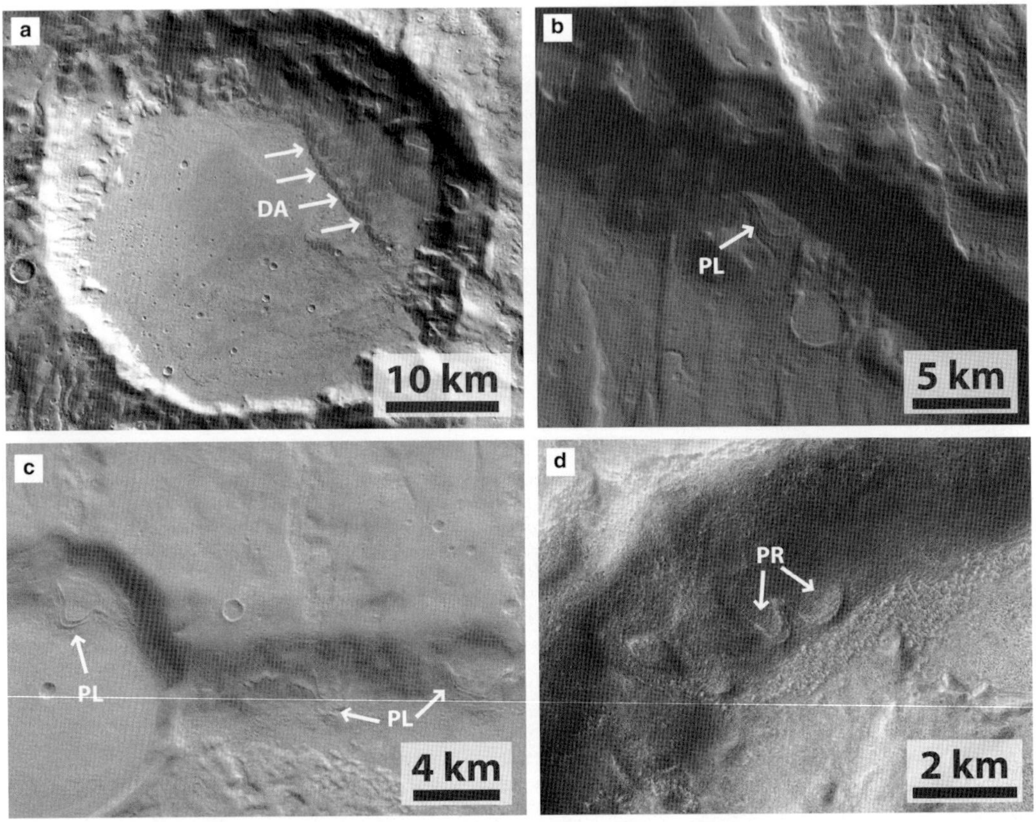

Fig. 5. Examples of debris aprons, protalus lobes and ramparts. North is up for all images. (**a**) Convex, *c*. 10 km-wide debris apron in a large crater: white arrows indicate its edge (HRSC nadir mosaic); in the same large crater protalus ramparts are also present. (**b**) Lobate, ridged protalus lobe (marked 'PL' in the figure) at the base of a very steep scarp (HRSC nadir from orbit 497). (**c**) Protalus lobes (an example labeled 'PL', other two on the scarp are unlabelled) at the base of a scarp and a crater rim (HRSC nadir from orbit 292). (**d**) Protalus ramparts (marked 'PR') on a crater rim (CTX P10_005110_1383_XI_41S095W). Image credit: ESA/DLR/FUB and NASA/JPL/MSSS.

the size of affected areas is rather small and high-resolution imagery is lacking in that particular spot, being able to date the fill would possibly also provide a strain-rate estimate.

Debris aprons

Landforms displaying similar morphologies to debris aprons observed elsewhere on Mars (e.g. Crown *et al.* 2003; Mangold 2003; van Gasselt *et al.* 2008) are present in some of the impact craters in the Thaumasia Highlands region (Fig. 5a). They are isolated or appear superimposed on lineated crater floors or fill materials (Fig. 5a).

Debris aprons in the Thaumasia Highlands region are characterized by smooth to moderately rough surface textures. In terms of scale, apparent chronology and degradational state, they appear to be transitional between larger older lineated fill materials and smaller protalus lobes. They tend to emanate from south-facing slopes, but with less distinct correlation to the general direction of exposure when compared with protalus lobes.

These landforms often show concave-upwards profiles on MOLA and HRSC DTM, in contrast to large-scale features observed at the dichotomy boundary that are usually convex upwards (e.g. Mangold 2003).

The thickness of the debris apron shown in Figure 5a is nearly 200 m, as estimated by prolonging the curvature of the slope/floor below the apron itself. Debris aprons in the Thaumasia Highlands region are far fewer in number compared with other parts of Mars, which is especially highlighted at northern latitudes (e.g. Crown *et al.* 2003; Mangold 2003; van Gasselt *et al.* 2008).

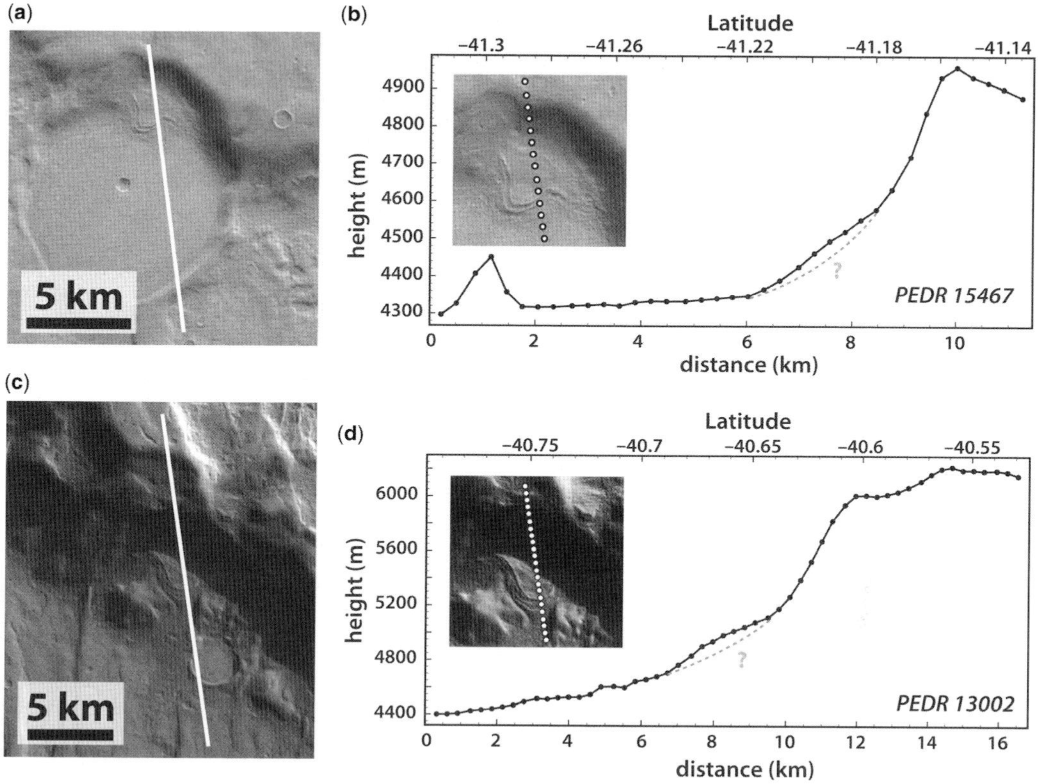

Fig. 6. Topography of protalus lobes (MOLA, HRSC profiles, contours, etc.). (**a**) MOLA PEDR profile location, over the HRSC nadir image (orbit 292). (**b**) PEDR profile 15467, showing a convex profile of the protalus lobe. Zoomed inset shows the actual MOLA shot location. (**c**) MOLA PEDR profile location, over the HRSC nadir image (orbit 497). (**d**) PEDR profile 13002. Image credit: ESA/DLR/FUB.

Protalus lobes and ramparts

Protalus lobes (Shakesby 1997) are commonly found in the study area in close association with impact-crater rims and, in general, fault scarps resulting from compressional or extensional deformation. The lobes are characterized by multiple lobate concentric ridges with a relatively simple geometry when compared with the fill materials described in the subsection on 'Lineated crater/depression-fill materials' or other more complex landforms (e.g. Fig. 8). Their width usually exceeds their length, thus forming broad features along foot-slopes. In most cases their total length is limited to a few kilometres. Their texture appears moderately rough when observed at the scales of HRSC (*c*. 15–20 m per pixel resolution) and MOC (*c*. 3–5 m per pixel resolution); where high-resolution MOC, CTX and HiRISE image data are available, their surface even appears blocky (Fig. 5d).

Protalus lobes tend to develop preferentially on south-facing slopes in the study area both in the case of linear south-facing scarps and inner rims of impact craters. The apparent flow direction is from north to south. The correlation between exposure and the development of protalus lobes is much greater than for any other landforms discussed here (Fig. 9).

Impact craters on these lobate landforms are scarce, with lower densities than any other feature described here. This is also consistent with the observed geometrical relationship between the different possible periglacial landforms in the study area, in that protalus lobes can be seen to be overlapping other landforms and demonstrate a young age.

The thickness of the protalus lobes can be evaluated using MOLA PEDR profiles; thicknesses range from a few tens of metres up to approximately 200 m (Fig. 6). Topographical profiling on these landforms has been performed using PEDR rather than HRSC DEM because of their relatively small size, and thus are not well resolved on stereo imagery but rather are detectable on single MOLA

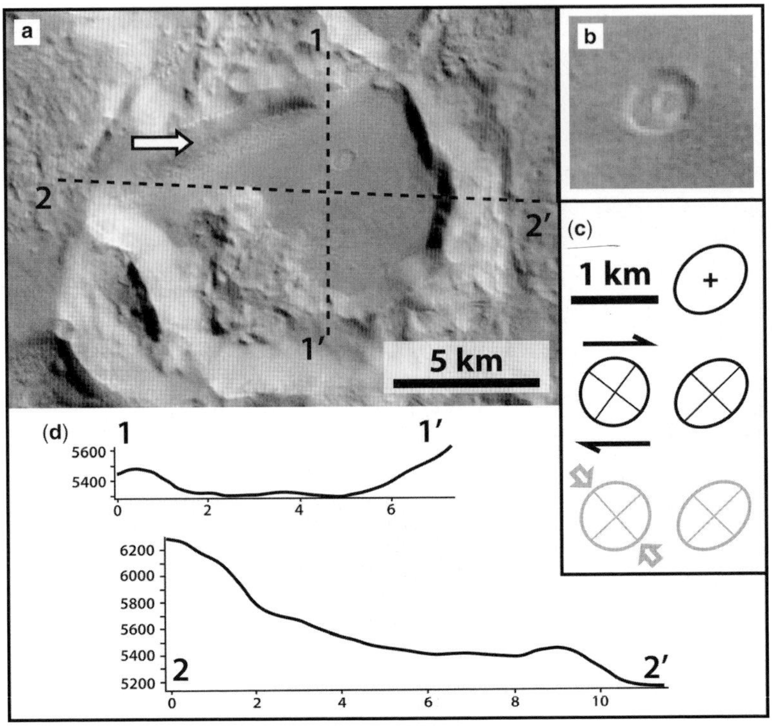

Fig. 7. Kinematic and finite deformation indicators in crater/valley fill. (**a**) Deformed possible impact crater and outline of the HRSC DEM profiles (HRSC nadir, orbit 497). (**b**) Enlargement of the elliptical crater-like feature. (**c**) Elliptical crater as the result of either simple (considered here more likely) or pure shear. (**d**) HRSC DEM profiles, as outlined in (a). Image credit: ESA/DLR/FUB.

shots (Fig. 6). Moreover, they often occur on impact-crater rims and other types of scarps, producing strong shadows (Fig. 5b). The greatest convex parts of these rock-glacier-like landforms have slopes of up to about 7°.

Association of landforms

At some point, probably much later than for previous fluvial activity (Dohm *et al.* 2001*b*; Ansan & Mangold 2006), extensive ice-rich mantling (Figs 3e, 4 & 8) appears to have occurred in extensive areas within the mountain range of the Thaumasia Highlands. This possibly occurred over a specific time or, perhaps, as multiple episodes. This may or may not have coincided with a large glacial cover, possibly associated with very different obliquity conditions to the present one (e.g. Berman *et al.* 2005, 2009; Dickson *et al.* 2006; Dickson *et al.* 2008) or with stages of Tharsis magmatism (Dohm *et al.* 2001*a*, *b*, 2007).

Our analysis shows distinct landform development through time in the Thaumasia Highlands region. A generic sequence of periglacial landform development in the study area includes (in chronological order) the following. (1) The poorly time constrained development of extensive lineated fill materials (Fig. 3a–c) (areas of *c.* 150–170 km^2), which presently exhibit possible thermokarst (Fig. 3a) and/or sublimation features together with possible impact craters; these landforms may be remnants of more extensive ice sheets and/or glaciers (Dickson *et al.* 2006, 2008). (2) Emplacement of moderately sized debris aprons (areas of several tens of kilometres), which appear to be 'deflated'. (3) The formation of isolated small protalus lobes (Fig. 10) marked by a few, if any, impact craters visible at available image resolutions (lengths of *c.* 2–5 km); protalus ramparts, which are several hundreds of metres to a few kilometres in extent, are also to be linked to this phase (Fig. 5). (4) Progressive sublimation and the deflation of landforms, possibly occurring under obliquity conditions similar to the present, leading to the currently observed association of landforms (Fig. 10). This sequence of events would be in general agreement with the notion of Dickson & Head (2009), who suggested that current morphologies in mid-latitude

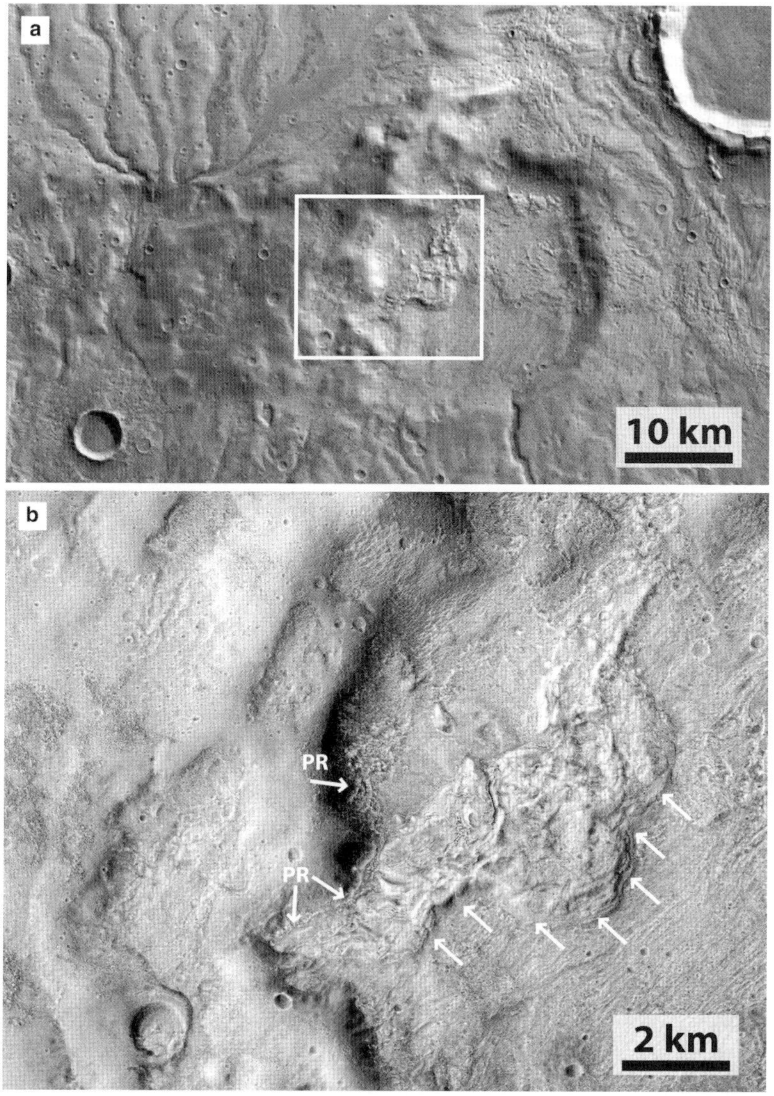

Fig. 8. Complex deformed possible multi-stage example of fill, with possibly exhumed craters within the deposit. (**a**) Local setting (HRSC nadir mosaic). (**b**) Detail of the complex convoluted, textured deposit. Small protalus ramparts are marked 'PR' in the figure. Parallel white arrows are pointing towards the edge of the complex deposit, which might be a degraded lobe: the edge marks the contact with the texture crater floor (CTX P08_004266_1377_XI_42S092W). Image credit: ESA/DLR/FUB and NASA/JPL/MSSS.

craters reflect a late-stage phase in the most recent ice age on Mars.

Inferred directions of movements for the various landforms are variable. Some of them are more clearly linked to topography (slope and aspect), probably related to the geological setting, while others (mostly the smaller features such as protalus lobes and ramparts) display a more direct link to local topography, aspect and exposure.

Discussion

Landforms in the ancient mountain range of the Thaumasia Highlands region are the result of a complex geological history, which includes magmatic, tectonic, erosional and aggradational processes (all of which may not be mutually exclusive), the latest of which is dominated by periglacial processes. Landforms such as lineated

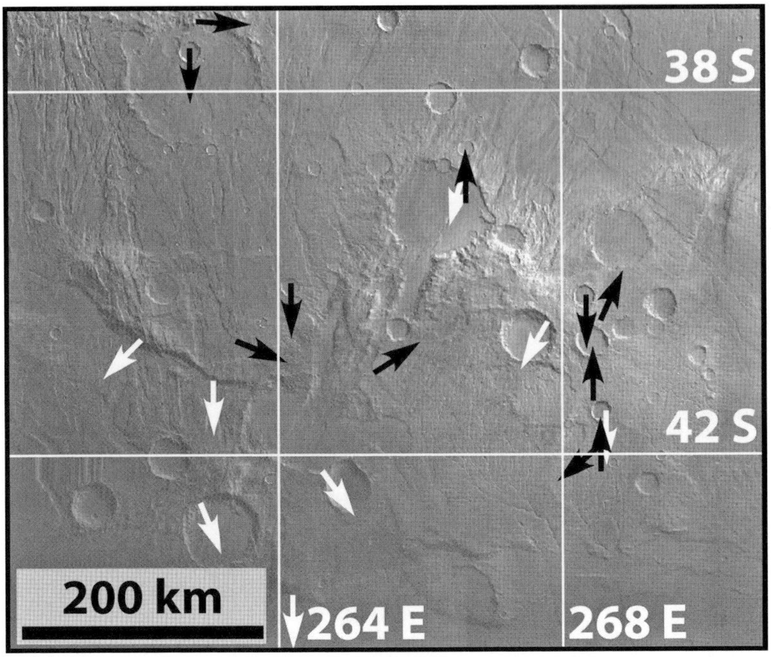

Fig. 9. Inferred direction of the flow map of the protalus lobes (white) and crater/valley fill (black) inferred direction of movements. Protalus lobes appear more pristine, and mostly showing southward exposure and flow direction. Lineated crater fills show more deflated morphologies, and a wider distribution of flow directions. Image credit: ESA/DLR/FUB.

fill materials, for example, may result from some form of climate-driven phenomena.

Landforms presented here are often spatially associated, co-occurring at close distances, but have developed over different time periods, often overlapping. Also, the size of these two main groups of landforms is very different: lineated fill materials are extensive, with areas as large as 100 km^2; while protalus lobes usually cover areas no larger than 5–10 km^2.

No clear distinction between rock glaciers and debris-covered glaciers can be made, and the difference between the two landforms appears to be subtle.

Linear and curvilinear fill materials have been interpreted as being related to the flow of ice (Squyres & Carr 1986) or eolian erosion (Zimbelman *et al.* 1989). Here, we favour the former hypothesis, although it is also possible that eolian erosion played some important role in modifying the landforms over long timescales, depleting the deposits of fine-grained material in association with the loss of volatiles through sublimation.

Relative-age relationships among the periglacial-like landforms can be determined in individual basins (e.g. Figs 3e, f & 5c), but it is more difficult to have a complete picture over a more regional extent, including the exact relationship between lineated fill materials and mantling. It is possible that both mantling (perhaps more extensive than observed at present) and the development of lineated fill materials occurred contemporaneously, and that both decreased with time, so producing the present residual landforms.

Crater-size frequency analyses to derive absolute ages will produce unreliable absolute ages due to both the limited extents of the features and the flow deformation of the materials. The small extent of the discovered protalus lobes, for example, makes it difficult to estimate absolute formational ages. Geometrical and stratigraphic relationships indicate a more recent age for the relatively small protalus lobes with respect to lineated fill materials (e.g. Fig. 5a, c). A few of the features, such as the lineated valley-fill materials, are often marked by circular or quasi-circular depressions. We have not attempted to date their surfaces because it is difficult to determine whether the origin of the often-deformed features is impact (Figs 3 & 4). Their diameters are typically a few hundreds of metres.

Some of the lineated fill materials, most protalus lobes and the debris aprons appear to be inflated, suggestive of subsurface ice. While others appear to be more deflated morphologically (Figs 4c–f &

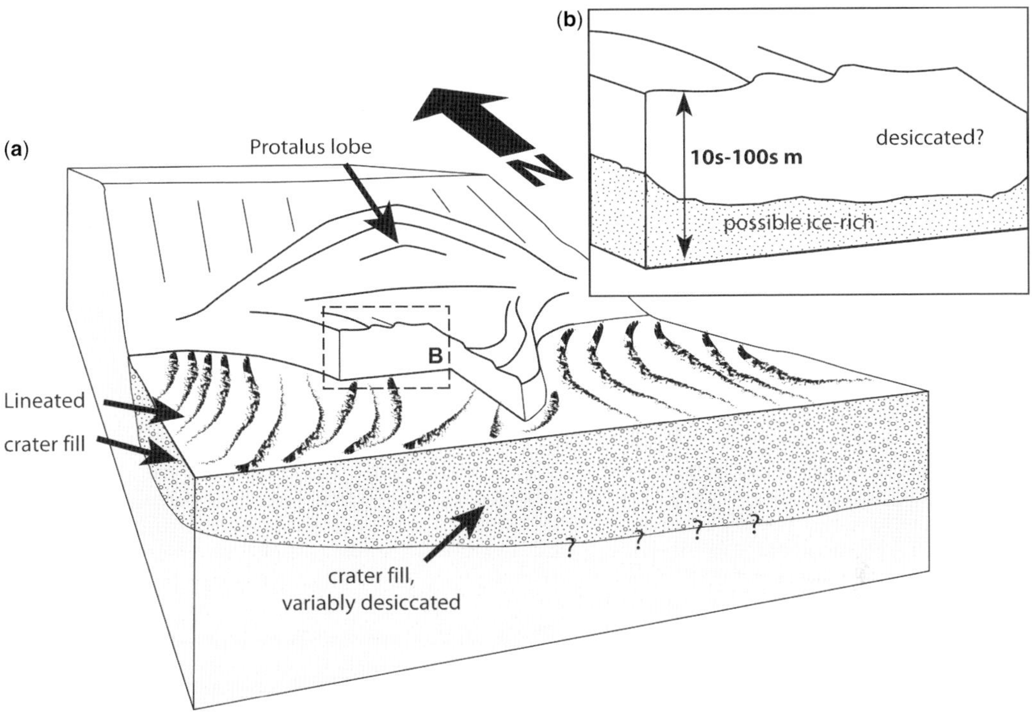

Fig. 10. Idealized block diagram (not to scale) of a protalus lobe. (**a**) Generalized location of small recent protalus lobes with respect to older gently dipping lineated crater fills. (**b**) Possible section across a protalus lobe. These lobes show a thickness of up to a few hundred metres.

6a, b), as observed in comparable deposits in different settings and locations on Mars (Dickson *et al.* 2008). The estimated amount of volume loss over a poorly constrained period of time can possibly be hinted at by the presence of terraces, ridges and moraine-like features (Dickson *et al.* 2008) at the edge of craters (Fig. 3b, c). The variation in thickness, based on these features, may be up to a few hundred metres (e.g. Dickson *et al.* 2008).

The presence of faults and their relationship with the emplacement, development and modification of lineated fill materials is also an issue; in Figure 3f a fault clearly deforms the textured floor and rim of an impact crater, but the extent of exhumation in such a setting is less clear. The fault could have a synsedimentary relationship with the fill materials of the impact crater or it could simply be that the infill materials partially infilled the fault and then was modified and later partially exhumed. We tend to favour pre(syn?)-tectonic development of lineated fill materials as both the textured crater fills and the rim bedrock display fault scarps with a comparable morphological and degradational level (Fig. 3f).

The flow direction inferred from the lineated fill materials show little to no correlation with slope orientation, being more correlated with local and regional topography. However, mostly concave-upwards debris aprons and mostly convex-upwards protalus lobes have developed preferentially on south-facing slopes (pole facing), suggesting a stronger and temporarily closer role of morphoclimatic conditions during their development (Fig. 9). This is consistent with previous observations of glacial–periglacial features at mid- to low latitudes (e.g. Berman *et al.* 2005, 2009; Dickson *et al.* 2008; Hauber *et al.* 2008). Therefore, smaller lobes are linked to present morphoclimatic conditions and could still be active. Indeed, based on modelling, ice accumulation is possible in the Thaumasia Highlands region (Levrard *et al.* 2004; Forget *et al.* 2007) given that past obliquity conditions have been different to those of the present. Thus, the Thaumasia Highlands region may have retained a record of climatic events complementary to that which have been recorded in the northern hemisphere (e.g. Dickson *et al.* 2008; Hauber *et al.* 2008; van Gasselt *et al.* 2008).

Some of the areas show complex fill materials that have possibly been deformed by more than one sequence of events, either with a different direction of movement superimposed (e.g. Figs 3c, 5a & 8) or with highly deformed fill materials – showing convoluted surface textures and the presence of possibly exhumed craters – being possibly linked to older or multiple different climatic periods. The evolution of the fill materials, which record multiple glacial–periglacial phases, is probably complex and may be difficult to unravel. However, it may prove useful to correlate and extend the observations over a wide range of locations on Mars (e.g. Dickson *et al.* 2008; Hauber *et al.* 2008; van Gasselt *et al.* 2008).

The location and development of the possible periglacial landforms may be strongly linked to the pre-existing topography because the ancient rugged mountain range of the Thaumasia Highlands region comprise impact craters of varying size and degradational states, extensional and contractional faults and folds, complex rift systems, and shield volcanoes that formed along the rift systems. All of these features would influence subsequent geological and geomorphological activities that include periglacial processes. Some of the lineated fill materials hosting impact craters appear to have been affected by fluvial erosion and deposition. In particular in the vicinity of Warrego Valles (Ansan & Mangold 2006), lobate deposits with lineated and convoluted textures occur in impact craters and other types of depressions that have been modified by fluvial activity, as evidenced by highly resurfaced channels which enter and/or exit the basins (Fig. 8). In fact, several of the more recent landforms, such as protalus lobes, appear to have a convex-upwards longitudinal profile, possibly suggesting (Clark *et al.* 1994) the active presence of an ice core.

In the Thaumasia Highlands region, the dominant concave-upwards profile of debris aprons, where present, is indicative of a past scenario involving the melting of ice cores of rock glaciers or debris-covered glaciers. In contrast, protalus lobes mostly show convex-upwards profiles (Fig. 6) that, together with their apparent relatively young age, is indicative of recent and/or current activity.

All landforms described here appear to be at moderately high altitude. Most of them are above 5000 m in altitude. Smaller ones, such as protalus lobes occurring on scarps with frequent shadows, are also found at elevations closer to 4000 m (Figs 2, 4d, f & 6b, d). Craters with sloping floors elsewhere in the southern hemisphere have also been described by Berman *et al.* (2005, 2009) at lower elevations. However, further observations on their latitudinal and altitudinal dependence across the hemisphere are still needed.

Although the nature and evolution of such landforms are often controversially discussed (e.g. Head *et al.* 2006; Hauber *et al.* 2008; van Gasselt *et al.* 2008), there is evidence that the existence and evolution of such landforms is related to climatic variations controlled by the orbital configuration of Mars (Levrard *et al.* 2004; Forget *et al.* 2007), which was responsible for the deposition of ice in the equatorial region during high-obliquity phases and the depletion of an ice reservoir during periods of low obliquities.

The search for subsurface ice or ice–rock mixture signatures with sounding radar data such as SHARAD (Shallow Radar on board MRO) has proved to be successful in the detection of ice in debris aprons in a few cases (Holt *et al.* 2008), but thus far unsuccessful in the Thaumasia Highlands region. This may be due either to the lack of subsurface ice-rich/ice-depleted interfaces or to the presence of a smooth gradient of ice content, which would not produce a sharp reflector in subsurface sounding radar data. Moreover, the small size of the landforms (apart from large craters with lineated fills), as well as the generally high surface roughness in the Thaumasia Highlands region, makes the analysis of subsurface radar data problematic.

Conclusions and future prospects

The Thaumasia Highlands region provides the geomorphological setting necessary for the formation of creep-related landforms caused by an abundance of high-relief slopes and a tectonically dissected terrain, which allows the accumulation and supply of wall-rock debris at footslopes.

In our survey, we identified flow and creep morphologies exhibiting a lobate to tongue-like shape, characterized by linear to curvilinear ridges and furrows closely resembling large-scale gelifluction lobes or terrestrial rock glaciers and protalus landforms indicative of periglacial environments. Larger, stratigraphically older lineated fill materials may have recorded older, more enhanced glacial phases (Dickson *et al.* 2006, 2008). The general lack of impact craters suggests relatively young surface ages. Although water ice is not considered to be presently stable at equatorial latitudes, there are morphological indicators suggestive of the reactivation and/or formation of such landforms in the transitional belt between equatorial latitudes and mid-latitudes on Mars during geologically recent times (e.g. Levrard *et al.* 2004; Forget *et al.* 2006).

Radar-based analysis of the Thaumasia Highlands region using SHARAD, although difficult due to the attenuation of the radar signature caused by the rugged, highly modified mountain-forming rock materials, may, with the aid of high-resolution stereo-derived topography (e.g. Gwinner *et al.* 2005; Dumke *et al.* 2008) for radar topographic clutter modelling (e.g. Cutigni *et al.* 2007), help constrain the presence of ice in the landforms described in this work.

Being one of the few high-altitude regions on Mars located within the southern mid-latitudes, the ancient mountain range of the Thaumasia Highlands may be key to unfolding the geological and geomorphological record of past climatic phases on Mars.

Our gratitude goes to A. Basilevsky and an anonymous reviewer, whose comments and suggestions greatly improved the manuscript. We thank the HRSC Experiment Teams at DLR Berlin and Freie Universitaet Berlin, as well as the *Mars Express* Project Teams at ESTEC and ESOC for their successful planning and acquisition of data, and for making the processed data available to the HRSC Team. We acknowledge the effort of the HRSC Co-Investigator Team members and their associates who have contributed to this investigation in the preparatory phase and in scientific discussions within the team. We thank MOLA, MOC, THEMIS, CTX and HiRISE respective teams for making data available to the public through PDS.

References

ANDERSON, J. A., SIDES, S. C., SOLTESZ, D. L., SUCHARSKI, T. L. & BECKER, K. J. 2004. Modernization of the integrated software for imagers and spectrometers. *In*: *Proceedings of the 35th Annual Lunar and Planetary Science Conference, March 15–19, 2004, League City, Texas*. Lunar and Planetary Institute, Houston, TX, Abstract 2039.

ANGUITA, F., FERNÁNDEZ, C. ET AL. 2006. Evidences for a Noachian-Hesperian orogeny in Mars. *Icarus*, **185**, 331–357.

ANSAN, V. & MANGOLD, N. 2006. New observations of Warrego Valles, Mars: evidence for precipitation and surface runoff. *Planetary and Space Science*, **54**, 219–242.

BAKER, V. R. 2001. Water and the Martian landscape. *Nature*, **412**, 228–236.

BAKER, V. R., MARUYAMA, S. & DOHM, J. M. 2007. *Tharsis Superplume and the Geological Evolution of Early Mars, Superplumes: Beyond Plate Tectonics*. Springer, Dordrecht, 507–522.

BERMAN, D. C., HARTMANN, W. K., CROWN, D. A. & BAKER, V. R. 2005. The role of arcuate ridges and gullies in the degradation of craters in the Newton Basin region of Mars. *Icarus*, **178**, 465–486.

BERMAN, D. C., CROWN, D. A. & BLEAMASTER, L. F., III. 2009. Degradation of mid-latitud craters on Mars. *Icarus*, **200**, 77–95.

BOYNTON, W. V., FELDMAN, W. C. ET AL. 2002. Distribution of hydrogen in the near-surface of Mars: evidence for subsurface ice deposits. *Science*, **297**, 81–85.

BOYNTON, W. V., FELDMAN, W. C. ET AL. 2004. The Mars Odyssey Gamma-Ray Spectrometer instrument suite. *Space Science Reviews*, **110**, 37–83.

CLARK, D. H., CLARK, M. M. & GILLESPIE, A. R. 1994. Debris-covered glaciers in the Sierra Nevada, California, and their implications for snowline reconstructions. *Quaternary Research*, **41**, 139–153.

COLAPRETE, A. & JAKOSKY, B. M. 1998. Ice flow and rock glaciers on Mars. *Journal of Geophysical Research*, **103**, 5897.

COSTARD, F. M. & KARGEL, J. S. 1995. Outwash plains and thermokarst on Mars. *Icarus*, **114**, 93–112.

CROWN, D. A., MCELFRESH, S. B. Z., PIERCE, T. L. & MEST, S. C. 2003. Geomorphology of debris aprons in the Eastern Hellas region of Mars. *In*: *Proceedings of the 34th Annual Lunar and Planetary Science Conference, March 17–21, 2003, League City, Texas*. Lunar and Planetary Institute, Houston, TX, Abstract **1126**.

CUTIGNI, M., RUSSO, F. ET AL. 2007. Incoherent simulator for Mars surface applied to the analysis of Sharad Radar Data. *AGU Fall Meeting Abstracts*, **11**, 0546.

DICKSON, J. L. & HEAD, J. W. 2009. The formation and evolution of youthful gullies on Mars: gullies as the late-stage phase of Mars most recent ice age. *Icarus*, **204**, 63–86.

DICKSON, J. L., HEAD, J. W., KRESLAVSKY, M. A. & MARCHANT, D. R. 2006. Linear lobate debris aprons, piedmont-like lobes, and crater fill in the Acheron Fossae graben region, Mars: evidence for debris-covered glacier formation and flow. *In*: *Proceedings of the 37th Annual Lunar and Planetary Science Conference, March 13–17, 2006, League City, Texas*. Lunar and Planetary Institute, Houston, TX, Abstract 1321.

DICKSON, J. L., HEAD, J. W. & MARCHANT, D. R. 2008. Late Amazonian glaciation at the dichotomy boundary on Mars: evidence for glacial thickness maxima and multiple glacial phases. *Geology*, **36**, 411–414.

DOHM, J. M. & TANAKA, K. L. 1999. Geology of the Thaumasia region, Mars: plateau development, valley origins, and magmatic evolution. *Planetary and Space Science*, **47**, 411–431.

DOHM, J. M., FERRIS, J. C. ET AL. 2001*a*. Ancient drainage basin of the Tharsis region, Mars: potential source for outflow channel systems and putative oceans or paleolakes. *Journal of Geophysical Research*, **106**, 32,943–32,958.

DOHM, J. M., TANAKA, K. L. & HARE, T. M. 2001*b*. *Geologic Map of the Thaumasia Region, Mars*. United States Department of the Interior, United States Geological Survey Map, **I-2650**.

DOHM, J. M., MARUYAMA, S., BAKER, V. R., ANDERSON, R. C., FERRIS, J. C. & HARE, T. M. 2002. Plate tectonism on early Mars: diverse geological and geophysical evidence. *In*: *Proceedings of the 33rd Annual Lunar and Planetary Science Conference, March 11–15, 2002, League City, Texas*. Lunar and Planetary Institute, Houston, TX, Abstract 1639.

DOHM, J. M., MARUYAMA, S., BAKER, V. R. & ANDERSON, R. C. 2007. Traits and evolution of the Tharsis superplume, Mars. *In*: YUEN, D. A., MARUYAMA, S., KARATO, S.-I. & WINDLEY, B. F. (eds) *Superplumes: Beyond Plate Tectonics*. Springer, Berlin, 523–537.

DOHM, J. M., ANDERSON, R. C. ET AL. 2009. Claritas rise: pre-Tharsis magmatism. *Journal of Volcanology and Geothermal Research*, **185**, 139–156, doi: 10.1016/j.jvolgeores.2009.03.012.

DUMKE, A., SPIEGEL, M., SCHMIDT, R. & NEUKUM, G. 2008. High-resolution digital terrain models and ortho-image mosaics of Mars: generation on the basis of Mars-Express HRSC Data. *In*: *Proceedings of the 39th Annual Lunar and Planetary Science Conference, March 10–14, 2008, League City, Texas*. Lunar and Planetary Institute, Houston, TX, Abstract 1910.

FELDMAN, W. C., BOYNTON, W. V. *ET AL*. 2002. Global distribution of Neutrons from Mars: results from Mars Odyssey. *Science*, **297**, 75–78.

FELDMAN, W. C., PRETTYMAN, T. H. *ET AL*. 2004. Global distribution of near-surface hydrogen on Mars. *Journal of Geophysical Research – Planets*, **109**, 09006, doi: 10.1029/2003JE002160.

FORGET, F., HABERLE, R. M., MONTMESSIN, F., LEVRARD, B. & HEAD, J. W. 2006. Formation of glaciers on Mars by atmospheric precipitation at high obliquity. *Science*, **311**, 368–371.

FORGET, F., MONTMESSIN, F., LEVRARD, B., HABERLE, R. M., HEAD, J. W. & MADELEINE, J.-B. 2007. Glaciers, polar caps and ice mantling: the effect of obliquity on Martian climate. *LPI Contributions*, **1353**, 3028.

GADDIS, L., ANDERSON, J. *ET AL*. 1997. An overview of the Integrated Software for Imaging Spectrometers (ISIS). *In*: *Proceedings of the 28th Annual Lunar and Planetary Science Conference, March 17–21, 1997, League City, Texas*. Lunar and Planetary Institute, Houston, TX, Abstract 387.

GAULT, D. E. & WEDEKIND, J. A. 1978. Experimental studies of oblique impact. *In*: *Proceedings of the ninth Lunar Science Conference, Houston, Texas, March 13–17, 1978, League City, Texas*. Lunar and Planetary Institute, Houston, TX, Abstract 374–376.

GROTT, M., HAUBER, E., WERNER, S. C., KRONBERG, P. & NEUKUM, G. 2005. High heat flux on ancient Mars: evidence from rift flank uplift at Coracis Fossae. *Geophysical Research Letters*, **32**, L21201, doi: 10.1029/2005GL023894.

GROTT, M., HAUBER, E., WERNER, S. C., KRONGBERG, P. & NUEKUM, G. 2007. Mechanical modeling of thrust faults in the Thaumasia region, Mars, and implications for the Noachian heat flux. *Icarus*, **186**, 517–526.

GWINNER, K., SCHOLTEN, F. *ET AL*. 2005. Hochauflösende Digitale Geländemodelle auf der Grundlage von Mars express HRSC-Daten. *Photogrammetrie–Fernerkundung–Geoinformation*, **5**, 387–394.

HAUBER, E. & KRONBERG, P. 2005. The large Thaumasia Graben on Mars: is it a rift? *Journal of Geophysical Research*, **110**, E07003, doi: 10.1029/2005JE002407.

HAUBER, E., VAN GASSELT, S., CHAPMAN, M. G. & NEUKUM, G. 2008. Geomorphic evidence for former lobate debris aprons at low latitudes on Mars: indicators of the Martian paleoclimate. *Journal of Geophysical Research*, **113**, E02007, doi: 10.1029/2007JE002897.

HEAD, J. W. & MARCHANT, D. R. 2003. Cold-based mountain glaciers on Mars: Western Arsia Mons. *Geology*, **31**, 641–644.

HEAD, J. W., MARCHANT, D. R., AGNEW, M. C., FASSETT, C. I. & KRESLAVSKY, M. A. 2006. Extensive valley glacier deposits in the northern mid-latitudes of Mars: evidence for Late Amazonian obliquity-driven climate change. *Earth and Planetary Science Letters*, **241**, 663–671.

HOLT, J. W., SAFAEINILI, A. *ET AL*. 2008. Radar sounding evidence for buried glaciers in the southern mid-latitudes of Mars. *Science*, **322**, 1235–1238.

KARGEL, J. S. & STROM, R. G. 1992. Ancient glaciation on Mars. *Geology*, **20**, 3–7.

KRESLAVSKY, M. A. & HEAD, J. W. 2003. North–south topographic slope asymmetry on Mars: evidence for insolation-related erosion at high obliquity. *Geophysical Research Letters*, **30**, 1815, doi 10.1029/2003GL017795.

LEVRARD, B., FORGET, F., MONTMESSIN, F. & LASKAR, J. 2004. Recent ice-rich deposits formed at high latitudes on Mars by sublimation of unstable equatorial ice during low obliquity. *Nature*, **431**, 1072–1075.

LUCCHITTA, B. K. 1981. Mars and Earth – comparison of cold-climate features. *Icarus*, **45**, 264–303.

LUCCHITTA, B. K. 1984. Ice and debris in the fretted terrain, Mars. *Journal of Geophysical Research*, **89**, 409.

MADELEINE, J. B., FORGET, F., HEAD, J. W., LEVRARD, B. & MONTMESSIN, F. 2007. Mars: a proposed climatic scenario for northern mid-latitude glaciation. *In*: *Proceedings of the 38th Annual Lunar and Planetary Science Conference, March 12–16, 2007, League City, Texas*. Lunar and Planetary Institute, Houston, TX, Abstract 1778.

MAHANEY, W., MIYAMOTO, H., DOHM, J. M., BAKER, V. R., CABROL, N. A., GRIN, E. A. & BERMAN, D. C. 2006. Rock glaciers on Mars: earth-based clues to Mars' recent paleoclimatic history. *Planetary and Space Sciences*, **55**, 181–192.

MANGOLD, N. 2003. Geomorphic analysis of lobate debris aprons on Mars at Mars Orbiter Camera scale: evidence for ice sublimation initiated by fractures. *Journal of Geophysical Research – Planets*, **108**, 8021.

MILLIKEN, R. E., MUSTARD, J. F. & GOLDSBY, D. L. 2003. Viscous flow features on the surface of Mars: observations from high-resolution Mars Orbiter Camera (MOC) images. *Journal of Geophysical Research – Planets*, **108**, 5057, doi: 10.1029/2002JE002005.

MONTGOMERY, D. R., JACKSON, M. P. A., SCHREIBER, B. C., GILLESPIE, A. R. & ADAMS, J. B. 2009. Continental-scale salt tectonics on Mars and the origin of Valles Marineris and associated outflow channels. *GSA Bulletin*, **121**, 117–133.

MUSTARD, J. F., COOPER, C. D. & RIFKIN, M. K. 2001. Evidence for recent climate change on Mars from the identification of youthful near-surface ground ice. *Nature*, **412**, 411–414.

ROSSBACHER, L. A. & JUDSON, S. 1981. Ground ice on Mars – Inventory, distribution, and resulting landforms. *Icarus*, **45**, 39–59.

ROSSI, A. P., KOMATSU, G. & KARGEL, J. S. 2000. Rock glacier-like landforms in Valles Marineris, Mars. *In*: *Proceedings of the 31st Annual Lunar and Planetary Science Conference, March 13–17, 2000, League City, Texas*. Lunar and Planetary Institute, Houston, TX, Abstract 1587.

ROSSI, A. P., CHICARRO, A. F. *ET AL*. 2006. Widespread periglacial landforms in Thaumasia Highland, Mars. *In*: *Proceedings of the 37th Annual Lunar and Planetary Science Conference, March 13–17, 2006,*

League City, Texas. Lunar and Planetary Institute, Houston, TX, Abstract 1568.

ROSSI, A. P., VAN GASSELT, S., PONDRELLI, M., ZEGERS, T., HAUBER, E. & NEUKUM, G. 2008. Periglacial landscape evolution at lower mid-latitudes on Mars: the Thaumasia Highlands. *In*: *The Ninth International Conference on Permafrost*. Institute of Northern Engineering, University of Alaska Fairbanks, 1531–1536.

RUIZ, J., WILLIAMS, J. P., DOHM, J. M., FERNÁNDEZ, C. & LÓPEZ, V. 2009. Ancieint heat flow and crustal thickness at Warrego rise, Thaumasia highlands, Mars: implications for a stratified crust. *Icarus*, **203**, 47–57.

SCOTT, D. H. & TANAKA, K. L. 1986. *Geologic Map of the Western Equatorial Region of Mars, scale 1:15,000,000*. United States Geological Survey Miscellaneous Investigations Series Map, **I-1802-A**.

SHAKESBY, R. A. 1997. Pronival (protalus) ramparts: a review of forms, processes, diagnostic criteria and palaeoenvironmental implications. *Progress in Physical Geography*, **21**, 394–418.

SQUYRES, S. W. 1978. Martian fretted terrain – flow of erosional debris. *Icarus*, **34**, 600–613.

SQUYRES, S. W. & CARR, M. H. 1986. Geomorphic evidence for the distribution of ground ice on Mars. *Science*, **231**, 249–252.

TORSON, J. M. & BECKER, K. J. 1997. ISIS – a software architecture for processing planetary images. *In*: *Proceedings of the 28th Annual Lunar and Planetary Science Conference, March 17–21, 1997, League City, Texas*. Lunar and Planetary Institute, Houston, TX, Abstract 1443.

VAN GASSELT, S., HAUBER, E., ROSSI, A. P. & NEUKUM, G. 2008. Emplacement of Lobate Rock–Glacier Landforms and Landscape Modification, Mareotis Fossae, Mars. *In*: *The Ninth International Conference on Permafrost*. Institute of Northern Engineering, University of Alaska Fairbanks, 487–492.

WAGNER, R., WOLF, U., NEUKUM, G. & THE GALILEO S.S.I.T 1999. Ages of individual craters on the Galilean Satellites Ganymede and Callisto. *In*: *Proceedings of the 30th Annual Lunar and Planetary Science Conference, March 15–19, 1999, League City, Texas*. Lunar and Planetary Institute, Houston, TX, Abstract 1818.

WHALLEY, W. B. & AZIZI, F. 2003. Rock glaciers and protalus landforms: analogous forms and ice sources on Earth and Mars. *Journal of Geophysical Research – Planets*, **108**, 8032, doi: 10.1029/2002JE001864.

ZIMBELMAN, J. R., CLIFFORD, S. M. & WILLIAMS, S. H. 1989. Concentric crater fill on Mars – an aeolian alternative to ice-rich mass wasting. *In*: *Proceedings of the 19th Annual Lunar and Planetary Science Conference, March 14–18, 1988, Houston, Texas*. Cambridge University Press, Cambridge/Lunar and Planetary Institute, Houston, TX, 397–407.

Landforms indicative of ground-ice thaw in the northern high latitudes of Mars

C. J. GALLAGHER[1]* & M. R. BALME[2,3]

[1]*UCD School of Geography, Planning and Environmental Policy, Newman Building, University College Dublin, Belfield, Dublin 4, Ireland*

[2]*Department of Earth and Environmental Sciences, Open University, Walton Hall, Milton Keynes MK7 6AA, UK*

[3]*Planetary Science Institute Tucson, 1700 E. Fort Lowell, Suite 106, Tucson, AZ 85719, USA*

**Corresponding author (e-mail: colman.gallagher@ucd.ie)*

Abstract: The confirmation of near-surface ground ice and perchlorates at the *Phoenix* landing site suggest that high-latitude ground-ice thaw may be more easily achieved than previously envisaged, providing the potential to drive significant, distinctive morphogenesis. We describe the results of a survey of 23 High Resolution Imaging Science Experiment (HiRISE) images covering 337° of longitude between latitudes 59°N and 79°N in which such morphogenesis is apparent, confirming that thaw has been a regionally important morphological agent. Some of the strongest geomorphological indicators of cyclical ground-ice thaw described are assemblages of sorted landforms, including clastic patterned ground resulting from cryoturbation of ice-rich regolith and lobate forms reflecting solifluction. Also described are braided gully-fan systems sourced at thermokarst pits and channels that have evolved from enlarged thermal contraction cracks. Not only are these landforms indicative of thaw and flowing liquid but the incision of solifluction lobes by thermokarst gullies demonstrates that thaw has been responsible for polycyclic morphogenesis. The presence of these landforms across the high northern latitudes of Mars indicates that the regional importance of thaw has been underestimated. This in turn has important implications for the development of better climate models and the search for life on Mars.

The direct observation of near-surface ground ice at the *Phoenix* landing site (Smith *et al.* 2009) confirmed the permafrost origin of patterned ground at Martian high latitudes. However, although this geomorphology is clearly associated with freezing conditions, the possibility of liquid being released through ground-ice thaw is an important consideration for our understanding of morphogenesis, cryosphere–climate interactions and the biological potential of Mars. Although very minor regional thaw was modelled by Mellon *et al.* (2008) prior to the landing of *Phoenix*, it was deemed insufficient to be geomorphically effective. Sorted rubble piles (cf. 'sorted islands'), a landform widely associated on Earth with freeze–thaw cryoturbation and observed throughout the *Phoenix* landing site region by Mellon *et al.* (2008), were attributed to clast collapse into thermal contraction cracks in a dry environment characterized by sublimation, not thaw. However, the observation of slowly moving liquid globules on the struts of the *Phoenix Lander* by Renno *et al.* (2009), the presence of $CaCO_3$ in the regolith of the *Phoenix Lander* site (Boynton *et al.* 2009) and the inferred recent activity of segregation ice resulting from the freezing of liquid water in the regolith (Smith *et al.* 2009) have raised the possibility of recent significant thaw, possibly involving brines or perchlorate solutions. Added to these findings, the observation of an active weak but complex hydrological cycle at the *Phoenix Lander* site involving seasonal and diurnal cycles of sublimation, condensation and precipitation suggests that significant high-latitude thaw could have occurred during recent periods of favourable obliquity when the Martian climate was warmer and wetter (Smith *et al.* 2009).

One of the strongest geomorphic indications of cyclic ground-ice thaw is the presence of texturally sorted patterned ground resulting from the cryoturbation of ice-rich regolith. Although Marchant *et al.* (2002) and Levy *et al.* (2006) have documented the production of clast-bordered thermal contraction polygons in Antarctica driven by the sublimation of debris-covered Miocene-age glacial ice, thaw is generally required to produce texturally sorted forms on Earth (Ballantyne & Harris 1994; Kessler & Werner 2003). Moreover, sorted patterned ground includes not only islands and clast-bordered circular or polygonal forms but also clastic garlands, stripes, lobes and terraces (Benedict 1970;

From: BALME, M. R., BARGERY, A. S., GALLAGHER, C. J. & GUPTA, S. (eds) *Martian Geomorphology*. Geological Society, London, Special Publications, **356**, 87–110.
DOI: 10.1144/SP356.6 0305-8719/11/$15.00

Werner & Hallet 1993; French 1996). All of these sorted forms reflect morphogenesis by the iterative segregation of regolith clasts from fines driven by the ice–water phase change, and modulated by clast concentration and slope (Kessler & Werner 2003).

A wide range of landforms including polygonal crack networks, polygonal troughs bordering mounds and narrow polygonal ridges bordering depressions are found at mid to high Martian latitudes (Seibert & Kargel 2001; Mangold 2005; Mellon *et al.* 2008), and have been imaged from the surface by the *Phoenix Mars Lander*. All reflect the modification of the surface by ground-ice processes and, although some may be relict forms, water-based ground ice is present beneath a very thin veneer of regolith at the *Phoenix* landing site (Smith *et al.* 2009). Shallow pits, and pitted mounds and cones in Martian mid-latitudes also have been attributed to periglacial processes (Soare *et al.* 2005; Dundas *et al.* 2008). Likewise, a periglacial genesis has been invoked for low-latitude, low-elevation, decametre-scale, patterned ground, small pitted cones and mounds within the outflow channel Athabasca Vallis, and in basins and channels in Elysium Planitia (Burr *et al.* 2005; Page & Murray 2006; Balme & Gallagher 2009). The precursor moisture in these situations was attributed to catastrophic flooding rather than condensation of volatiles from the atmosphere. However, the discovery of decametre-scale sorted stone circles in deposits associated with recent floods in Athabasca Vallis and the Elysium Basin by Balme *et al.* (2009) points to cyclic cryoturbation and, therefore, the repeated presence of ground moisture in the past approximately 10 Ma. Terrestrial analogues of similar scale to these sorted clastic forms are associated with deep permafrost (Ballantyne & Harris 1994) subjected to several tens to hundreds of cycles of freeze–thaw (cf. Corte 1962*a*, *b*; Ballantyne & Matthews 1982) extending to depths of several metres below the surface. In addition, Page (2007) presented evidence for recent thaw in low-latitude terrains in Amazonis Planitia that are similar to those in the Elysium Basin, and Soare *et al.* (2008) presented observations of mid-latitude pits and cones that might also imply recent thaw. Balme & Gallagher (2009) concluded that the geomorphology of the head of Athabasca Vallis, especially the presence of retrogressive thermokarst slumps and gullies fringing alas-like basins containing epigenetic pingos and polygons, indicates recent persistent standing water associated with thaw degradation of ice-rich sediments.

Malin & Edgett (2000) also inferred the recent action of surface flows of water in the incision of fluviatile gullies. However, rather than being attributed to pervasive, stable, near-surface moisture indicative of a wetter environment, these forms have been attributed to the melting of water-ice precipitated from the atmosphere at high obliquity (e.g. Costard *et al.* 2002; Reiss *et al.* 2009) or to breakouts from subsurface aquifers (e.g. Malin & Edgett 2000; Heldmann *et al.* 2005). Although observations of the distribution and orientation of the gullies support the melting hypothesis (e.g. Balme *et al.* 2006; Dickson *et al.* 2007), models favour sublimation of ice and snow over direct melting (Mellon & Phillips 2001), except perhaps in isolated microclimates (Head *et al.* 2008). However, the observation by Balme & Gallagher (2009) of polycyclic gullies associated with thaw degradation of patterned ground shows that gullies can reflect both deep and areally extensive ground-ice thaw. Hence, although individual landforms are often genetically ambiguous, landform assemblages are more often genetically specific and indicate that liquid water, or a water-based thaw-fluid, has been an important morphogenetic agent at low Martian latitudes.

While liquid water is only metastable under current Martian conditions (Hecht 2002), the upper 50–100 cm of regolith poleward of approximately 60° latitude contains several tens of weight-percentage (wt%) water ice (e.g. Feldman *et al.* 2004). Even a sublimed lag analysed by the TEGA (Thermal Evolved Gas Analyser) mass spectrometer on board the *Phoenix Lander* had a water content of approximately 2% (Smith *et al.* 2009). The stability of near-surface ice and the possible presence of liquid water depends largely on insolation and, consequently, axial obliquity (Laskar *et al.* 2004). When Mars' obliquity exceeds 30°, the vapour pressure of the atmosphere could be raised significantly by the destabilization of the north polar cap, offering the possibility of significant precipitation of atmospheric water at the surface at high northern latitudes (Smith *et al.* 2009). In addition, the likely presence of perchlorates in the regolith, some of which are eutectic even within the range of present temperatures (Hecht *et al.* 2009), suggests that the thawing of high-latitude ground ice has the potential to drive significant, distinctive morphogenesis. We describe such a landform assemblage here.

Reconnaissance survey and initial findings

A reconnaissance-level survey of High Resolution Imaging Science Experiment (HiRISE) images in a latitudinal band approximately 5° wide centred on the *Phoenix* landing site was performed. This revealed that sorted clastic circles, stripes and lobes were common forms, and especially evident on the slopes of craters. Arising from this result, a more detailed analysis of 24 HiRISE images was completed (Fig. 1). This confirmed that landform

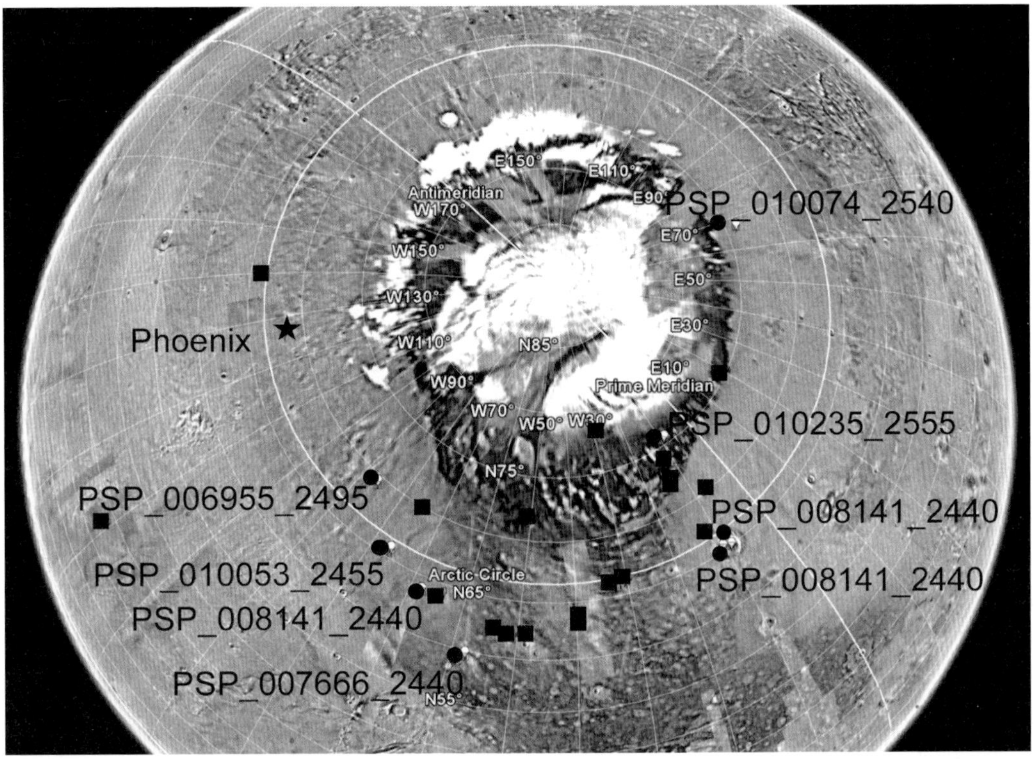

Fig. 1. Part of the northern hemisphere of Mars showing the locations of the HiRISE images referred to in the text as well as major albedo features and the north polar cap. Circles represent images discussed in detail, squares are images referred to only in Table 2. The location of the *Phoenix* landing site is marked with a star. The bold white circle is the Arctic Circle and the bold lines of latitude are the prime meridian and anti-meridian. Lines of latitude and longitude are at 5° spacings. The image is intended only to show the relative locations and latitudinal range of the HiRISE images referred to in the text. Image credit: Google Mars.

assemblages associated with freeze–thaw and the action of near-surface liquids were evident in at least 22 of the 24 images. Detailed image interpretation was performed on eight images (Table 1) and forms the observational basis of this paper. However, Table 2 summarizes the dominant morphological attributes of the 16 images not selected for detailed analysis here. In only two of these 16 images was morphological evidence of freeze–thaw either equivocal (PSP_007508_2440) or not evident (PSP_007573_2435), supporting the inference that the landform descriptions that follow relate to typical rather than anomalous landscape features of the study area.

Table 1. *Location of HiRISE images and landforms discussed in detail in the text*

Image	Longitude	Latitude	Attributes
PSP_010235_2555	350.8	75.3	Sorted clastic islands outside crater context
PSP_007666_2400	302.4	59.5	Sorted clastic circles; gullies, fans; lobe incision by gullies
PSP_010644_2455	349.5	65.3	Sorted circles, stripes, lobes
PSP_007440_2455	351.0	65.0	Clastic circles evolution to clastic stripes
PSP_010053_2455	284.1	65.5	Solifluction lobes, terraces
PSP_006955_2495	274.0	69.3	Dendritic gullies up to crater rim; possible vein ice on flat crater bottom
PSP_008141_2440	292.3	63.8	Gullies; solifluction lobes; RTS-like form in gully

Table 2. *Location and morphological attributes of images only surveyed*

Image	Longitude	Latitude	Attributes
PSP_008416_2585	331.4	78.6	Gullies incised in sand dune faces; clastic stripes; dark possibly clastic polygons under sandy surface
PSP_009087_2550	14.9	74.9	Polycyclic gullies in sand dune faces; buried polygons
PSP_009233_2535	348.2	73.5	Gullies incised in sand dune faces
PSP_007572_2520	344.8	71.9	Braided gullies and fans; solifluction lobes; clastic stripes
PSP_008628_2515	309.1	71.2	Clastic stripes; interaction between stripes and dunes
PSP_009457_2505	352.1	70.0	Sorted clastic forms
PSP_008352_2500	285.1	69.8	Blockfield; clastic stripes; lobes; circles
PSP_010011_2460	350.7	65.9	Possible thermokarst channels; lobate viscous flow features
PSP_008456_2460	329.6	65.6	Slope-controlled cracks; solifluction lobes; faint clastic polygons
PSP_010236_2460	329.4	65.6	Thermokarst-like collapse forms; some sorted clastic forms and albedo patterning
PSP_007508_2440	295.8	63.8	Sorted clastic circles but dominated by crack-filling; thermal contraction cracking dominant
PSP_007574_2440	292.1	63.8	Gullies; sorted clastic circles and stripes
PSP_007573_2435	320.2	63.0	Dominated by complex polygonal cracks, many with cracked uplifted shoulders; no obvious ground patterns independent of cracks; thaw not indicated
PSP_007613_2420	308.0	61.6	Clastic lobes breached by faint gullies; garlands, lobes, faint stripes; lobate viscous flow feature; elongate clast-lobed flow feature; sorted circles; some trough-lining clast fields
PSP_007547_2415	311.5	61.3	Sorted circles, faint stripes and fine textured solifluction lobes
PSP_007376_2395	302.1	59.3	Possible ground ice exposed in gullies; sorted clastic circles; faint fine stripes; fine textured lobes

Detailed morphological observations

Sorted islands, circles, polygons, garlands and stripes

On low-gradient surfaces of the Northern Plain outside craters, clastic islands (cf. 'clastic piles') are the typical sorted landform. For example, the surface shown in Figure 2 is characterized by polygons, near-circular forms and ellipses approximately 3–10 m across. These forms occur in groups delimited by zigzagging cracks, which are the outside edges of outlying polygons. Well-defined clast islands, some also displaying a lower

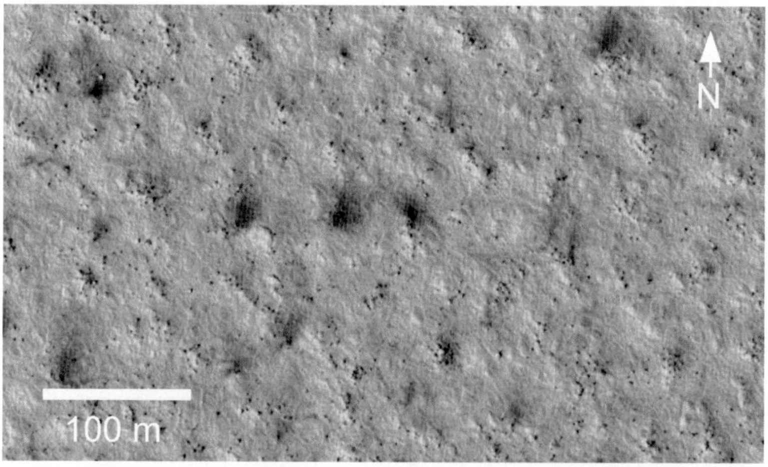

Fig. 2. Sorted clastic islands here are approximately 15–30 m wide and spaced at approximately 20–30 m intervals in both spatial dimensions. They are defined by clasts of about 1.8–4 m in diameter. Importantly, the clastics are not filling cracks but, rather, are positive relief features indicative of either clast (cf. stone domain) uplift or fine domain downwearing. Part of HiRISE image PSP_010235_2555 (63.5 cm per pixel, 2 × 2 binning, 191 cm object resolution, image centre 75.3°N, 350.8°E). Image credit: NASA/JPL/University of Arizona (UofA); see prelim viii for acronym definitions.

albedo than their surroundings, dominate the fabric of the surface. They are about 20 m across, spaced up to 115 m apart and contain clasts of approximately 1.8–4 m. These clast islands occur within crack-delimited polygon assemblages that have positive relative relief. Polygon assemblages of negative relief tend to be fine domains (Kessler & Werner 2003) devoid of islands and largely even clast-free. Hence, these sorted clastic islands appear to be associated with a mechanism involving either clast uplift or fine domain downwearing, but not collapse of clasts into cracks or troughs.

While sorted clastic islands occur also on low-gradient surfaces within craters, the dominant within-crater sorted forms are clastic nets and circles. Figure 3 shows clastic nets consisting of polygonal arrangements of clasts up to 4 m across, bordering clast-deficient centres up to 13 m across. While some polygon centres are slightly domed, and the clastic borders occupy bounding polygonal troughs, the clasts here appear to stand higher than the polygon centres. Hence, although cracking is apparent in the fine textured surface, a further mechanism is required to either lift the clasts at the polygon edges or to lower the fine regolith at the polygon edges without subducting the clastic borders. The presence of slightly raised dark polygons, themselves containing clastic polygons, bounding fine textured centres indicates that the clastic sorting is associated with active uplift at polygon borders rather than just passive clast collapse into border troughs. Figure 4a, b appear to confirm this conclusion, for they show situations in which clastic polygons within dark albedo-defined polygons surround bright centres but with little (Fig. 4a) or no (Fig. 4b) apparent relief between

Fig. 3. Clastic nets on a near-flat surface at the bottom of a slope running out from a crater rim (towards the top left of the inset). The nets consist of polygonal arrangements of clasts, bordering clast-deficient centres up to 13 m across. Polygon centres are slightly domed, with clastic borders occupying bounding polygonal troughs, but the clasts stand higher than the polygon centres. Hence, these forms reflect both cracking of the fine textured regolith domain and a mechanism of either clast uplift at the polygon edges or fine domain lowering without clast subduction at polygon edges. Part of HiRISE image PSP_007666_2400 (31.4 cm per pixel, 1 × 1 binning, 94 cm object resolution, image centre 59.5°N, 302.4°E). Clasts are up to 4 m, fine domains approximately 3 m wide, albedo border 4–7 m wide. Inset shows the context of Figure 3 and relative locations of Figure 4a, b. Image credit: NASA/JPL/UofA.

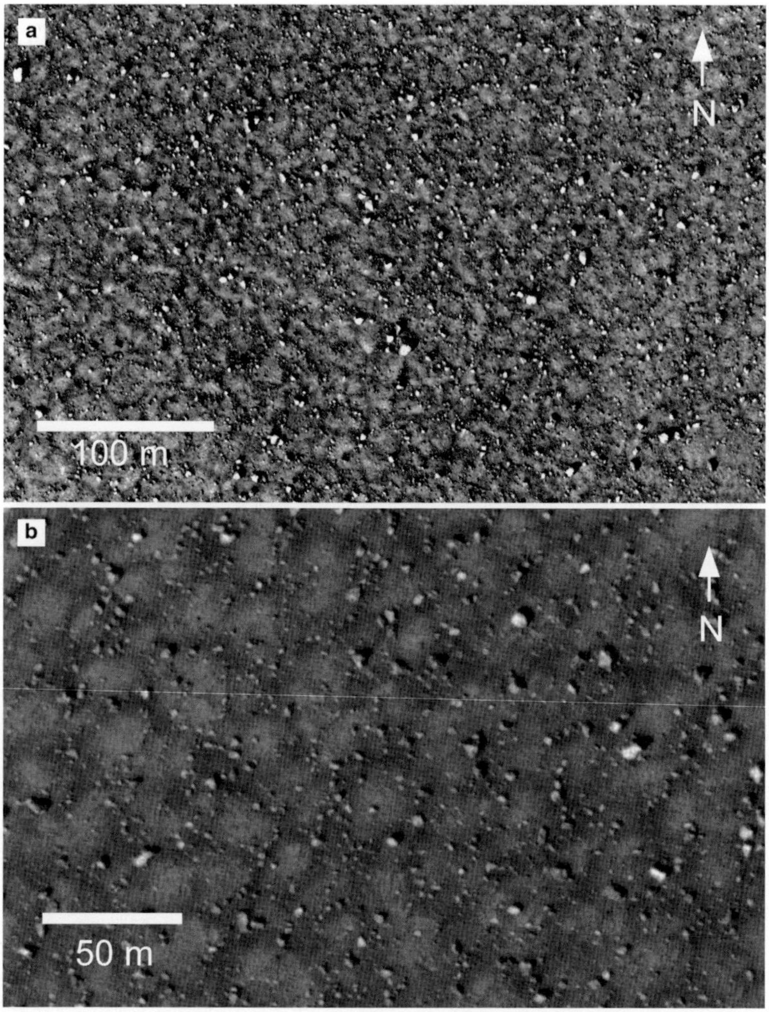

Fig. 4. (**a**) Clastic nets within albedo-defined polygons surround flat to slightly domed centres. There is little apparent relief between the polygon centres and borders. (**b**) shows clastic nets within albedo-defined polygons with no relief between the polygon centres and borders. The conclusion is that clastic sorting does not depend on initial relief but instead produces relief, here in the form of raised clastic borders. Parts of HiRISE image PSP_007666_2400. Maximum clast size is 6 m, albedo border width 4–12 m. Image credit: NASA/JPL/UofA.

the bright centres and the dark albedo polygons. Clastic sorting does not depend on relief but instead produces relief, whether in the form of clastic islands or of raised clastic polygons. The prevalence of sorted islands outside craters, but sorted polygons and circles within craters, probably reflects a greater concentration of available clasts inside craters than outside. A consequence of, perhaps, both the abundance of impact-created breccias and joint exposure within craters and the build-up of a thick weathered mantle of fines outside craters.

With increasing ground slope, blockfields develop a lineated 'grain' (Fig. 5) and sorted circles become elongated into downslope-oriented ellipses, garlands (i.e. ellipses open at their downslope limb) and stripes (Benedict 1970; Werner & Hallet 1993; French 1996; Kessler & Werner 2003). Figure 6 shows an example of this slope-controlled evolution in the Lomonsov Crater in which sorting of clasts, ranging from 2 to 9 m across, from fines takes the form of closed clastic ellipses (up to 50 m in extent along their major axis), garlands and stripes. The marbling of the

Fig. 5. This blockfield exhibits a subtle slope-modified textural grain, including slope-parallel clastic lineations and slope-normal clastic banding (e.g. to the right of the double-headed slope-direction arrow). Although slope-parallel cracks are evident (e.g. below point of double-headed arrow), clasts do not simply occupy the cracks. Hence, this image reflects the well-documented transition from sorted clastic polygons and circles on flat surfaces to lineated clastic forms, particularly on debris-limited slopes. Part of HiRISE image PSP_010644_2455 (31.7 cm per pixel, 1 × 1 binning, 95 cm object resolution, image centre 65.3°N, 349.5°E) Clasts up to 4.5 m diameter, fine domains up to 15 m across. Inset image shows relative locations and topographical situations of Figures 5, 7a, b, 8 & 9b. Image credit: NASA/JPL/UofA.

fine textured surface here appears to reflect the filling of fissures by a secondary material – possible candidates including salts and vein ice. If the latter, this is important because vein ice forms by surface-water penetration into narrow thermal contraction fissures. Vein ice forms reticulate networks in very-fine-grained regolith and, if water penetration is repeated over many years, ultimately develops into ice wedges (French 1996). Similar networks characterized by feather-edged fissure and crack fills are very common on crater bottoms throughout the Martian high north, a point that will be returned to later.

Figure 7a, b demonstrate that sorted clastic slope forms often originate in blockfields, themselves a product of rock disintegration broadly associated with freeze–thaw weathering (Ballantyne & Harris 1994). Here a hill-summit blockfield, consisting of clasts up to 6 m across, is seen to develop into sorted clastic stripes 6–10 m wide, spaced up to 17 m apart and oriented radially downslope on the SE flank of the hill. Figure 8 exemplifies the downslope fining typical of sorted stone stripes. Although stripes are a product of lateral sorting coupled with solifluction, fine clasts move mainly by gelifluction involving liquefaction of the regolith, while coarse clasts move downslope by frost creep dependent on the angular difference between slope-normal heave during freezing and vertical settling under gravity during thaw.

Lobes, stone banked lobes and terraces

Hillslopes patterned by sorted clastic forms often also exhibit a range of lobate forms ranging from transverse clastic bars linking stripes, to stone-banked lobes and terraces, to texturally fine lobes lacking surface clasts. Figure 9a shows clastic stripes adjacent to texturally fine lobes. However, clastic stripes, lobes and terraces in periglacial environments on Earth are not only spatially associated but genetically too, all being products of solifluction involving a varying balance between frost creep and gelifluction. Figure 9b demonstrates a subtle but typical internal variability in stripe morphology characterized by the local evolution from clastic stripes to clastic lobes. An evolution that Benedict (1970) demonstrated was a reflection of

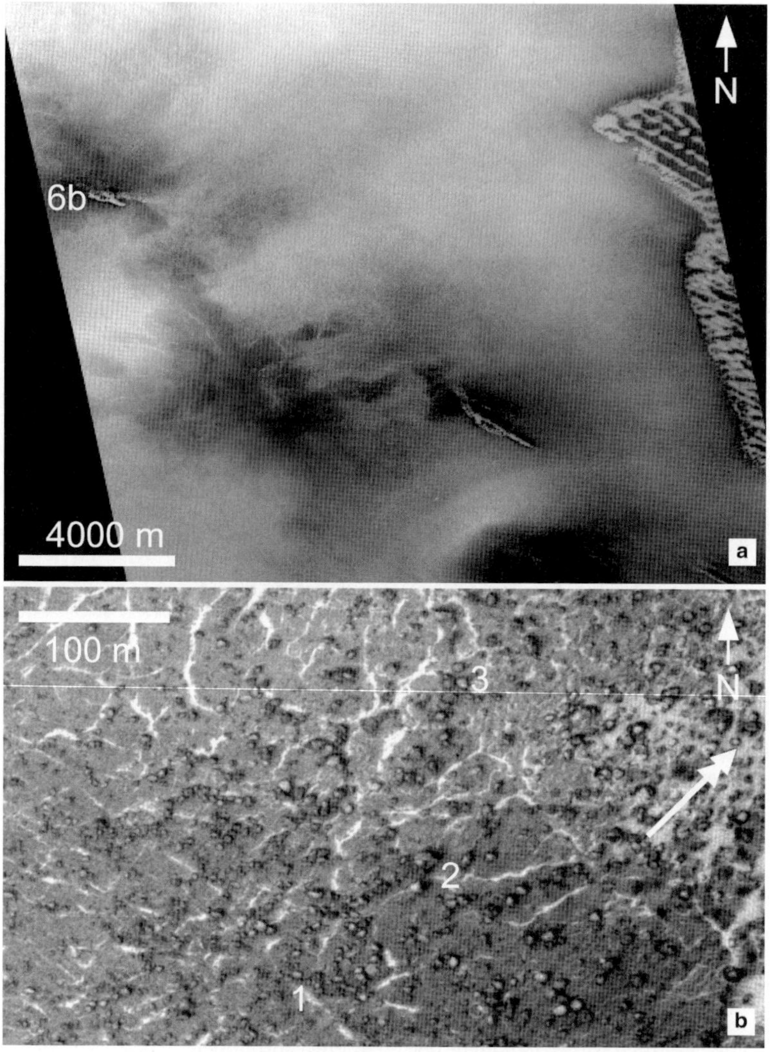

Fig. 6. (**a**) Part of Lomonsov Crater showing dark clastic accumulations on convex summits and long clastic aprons running-out downslope from the summits. (**b**) (relative location indicated in a) shows the evolution of sorted clastic circles near a summit to slope-parallel clastic stripes further downslope. Circles segue into ellipses (1), open ellipses or garlands (2) and ultimately stripes (3). Maximum clast size is 9 m, maximum ellipse major axis 50 m. PSP_007440_2455 (63.8 cm per pixel, 2 × 2, 191 cm object resolution, image centre 65°N, 351°E). Image credit: NASA/JPL/UofA.

slope transitions from supply-limited units to transport-limited units; as clasts decelerate (e.g. at the foot of slopes or on slope undulations) and clast concentration rises, clasts diffuse laterally out of the stripes and accumulate in lobes normal to the slope. However, Figures 10a, b and 11a, b demonstrate that the modification of slopes by periglacial processes involving solifluction in the high northern latitudes of Mars is far from subtle. The almost complete modification of both the interior and exterior slopes of this crater by forms involving solifluction echo the observation of Holdgate *et al.* (1967) that solifluction in the Antarctic is extremely extensive and can mobilize entire hillsides.

Gullies and fans

Fine textured clastic and non-clastic solifluction lobes reflect solifluction involving a dominance of gelifluction over frost creep. Hence, they reflect

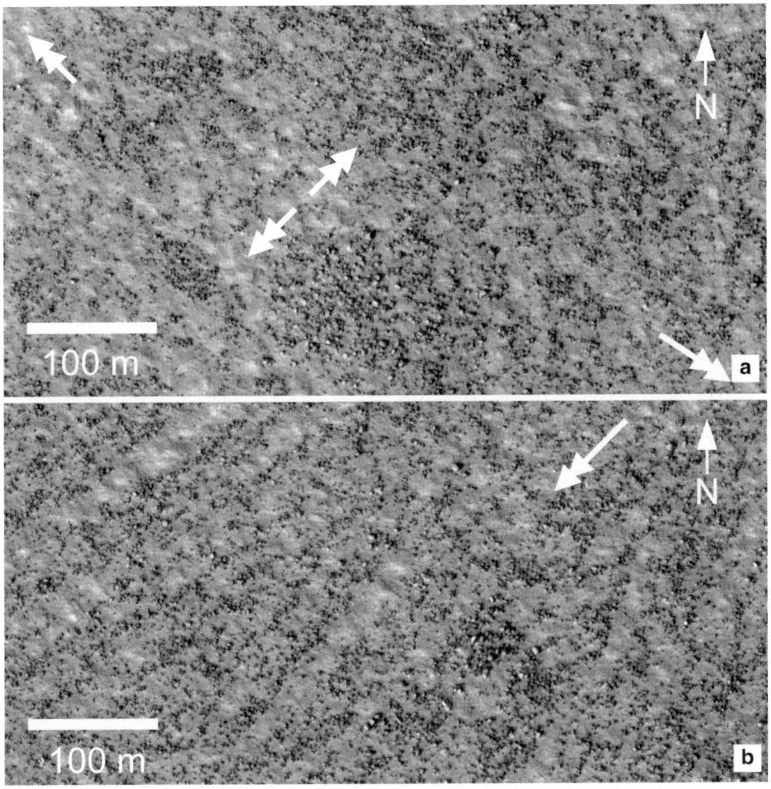

Fig. 7. (**a**) Clastic sorting and slope-controlled morphogenesis of a hill-summit blockfield consisting of clasts up to 6 m across, organized into sorted stripes and bands 6–10 m wide and spaced up to 17 m apart. Slopes are divergent. (**b**) Clastic stripes oriented radially downslope on the SE flank of the same hill. Parts of HiRISE image PSP_010644_2455. For context see Figure 5 inset. Image credit: NASA/JPL/UofA.

Fig. 8. Downslope fining of stripes reflecting the greater mobility of fine compared with coarse clasts. This reflects the increasing dominance downslope of gelifluction, involving liquefaction, over frost creep. Part of HiRISE image PSP_010644_2455. Image credit: NASA/JPL/UofA.

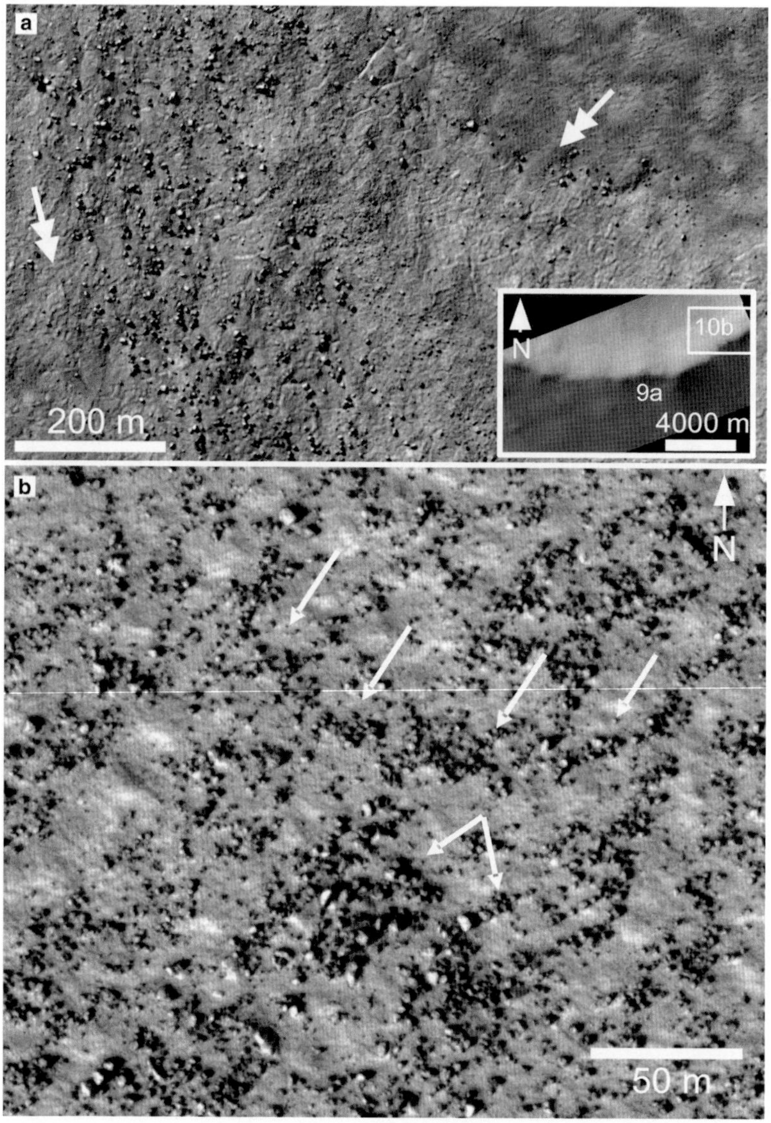

Fig. 9. (**a**) Clastic stripes formed adjacent to texturally fine lobes (at the top right-hand corner of image), here morphologically defined by albedo. Clastic stripes, lobes, clastic lobes and terraces are all products of solifluction involving a varying balance between frost creep and gelifluction. The inset shows both context of Figure 9a on the exterior slope of a crater and the relative location of Figure 10b. See Figure 10a for a larger contextual image. Parts of HiRISE image PSP_010053_2455 (31.4–62.8 cm per pixel with $1 \times 1 - 2 \times 2$ binning, image centre 65.5°N, 284.1°E). Image credit: NASA/JPL/UofA. (**b**) Local evolution from clastic stripes to clastic lobes, reflecting transitions from supply-limited to transport-limited slope units. Where clasts decelerate and clast concentration rises, clasts diffuse laterally out of the stripes, accumulating in lobes normal to the slope. Part of HiRISE image PSP_010644_2455. Image credit: NASA/JPL/UofA.

morphogenesis driven by liquefaction of the regolith. The presence of permafrost near the surface can retard the downward percolation/permeation of water, meaning that only a relatively thin regolith layer is left to liquefy and which could occur with only small absolute amounts of water; for example, during rare but repeated thaw events. However, the widespread occurrence of fluviatile gullies and braided fans on the inner walls of many of the high-latitude Martian craters surveyed for this research

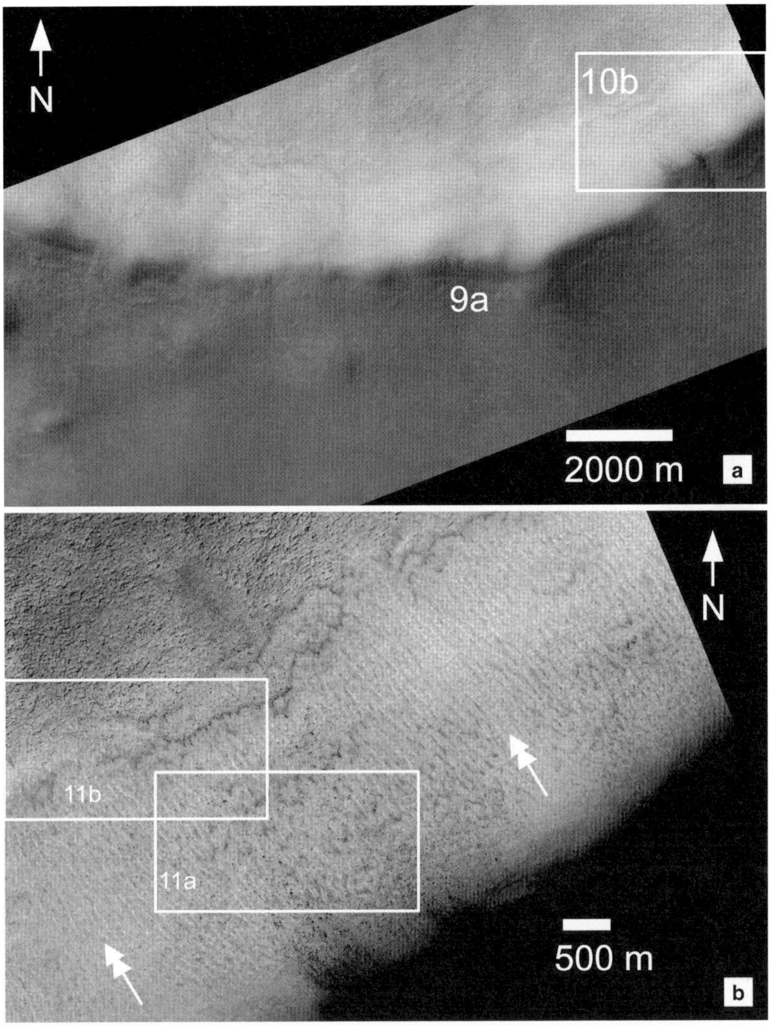

Fig. 10. (**a**) Clastic stripes evolve to lobes and terraces on both interior and exterior crater slopes. Both the interior and exterior slopes of this crater are dominated by landforms reflecting solifluction. The location of Figure 9a and 10b is shown. (**b**) Mid-slope units are dominated by clastic lobes (e.g. between double-headed slope direction arrows) but the slope termination is marked by an extensive pair of stone-banked terraces (e.g. running across and extending beyond the box labelled 11b). Boxes show the context of Figure 11a, b. Part of HiRISE image PSP_010053_2455. Image credit: NASA/JPL/UofA.

reflects the generation and surface flow of significant quantities of water. Crucially, as shown in Figure 12a–d, the gullies have incised into stone-banked solifluction lobes, demonstrating that the gullies and the surface flows of liquid they reflect are either coeval with or post-date the solifluction forms. In most instances, gully erosion has left a lag of coarse clasts, partially preserving the lobe and showing that, rather than experiencing total erosion, the lobes have been deflated by the winnowing of fines that tend to dominate the upper, horizontal, tread of solifluction lobes. For example, Figure 12c shows a fluviatile system comprising an axial gully and subordinate rills that have deflated a solifluction lobe. The development of a braided fan just below the deflated lobe demonstrates not only that the fines winnowed from the lobe caused the transport capacity of the gully system to be exceeded, triggering deposition of the fan, but also that the gully post-dates this lobe. Hence, locally at least, morphogenesis driven by gelifluction, but perhaps involving minimal thaw, was succeeded

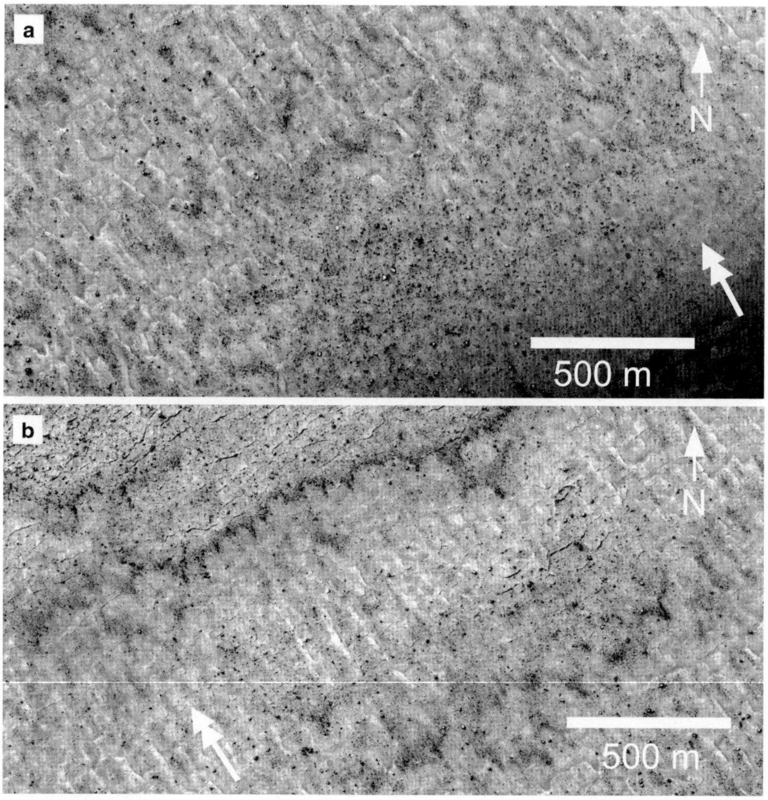

Fig. 11. (**a**) Subtly lineated blockfield near the head of the slope evolves to clastic stripes and stone-banked lobes. (**b**) Near the slope termination, stone-banked lobes are drawn out into very faint stripes across a convex–concave slope transition (e.g. transition along the double-headed slope-arrow). The faint stripes terminate at a stone-banked terrace composed of linked lobes, each supplied by a pair of stripes; hence, the terrace resembles a waveform with peaks oriented upslope. Part of HiRISE image PSP_010053_2455. Image credit: NASA/JPL/UofA.

by morphogenesis reflecting the action of flowing liquid.

The gullies shown in Figures 13 and 14 have incised crater interior slopes extensively modified by non-clastic lobes indicative of relatively fine regolith compared with that of the crater shown in Figure 12. The inferred fine texture is consistent with the well-developed drainage hierarchy developed by headward erosion of individual gullies, the absence of channel terminations at mid-slope fans and, consequently, the complete run-out of the gully systems from the crater rim to its bottom, ultimately via fans that originate below mid-slope (Fig. 13b). The gullies that have eroded all the way back to the crater rim (Fig. 13c) suggest that, unless a groundwater source on the outside slope slumped away from the rim en masse after the gullies formed, these fluviatile systems are the product of incision by thaw fluids seeping from solifluction lobes, frost melt or even snow melt (Smith *et al.* 2009). Rim flattening, in places, and extensive loss of crater-rim ground-ice reservoirs cannot be precluded (Fig. 13c). Hence, although these need not necessarily be mutually exclusive drivers, the complete presence of small lobes all the way up to the crater rim and constraining the alcoves of incipient gullies (Fig. 14a) indicates that seepage of thaw fluids from solifluction lobes is a strong likelihood. Also, the beaded appearance of gullies surrounded by downslope-elongated dimples (Fig. 14b) indicates that these gullies have extended by the headward linking of the dimples, strongly suggestive of a process analogous to gully propagation in patterned ground experiencing thermokarst degradation. However triggered, once some gullies form, frost will accumulate in them, both out of the atmosphere and blown into them from initial settling points. If this process ends in thaw, the systems will deepen and back erode, eventually extending via gully piracy and areal complexification achieved solely by headward extension of self-similar linear systems (Fig. 14c).

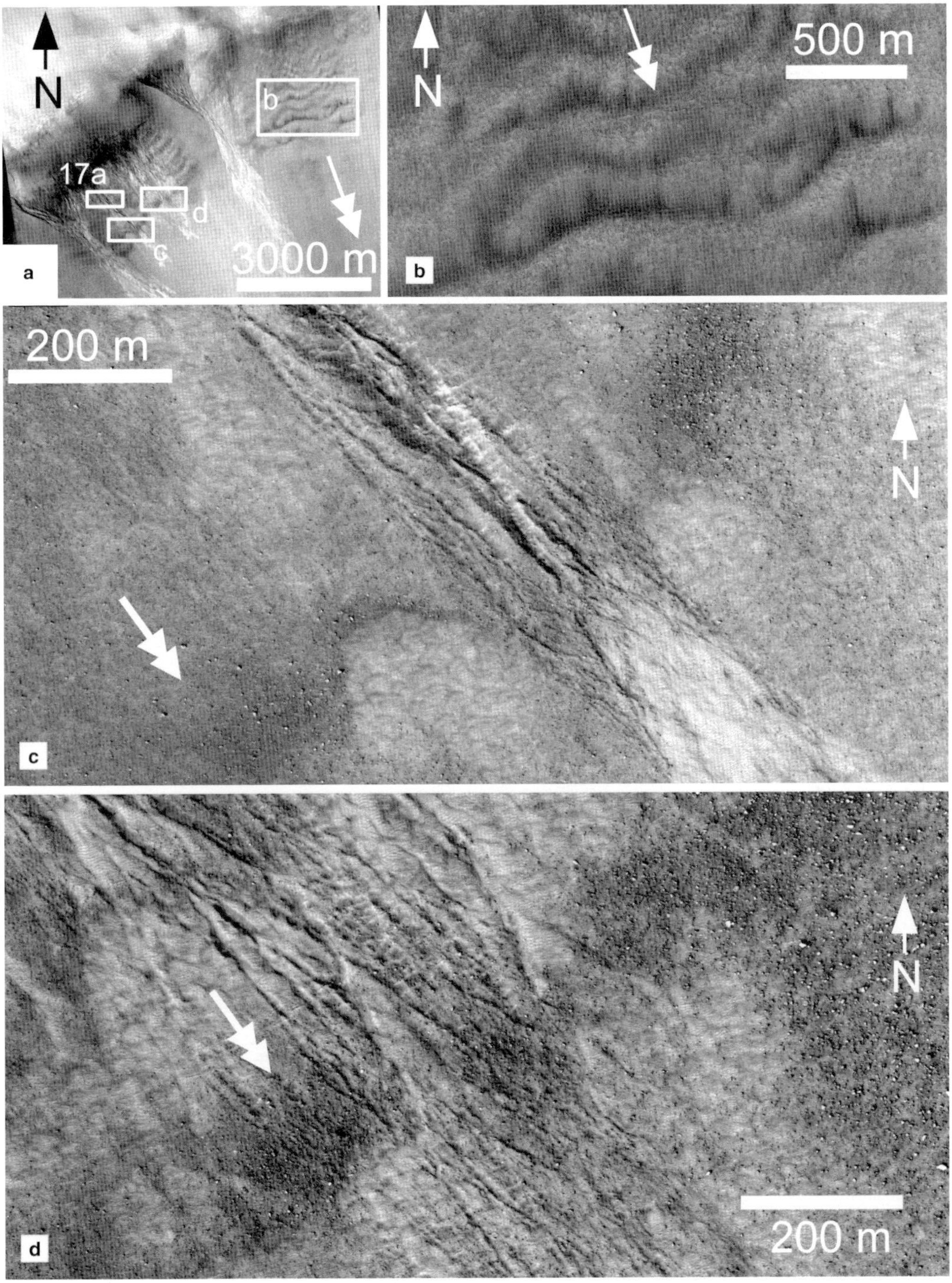

Fig. 12. (**a**) Stone-banked solifluction lobes incised by gullies on the interior slopes of a crater, context. (**b**) Lobes exhibiting bright, fine-textured low-angle treads bounded by dark, steeper clastic risers. (**c**) Axial gully and rills have incised and partly deflated a clastic solifluction lobe by winnowing. Fines winnowed from the lobe caused the capacity of the gully system to be exceeded, triggering deposition of the fan below the lobe. (**d**) Lobe deflation by younger gully rills, demonstrating that solifluction was succeeded by liquid flow across the surface. Parts of HiRISE image PSP_007666_2400. Image credit: NASA/JPL/UofA.

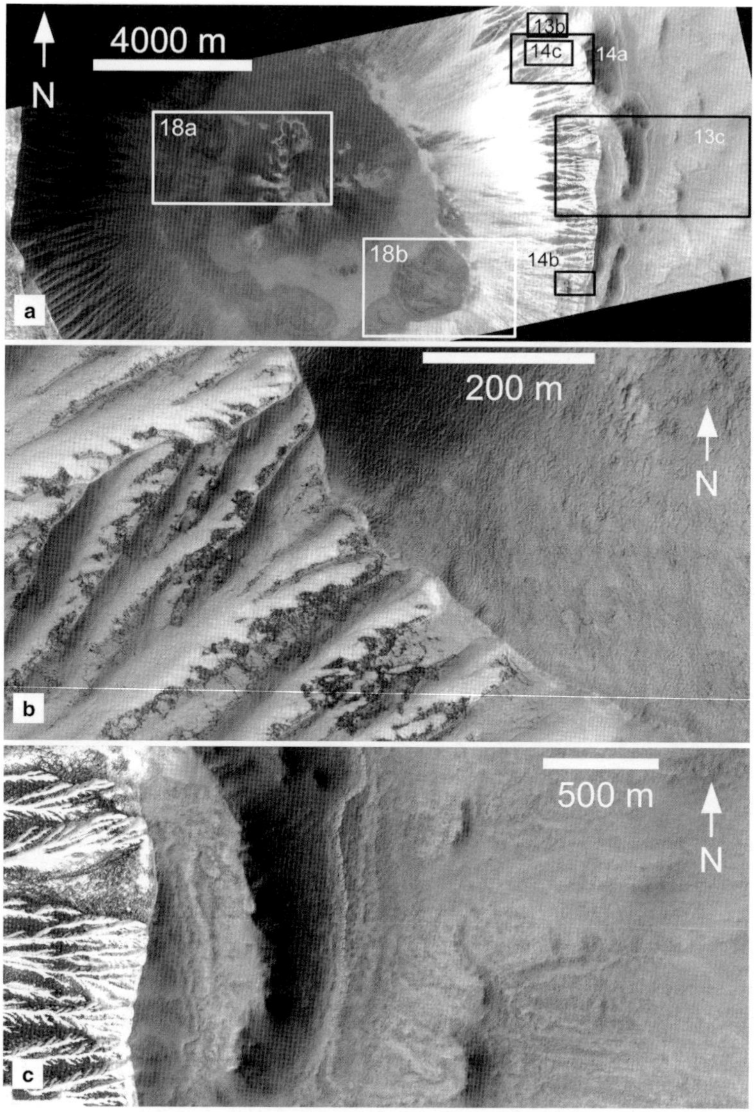

Fig. 13. (**a**) Contextual image showing a crater intensely incised by a dendritic gully network on interior crater slopes, surrounding a flat, although morphologically complex, crater bottom. Boxes show locations of (b), (c), and Figures 14a–c & 18a, b. (**b**) Gullies have eroded headward as far as the crater rim, demonstrating network evolution involving the entire interior slope from rim to crater bottom. Figure 12c shows that the exterior crater rim has experienced periglacial mass wasting and rim flattening associated with multiple generations of translational slides on the exterior slopes. Parts of HiRISE image PSP_006955_2495 (31.8 cm per pixel, 1 × 1 binning, 95 cm object resolution, image centre 69.3°N, 274.0°E). Image credit: NASA/JPL/UofA.

This relationship between possible thermokarst downwearing (Czudek & Demek 1970) of the surface and the evolution of fluvioperiglacial channel systems is further exemplified by the development of a complex incipient gully head formed within a pre-existing gully system (Fig. 15a–c). The morphology of this incipient system is based on downwearing of individual slope-elongated, domed polygons. Domed polygons like these are epigenetic forms arising from the refreezing of accumulated thaw fluids, and have previously been observed at the head of Mars' Athabasca Vallis in a thermokarst landform assemblage (Balme & Gallagher 2009). Those in Figure 15 occur only

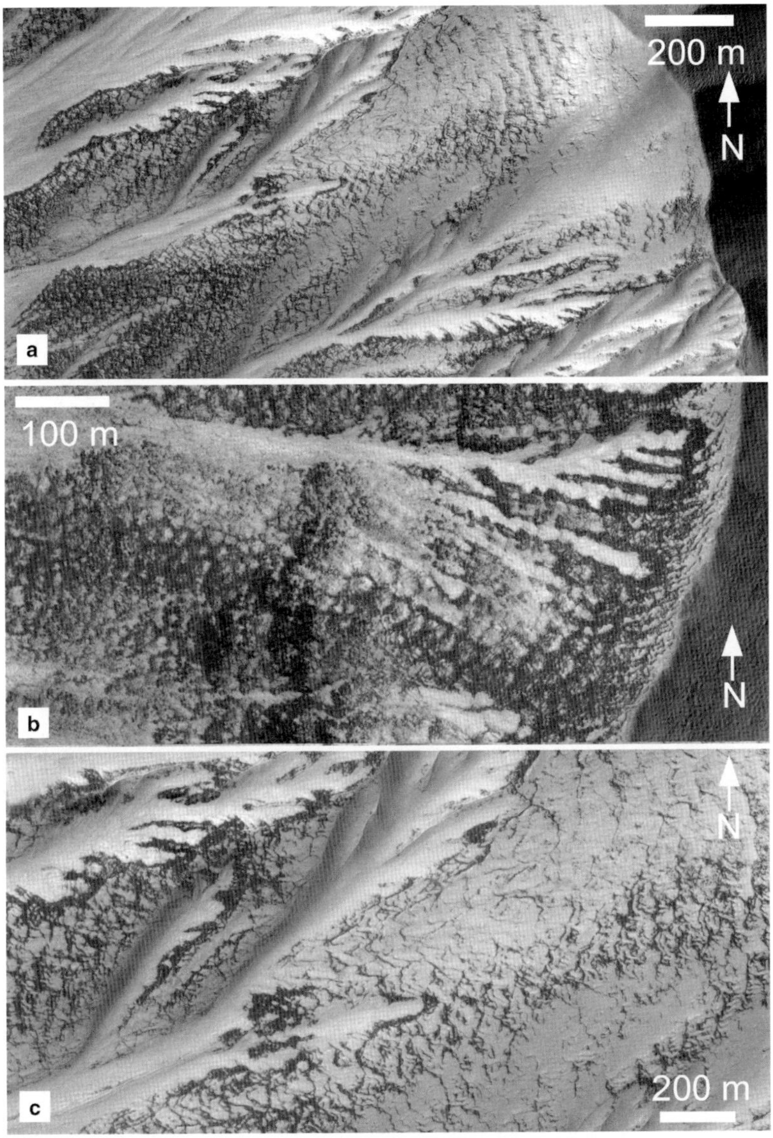

Fig. 14. (**a**) As well as gullies, solifluction lobes occur all the way up to the rim of this crater (see Fig. 13a for context), demonstrating at least a spatial association between landforms associated with flowing liquid and regolith liquefaction, respectively. (**b**) Dimples that cover parts of interior slopes of this crater reflect the presence of small pits, garlands and lobes. Many of the gullies in this crater have a beaded appearance that is clearly due to the linkage of dimples by headward gully erosion. (**c**) Gullies are evidently effective frost traps and could be incised both by seepage from surrounding solifluction lobes and the action of frost-melt liquids. Parts of HiRISE image PSP_006955_2495. Image credit: NASA/JPL/UofA.

along the gully axis, the ground patterning outside the gully taking the form of clastic lobes, stripes and reticulate cracks. The entire feature exhibits a dimensional hierarchy of forms based initially on downwearing of individual polygons followed by the coalescence of many individuals as their backwalls retreated, characteristics typical of thermokarst degradation of patterned ground. The regolith mobilized by this process appears to have flowed out of compound alcoves, joining an axial accumulation reflecting debris transport out of the complex into the axis of the pre-existing gully

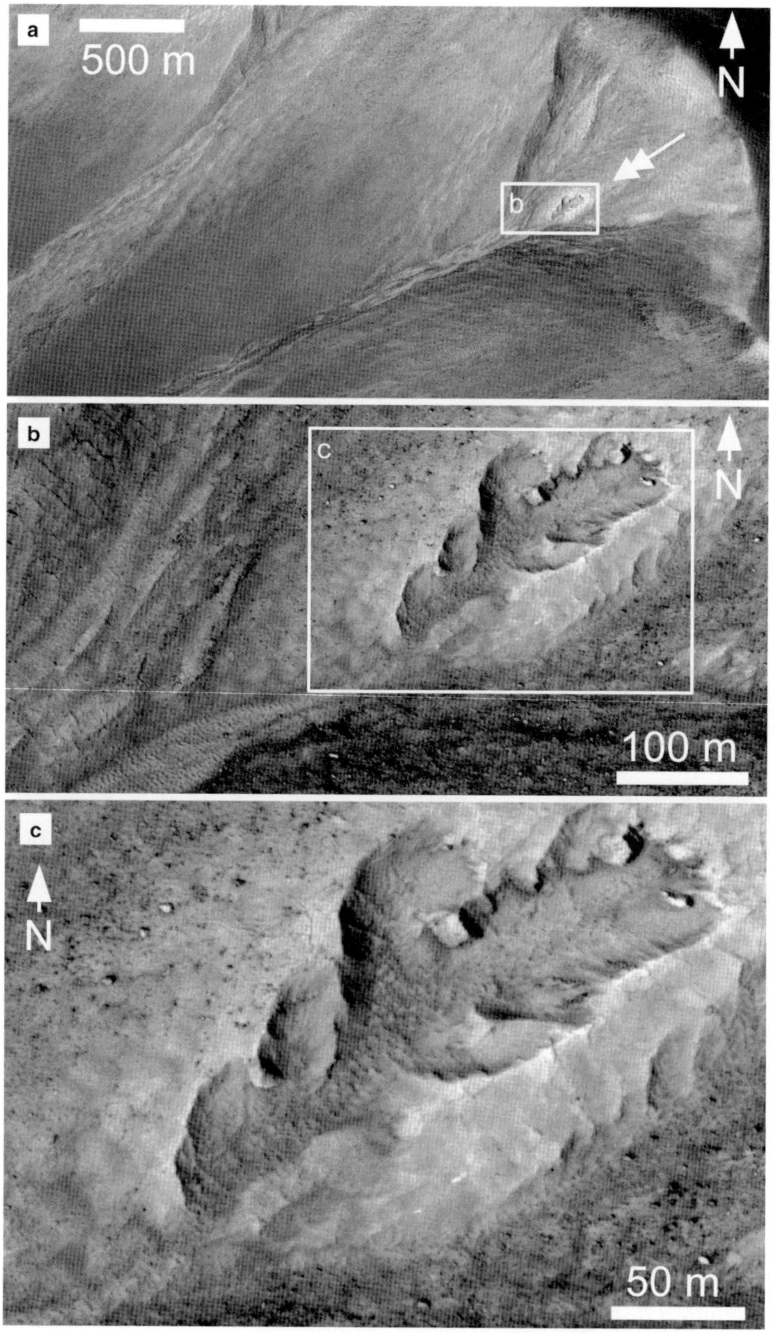

Fig. 15. (**a**) Gullies on the interior slopes of this crater are associated with sorted stripes and clastic lobes, and the crater exterior exhibits extensive translational mass wasting, all forms associated with liquefaction. (**b**) Overdeepening of a pre-existing gully based on alcove formation by downwearing of individual domed polygons followed by the coalescence of many individuals by alcove-backwall retreat. These are characteristics typical of thermokarst degradation of patterned ground. (**c**) Regolith has been mobilized by this process and flowed out of alcoves and compound alcoves, joining an axial debris accumulation. Altogether, this form has the characteristics of complex retrogressive thaw slumps (RTS) that develop through the thermokarst degradation of ground ice. Parts of HiRISE image PSP_008141_2440 (31.5 cm per pixel, 1 × 1 binning, 95 cm object resolution, image centre 63.8°N, 292.3°E). Image credit: NASA/JPL/UofA.

(Fig. 15b, c). Altogether, this form has the characteristics of complex retrogressive thaw slumps (RTS) that develop through the thermokarst degradation of ground ice (Lantuit & Pollard 2005). Sublimation karst is common on the residual south polar cap of Mars and has some components of RTSs (Thomas *et al.* 2000; Malin *et al.* 2001; Byrne & Ingersoll 2003), but notably lacks features associated with liquid outflows that are integral to terrestrial RTSs. Hence, the key component of the feature in Figure 15b, c that supports the inference of thaw, but which is absent in polar sublimation karst, is the hierarchy of debris flows sourced in individual and compound alcoves that exit the system via a single, axial outlet. Gullies elsewhere in the region are sourced from similar, although sometimes simpler, RTS-like forms (Fig. 16a–c). For example, the slope-elongated depression in Figure 16b reflects incipient coalescence both with a shallow circular pit, via backwearing (Czudek & Demek 1970) of the main headwall, and with an adjacent gully via backwearing of its headwall. Both are examples of channel extension and network complexification by gully piracy. Figure 16c shows a braided gully system sourced at a narrow linear pit, the presence of levees bounding the gully indicative of breakouts by viscous debris flows and, therefore, symptomatic of liquefaction of the regolith associated with thaw.

Figure 17a–c demonstrates that gully evolution reflects not only headward extension through the backwearing of thermokarst depressions at the head of gully networks but also by the deepening and widening of existing channels. Figure 17b shows a very small depression from which debris has flowed into a gully channel, some of the debris accumulating just below the breached lower rim of the depression. Further downslope, however, the channel is characterized by a deeply incised thalweg that reflects overdeepening of the pre-existing channel axis by a single, over-competent flow sourced in the depression. Figure 17c shows that at least some channels in the same gully system owe their form to the widening of pre-existing longitudinal cracks. While such crack-widening on its own could reflect sublimation, its location adjacent to a gully that has experienced secondary incision by flows clearly sourced at a single depression demonstrates the thermokarst nature of this gully complex. Ground-ice thaw was the fundamental morphogenetic driver of the system through the extended production and action of thaw liquids.

Given the abundance of slope forms associated both with liquefaction of the regolith and with flowing liquids, it is not unreasonable to infer that thaw fluids have collected in crater bottoms. The presence of lobate, fountain-like flows horizontally intruding crater-bottom debris from distal fans on crater slopes supports this inference (Fig. 18a, b). In addition, the termination of the interior crater slope at a series of cirque-shaped indentations in polygonized ground and the exhibition of backwearing by the collapse of individual polygonal blocks (Fig. 18a) is analogous to shoreline retreat by thermo-abrasional collapse in terrestrial thermokarst environments (Are 1988). Similarly, the frequent presence of vein-like fissures (Fig. 18a, c) may be indicative of ground intrusion by surface water. These features are branching – but generally not polygonal – and exhibit structures that may be the surface expression of internal vertical foliations, usually associated with discrete cycles of intrusion by a secondary material. Analogous terrestrial forms that develop in association with other indicators of freeze–thaw are attributed to vein-ice growth in narrow thermal contraction cracks and are a precursor to ice wedges and fully polygonized ground. Hence, if these forms are ice veins rather than salts, seasonal frost pockets or fine textured aeolian sedimentary fills, the intrusion of extensive, foliated veins into ground patterned with smaller crack networks would indicate polycyclic ground patterning and the repeated presence of surface liquids.

Discussion

Cyclical freeze–thaw cryoturbation results initially in the vertical sorting of regolith clasts, including large boulders (Harris & Matthews 1984), from fines. Kessler & Werner (2003) successfully modelled the development of all forms of sorted patterned ground on the basis that morphogenesis resulted initially from the expansion of fines (the soil domain) during freezing, causing clast-dominated areas (the stone domain) to be displaced outwards from the soil domain centres. Expansion of individual cells results in stone domain squeezing, constraining clast migration into narrow lines normal to the compression. Consequently, where linear stone domains lengthen and intersect, sorted polygons emerge.

Sorted stripes are a subset of sorted patterned ground and form by a combination of freeze–thaw and mass-wasting processes. The combination of the lateral clast sorting away from fine 'soil' domains (Kessler & Werner 2003) and downhill movement constrains clasts within stripes. Sorted circles, garlands, stripes and lobes form a gradational series of landforms, the exact morphology being dependent, through solifluction, on slope and clast concentration (cf. Benedict 1970). Similarly, solifluction sheets (cf. blockfields), lobes and terraces develop on slopes that experience cyclical thaw, and which are characterized by

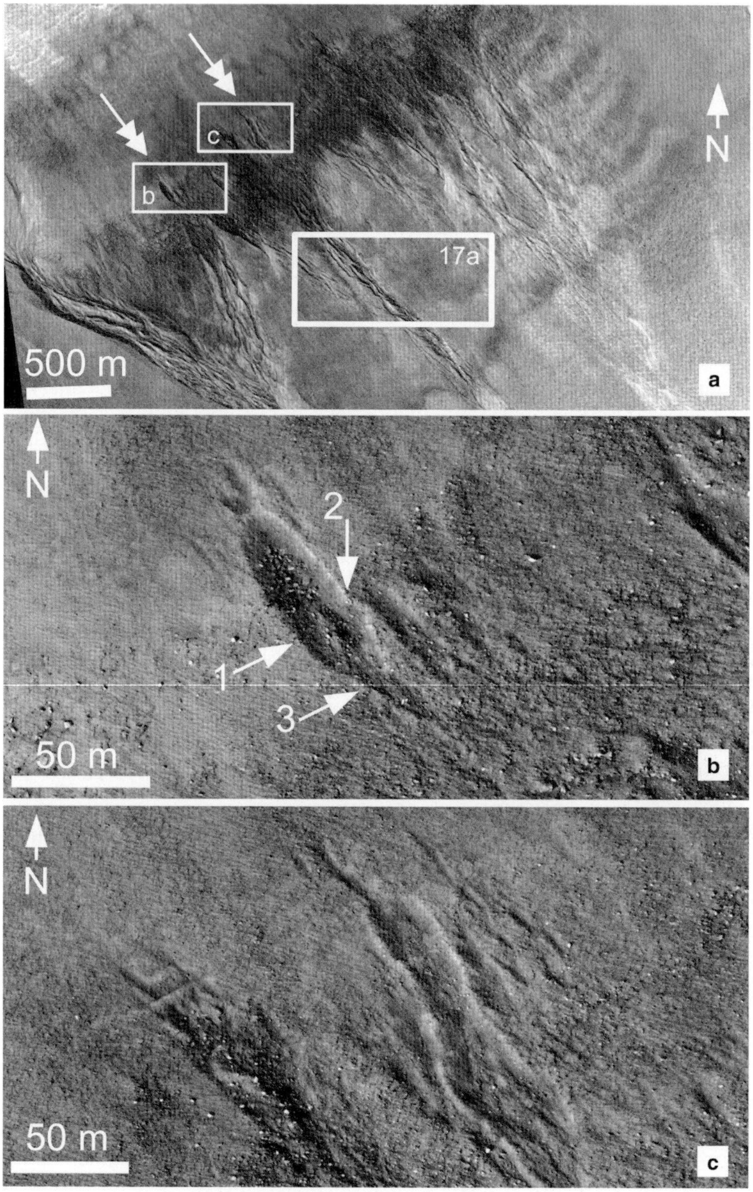

Fig. 16. (**a**) Downslope-elongated pits with RTS-like morphology are the source of headward-extending braided gullies. Boxes show the context of (b), (c) and (a). (**b**) A slope-elongated RTS-like pit exhibits an axial debris island isolated by marginal channels (arrows 1 and 2). The debris island lies just above the efflux point of the form (arrow 3), beyond which channel incision is evident. Also apparent is incipient coalescence with both a higher pit and an adjacent slope-elongated pit, the interfluve between the latter and the main form already showing lowering and, therefore, the early stages of piracy. (**c**) Gully sourced in a narrow slope-elongated pit is bounded by bright levees, consistent with breakouts in a viscous debris flow. Parts of HiRISE image PSP_007666_2400. Image credit: NASA/JPL/UofA.

variable rates of downslope debris movement. Solifluction is often most clearly manifested by the development of 'stone-banked' terraces or lobes. Benedict (1970) observed that rapid downslope movement of fine debris relative to coarse lobate material is characteristic of a solifluction environment. Thus, the well-developed vertical sorting typical of clastic lobes and terraces, together with

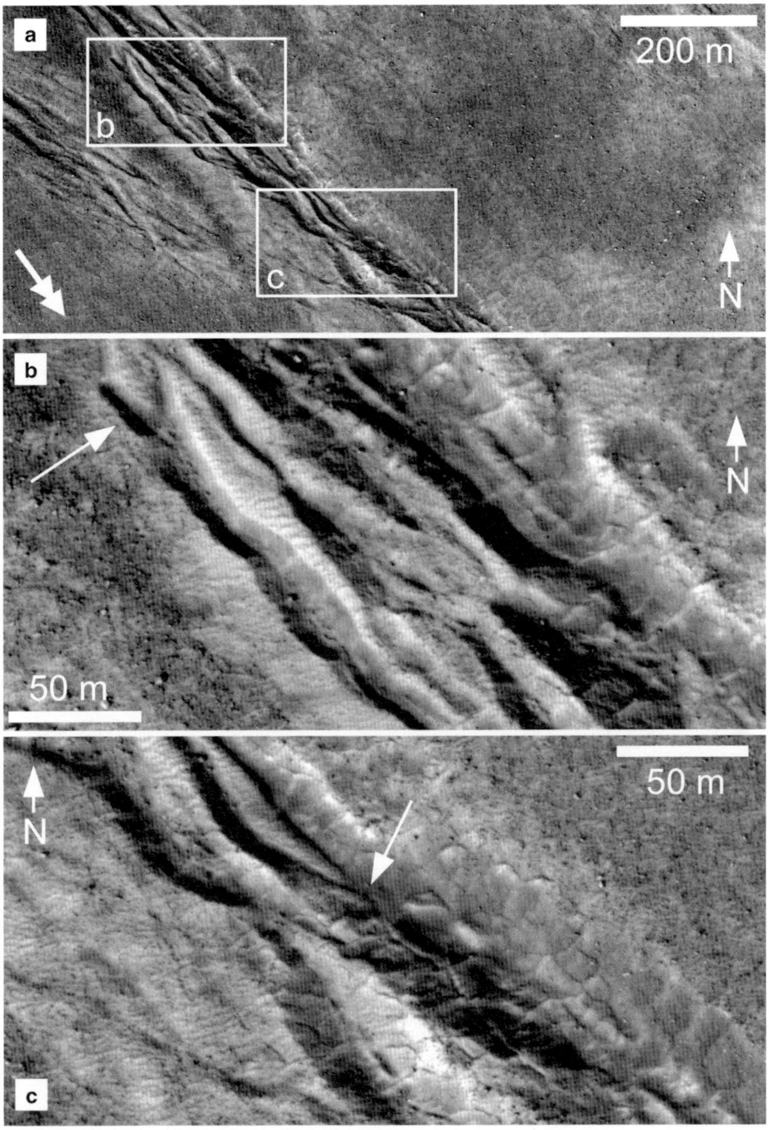

Fig. 17. (**a**) Shows the context of a gully system that has incised a suite of clastic lobes. (**b**) The gullies shown here evidently are thermokarst forms arising from headward extension, and polycyclic overdeepening of existing channels by the degradation of epigenetic pits. The example arrowed shows that liquid was released during this process and overdeepened an existing channel thalweg. (**c**) Channel evolution also occurs by flows exploiting and widening longitudinal cracks in the surface (arrowed). Parts of HiRISE image PSP_007666_2400. Image credit: NASA/JPL/UofA.

their best development on steep upper slopes, indicates that these forms develop only where frost creep is the dominant movement process. On gentler, moister, lower slopes, gelifluction involving liquefaction dominates. Consequently, the relative velocity of fine-textured regolith increases downslope, and stone-banked lobes and terraces become increasingly unlikely to develop down a slope profile.

Benedict (1970) also observed that stone stripes evolved into stone-banked lobes where clasts decelerated and spread laterally, and clast concentration increased as the slope became transport limited. Similarly, stone-banked terraces formed at the

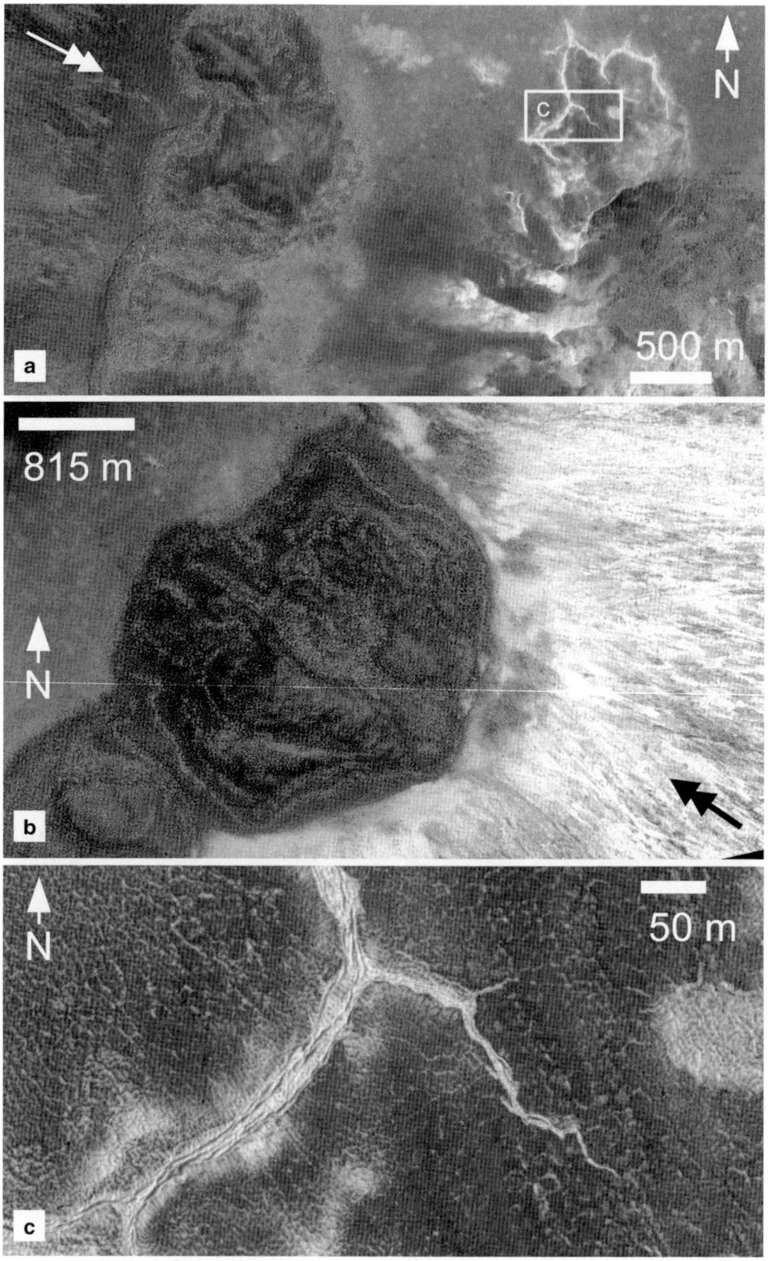

Fig. 18. (**a**) The termination of the interior crater slope at a series of cirque-shaped indentations in polygonized ground, backworn by the collapse of individual polygonal blocks, is strongly reminiscent of shoreline retreat by thermo-abrasional collapse in terrestrial thermokarst environments that border water bodies. (**b**) Possible fountain flows beyond distal fans on the crater bottom may reflect the injection of flowing liquid sourced on the crater slopes into the liquefied debris lining the crater bottom. (**c**) Vein-like fissures with possible internal vertical foliations may be indicative of polycyclic ground intrusion by surface water. Analogous terrestrial forms associated with other indicators of freeze–thaw reflect ice growth in thermal contraction cracks, and are a precursor to ice wedges and fully polygonized ground. Parts of HiRISE image PSP_006955_2495. Image credit: NASA/JPL/UofA.

downslope margin of decelerating blockfields or very closely spaced stone stripes. Although stone-banked lobes and terraces largely retain the textural characteristics of their clast-dominated sources, continued frost sorting accentuates clast concentration, and lobes and terraces develop a bimodal morphology characterized by steep clastic risers bounding gently sloping texturally fine treads (Ballantyne & Harris 1994).

Braided fluvioperiglacial channels and gullies often form in association with patterned ground and solifluction landforms, and are the dominant fluviatile landform of periglacial areas (French 1996). Terrestrial fluvioperiglacial braiding is associated with intermittent, often catastrophic, nival or proglacial water discharge and high sediment loading. However, steep gullies and braided fans are frequently a response to singular triggers, such as extreme rainfall (e.g. Øygarden 2003), snowmelt (e.g. Theakstone 1982), thermokarst collapse of tunnels developed in ice-wedge networks and ice-rich permafrost (e.g. Fortier *et al.* 2007) or catastrophic failure of destabilized saturated slope deposits (e.g. Bovis & Jakob 2000).

Braided gullies are common on Mars (Levy *et al.* 2009) and probably result from the melting of ice-rich regolith (e.g. Costard *et al.* 2002) and/or seepage or outflow from subsurface liquid reservoirs (e.g. Malin & Edgett 2000). Recent studies of morphology favour a melting–insolation control process (Levy *et al.* 2009). Balme *et al.* (2006) and Dickson *et al.* (2007) concluded from the distribution and orientation of gullies that melting and insolation played a dominant role, and linked Martian gullies to periglacial–permafrost landform assemblages including polygonally patterned ground (Levy *et al.* 2008). The clear association of the gullies presented in Figures 15–17 with thermokarst pits confirms that melting is the dominant driver of gully formation at high northern latitudes on Mars. However, not only are these landforms indicative of the action of thaw and flowing liquid but the incision of solifluction lobes by thermokarst gullies demonstrates that the thaw has been responsible for polycyclic morphogenesis.

Levy *et al.* (2009) concluded that active layer processes are unlikely to have been significant morphogenetic agencies on Mars. However, the regional ubiquity in the Martian high north of landforms produced by cryoturbation involving cyclic thaw and associated closely with landforms produced by surface flows of liquid – some probably sourced from thaw fluids seeping from solifluction lobes in addition to sources in thermokarst pits – suggests that active layer processes are instead one of the dominant modes of surface modification. This raises the question of how much liquid would be required to liquefy the slopes encountered in this survey and whether sufficient liquid could have been available.

Liquefaction of regolith in periglacial environments depends both on the production of thaw liquid through the melting of ground ice and on thaw consolidation of the regolith. It is cyclic disturbance of the regolith during thaw consolidation that triggers liquefaction through the concentration of regolith into smaller volumes and the consequent increase in pore-water pressure, in the same way that earthquakes trigger fluidization of previously stable hillsides. The amount of liquid (%wt) required to cause liquefaction is dictated by the liquid limit (LL) of the regolith; sands and silts have LLs ranging from 15 to 20% and from 30 to 40%, respectively. Even a LL of 15% might seem a large requirement in the Martian environment, but if permafrost underlies the regolith and retards downward percolation/permeation of water only a relatively thin regolith layer is left to liquefy, and this could occur relatively easily. Gelifluction occurs when the moisture content of regolith is very close to its liquid limit (i.e. at a Liquid Index just greater than 1). A cubic metre (1 m^3) of sand, typically weighing 1700 kg (or experiencing a gravitational force of *c.* 6671N due to Mars' 0.4 g) at its LL, will have a water mass of approximately 255 kg, equivalent to 255 l m^{-3}. On a slope profile, with each cubic metre contributing even a small amount of thaw liquid, only short slope runs would be required for the liquid limit of the regolith to be reached. For example, with a water content of only 2% or 20 l m^{-3}, as recently determined for a sublimed lag at the *Phoenix Lander* site and, therefore, an underestimate of the prevailing water content (Smith *et al.* 2009) estimated to be several tens of per cent (Feldman *et al.* 2004), and a regolith depth of only 0.05 m above permafrost (as at the *Phoenix* landing site), only 255 linear metres of upper slope would be required to supply enough water to exceed the LL of the rest of the slope regolith (assuming no losses downslope) and for gelifluction to become possible.

Although liquid water is at present only metastable on Mars (Hecht 2002), near-surface ground ice clearly comprises a significant potential reservoir, particularly if perchlorate solutions are a component of the Martian cryosphere. This raises the question of whether Mars' recent climate has ever been such that sublimation of water ice has been suppressed in favour of melt. The landform assemblages described in this paper indicate that this has been the case but may be indicative of environmental controls operating at two spatial scales. First, regional thaw could be initiated by significant increases in the temperature and vapour pressure of the atmosphere, owing to the destabilization of the north polar cap, during periods of high obliquity

(Smith *et al.* 2009). Moreover, given that perchlorates are present in the regolith (Hecht *et al.* 2009) and that some of these are eutectic within the range of present temperatures, it is possible that a significant transition from sublimation to regional thaw of ground ice could occur as a consequence of a more moderate climate change than previously envisaged. Secondly, because a minor degree of ground-ice thaw is a characteristic of the present high-latitude regolith (Mellon *et al.* 2008) and because liquefaction is largely controlled by regolith consolidation, it may be that sublimation consolidation could enhance liquefaction on very small to local scales, particularly in regolith with high perchlorate concentrations.

Conclusions

Given the abundance of near-surface ice in the northern plains of Mars, and the short slope runs required to provide sufficient thaw water for liquefaction to occur under favourable climatic conditions, it is not surprising that landforms associated with freeze–thaw cryoturbation, liquefaction and surface liquid flows are widespread at high northern latitudes on Mars. Table 2 lists the dominant landforms observed in 16 HiRISE images surveyed but not described here in detail. Together with the eight images analysed in detail for this paper (Fig. 1 shows the locations of both image categories), these show that landforms associated with thaw and, therefore, the activity of surface and near-surface liquids are probably ubiquitous; in only two images of these 24 were morphological landforms indicative of freeze–thaw either equivocal (PSP_007508_2440) or not evident (PSP_007573 _2435). At the very least, therefore, processes driven by freeze–thaw appear to be the dominant mass-wasting process of crater slopes, especially interior slopes. However, this association probably reflects the greater abundance of coarse, near-surface clasts on interior crater slopes compared with exterior slopes and surrounding plains. As Figure 2 indicates, clastic islands – on Earth a self-organized pattern associated with freeze–thaw cryoturbation of regolith with low clast concentrations (Kessler & Werner 2003) – occur on surfaces away from the influence of craters. Moreover, despite the exquisite resolution of HiRISE, the presence of patterned ground manifested by albedo but associated with diffuse overlying or neighbouring clastic patterns (Fig. 4a, b and, possibly, Fig. 2), suggests that sub-metre sorted patterned ground is also present beyond the resolution of imaging instruments currently in orbit. Hence, landforms associated with thaw are apparent or can reasonably be inferred to exist throughout the high northern latitudes of Mars in many morphological and topographical settings. This demonstrates that thaw has played not only a more important role than previously envisaged but also a dominant role in shaping the geomorphology of the high latitudes of Mars.

Perhaps of greatest significance, though, is the presence of crater-bound landform assemblages indicative of cryoturbation, frost creep and gelifluction (Figs 2–11) that have been superceded by landforms indicative of flows of surface liquids, surface liquid accumulations and possible ground-ice regrowth (Figs 12–18). These landform assemblages suggest that many high-latitude crater bottoms have been wet environments and this raises their importance as targets for future landers that will be sent to Mars to search for evidence of life.

References

ARE, F. E. 1988. Thermal abrasion of Sea Coasts (Part 1). *Polar Geography and Geology*, **12**, 1–86.

BALLANTYNE, C. K. & HARRIS, C. 1994. *The Periglaciation of Great Britain.* Cambridge University Press, Cambridge.

BALLANTYNE, C. K. & MATTHEWS, J. 1982. The development of sorted circles on recently deglaciated terrain, Jotunheimen, Norway. *Arctic and Alpine Research* **14**, 341–354.

BALME, M. R. & GALLAGHER, C. 2009. An equatorial periglacial landscape on Mars. *Earth and Planetary Science Letters*, **285**, 1–15.

BALME, M. R., GALLAGHER, C., PAGE, D. P., MURRAY, J. B. & MULLER, J.-P. 2009. Sorted stone circles in Elysium Planitia, Mars: implications for recent martian climate. *Icarus*, **200**, 30–38.

BALME, M. R., MANGOLD, N. *ET AL.* 2006. Orientation and distribution of recent gullies in the southern hemisphere of Mars: observations from HRSC/MEX and MOC/MGS data. *Journal of Geophysical Research*, **111**, E05001, doi: 10.1029/2005JE002607.

BENEDICT, J. B. 1970. Downslope soil movement in a Colorado Alpine region: rates, processes and climatic significance. *Arctic and Alpine Research*, **2**, 165.

BOVIS, M. J. & JAKOB, M. 2000. The July 29, 1998, debris flow and landslide dam at Capricorn Creek, Mount Meager Volcanic Complex, southern Coast Mountains, British Columbia. *Canadian Journal of Earth Sciences*, **37**, 1321–1334.

BOYNTON, W. V., MING, D. W. *ET AL.* 2009. Evidence for calcium carbonate at the Mars Phoenix Landing Site. *Science*, **305**, 61–64.

BURR, D. M., SOARE, R. J., WAN BUN TSEUNG, J.-M. & EMERY, J. P. 2005. Young (late Amazonian), near-surface, ground ice features near the equator, Athabasca Valles, Mars. *Icarus*, **178**, 56–73.

BYRNE, S. & INGERSOLL, A. P. 2003. A sublimation model for Martian south polar ice features. *Science*, **14**, 1051, 1053.

CORTE, A. E. 1962*a*. Vertical migration of particles in front of a moving freezing plane. *Journal of Geophysical Research*, **67**, 1085–1090.

CORTE, A. E. 1962*b*. *The Frost Behaviour of Soils: Laboratory and Field Data for a New Concept. I: Vertical Sorting*. United States Army Corps of Engineers, Cold Regions Research and Engineering Laboratory, Research Report, **85**.

COSTARD, F., FORGET, F., MANGOLD, N. & PEULVAST, J. P. 2002. Formation of recent martian debris flows by melting of near-surface ground ice at high obliquity. *Science*, **295**, 110–113.

CZUDEK, T. & DEMEK, J. 1970. Thermokarst in Siberia and its influence on the development of lowland relief. *Quaternary Research*, **1**, 103–120.

DICKSON, J. L., HEAD, J. W. & KRESLAVSKY, M. A. 2007. Martian gullies in the southern mid-latitudes of Mars: evidence for climate-controlled formation of young fluvial features based upon local and global topography. *Icarus*, **188**, 315–323.

DUNDAS, C. M., MELLON, M. T., MCEWEN, A. S., LEFORT, A., KESZTHELYI, L. P. & THOMAS, N. 2008. HiRISE observations of fractured mounds: possible Martian pingos. *Journal of Geophysical Research*, **35**, L04201.

FELDMAN, W. C., PRETTYMAN, T. H. *ET AL*. 2004. The global distribution of near-surface hydrogen on Mars. *Journal of Geophysical Research*, 109.

FORTIER, D., ALLARD, M. & SHUR, Y. 2007. Observation of rapid drainage system development by thermal erosion of ice wedges on Bylot Island, Canadian Arctic Archipelago. *Permafrost and Periglacial Process*, **18**, 229–243.

FRENCH, H. M. 1996. *The Periglacial Environment*, 2nd edn. Addison Wesley Longman, Harlow.

HARRIS, C. & MATTHEWS, J. A. 1984. Some observations on Boulder-Cored Frost Boils. *The Geographical Journal*, **150**, 63–73.

HEAD, J. W., MARCHANT, D. R & KRESLAVSKY, M. A. 2008. Formation of gullies on Mars: Link to recent climate history and insolation microenvironments implicate surface water flow origin. *Proceedings of the National Academy of Science USA*, **105**, 13 258–13 263.

HECHT, M. H. 2002. Metastability of water on Mars. *Icarus*, **156**, 373–386, doi: 10.1006/icar.2001.6794.

HECHT, M. H., KOUNAVES, S. P. *ET AL*. 2009. Detection of perchlorate and the soluble chemistry of martian soil at the Phoenix Lander Site. *Science*, **325**, 64–67.

HELDMANN, J. L., TOON, O. B., POLLARD, W. H., MELLON, M. T., PITLICK, J., MCKAY, C. P. & ANDERSEN, D. T. 2005. Formation of martian gullies by the action of liquid water flowing under current martian environmental conditions. *Journal of Geophysical Research*, 110, E05004, doi: 10.1029/2004JE002261.

HOLDGATE, M. W., ALLEN, S. E. & CHAMBERS, M. J. G. 1967. A preliminary investigation of the soils of Signy Island, South Orkney Islands. *British Antarctic Survey Bulletin*, **12**, 53–71.

KESSLER, M. A. & WERNER, B. T. 2003. Self-organization of sorted patterned ground. *Science*, **299**, 380.

LANTUIT, H. & POLLARD, W. H. 2005. Fifty years of coastal erosion and retrogressive thaw slump activity on Herschel Island, southern Beaufort Sea, Yukon Territory, Canada. *Geomorphology.*, **95**, 84–102.

LASKAR, J., CORREIA, A. C. M., GASTINEAU, M., JOUTEL, F., LEVRARD, B. & ROBUTEL, P. 2004. Long term evolution and chaotic diffusion of the insolation quantities of Mars. *Icarus*, **170**, 343–364.

LEVY, J. S., HEAD, J. W., MARCHANT, D. R., DICKSON, J. L. & MORGAN, G. A. 2009. Geologically recent gully–polygon relationships on Mars: Insights from the Antarctic Dry Valleys on the roles of permafrost, microclimates, and water sources for surface flow. *Icarus*, **201**, 113–126.

LEVY, J. S., HEAD, J. W., MARCHANT, D. R. & KOWALEWSKI, D. E. 2008. Identification of sublimation-type thermal contraction crack polygons at the proposed NASA Phoenix landing site: Implications for substrate properties and climate-driven morphological evolution. *Geophysical Research Letters*, **35**, L04202, doi: 10.1029/2007GL032813.

LEVY, J. S., MARCHANT, M. R. & HEAD, J. W. 2006. Distribution and origin of patterned ground on Mullins Valley debris covered glacier, Antarctica: the roles of ice flow and sublimation. *Antarctic Science*, **18**, 385–397.

MALIN, M. C. & EDGETT, K. S. 2000. Evidence for recent groundwater seepage and surface runoff on Mars. *Science*, **288**, 2330–2335.

MALIN, M. C., CAPLINGER, M. A. & DAVIS, S. D. 2001. Observational evidence for an active surface reservoir of solid carbon dioxide on Mars. *Science*, **294**, 2146–2148.

MANGOLD, N. 2005. High latitude patterned grounds on Mars: Classification, distribution and climatic control. *Icarus*, **174**, 336–359.

MARCHANT, D. R., LEWIS, A., PHILLIPS, W. C., MOORE, E. J., SOUCHEZ, R. & LANDIS, G. A. 2002. Formation of patterned-ground and sublimation till over Miocene glacier ice in Beacon Valley, Antarctica. *Geological Society of America Bulletin*, **114**, 718–730.

MELLON, M. T. & PHILLIPS, R. J. 2001. Recent gullies on Mars and the source of liquid water. *Journal of Geophysical Research*, **106**, 23,165–23,180.

MELLON, M. T., ARVIDSON, R. E., MARLOW, J. J., PHILLIPS, R. J. & ASPHAUG, E. 2008. Periglacial landforms at the Phoenix landing site and the northern plains of Mars. *Journal of Geophysical Research*, 113.

ØYGARDEN, L. 2003. Rill and gully development during an extreme winter runoff event in Norway. *Catena*, **50**, 217–242.

PAGE, D. P. 2007. Recent low-latitude freeze–thaw on Mars. *Icarus*, **189**, 83–117.

PAGE, D. P. & MURRAY, J. B. 2006. Stratigraphical and morphological evidence for pingo genesis in the Cerberus plains. *Icarus*, **183**, 46–54.

REISS, D., HIESINGER, H., HAUBER, E. & GWINNER, K. 2009. Regional differences in gully occurrence on Mars: A comparison between the Hale and Bond craters. *In*: PIO ROSSI, A. & WITASSE, O. (eds) *European Mars Science and Exploration Conference (EMSEC) – Mars Express & ExoMars, European Mars Science and Exploration Conference*, July 2009. *Planetary and Space Science*, **57**, 958–974.

RENNO, N. O., BOS, B. J. *ET AL*. 2009. Physical and thermodynamical evidence for liquid water on Mars? *In*: *Proceedings of the 39th Annual Lunar and Planetary Science Conference, March 23–27, 2009, The Woodlands, Texas*. Lunar and Planetary Institute, Houston, TX, 1440.

Seibert, N. M. & Kargel, J. S. 2001. Small-scale Martian polygonal terrain: implications for liquid surface water. *Geophysical Research Leters*, **28**, 899–902.

Smith, P. H., Tamppari, L. K. *et al.* 2009. H_2O at the Phoenix Landing Site. *Science*, **325**, 58–61.

Soare, R. J., Burr, D. M. & Wan Bun Tseung, J. M. 2005. Pingos and a possible periglacial landscape in northwest Utopia Planitia, Mars. *Icarus*, **174**, 373–382.

Soare, R. J., Osinski, G. R. & Roehm, C. L. 2008. Thermokarst lakes and ponds on Mars in the very recent (late Amazonian) past. *Earth and Planetary Science Letters*, **272**, 382–393.

Theakstone, W. H. 1982. Sediment fans and sediment flows generated by snowmelt: observations at Austerdalsisen, Norway. *Journal of Geology*, **90**, 583–588.

Thomas, P. C., Malin, M. C. *et al.* 2000. North–south geological differences between the residual polar caps on Mars. *Nature*, **404**, 161–164.

Werner, B. T. & Hallet, B. 1993. Numerical simulation of self-organised stone stripes. *Nature*, **361**, 142–145.

Landscape evolution in Martian mid-latitude regions: insights from analogous periglacial landforms in Svalbard

E. HAUBER[1]*, D. REISS[2], M. ULRICH[3], F. PREUSKER[1], F. TRAUTHAN[1], M. ZANETTI[2], H. HIESINGER[2], R. JAUMANN[1], L. JOHANSSON[4], A. JOHNSSON[4], S. VAN GASSELT[5] & M. OLVMO[4]

[1]*Institut für Planetenforschung, DLR, Rutherfordstrasse 2, 12489 Berlin, Germany*

[2]*Institut für Planetologie, Westfälische Wilhelms-Universität, 48149 Münster, Germany*

[3]*Alfred-Wegener-Institut, 14473 Potsdam, Germany*

[4]*Department of Earth Sciences, University of Gothenburg, Box 460, SE-405 30 Göteborg, Sweden*

[5]*Institut für Geologische Wissenschaften, Freie Universität Berlin, Malteserstrasse 74-100, 12249 Berlin, Germany*

**Corresponding author (e-mail: Ernst.Hauber@dlr.de)*

Abstract: Periglacial landforms on Spitsbergen (Svalbard, Norway) are morphologically similar to landforms on Mars that are probably related to the past and/or present existence of ice at or near the surface. Many of these landforms, such as gullies, debris-flow fans, polygonal terrain, fractured mounds and rock-glacier-like features, are observed in close spatial proximity in mid-latitude craters on Mars. On Svalbard, analogous landforms occur in strikingly similar proximity, which makes them useful study cases to infer the spatial and chronological evolution of Martian cold-climate surface processes. The analysis of the morphological inventory of analogous landforms on Svalbard and Mars allows the processes operating on Mars to be constrained. Different qualitative scenarios of landscape evolution on Mars help to better understand the action of periglacial processes on Mars in the recent past.

Many young landforms on Mars that were probably formed by exogenic processes show a latitude-dependent geographical distribution. They include surface mantling (Kreslavsky & Head 2000; Mustard *et al.* 2001; Morgenstern *et al.* 2007), lobate debris aprons, lineated valley fill and concentric crater fill (e.g. Squyres 1978), viscous flow features (Milliken *et al.* 2003), gullies (Balme *et al.* 2006; Kneissl *et al.* 2010) and patterned ground (Mangold 2005). Other landforms, such as pedestal craters, seem to indicate a preservation of near-surface ice and are also latitude-dependent (Kadish *et al.* 2009). Collectively, these landforms are hypothesized to represent the surface records of Martian ice ages (e.g. Head *et al.* 2003) that were induced by astronomical forcing (Laskar *et al.* 2004) and associated climate changes (Toon *et al.* 1980; Jakosky & Carr 1985; Mischna *et al.* 2003; Forget *et al.* 2006; Schorghofer 2007). Previous authors often considered only one of such feature classes in isolation (e.g. gullies), without taking into account the geomorphological context. It was not until the recent advent of high-resolution data from orbit and the *in situ* investigation of Martian high-latitude terrain by the *Phoenix Lander* that a more integrated view of diverse landforms into a landscape evolution model were allowed to develop (e.g. Balme & Gallagher 2009; Levy *et al.* 2009*a*). A more comprehensive investigation of the full assemblage of landforms by means of landscape analysis, however, has the potential to reduce the ambiguity in interpreting landforms and to reveal the evolution of the climatic environment in more detail. The phenomenon of equifinality (i.e. similar-looking landforms resulting from diverse processes) is particularly problematic in planetary geomorphology, where the morphology as inferred from remote-sensing data such as images and digital elevation models (DEM) is the only observable component. An instructive example is the case of pitted mounds on Mars, which have been interpreted in the past as modified impact craters, rootless cones, cinder cones and pingos. In some of the studies that favoured pingos, the interpretations were based on poor evidence and attracted criticism from terrestrial permafrost researchers (Humlum & Christiansen 2008). Here we present permafrost landforms from Svalbard

From: Balme, M. R., Bargery, A. S., Gallagher, C. J. & Gupta, S. (eds) *Martian Geomorphology*. Geological Society, London, Special Publications, **356**, 111–131.
DOI: 10.1144/SP356.7 0305-8719/11/$15.00

(Norway) as useful terrestrial analogues for the suite of possible periglacial landforms that are typically found at mid-latitudes on Mars. We build on our previous investigations of gullies and fans (Hauber *et al.* 2009), and include a number of classical periglacial landforms (patterned ground, rock glaciers, pingos) that all have close morphological analogues on Mars. Based on this comparison, we propose several evolutionary scenarios that could help to develop a better understanding of the sequential formation of the Martian landforms.

Permafrost and periglacial features on Mars and Svalbard

Mars may be regarded as a permafrost planet, following the definition of permafrost given by van Everdingen (2005): 'Ground (soil or rock [...]) that remains at or below 0 °C for at least two consecutive years, regardless of the water content'. In fact, the shallow subsurface of Mars probably experienced temperatures that were continuously below 0 °C for most of its history (e.g. Shuster & Weiss 2005). In the current Martian climate, ground ice is thought to be stable only at higher latitudes (e.g. Leighton & Murray 1966; Smoluchowski 1968; Fanale *et al.* 1986; Mellon & Jakosky 1993) and, indeed, the *Phoenix* mission has provided unambiguous evidence for very shallow and rather pure ground ice at a latitude of 68.2°N (Smith *et al.* 2009). The latitudinal range of ice stability is, however, a function of the planet's obliquity (i.e. the tilt of the rotational axis). Mars' obliquity is assumed to vary widely (Ward 1973; Touma & Wisdom 1993), and at an obliquity exceeding 32° (today *c.* 25°) ground ice becomes globally stable (Mellon & Jakosky 1995). An obliquity exceeding about 27° is required for ice to be stable at latitudes of 30° and higher (Mellon & Jakosky 1995, their fig. 10d). Other factors that affect the stability of ground ice are geographical variability, soil properties, rocks and local slopes (see Mellon *et al.* 2009 and references therein). The large and frequent oscillations of Mars' obliquity (an obliquity cycle spans 117 000 years: Laskar *et al.* 2004) should have a significant influence on the volatile distribution on the surface (Jakosky *et al.* 1995), and climate modelling using global circulation models (GCM) confirms this view (Levrard *et al.* 2004; Forget *et al.* 2006; Madeleine *et al.* 2009). It appears likely that water ice was frequently driven from the poles towards lower latitudes during periods of higher obliquities, when the polar regions received more incoming solar energy (Forget *et al.* 2006). In contrast, water ice was redistributed towards higher latitudes during the following periods of lower obliquities (Levrard *et al.* 2004). Ground ice can thus be expected to be a significant factor in Martian landscape evolution. Recent observations, indeed, showed that near-surface water ice is present even in mid- and low-latitude regions (Holt *et al.* 2008; Byrne *et al.* 2009; Vincendon *et al.* 2010*a*, *b*), in contrast to expectations from theoretical modelling (see above).

To complement theoretical modelling, comparisons with terrestrial analogues are mandatory to constrain the action of periglacial processes and the corresponding landscape evolution on Mars. Present-day Mars is cold and dry, so surface processes acting in terrestrial cold deserts should be considered as useful analogues. The closest cold-climate analogue to Mars on Earth are the Antarctic Dry Valleys (Anderson *et al.* 1972; Marchant & Head 2007, 2010), a polar desert environment with exceptionally cold and dry conditions (Doran *et al.* 2002) and correspondingly small active layer depth (Bockheim *et al.* 2007). Other polar regions also display morphological analogues to Mars, however, and the archipelago of Svalbard and its largest island, Spitsbergen (Fig. 1a), offer a diverse inventory of periglacial landforms in close spatial proximity. Terrain phenomena such as pingos, ice-wedge polygons and rock glaciers are widespread, especially in the dry central regions of Spitsbergen. Periglacial features such as solifluction lobes occur primarily in the more humid western regions. Various forms of patterned ground, such as stone circles and stripes, are widespread and well developed (see Åkerman 1987 for a review of periglacial landforms of Svalbard). Examples of periglacial morphologies are closely located to the settlements of Longyearbyen and Ny Ålesund on the main island of Spitsbergen, making them very useful morphological analogues to Martian cold-climate landforms. Major controls on permafrost aggradation are wind, snow and avalanches (Humlum 2005). A particularly interesting aspect of permafrost on Svalbard is its interaction with glaciers (Etzelmüller & Hagen 2005) because such interaction is often neglected in the literature (Haeberli 2005) but may be highly important on Mars.

Data

Martian surface features were analysed using high-resolution images of the CTX (Context Camera) and HiRISE (High Resolution Imaging Science Experiment) cameras, which have spatial resolutions of 5–6 m per pixel and approximately 30 cm per pixel, respectively. An airborne version of the HRSC (High Resolution Stereo Camera) was used for the acquisition of stereo and colour

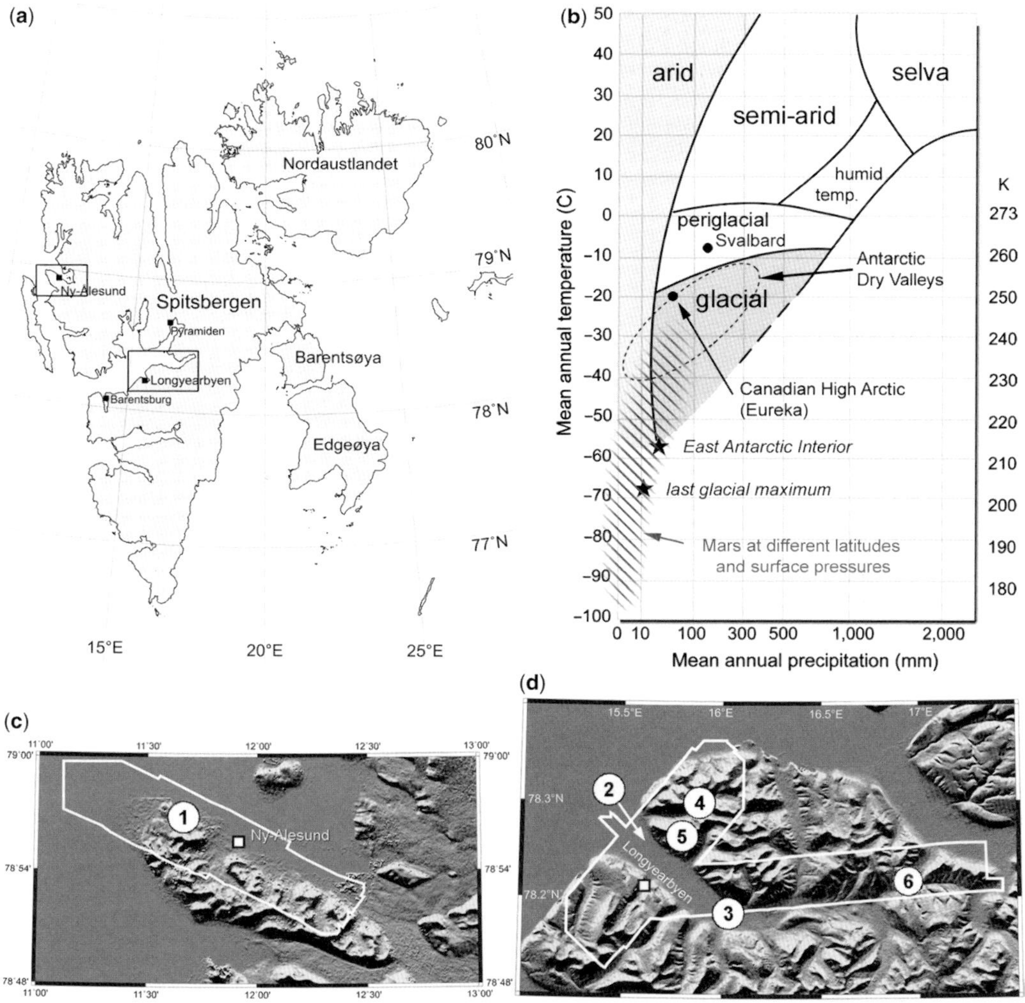

Fig. 1. Location and climate of study areas on Svalbard. (**a**) Map of Svalbard with study areas (boxes, see c and d). (**b**) Climate zones and morphogenetic regions on Earth; modified from Baker (2001) and Head & Marchant (2007). The climatic conditions on Mars (present and inferred past) are indicated by the hatched area. (**c**) Study area on the Brøgger Peninsula (shaded elevation model derived from ASTER data). (**d**) Study area in Adventdalen (shaded elevation model derived from Advanced Spaceborne Thermal Emission and Reflection Radiometer (ASTER) data). Numbers in (c) and (d) mark the geographical locations mentioned in the text: 1, Stuphallet; 2, Adventfjord; 3, Adventdalen; 4, Hannaskogdalen; 5, Hiorthfjellet; 6, Eskerdalen. Image credit: Univ. Münster/NASA/GSFC/METI/ERSDAC/JAROS/US-Japan ASTER Science Team; see prelim viii for acronym definitions.

images of Spitsbergen. HRSC-AX is a multi-sensor push broom instrument with nine CCD (charge-coupled device) line sensors mounted in parallel. It simultaneously obtains high-resolution stereo, multi-colour and multi-phase images. The particular value of HRSC-AX is the stereo capability, which allows it to systematically produce high-resolution digital elevation models (DEM) with grid sizes of between 50 cm and 1 m (Wewel *et al.* 2000; Scholten & Gwinner 2004; Gwinner *et al.* 2005, 2006, 2009, 2010; Scholten *et al.* 2005). The HRSC-AX flight campaign in July–August 2008 covered a total of seven regions in Svalbard: (i) Longyearbyen and the surrounding area of Adventfjorden (all place names on Svalbard are as given as in the topographic map series, scale 1:100 000, published by the Norsk Polarinstitutt, Tromsø, Sheets C9 and A7); (ii) large parts of Adventdalen; (iii) large parts of the Brøggerhalvøya (halvøya means peninsula) in western Spitsbergen; (iv) the Bockfjorden area in northern Spitsbergen; (v) the NE shore of the Palanderbukta and the margin of the adjacent ice cap in

Nordaustlandet; (vi) an area on Prins Karls Forland; and (vii) the area of the abandoned Russian mining settlement of Pyramiden together with the nearby Ebbedalen. The landforms discussed in this study are located on the Brøgger Peninsula and in Adventdalen and its vicinity (Fig. 1a). In two field campaigns in 2008 and 2009, both of the areas covered by HRSC-AX were visited.

The Svalbard climate

The present climate of Svalbard is arctic (Fig. 1b). The mean annual air temperature at the airport in Longyearbyen, which is located only a few kilometres from the study area of Adventdalen, ranges between about −6 °C at sea level and −15 °C in the high mountains (Hanssen-Bauer & Førland 1998). Annual precipitation is low and reaches only about 180 mm in central Spitsbergen (Table 1). The central part of Spitsbergen can, therefore, be considered to be a polar (semi)desert, which is defined as an area with an annual precipitation of less than 250 mm and a mean temperature during the warmest month of less than 10 °C (Walker 1997). About 60% of Svalbard is covered by glaciers and ice caps, and relatively small glaciers and ice caps are situated on many massifs and valleys around Adventdalen. The unglaciated part of Svalbard is characterized by continuous permafrost, which has a thickness of 10–40 m in coastal regions, about 100 m in the major valleys and more than 450 m in the highlands (Liestøl 1976; Isaksen *et al.* 2000; Sollid *et al.* 2000).

Morphological comparisons between Mars and Svalbard

Many possible glacial and periglacial landforms are located in mid-latitude impact craters on Mars. This specific geological setting provides ideal study cases because there is high relief present at the crater walls, and the opportunity to study the effects of insolation variations because craters are axisymmetric features and their inner walls have an azimuthal range of the entire 360°. It has been found by many previous researchers that the pole-facing walls of impact craters are particularly prone to be shaped by glacial and periglacial processes (e.g. Dickson *et al.* 2007). In this section the inventory of such landforms is briefly reviewed and compared to analogous landforms on Svalbard. We note here that all of these features have been found in craters on Mars, sometimes several of them in the same crater but, so far, no crater has been found that hosts all of them together.

Martian landforms

Many landforms on Mars that are morphologically similar to terrestrial glacial and periglacial landforms occur in the middle latitudes, between about 30° and 60° (Fig. 2). They are situated along the high-relief belt of the Martian dichotomy boundary and other regions of high relief (e.g. Pierce & Crown 2003; Chuang & Crown 2005; Head *et al.* 2006; van Gasselt *et al.* 2010), as well as in flat-lying regions such as Utopia Planitia (Soare *et al.* 2005; Morgenstern *et al.* 2007; Lefort *et al.* 2009). A

Table 1. *Climate at Svalbard Airport. For the series of observed and modelled annual and seasonal temperature means and precipitation sums from 1912 to 1993 the following values are given: mean, standard deviation, absolute minimum and absolute maximum. SD, standard deviation; Corr., correlation coefficient between observed and modelled temperature and precipitation series (data from Hanssen-Bauer & Førland 1998). For comparison, the mean annual air temperature at the floor of the Dry Valleys in Antarctica ranges from −14.8 to −30 °C, and the mean annual precipitation is ≪100 mm, but can be as low as 13 mm (Doran* et al. *2002; Campbell & Claridge 2004)*

Season		Mean	SD	Min.	Max.	Mean	SD	Min.	Max.	Corr.
		Observed *T* (°C)				**Modelled *T* (°C)**				
Year		−6.3	1.7	−12.2	−3.1	−6.4	1.0	−8.9	−4.0	0.61
Winter	(DJF)	−14.0	3.6	−23.2	−7.6	−14.1	2.4	−19.1	−9.1	0.62
Spring	(MAM)	−10.8	2.4	−19.3	−6.7	−10.8	1.7	−15.2	−7.5	0.58
Summer	(JJA)	4.3	0.7	2.5	6.1	4.2	0.5	3.2	5.4	0.54
Autumn	(SON)	−4.8	2.0	−11.3	−1.3	−4.9	1.5	−8.7	−1.8	0.66
		Observed *P* (mm)				**Modelled *P* (mm)**				
Year		180.7	49.8	86.4	317.0	178.7	33.5	93.5	286.6	0.54
Winter	(DJF)	53.4	24.3	16.8	140.0	52.8	11.5	24.5	86.8	0.40
Spring	(MAM)	35.6	10.4	6.4	125.9	34.3	13.6	10.6	65.5	0.60
Summer	(JJA)	43.7	21.2	3.0	114.0	43.7	18.7	8.3	100.8	0.57
Autumn	(SON)	48.1	17.0	18.4	109.0	47.9	13.1	21.5	79.1	0.54

DJF, December, January, February; MAM, March, April, May; JJA, June, July, August; SON, September, October, November.

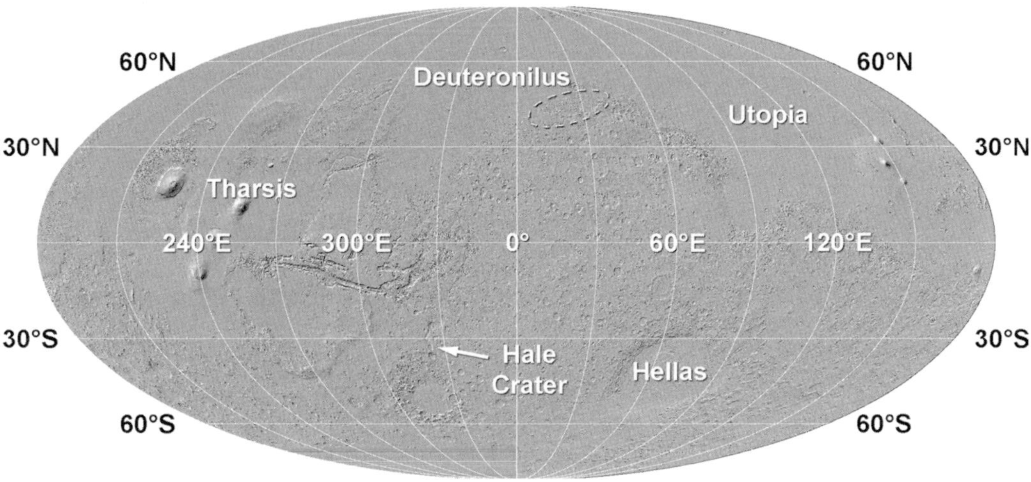

Fig. 2. Locations of regional features on Mars mentioned in the text (shaded version of the MOLA DEM). Image credit: DLR/NASA/JPL/MOLA Science Team.

particularly interesting setting is the pole-facing inner wall of impact craters. Most gullies (Fig. 3a) have been found on such walls, especially in the southern hemisphere (Dickson *et al.* 2007). On the base of some gullies, spatulate depressions are delineated towards the inner crater floor by arcuate ridges, which have been compared to moraines (e.g. Berman *et al.* 2005, fig. 1). Other landforms of possible periglacial origin have been observed in close spatial association with the crater-wall gullies, including polygons (Fig. 4a) (Levy *et al.* 2009*c*), patterned ground (Fig. 5a) (Mangold 2005), lobate features (Figs 6a & 7a) (Milliken *et al.* 2003) and fractured mounds (Fig. 8a) (Dundas *et al.* 2008). The unique occurrence of diverse possible periglacial landforms within a small area with considerable relief makes such craters an ideal study case for the action of periglacial processes on Mars. In the following, they will be compared with terrestrial analogues on Spitsbergen. Based on this comparison, possible scenarios of landscape evolution on Mars will be outlined.

Svalbard landforms

The main study site is Adventdalen, an approximately 40 km-long and up to about 3 km-wide valley in central Spitsbergen, that was deglaciated

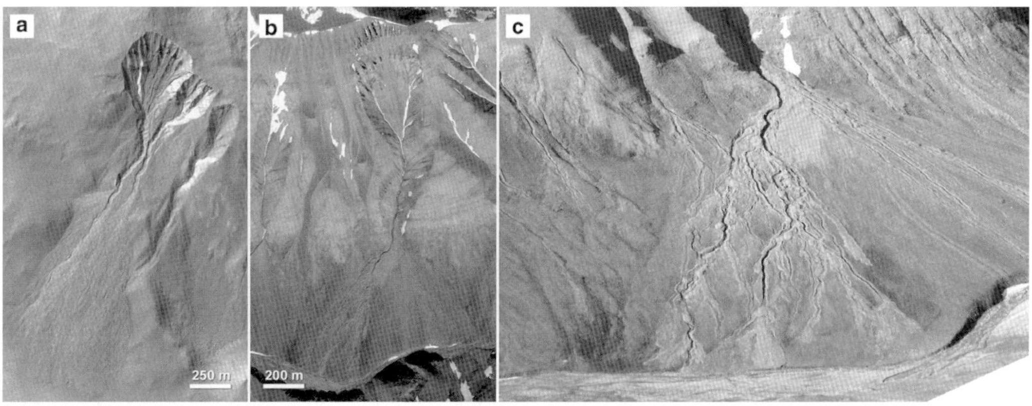

Fig. 3. Gullies and fans on Mars and Svalbard. (**a**) Gully in Martian crater at 38.5°S, 319.8°E (HiRISE PSP_006888_1410). (**b**) Gully and debris-flow fan in Hannaskogdalen, Svalbard. Note the similarity in morphology and scale between the two systems. (**c**) Close-up field photograph (taken from the opposite mountain) of the fan surface shown in (b). Note the morphological indicators of debris flows, such as large lateral levees and flow tongues. Image credit: NASA/JPL/UofA, DLR and Univ. Münster/Mike Zanetti.

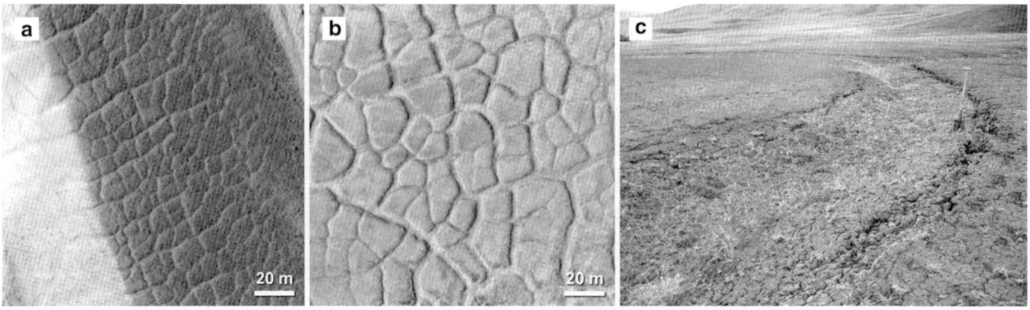

Fig. 4. Polygons on Mars and Svalbard. (**a**) Oriented-orthogonal polygons pattern on a ridge between two gullies on the northern wall of Hale Crater, Mars. The polygons have high centres and diameters between about 5 and 10 m (HiRISE image PSP_004072_1845; near 34.6°S, 323.1°E). (**b**) High-centre orthogonal polygons in central Adventdalen (HRSC-AX image). The polygons have high centres, and diameters between approximately 10 and 20 m. A trough that is typical for this type of polygon is shown in panel (c). (**c**) Trough between high-centre polygons in central Adventdalen. Note the fractured and degraded appearance of the trough shoulders. Spade for scale. Image credit: NASA/JPL/UofA, DLR and AWI/Mathias Ulrich.

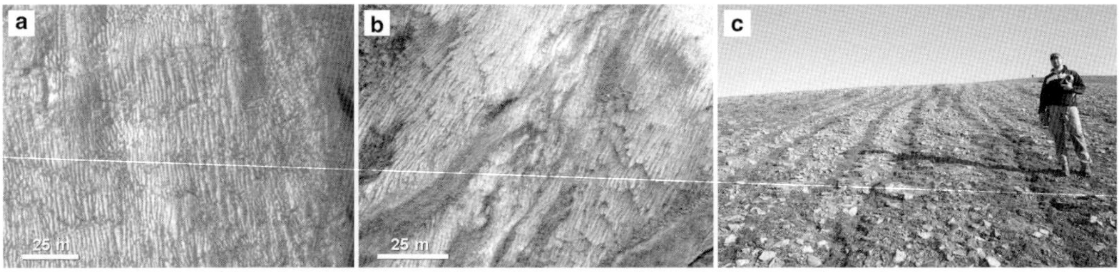

Fig. 5. Comparison between alternating bright and dark stripes on Mars and sorted stripes on Svalbard. (**a**) Alternating dark and bright stripes near gullies on the inner wall of a Martian impact crater (HiRISE image PSP_001684_1410; near 38.9°S, 196.0°E). The orientation of the stripes is approximately downslope. (**b**) Sorted stripes on the western slopes of the Hiorthfjellet massif (east of Adventfjorden, Spitsbergen). Note the striking similarity in scale between (a) and (b). (**c**) Sorted stripes in Adventdalen (Spitsbergen). Coarser and slightly elevated unvegetated stripes alternate with finer-grained and vegetated stripes (person for scale). Image credit: NASA/JPL/UofA, DLR and DLR/Ernst Hauber.

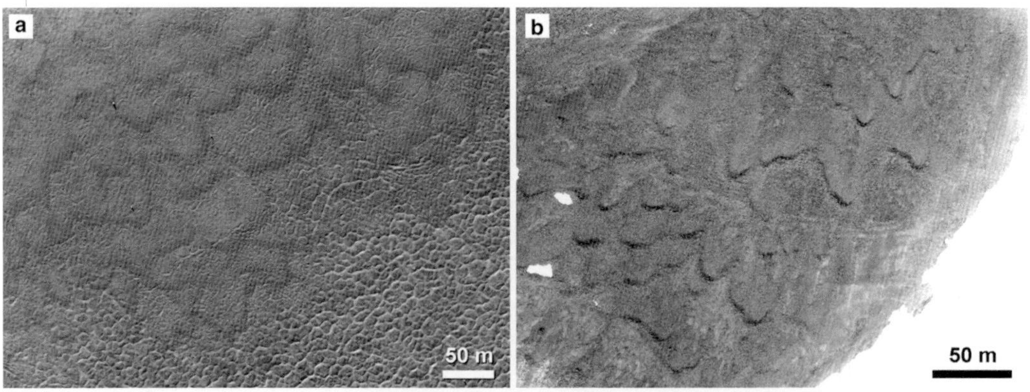

Fig. 6. Comparison between lobate structures on Martian slopes and solifluction features on Svalbard. (**a**) Lobate features on the inner wall of an impact crater on Mars (near 71.9°N, 344.5°E; HiRISE PSP_010077_2520). The morphology is identical to that of lobate solifluction sheets (cf. Ballantyne & Harris 1994, fig. 11.1). Although this particular example is on the wall of a crater in high latitudes, it is expected that such features might also be found in mid-latitude craters. (**b**) Solifluction lobes on the slopes of Louisfjellet (central Spitsbergen, Svalbard). Note the striking similarity in scale and morphology between (a) and (b). Image credit: NASA/JPL/UofA and DLR.

Fig. 7. Possible protalus ramparts on Mars (left) and Svalbard (right). (**a**) Protalus lobe-like structures at the base of a large scarp on the northern wall of Hale Crater (CTX image P15_006756_1454; near 34.6°S, 323.1°E; north is up). The steep front is characterized by polygons (see Reiss *et al.* 2009; their fig. 10b, c). (**b**) Protalus lobes on the northern tip of Prins Karls Forland, Svalbard (see Berthling *et al.* 1998), at the western foot of the Fuglehukfjellet massif (aerial photograph S704128, Norsk Polarinstitutt, Oslo, Norway; from André (1994); north is towards the left). (**c**) Close-up image of a protalus rampart at Stuphallet, Brøgger Peninsula (see person for scale). The surface of the steep front consists of very coarse blocks (diameters of up to tens of centimetres). Image credit: NASA/JPL/MSSS and Univ. Münster/Dennis Reiss.

about 10 000 years ago (Mangerud *et al.* 1992). The valley hosts a large number of periglacial landforms, both on the valley flanks and on the valley floor. The mountain massifs and the upper parts of many valleys are still partly covered by polythermal or cold-based glaciers, which can be partly debris-covered (Tolgensbakk *et al.* 2000). Distinctive end moraines, which may be ice-cored (Lukas *et al.* 2005), mark the former larger extent of the glaciers. Some tongue-shaped rock glaciers are perched in cirques and broad alcoves (Isaksen *et al.* 2000; Ødegård *et al.* 2003). Protalus ramparts, defined as 'ridges or ramps of debris formed at the downslope margin of a snowbed or firn field' (Shakesby 1997, p. 395), are well developed on the foot of high cliffs on the Brøgger Peninsula (Fig. 7b, c). Rock fall is frequent from the steep cliffs that mark the flat-topped summits of the mountains (André 1995). The flanks of the massifs bordering the valley are dissected by numerous gullies (Fig. 3b, c), which are the transport pathways for debris flows. Debris flows can reach volumes of 50–500 m^3 in the Longyearbyen Valley (Larsson 1982), and their recurrence interval is 80–500 years (André 1990). Between the gullies, many slopes display evidence of solifluction (Fig. 6b) and sorting processes (sorted and non-sorted nets and stripes: Fig. 5b, c) (Sørbel & Tolgensbakk 2002). The debris flows build up fans, characterized by channels with lateral levees, flow tongues and coarse sediment (for a description of an alluvial fan in a permafrost region see Catto 1993). Where fans extend to the shore of the estuary at the mouth of Adventdalen, they can form an arctic fan delta (Lønne & Nemec 2004). In the inland, debris-flow fans at the downstream-end of the gullies coalesce

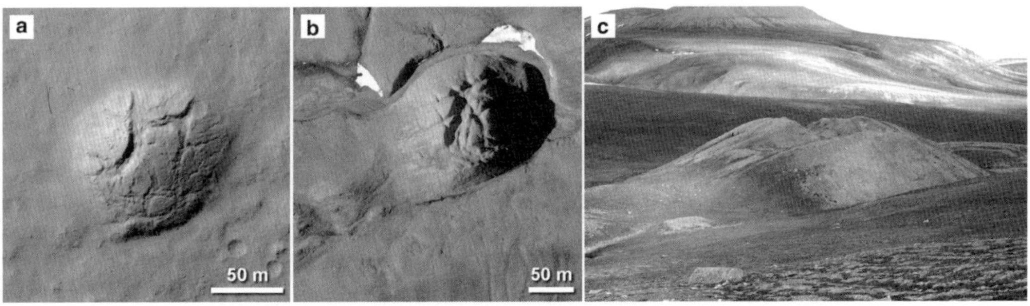

Fig. 8. Comparison between a fractured mound on Mars and pingo on Svalbard. (**a**) Fractured mound on the floor of a crater in the southern hemisphere (detail of HiRISE image PSP_007533_1420; near 37.9°S, 347.2°E: see Dundas & McEwen 2010). (**b**) Pingo in upper Eskerdalen (central Spitsbergen) with fractures on its top. HRSC-AX image, acquired in July 2008. Note the morphological similarity to the shallow fractured mound shown in (a). (**c**) Field photograph of pingo shown in (b). North is up in panels (a) and (b). View towards the NE in panel (c). Image credit: NASA/JPL/UofA, DLR and DLR/Ernst Hauber.

along the valley to form bajadas. The valley floor is occupied by the large braided river, Adventelva, which often cuts the toes of the fans. Several open-system (hydraulic) pingos are located near the fans on the valley floor (Fig. 8b, c) (Liestøl 1976; Yoshikawa 1993; Yoshikawa & Harada 1995). River terraces are overprinted by thermal-contraction cracks that form widespread nets of ice-wedge polygons (Fig. 4b, c) (Christiansen 2005). Most of the landforms on the valley flanks can be considered to be part of an ice-debris transport system, where mass wasting takes place both by steady-state processes (small-scale rockfall, avalanches, glacial and fluvial transport and solifluction) and by more extreme short-lived processes (large-scale rockfall, landslides, debris flows) (Haeberli 1985). Figure 9 demonstrates the spatial arrangement of the landforms in Adventdalen in an idealized sketch, and Figure 10 shows a three-dimensional perspective view of the Hiorthfjellet massif exhibiting some of the features in their real setting.

Discussion

The above comparisons suggest that periglacial processes might have operated in Martian mid-latitude craters. However, the exact nature, intensity and sequence of these processes are unclear. We present three different models that outline in a qualitative way some possible scenarios of how mid-latitude craters were shaped in the recent Martian history by processes involving water ice and, to a lesser degree, liquid water. It is important to note that these models are not thought to be mutually exclusive, nor do they necessarily include all processes that operated on Mars. Instead, they are suggested as examples of how planetary landform analysis guided by terrestrial knowledge can yield improved insight into the evolution of complex landscapes.

The premise of the models is that during higher obliquity water ice is driven from the poles towards lower latitudes where it is precipitated as snow. During periods of lower obliquities, the precipitated snow would sublime or melt, and water vapour would be redistributed at higher latitudes. This basic pattern of volatile transport through the atmosphere as a function of obliquity has been modelled with GCM (Mischna *et al.* 2003; Levrard *et al.* 2004; Forget *et al.* 2006; Madeleine *et al.* 2009), and the modelling results successfully predict ice accumulation in places where, indeed, an increased frequency of possible glacial landforms have been

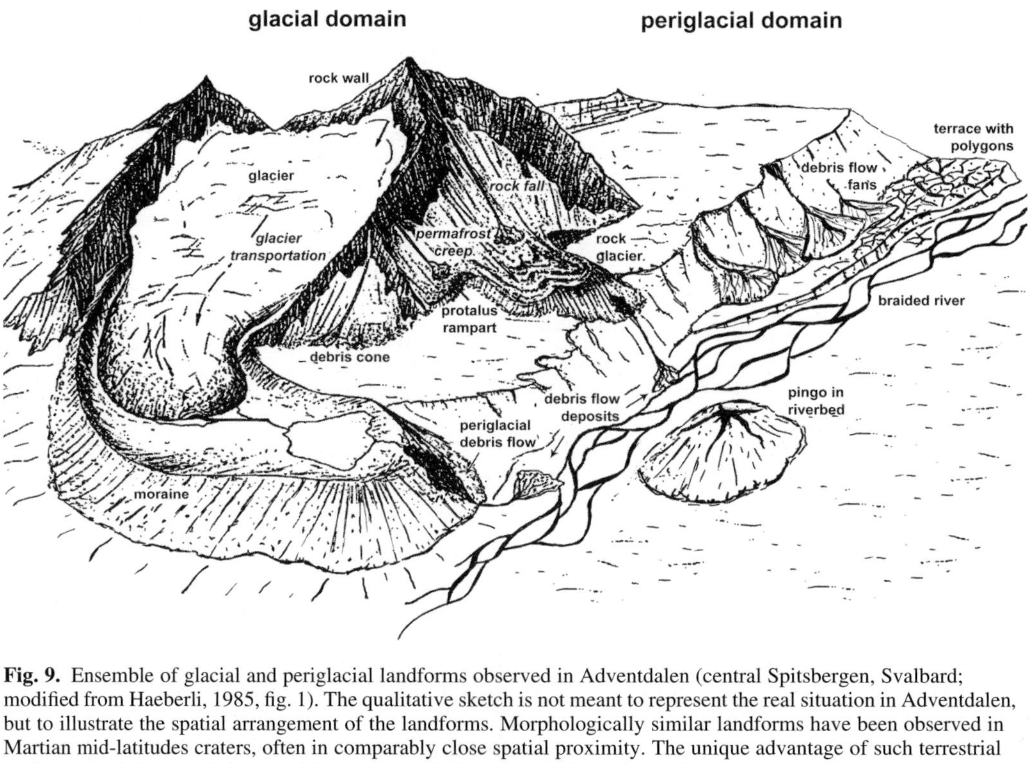

Fig. 9. Ensemble of glacial and periglacial landforms observed in Adventdalen (central Spitsbergen, Svalbard; modified from Haeberli, 1985, fig. 1). The qualitative sketch is not meant to represent the real situation in Adventdalen, but to illustrate the spatial arrangement of the landforms. Morphologically similar landforms have been observed in Martian mid-latitudes craters, often in comparably close spatial proximity. The unique advantage of such terrestrial analogues is their potential to provide constraints in the interpretation of planetary surface morphologies.

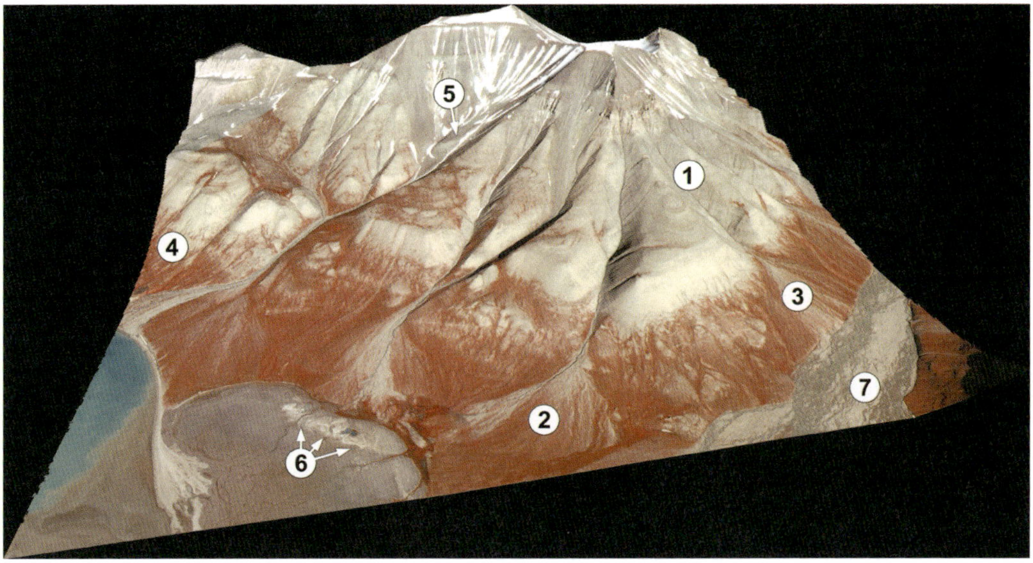

Fig. 10. Example of the close spatial proximity of glacial and periglacial landforms on Svalbard. The scene (width *c.* 3.7 km; north is towards the background) was computed from HRSC-AX false-colour stereo images and shows the Hiorthfjellet mountain massif on the northern side of Adventfjord. Numbers refer to specific landforms: 1, gullies; 2, alluvial fan; 3, debris-flow fans merging along the valley wall into a bajada; 4, slope stripes; 5, rock glacier; 6, pingo; 7, braided river. All of these landforms with the exception of the braided river have close morphological analogues in Martian mid-latitude craters. Image credit: DLR.

observed (e.g. east of the Hellas Basin, west of the Tharsis Montes and at the Deuteronilus Mensae region). If this premise is accepted, it implies that the pattern of deposition and degradation of snow and the associated periglacial processes operate in cycles, as the obliquity varies cyclically. One of such cycles is discussed in the following for each of the scenarios.

Following the scheme of landscape evolution proposed by Morgenstern *et al.* (2007) for the lowlands of Utopia Planitia, the initial process in the cycle of deposition and degradation is the subaerial deposition of a volatile-rich mantle consisting of a layered mixture of dust and snow. Martian dust is suggested to originate from volcanic sources, meteoritic impact and rock erosion, and is redistributed by global dust storms (Kahn *et al.* 1992). The dust particles act as condensation nuclei for water ice (H_2O-ice: e.g. Gooding 1986). The dusty snow mantle would be thicker at the pole-facing wall, but would also cover the crater interior and smaller crater therein. This stage is common to all three scenarios (Fig. 11a–c, stage I). Such a mantling deposit had already been suggested on the basis of *Mariner 9* data (Soderblom *et al.* 1973), and was later revealed in detail by high-resolution topography (Kreslavsky & Head 2000) and images (Mustard *et al.* 2001). This mantling layer has a thickness of the order of tens of metres in lowlands (Morgenstern *et al.* 2007), but it is not clear how much of this thickness is deposited during one obliquity cycle. The microclimatic conditions at pole-facing (inner) walls of craters are such that ice is preferentially accumulated and preserved in these locations, that is, they function as cold traps for atmospheric water ice (Hecht 2002; Schorghofer & Edgett 2006; Head *et al.* 2008).

The 'dry' scenario

Over time, the accumulated snowpack would increase in thickness and eventually the lower portions would transform into glacier ice (Fig. 11a, stage II). This glacier would probably contain a significant amount of dust (and perhaps wind-blown sand, but no or very few rock fragments) and we tentatively suggest the term 'dust glacier'. The plan-view shape of such glaciers would typically be tongue-shaped (length > width), as it is commonly observed on Earth (for a comparison between these shapes on Mars and Earth, see Arfstrom & Hartmann 2005, their fig. 2). If it were cold enough, this glacier would freeze onto the underlying crater wall and be a cold-based glacier, as has previously been suggested for Mars (Head & Marchant 2003). A cold-based glacier would cause little or no erosion of the underlying crater wall and, therefore, the slope of this wall might remain

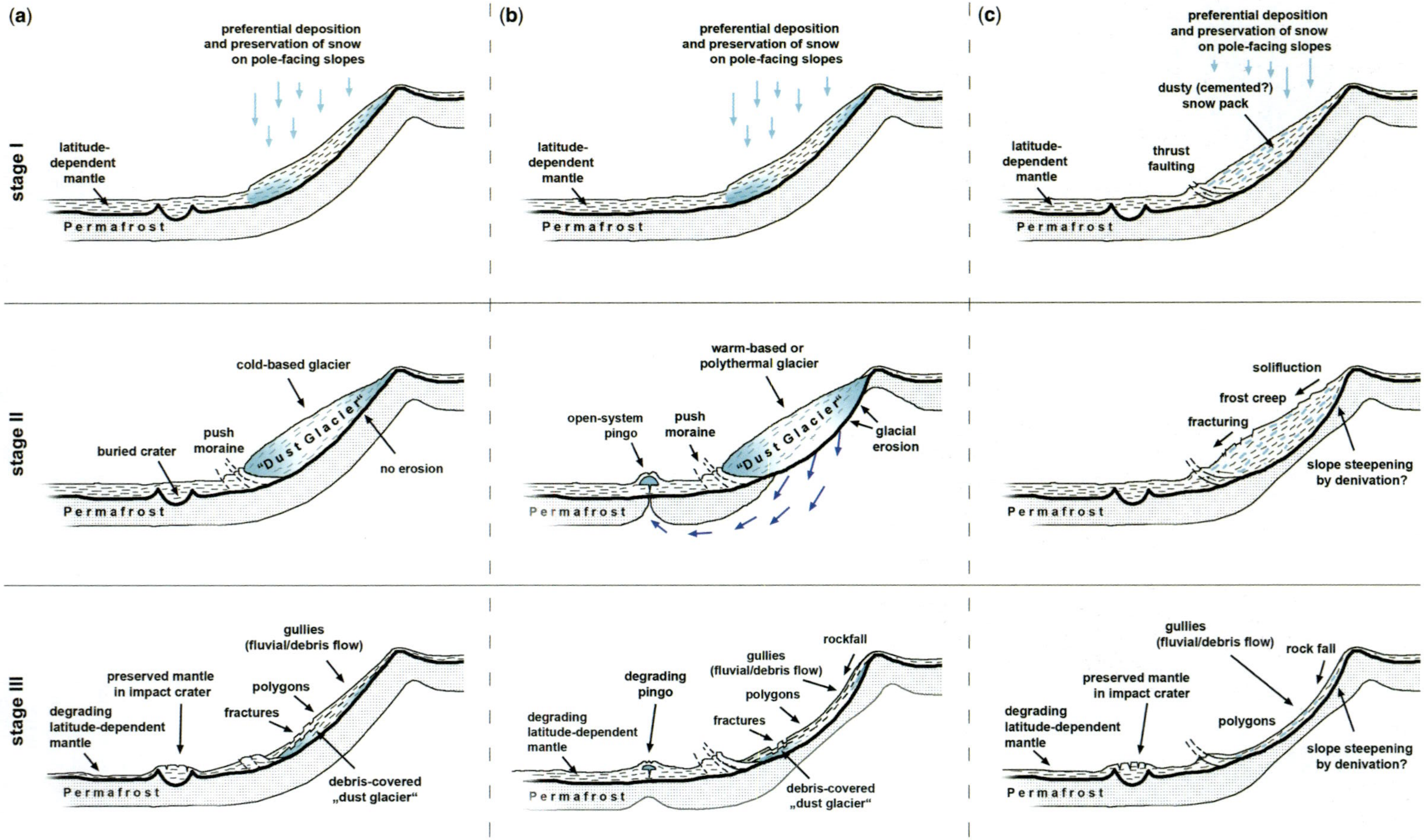

Fig. 11. Qualitative scenarios of landscape evolution in Martian mid-latitude craters. See the text for details.

more or less unchanged. At the downslope termination of the glacier, thrust or push moraines could develop (e.g. Berman *et al.* 2005) because even if there was no basal sliding of the glacier it would deform internally and move downslope. The presence of push moraines in front of cold-based glaciers has been well documented on Earth (e.g. Haeberli 1979), where push moraines are the morphological result of permafrost deformation. These moraines would be piled up to form ridges, which might contain some ice. In some cases, a lobate body might form at the base of the crater wall that has a width larger than its length (Fig. 7a). This class of flow feature exhibits a striking large-scale similarity to protalus ramparts on Svalbard (Fig. 7b). The spatial proximity of 'dust glaciers' (ice-cored), moraines and permafrost features, such as protalus ramparts, would not be surprising because it was suggested that these landforms might be part of a morphological and developmental continuum (Shakesby *et al.* 1987). At smaller scales of observation, however, significant differences become obvious between the Martian and terrestrial features shown in Figure 7. The steep distal front of the Martian flow feature is overprinted by polygons (cf. Reiss *et al.* 2009, fig. 10), which are likely to have been developed as thermal contraction cracks in fine-grained material. In contrast, the distal fronts of the protalus ramparts on Svalbard consist of coarse, decimetre-sized rocks derived from steep cliffs and mountain slopes. The difference is easily explained, however, if the relief above the features is taken into account. On Svalbard, the slopes are steep and frequent mass wasting delivers copious amounts of coarse particles, which form the rocky part of the rock glacier. Conversely, the lower slopes of large and old craters on Mars (such as Hale Crater, in the example of Fig. 7a) are much gentler, and the material being mixed with ice to form the protalus rampart-like feature would be fine-grained airborne dust. On the surface of such a body, it would be reasonable to expect the formation of sublimation polygons.

After the obliquity decreases, the ice would slowly become unstable and begin to sublimate. A lag deposit of dust and sand would form at the top of the glacier, decreasing the rate of sublimation (Mellon & Jakosky 1993; Chevrier *et al.* 2007). Internally, the glacier might still be deformed. If the lag deposit has some cohesion (e.g. from cementation), the ongoing internal deformation of the glacier body might crack the lag deposit and form tension fractures, normal faults and grabens paralleling the topographical contours. When sublimation would have removed most of the ice, a thick and very fine-grained lag deposit (dominated by dust-sized particles) would remain above a thinned body of buried glacier ice. At the same time, the mantling deposit in the crater interior would also degrade and become thinner. Where this mantling filled a smaller impact crater, it might be preferentially preserved, leaving a high-standing mound of the mantling that could develop fractures at its top. As Dundas & McEwen (2010) have already discussed, such a fractured mound could easily be misinterpreted as a pingo (Fig. 11a, stage III). Thermal contraction polygons could develop in the sublimation lag. By analogy to the McMurdo Dry Valleys in the Antarctica, these polygons could be sublimation polygons (Marchant *et al.* 2002), as suggested for Mars by Levy *et al.* (2009*b*). It has to be noted, however, that the exact nature of the polygons (ice-wedge polygons, sand-wedge polygons or sublimation polygons) remains an open question as the morphology alone does not allow for an unambiguous identification of either of these forms (e.g. van Gasselt *et al.* 2005). For example, degraded ice-wedge polygons in Adventdalen (Fig. 4b, c) display a morphology that can hardly be distinguished from sublimation polygons in remote-sensing imagery. With continual degradation, the volume of the remaining ice would be so small that scalloped depressions would form between the thrust moraines, left behind as arcuate ridges, and the remaining lag deposit on the crater walls (Fig. 11a, stage III). Remnant thicker patches of near-surface ground ice (Costard *et al.* 2002) or snow perched high in alcoves on the crater rim (Head *et al.* 2008) might finally melt (Hecht 2002; Kossacki & Markiewicz 2004). The meltwater could run off surficially and initiate fluvial transport and downstream deposition, where a resulting alluvial fan would form (Fig. 11a, stage III). Alternatively, the meltwater could infiltrate into the lag deposit, saturate it, increase the pore pressure and thus reduce its shear strength, which would increase the susceptibility of the material to gravity-driven failure and debris flows (e.g. Iverson *et al.* 1997). The degree of saturation is commonly increased if a low-permeability layer in the subsurface is present, which leads to the transient perching of the water table (Reid *et al.* 1988), and the frozen underground would be such a hydrological barrier. Another factor favouring the development of debris flows in this setting on Mars is the small grain size of the lag deposit because clay-sized material is required to maintain the high pore pressures needed during the flow (Iverson 1997). This mechanism of debris-flow initiation has also been proposed by Lanza *et al.* (2010). Unambiguous evidence for debris or mud flows on Mars has, indeed, been found by Levy *et al.* (2010). The debris flows and the fluvial processes would form a downstream fan, as is typical for Earth. The fans have been dated by crater counting and have ages of the order of 10^5–10^6 years (Reiss *et al.* 2004; Schon *et al.* 2009*a*). In the 'dry' model, a transition takes place from glacial to periglacial processes, and the formation of gullies and fans from and on the lag deposit would be the final stage (Dickson & Head 2009).

The 'wet' scenario

The second scenario starts as the 'dry' one, except that a warm-based or polythermal 'dust glacier' would form. This glacier would experience basal melting and, therefore, the ice and subglacial meltwater would erode and steepen the crater wall (Fig. 11b, stage II). Another difference to the first scenario would be the extent of the permafrost layer. Beneath the warm-based glacier, the permafrost would disappear and liquid water generated by the basal melting of the glacier would infiltrate into the substrate. A similar scheme was proposed by Carr & Head (2003) and Fassett & Head (2006). The groundwater would migrate down towards the interior of the crater. In the subsurface of the crater floor, beyond the extent of the

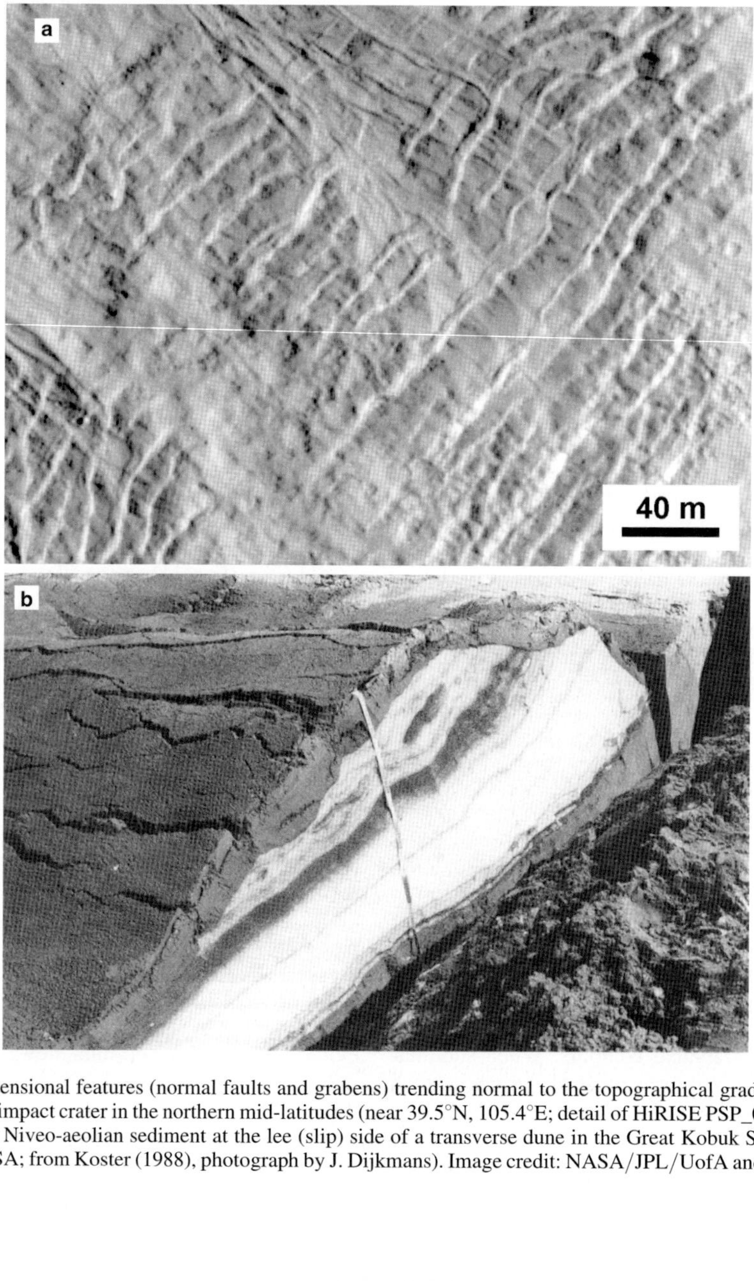

Fig. 12. (**a**) Extensional features (normal faults and grabens) trending normal to the topographical gradient of the inner wall of an impact crater in the northern mid-latitudes (near 39.5°N, 105.4°E; detail of HiRISE PSP_001357_2200, north is up). (**b**) Niveo-aeolian sediment at the lee (slip) side of a transverse dune in the Great Kobuk Sand Dunes (NW Alaska, USA; from Koster (1988), photograph by J. Dijkmans). Image credit: NASA/JPL/UofA and J. Dijkmans.

glacier, there would be an impermeable permafrost layer above the groundwater, and the hydraulic head would pressurize the groundwater. At weak spots in the crater floor, which would be abundant owing to the fracturing that was created at the impact, this groundwater could ascend as artesian water. Reaching the near surface, it would freeze and build a growing ice core. With time, a mound consisting of this ice core and some overlying mantle deposit would rise. This is how hydraulic (open-system) pingos grow on Earth, except that they do not form in craters, but rather in valleys where the hydraulic head has its source in nearby mountains (Müller 1959; Worsley & Gurney 1996; Mackay 1998). The pingos in the study area in Spitsbergen are also thought to form by this mechanism (Liestøl 1976, fig. 2). If the same process applies to Mars, it would represent an example of glacier–permafrost interaction, which is also considered to be an important factor in landform evolution on Svalbard (Etzelmüller & Hagen 2005). The steepening of the crater wall by glacial erosion would increase the probability for rockfall, which was suggested as a triggering mechanism for debris flows on Earth if the other requirements (saturated soil, positive pore pressure) are met (Hsu 1975; Johnson 1995). Apart from these differences, this 'wet' scenario would otherwise be very similar to the 'dry' scenario, and glacial processes (including surficial meltwater production and runoff: Fassett *et al.* 2010) would be followed by the formation of periglacial landforms (polygons, solifluction lobes, rock glaciers and pingos) and, finally, paraglacial processes (avalanches, rock falls, debris flows, chemical denudation, mechanical fluvial denudation and surface movements/creep) (Fig. 11b, stage III).

The 'snow' scenario

A thick snowpack might form instead of a glacier in the 'snow' scenario (Fig. 11c, stage I). The transition of snow (or rather firn) to glacier ice is defined by density and starts at approximately 830 kg m^{-3}, where interconnecting air passages between ice grains become sealed off (Paterson 1994) and reach a final value of 917 kg m^{-3} (Knight 1999, table 3.3; for an extended discussion of ice metamorphism, firnification and ice formation see Shumskii 1964, pp. 240–303). While it is known that many factors (e.g. vapour transport, and the diurnal and seasonal temperature variations) control the snow densification on Mars (Arthern *et al.* 2000), a clear difference between Earth and Mars is the rate of gravity-driven snow densification (sintering). Other factors being equal, the transition from snow (or rather firn) to glacier ice should, therefore, occur on Mars at a greater depth than on Earth (the Martian gravitational acceleration at its surface is about 38% of that on the Earth's surface). Typical values for this depth on Earth are approximately 10–20 m in temperate areas and much less than 50 m in cold continental areas (e.g. Shumskii 1964, p. 275). The timescales of this transformation are also vastly different, depending on the climate. In cold and dry climates, such as in Antarctica, the transformation may require up to 2500 years (Paterson 1994, table 2.2), whereas it can be as short as only a few years in

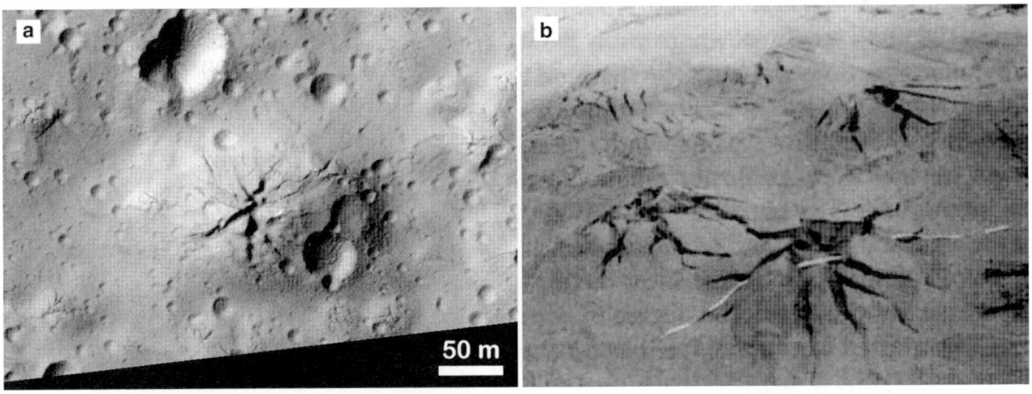

Fig. 13. Comparison between fractured mound on Mars and niveo-eolian features on Earth. (**a**) Mound with radial fractures on the floor of an impact crater in the southern mid-latitudes. The surface of the mound is superposed by several round depressions that might be due to collapse and/or impact cratering (near 33.6°S, 124°E; detail of HiRISE PSP_002135_1460; north is up). (**b**) Snow hummock with radial tensional cracks on the Great Kobuk Sand Dunes (Alaska, USA; from Koster & Dijkmans 1988). The hummock is a denivation form that developed in niveo-eolian beds. Note the morphological similarity to (a), but also note the large difference in scale (these hummocks are only a few decimetres to 1 m wide). See the text for details. Image credit: NASA/JPL/UofA and J. Dijkmans.

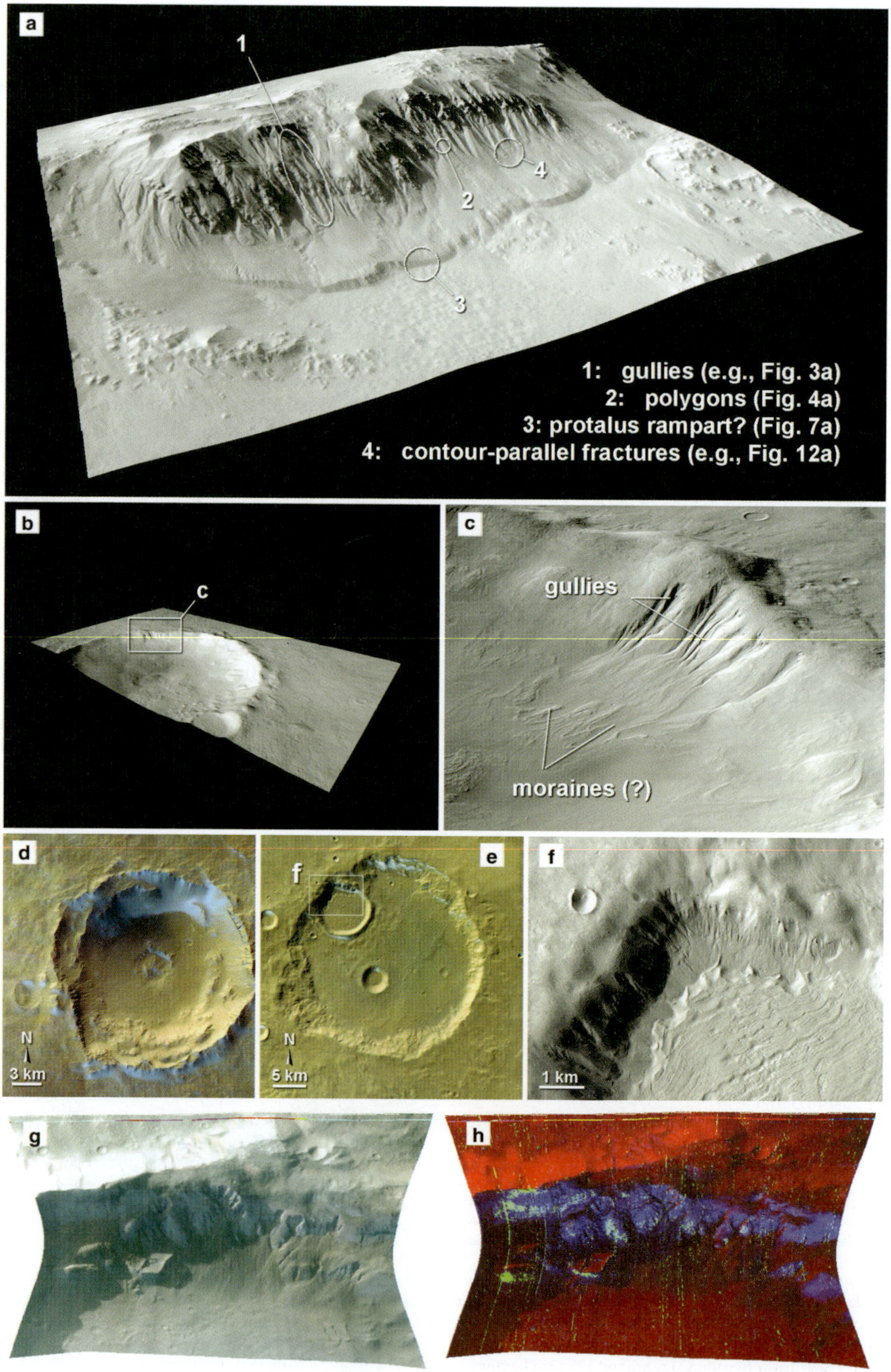
a
1
2
3
4
1: gullies (e.g., Fig. 3a)
2: polygons (Fig. 4a)
3: protalus rampart? (Fig. 7a)
4: contour-parallel fractures (e.g., Fig. 12a)
b
c
c
gullies
moraines (?)
d
N
3 km
e
f
N
5 km
f
1 km
g
h

Fig. 14.

more temperate regions such as in NW Canada. In summary, it can be expected that it takes longer in a very cold and, presumably, rather dry climate on Mars to transform snow to firn and finally to ice than on Earth. This should be true even for recent periods of higher obliquity. Similarly, one might expect snowpacks on Mars to reach larger thicknesses than on Earth before they transform to glacier ice. Based on these qualitative considerations, it seems likely that in many cases the accumulation of snow did not result in a glacier, but in a thick snowpack with intercalated layers of dust and, perhaps, wind-blown sand (cf. Williams *et al.* 2008, fig. 3). The snow scenario is perfectly in agreement with an interpretation of the features shown in Figure 7a as protalus ramparts because such landforms on Earth are evidence of snow accumulation. Sublimation of snow would, again, favour the formation of a lag deposit on top of the snowpack. The slow downward creep (Perron *et al.* 2003) in combination with compaction and sublimation of snow could induce fracturing of the overlying lag deposit (Fig. 12a). A terrestrial analogue for this process was described by Koster (1988), who investigated niveo-aeolian forms in Alaska. He found that denivation of sand-covered snow on dunes can produce deformational structures such as tensional cracks and compressional features (Fig. 12b) (see also Dijkmans 1990, fig. 3b), which are morphologically similar to the contour-parallel fractures and grabens commonly seen on the lower slopes of mantling deposits and fans on Mars (cf. Fig. 12a). The creep of the snowpack might also pile up some permafrost material at the base, analogous to the moraines in the dry and wet scenarios (Fig. 11c, stage II).

The melting of a dusty snowpack has long been recognized as a potential source of liquid water on Mars (Clow 1987), and more recent studies confirmed this possibility (Williams *et al.* 2008). While the results of Williams *et al.* (2008) apply only to periods of obliquity higher than that of today, Möhlmann (2010) emphasized the outcome of the 'solid-state greenhouse effect' in generating liquid water in snowpacks and concluded that, even in the current climate of Mars, liquid water can be produced. Williams *et al.* (2009) modelled snow melt at mid-latitudes on Mars and found that enough meltwater can be generated to produce gullies, an idea that had been previously suggested by Christensen (2003). Whenever the snow melting occurred exactly, it would be a viable process to provide the required liquid water for gully and fan formation in the 'snow' scenario. If an active layer existed in the past (Kreslavsky *et al.* 2008), solifluction might occur in the form of frost creep or gelifluction, although the period of the freeze–thaw cycles is difficult to constrain (day–night or seasonal cycles). Fractured mounds would form as erosional forms, not as pingos. Where all the snow in the surrounding has decreased in height or disappeared, snow hummocks would remain, consisting of residual snow patches or ridges (Koster & Dijkmans 1988). When the tops of these denivation forms are broken up into radial patterns, they display a strikingly similar morphology to Martian fractured mounds (Fig. 13). There is a huge difference in scale between the two types of fractured mounds shown in Figure 13, but the principle should work for the larger fractured mounds on Mars as well. The other landforms would form very similarly as in

Fig. 14. (*Continued*) Assemblages of possible periglacial landforms and water ice on Martian pole-facing crater walls. (**a**) Part of south-facing inner wall of Hale Crater, displaying several landforms that resemble periglacial landforms on Svalbard. CTX image P15_006756_1454 superposed on HRSC DEM (HRSC h0533_0000). View is towards the NE, no vertical exaggeration, image width is about 12 km. (**b**) Perspective view of a crater in the southern mid-latitudes (at 45.66°S, 238.11°E). CTX image B05_011519_1341 superposed on HRSC DEM (HRSC h0424_0000). View is towards the NE, no vertical exaggeration, crater diameter is 26 km. (**c**) Slightly rotated detail of the scene shown in (b), with gullies and possible moraines at the downward termination of the inferred location of former glaciers (view towards the north). (**d**) Snow and frost on pole-facing slopes (crater centre at 46.05°S, 183.85°E; detail of HRSC h8569_0000; image acquired during the southern winter at solar longitude (L_S) 147.8°). The bright material is likely to be water ice, as it was found by the Compact Reconnaissance Imaging Spectrometer (CRISM) at a similar latitude during the same season (see panels g and h). (**e**) Another example of snow and frost preferentially accumulated on pole-facing slopes (crater centre at 39.6°S, 158.32°E; detail of HRSC h8527_0000; image acquired during the southern winter at L_S 141.6°). The white box marks the location of panel (f) and corresponds to an area where bright material accumulated on the inner wall of a smaller impact crater. (**f**) Detail of the previous image. The area of snow accumulation corresponds exactly to sites where gullies, fans and moraine-like landforms are observed (detail of CTX B05_011746_1401). (**g**) CRISM false-colour image of a crater rim in Terra Sirenum (near 38.9°S, 195.9°E). Frost is characterized by a 'bluer' colour than the rock and soil. The image was taken during the Martian winter at L_S 140.6° (image source: NASA PlanetaryPhotojournal, #PIA09101). (**h**) Same scene as (g), with the colour indicating the depths of absorption bands of H_2O-frost at 1.50 μm (blue) and CO_2-frost at 1.45 μm (green). While CO_2-frost occurs only at the coldest, most shaded areas, water ice is more widespread and occurs on slopes incised by many gullies (image source: NASA PlanetaryPhotojournal, #PIA09101; see also Vincendon *et al.* 2009). Image credit: ESA/DLR/FUB, NASA/JPL/MSSS and NASA/JPI/JHUAPL.

the other scenarios (Fig. 11c, stage III). An important aspect of this scenario and the associated snow melting would be intensified chemical weathering, the role of which has been underestimated in the past even on Earth (Thorn 1988).

Conclusions

Despite significant differences in the climates of Mars and Svalbard, a suite of very analogous landforms has developed, although perhaps over enormously different timescales. Attempts to reconstruct palaeo-climates on Mars have to take into account the fact that different processes acting in different environments can produce similar results (equifinality). The integrated analysis of landscapes can reduce such ambiguities.

The landform inventory associated with pole-facing inner walls of impact craters in the Martian mid-latitudes (Fig. 14) suggests the geologically recent action and interaction of glacial and periglacial processes. Based on terrestrial analogue landforms in similarly close spatial proximity on Svalbard, three scenarios of sequential landscape evolution are presented for Mars. All scenarios start with initial snowfall and the deposition of a dusty snowpack, and they all end with recent gully and fan formation. These scenarios are qualitative in the sense that none of them is expected to exactly represent the real situation on Mars. In fact, the scenarios are not mutually exclusive, and mixed cases (e.g. the dry and the snow scenarios) are very plausible. Dependent on latitude and insolation, some craters might have been shaped by the dry scenario, while craters at other latitudes might have been shaped by the wet scenario. The different scenarios also have different implications for the interpretation of certain landforms. For example, fractured mounds are unlikely to be open-system pingos in the dry scenario because that does not predict liquid water in the subsurface, a prerequisite for the growth of hydraulic pingos. However, basal melting of snow in the snow scenario could lead to infiltration of liquid water into the subsurface and the formation of a hydraulic pingo as in the wet scenario.

The landscape evolution proposed here would be controlled by obliquity and/or orbital parameters such as eccentricity or the position of perihelion, and is therefore assumed to be cyclic. Several successive episodes of deposition and removal have already been suggested by, for example, Kreslavsky & Head (2002), Schon *et al.* (2009*b*) and Morgan *et al.* (2010). Processes implying an active layer might have operated in the past, although an active layer does not exist today (Kreslavsky *et al.* 2008). It is thus important to realize that the Martian mid-latitude morphologies do not represent a stable situation over long periods. Instead, this is a dynamic landscape in constant, although perhaps very slow, transition, and patterns of sedimentation and erosion overprint each other repeatedly. Nevertheless, the associated rates for erosion (e.g. in the dry scenario) are likely to be very low, and not all traces of former ice ages are extinguished by later glaciations. Therefore, the spatial extent of former and more widespread glaciations can be identified by careful morphological analysis (Hauber *et al.* 2008; Dickson *et al.* 2008, 2010; Head *et al.* 2010).

Not all craters are necessarily expected to be exactly in the same stage of this landscape evolution. In general, however, the observations of gullies with very recent activity (e.g. Diniega *et al.* 2010; Dundas *et al.* 2010; Reiss *et al.* 2010) point to a late-stage situation for most mid-latitude craters at the present time. This is also in agreement with observations of current degradation of the mantling deposit in mid-latitudes (Mustard *et al.* 2001; Morgenstern *et al.* 2007; Lefort *et al.* 2009; Zanetti *et al.* 2010) and with theoretical modelling of ground-ice stability in the recent history of Mars (Chamberlain & Boynton 2007).

The importance of snow (Figs 12 & 13) should not be neglected in assessing the relative importance of glacial and periglacial processes on Mars. Snow and nivation processes are important factors in the geomorphology of polar and cold-climate regions (e.g. Thorn 1978; Christiansen 1998), and snowpacks might be viable alternatives to glacial interpretations of some Martian surface features. Wind should also be an important factor, as it can transport snow and accumulate it in protected regions (Head *et al.* 2008) where it could act as a landscape-forming agent.

This study would not have been possible without the logistical support by the German–French research station AWIPEV and the kind hospitality of their staff, in particular M. Schumacher and D. Isambert. The generous help from UNIS and the Norwegian Polar Institute with transport and safety equipment for the field campaigns is highly appreciated. E. Carlsson, H. Johansson and S. McDaniel joined the first field trip, and their companionship made it a wonderful experience. We thank the HiRISE and CTX teams for making their data publicly available. This research has been partly supported by the Helmholtz Association through the research alliance 'Planetary Evolution and Life'. Constructive comments by G. Morgan and an anonymous reviewer are greatly appreciated.

References

Åkerman, J. 1987. Periglacial forms of Svalbard: a review. *In*: Boardman, J. (ed.) *Periglacial Processes and Landforms in Britain and Ireland*. Cambridge University Press, Cambridge, 9–25.

ANDERSON, D. M., GATTO, L. W. & UGOLINI, F. C. 1972. An Antarctic analog of Martian permafrost terrain. *Antarctic Journal of the United States*, **7**, 114–116.

ANDRÉ, M.-F. 1990. Frequency of debris flows and slush avalanches in Spitsbergen: a tentative evaluation from lichenometry. *Polish Polar Research*, **11**, 345–363.

ANDRÉ, M.-F. 1994. Rock glaciers in Svalbard – tentative dating and inferred long-term velocities. *Geografiska Annaler*, **76A**, 235–245.

ANDRÉ, M.-F. 1995. Holocene climate fluctuations and geomorphic impact of extreme events in Svalbard. *Geografiska Annaler*, **77A**, 241–250.

ARFSTROM, J. & HARTMANN, W. K. 2005. Martian flow features, moraine-like ridges, and gullies: terrestrial analogues and interrelationships. *Icarus*, **174**, 321–335.

ARTHERN, R. J., WINEBRENNER, D. P. & WADDINGTON, E. D. 2000. Densification of water ice deposits on the residual North Polar Cap of Mars. *Icarus*, **144**, 367–381.

BAKER, V. R. 2001. Water and the Martian landscape. *Nature*, **412**, 228–236.

BALLANTYNE, C. K. & HARRIS, C. 1994. *The Periglaciation of Great Britain*. Cambridge University Press, Cambridge.

BALME, M., MANGOLD, N. *ET AL*. 2006. Orientation and distribution of recent gullies in the southern hemisphere of Mars: observations from High Resolution Stereo Camera/Mars Express (HRSC/MEX) and Mars Orbiter Camera/Mars Global Surveyor (MOC/MGS) data. *Journal of Geophysical Research*, **111**, E05001, doi: 10.1029/2005JE002607.

BALME, M. R. & GALLAGHER, C. 2009. An equatorial periglacial landscape on Mars. *Earth and Planetary Science Letters*, **285**, 1–15.

BERMAN, D. C., HARTMANN, W. K., CROWN, D. A. & BAKER, V. R. 2005. The role of arcuate ridges and gullies in the degradation of craters in the Newton Basin region of Mars. *Icarus*, **178**, 465–486.

BERTHLING, I., ETZELMÜLLER, B., EIKEN, T. & SOLLID, J. L. 1998. Rock glaciers on Prins Karls Forland, Svalbard. I: internal structure, flow velocity and morphology. *Permafrost and Periglacial Processes*, **9**, 135–145.

BOCKHEIM, J. G., CAMPBELL, I. B. & MCLEOD, M. 2007. Permafrost distribution and active-layer depth in the McMurdo Dry Valleys, Antarctica. *Permafrost and Periglacial Processes*, **18**, 217–227.

BYRNE, S., DUNDAS, C. M. *ET AL*. 2009. Distribution of mid-latitude ground ice on Mars from new impact craters. *Science*, **325**, 1674–1676.

CAMPBELL, I. B. & CLARIDGE, G. G. C. 2004. Cryosols in the extremely arid Transantarctic Mountains region of Antarctica. *In*: KIMBLE, J. M. (ed.) *Cryosols*. Springer, Berlin, 415–426.

CARR, M. H. & HEAD, J. W. 2003. Basal melting of snow on early Mars: a possible origin of some valley networks. *Geophysical Research Letters*, **30**, 2245, doi: 10.1029/2003GL018575.

CATTO, N. R. 1993. Morphology and development of an alluvial fan in a permafrost region, Aklavik Range, Canada. *Geografiska Annaler*, **75A**, 83–93.

CHAMBERLAIN, M. A. & BOYNTON, W. V. 2007. Response of Martian ground ice to orbit-induced climate change. *Journal of Geophysical Research*, **112**, E06009, doi: 10.1029/2006JE002801.

CHEVRIER, V., SEARS, D. W. G. *ET AL*. 2007. Sublimation rate of ice under simulated Mars conditions and the effect of layers of mock regolith JSC Mars-1. *Geophysical Research Letters*, **34**, L02203, doi: 10.1029/2006GL028401.

CHRISTENSEN, P. R. 2003. Formation of recent Martian gullies through melting of extensive water-rich snow deposits. *Nature*, **422**, 45–48.

CHRISTIANSEN, H. H. 1998. Nivation forms and processes in unconsolidated sediments, NE Greenland. *Earth Surface Processes and Landforms*, **23**, 751–760.

CHRISTIANSEN, H. H. 2005. Thermal regime of ice-wedge cracking in Adventdalen, Svalbard. *Permafrost and Periglacial Processes*, **16**, 87–98.

CHUANG, F. C. & CROWN, D. A. 2005. Surface characteristics and degradational history of debris aprons in the Tempe Terra/Mareotis fossae region of Mars. *Icarus*, **179**, 24–42, doi: 10.1016/j.icarus.2005.05.014.

CLOW, G. D. 1987. Generation of liquid water on Mars through the melting of a dusty snow pack. *Icarus*, **72**, 95–127.

COSTARD, F., FORGET, F., MANGOLD, N. & PEULVAST, J. P. 2002. Formation of recent Martian debris flows by melting of near-surface ground ice at high obliquity. *Science*, **295**, 110–113.

DICKSON, J. L. & HEAD, J. W. 2009. The formation and evolution of youthful gullies on Mars: gullies as the late-stage phase of Mars' most recent ice age. *Icarus*, **204**, 63–86.

DICKSON, J. L., HEAD, J. W. & KRESLAVSKY, M. 2007. Martian gullies in the southern mid-latitudes of Mars: evidence for climate-controlled formation of young fluvial features based upon local and global topography. *Icarus*, **188**, 315–323.

DICKSON, J. L., HEAD, J. W. & MARCHANT, D. R. 2008. Late Amazonian glaciation at the dichotomy boundary on Mars: evidence for glacial thickness maxima and multiple glacial phases. *Geology*, **36**, 411–414.

DICKSON, J. L., HEAD, J. W. & MARCHANT, D. R. 2010. Kilometer-thick ice accumulation and glaciation in the northern mid-latitudes of Mars: evidence for crater-filling events in the Late Amazonian at the Phlegra Montes. *Earth and Planetary Science Letters*, **294**, 332–342.

DIJKMANS, J. W. A. 1990. Niveo-aeolian sedimentation and resulting sedimentary structures; Søndre Strømfjord area, Western Greenland. *Permafrost and Periglacial Processes*, **1**, 83–96.

DINIEGA, S., BYRNE, S., BRIDGES, N. T., DUNDAS, C. M. & MCEWEN, A. S. 2010. Present-day Martian dune gully activity. *In*: *Proceedings of the 41st Annual Lunar and Planetary Science Conference, March 1–5, 2010, The Woodlands, Texas*. Lunar and Planetary Institute, Houston, TX, Abstract 2216.

DORAN, P. T., MCKAY, C. P., CLOW, G. D., DANA, G. L., FOUNTAIN, A. G., NYLEN, T. & LYONS, W. B. 2002. Valley floor climate observations from the McMurdo Dry Valleys, Antarctica, 1986–2000. *Journal of Geophysical Research*, **107**, 4772, doi: 10.1029/2001JD002045.

DUNDAS, C. M. & MCEWEN, A. S. 2010. An assessment of evidence for pingos on Mars using HiRISE. *Icarus*, **205** 244–258.

DUNDAS, C. M., MCEWEN, A. S., DINIEGA, S., BYRNE, S. & MARTINEZ-ALONSO, S. 2010. New and recent gully activity on Mars as seen by HiRISE. *Geophysical Research Letters*, **37**, L07202, doi: 10.1029/2009GL041351.

DUNDAS, C. M., MELLON, M. T., MCEWEN, A. S., LEFORT, A., KESZTHELYI, L. P. & THOMAS, N. 2008. HiRISE observations of fractured mounds: possible Martian pingos. *Geophysical Research Letters*, **35**, L04201, doi: 10.1029/2007GL031798.

ETZELMÜLLER, B. & HAGEN, J. O. 2005. Glacier-permafrost interaction in Arctic and alpine mountain environments with examples from southern Norway and Svalbard. *In*: HARRIS, C. & MURTON, J. B. (eds) *Cryospheric Systems: Glaciers and Permafrost*. Geological Society, London, Special Publications, **242**, 11–27.

FANALE, F. P., SALVAIL, J. R., ZENT, A. P. & POSTAWKO, S. E. 1986. Global distribution and migration of subsurface ice on Mars. *Icarus*, **67**, 1–18.

FASSETT, C. I. & HEAD, J. W. 2006. Valleys on Hecates Tholus, Mars: origin by basal melting of summit snowpack. *Planetary and Space Science*, **54**, 370–378.

FASSETT, C. I., DICKSON, J. L., HEAD, J. W., LEVY, J. S. & MARCHANT, D. R. 2010. Supraglacial and proglacial valleys on Amazonian Mars. *Icarus*, **208**, 86–100.

FORGET, F., HABERLE, R. M., MONTMESSIN, F., LEVRARD, B. & HEAD, J. W. 2006. Formation of Glaciers on Mars by Atmospheric Precipitation at High Obliquity. *Science*, **311**, 368–371.

GOODING, J. L. 1986. Martian dust particles as condensation nuclei – A preliminary assessment of mineralogical factors. *Icarus*, **66**, 56–74.

GWINNER, K., COLTELLI, M. ET AL. 2006. *The HRSC-AX Mt Etna Project: High-Resolution Orthoimages and 1 m DEM at Regional Scale*. International Archives of Photogrammetry and Remote Sensing, XXXVI (Part 1), http://isprs.free.fr/documents/Papers/T05-23.pdf.

GWINNER, K., SCHOLTEN, F. ET AL. & THE HRSC CO-INVESTIGATOR TEAM 2005. Hochauflösende Digitale Geländemodelle auf der Grundlage von Mars Express HRSC-Daten. *Photogrammetrie, Fernerkundung, Geoinformation*, **5/2005**, 387–394.

GWINNER, K., SCHOLTEN, F. ET AL. 2009. Derivation and validation of high-resolution digital elevation models from Mars Express HRSC-data. *Photogrammetric Engineering & Remote Sensing*, **75**, 1127–1142.

GWINNER, K., SCHOLTEN, F. ET AL. 2010. Topography of Mars from global mapping by HRSC high-resolution digital elevation models and orthoimages: characteristics and performance. *Earth and Planetary Science Letters*, **294**, 506–519.

HAEBERLI, W. 1979. Holocene push-moraines in Alpine permafrost. *Geografiska Annaler*, **61A**, 43–48.

HAEBERLI, W. 1985. On the morphodynamics of ice/debris-transport systems in cold mountain areas. *Norsk Geografisk Tidsskrift*, **50**, 3–9.

HAEBERLI, W. 2005. Investigating glacier-permafrost relationships in high-mountain areas: historical background, selected examples and research needs. *In*: HARRIS, C. & MURTON, J. B. (eds) *Cryospheric Systems: Glaciers and Permafrost*. Geological Society, London, Special Publications, **242**, 29–37.

HANSSEN-BAUER, I. & FØRLAND, E. J. 1998. Long-term trends in precipitation and temperature in the Norwegian Arctic: can they be explained by changes in atmospheric circulation patterns? *Climate Research*, **10**, 143–153.

HAUBER, E., PREUSKER, F. ET AL. 2009. Morphometry of alluvial fans in a polar desert (Svalbard, Norway): implications for Interpreting Martian Fans. *In*: *Proceedings of the 41st Annual Lunar and Planetary Science Conference, March 1–5, 2010, The Woodlands, Texas*. Lunar and Planetary Institute, Houston, TX, Abstract 1468.

HAUBER, E., VAN GASSELT, S., CHAPMAN, M. G. & NEUKUM, G. 2008. Geomorphic evidence for former lobate debris aprons at low latitudes on Mars: indicators of the Martian paleoclimate. *Journal of Geophysical Research*, **113**, E02007, doi: 10.1029/2007JE002897.

HEAD, J. W. & MARCHANT, D. R. 2003. Cold-based mountain glaciers on Mars: Western Arsia Mons. *Geology*, **31**, 641–644.

HEAD, J. W., MARCHANT, D. R., AGNEW, M. C., FASSETT, C. I. & KRESLAVSKY, M. A. 2006. Extensive valley glacier deposits in the northern mid-latitudes of Mars: evidence for Late Amazonian obliquity-driven climate change. *Earth and Planetary Science Letters*, **241**, 663–671.

HEAD, J. W., MARCHANT, D. R., DICKSON, J. L., KRESS, A. M. & BAKER, D. M. 2010. Northern mid-latitude glaciation in the Amazonian period of Mars: criteria for the recognition of debris-covered glacier and valley glacier landsystem deposits. *Earth and Planetary Science Letters*, **294**, 306–320.

HEAD, J. W., MARCHANT, D. R. & KRESLAVSKY, M. A. 2008. Formation of gullies on Mars: link to recent climate history and insolation microenvironments implicate surface water flow origin. *Proceedings of the National Academy of Sciences*, **105**, 13,258–13,263.

HEAD, J. W., MUSTARD, J. F., KRESLAVSKY, M. A., MILLIKEN, R. E. & MARCHANT, D. R. 2003. Recent ice ages on Mars. *Nature*, **426**, 797–802.

HECHT, M. H. 2002. Metastability of liquid water on Mars. *Icarus*, **156**, 373–386.

HOLT, J. W., SAFAEINILI, A. ET AL. 2008. Radar sounding evidence for buried glaciers in the southern mid-latitudes of Mars. *Science*, **322**, 1235–1238.

HSU, K. J. 1975. Catastrophic debris streams (sturzstroms) generated by rockfalls. *Geological Society of America Bulletin*, **86**, 129–140.

HUMLUM, O. 2005. Holocene permafrost aggradation in Svalbard. *In*: HARRIS, C. & MURTON, J. B. (eds) *Cryospheric Systems: Glaciers and Permafrost*. Geological Society, London, Special Publications, **242**, 11–27.

HUMLUM, O. & CHRISTIANSEN, H. H. 2008. Lowland periglacial research: a review of published advances 2003–2007. *Permafrost and Periglacial Processes*, **19**, 211–235.

ISAKSEN, K., ØDEGÅRD, R. S., EIKEN, T. & SOLLID, J. L. 2000. Composition, flow and development of two tongue-shaped rock glaciers in the permafrost of Svalbard. *Permafrost and Periglacial Processes*, **11**, 241–257.

IVERSON, R. M. 1997. The physics of debris flows. *Reviews of Geophysics*, **35**, 245–296.

IVERSON, R. M., REID, M. E. & LAHUSEN, R. G. 1997. Debris-flow mobilization from landslides. *Annual Review of Earth and Planetary Sciences*, **25**, 85–138.

JAKOSKY, B. M. & CARR, M. H. 1985. Possible precipitation of ice at low latitudes of Mars during periods of high obliquity. *Nature*, **315**, 559–561.

JAKOSKY, B. M., HENDERSON, B. G. & MELLON, M. T. 1995. Chaotic obliquity and the nature of the Martian climate. *Journal of Geophysical Research*, **100**, 1579–1584.

JOHNSON, P. G. 1995. Debris transfer and sedimentary environments: Alpine glaciated areas. *In*: SLAYMAKER, O. (ed.) *Steepland Geomorphology*. Wiley, Chichester, 27–44.

KADISH, S. J., BARLOW, N. G. & HEAD, J. W. 2009. Latitude dependence of Martian pedestal craters: evidence for a sublimation-driven formation mechanism. *Journal of Geophysical Research*, **114**, E10001, doi: 10.1029/2008JE003318.

KAHN, R. A., MARTIN, T. Z., ZUREK, R. W. & LEE, S. W. 1992. The Martian dust cycle. *In*: KIEFFER, H. H., JAKOWSKY, B. M., SNYDER, C. & MATTHEWS, M. (eds) *Mars*. University of Arizona Press, Tucson, AZ, 1017–1053.

KNEISSL, T., REISS, D., VAN GASSELT, S. & NEUKUM, G. 2010. Distribution and orientation of northern-hemisphere gullies on Mars from the evaluation of HRSC and MOC-NA data. *Earth and Planetary Science Letters*, **294**, 357–367.

KOSSACKI, K. J. & MARKIEWICZ, W. J. 2004. Seasonal melting of surface water ice condensing in Martian gullies. *Icarus*, **171**, 272–283.

KOSTER, E. A. 1988. Ancient and modern cold-climate aeolian sand deposition: a review. *Journal of Quaternary Science*, **3**, 69–83.

KOSTER, E. A. & DIJKMANS, J. W. A. 1988. Niveo-aeolian deposits and denivation forms, with special reference to the Great Kobuk Sand Dunes, Northwestern Alaska. *Earth Surface Processes and Landforms*, **13**, 153–170.

KRESLAVSKY, M. A. & HEAD, J. W. 2000. Kilometer-scale roughness of Mars: results from MOLA data analysis. *Journal of Geophysical Research*, **105**, 26695–26712.

KRESLAVSKY, M. A. & HEAD, J. W. 2002. Mars: nature and evolution of young latitude-dependent water-ice-rich mantle. *Geophysical Research Letters*, **29**, 1719, doi: 10.1029/2002GL015392.

KRESLAVSKY, M. A., HEAD, J. W. & MARCHANT, D. R. 2008. Periods of active permafrost layer formation during the geological history of Mars: implications for circum-polar and mid-latitude surface processes. *Planetary and Space Science*, **56**, 289–302.

LANZA, N. L., MEYER, G. A., OKUBO, C. H., NEWSOM, H. E. & WIENS, R. C. 2010. Evidence for debris flow gully formation initiated by shallow subsurface water on Mars. *Icarus*, **205**, 103–112.

LARSSON, S. 1982. Geomorphological effects on the slopes of Longyear Valley, Spitsbergen, after a heavy rainstorm in July 1972. *Geografiska Annaler*, **64A**, 105–125.

LASKAR, J., CORREIA, A. C. M., GASTINEAU, M., JOUTEL, F., LEVRARD, B. & ROBUTEL, P. 2004. Long term evolution and chaotic diffusion of the insolation quantities of Mars. *Icarus*, **170**, 343–364.

LEFORT, A., RUSSELL, P. S., THOMAS, N., MCEWEN, A. S., DUNDAS, C. M. & KIRK, R. L. 2009. Observations of periglacial landforms in Utopia Planitia with the High Resolution Imaging Science Experiment (HiRISE). *Journal of Geophysical Research*, **114**, E04005, doi: 10.1029/2008JE003264.

LEIGHTON, R. R. & MURRAY, B. C. 1966. Behavior of carbon dioxide and other volatiles on Mars. *Science*, **153**, 136–144.

LEVRARD, B., FORGET, F., MONTMESSIN, F. & LASKAR, J. 2004. Recent ice-rich deposits formed at high latitudes on Mars by sublimation of unstable equatorial ice during low obliquity. *Nature*, **431**, 1072–1075.

LEVY, J. S., HEAD, J. W. & MARCHANT, D. R. 2009*a*. Cold and dry processes in the Martian Arctic: Geomorphic observations at the Phoenix landing site and comparisons with terrestrial cold desert landforms. *Geophysical Research Letters*, **36**, L21203, doi: 10.1029/2009GL040634.

LEVY, J. S., HEAD, J. W. & MARCHANT, D. R. 2009*b*. Thermal contraction crack polygons on Mars: classification, distribution, and climate implications from HiRISE observations. *Journal of Geophysical Research*, **114**, E01007, doi: 10.1029/2008JE003273.

LEVY, J. S., HEAD, J. W., MARCHANT, D. R., DICKSON, J. L. & MORGAN, G. A. 2009*c*. Geologically recent gully–polygon relationships on Mars: insights from the Antarctic Dry Valleys on the roles of permafrost, microclimates, and water sources for surface flow. *Icarus*, **201**, 113–126.

LEVY, J. S., HEAD, J. W., DICKSON, J. L., FASSETT, C. I., MORGAN, G. A. & SCHON, S. C. 2010. Identification of gully debris flow deposits in Protonilus Mensae, Mars: characterization of a water-related, energetic gully-forming process. *Earth and Planetary Science Letters*, **294**, 368–377.

LIESTØL, O. 1976. Pingos, springs and permafrost in Spitsbergen. *Norsk Polarinstitutt Årbok*, **1975**, 7–29.

LØNNE, I. & NEMEC, W. 2004. High-arctic fan delta recording deglaciation and environment disequilibrium. *Sedimentology*, **51**, 553–589.

LUKAS, S., NICHOLSON, L. I., ROSS, F. H. & HUMLUM, O. 2005. Formation, meltout processes and landscape alteration of high-arctic ice-cored moraines – examples from Nordenskiöld land, Central Spitsbergen. *Polar Geography*, **29**, 157–187.

MACKAY, J. R. 1998. Pingo growth and collapse, Tuktoyaktuk Peninsula area, Western Arctic Coast, Canada: a long term field study. *Geographique physique et Quaternaire*, **52**, 271–323.

MADELEINE, J.-B., FORGET, F., HEAD, J. W., LEVRARD, B., MONTMESSIN, F. & MILLOUR, E. 2009. Amazonian northern mid-latitude glaciation on Mars: a proposed climate scenario. *Icarus*, **203**, 390–405, doi: 10.1016/j.icarus.2009.04.037.

MANGERUD, J., BOLSTAD, M. *ET AL.* 1992. The last glacial maximum on Spitsbergen, Svalbard. *Quaternary Research*, **38**, 1–39.

MANGOLD, N. 2005. High latitude patterned grounds on Mars: classification, distribution and climatic control. *Icarus*, **174**, 336–359.

MARCHANT, D. R. & HEAD, J. W. 2007. Antarctic Dry Valleys: microclimate zonation, variable geomorphic processes, and implications for assessing climate change on Mars. *Icarus*, **192**, 187–222.

MARCHANT, D. R. & HEAD, J. W. 2010. Geologic analogies between the surface of Mars and the McMurdo Dry Valleys: microclimate-related geomorphic features and evidence for climate change. *In*: DORAN, P. T., LYONS, W. B. & MCKNIGHT, D. M. (eds) *Life in Antarctic Deserts and Other Cold Dry Environments*. Cambridge University Press, Cambridge, 9–77.

MARCHANT, D. R., LEWIS, A., PHILLIPS, W. C., MOORE, E. J., SOUCHEZ, R. & LANDIS, G. P. 2002. Formation of patterned-ground and sublimation till over Miocene glacier ice in Beacon Valley, Antarctica. *Geological Society of America Bulletin*, **114**, 718–730.

MELLON, M. T. & JAKOSKY, B. M. 1993. Geographic variations in the thermal and diffusive stability of ground ice on Mars. *Journal of Geophysical Research*, **98**, 3345–3364.

MELLON, M. T. & JAKOSKY, B. M. 1995. The distribution and behavior of Martian ground ice during past and present epochs. *Journal of Geophysical Research*, **100**, 11 781–11 799.

MELLON, M. T., ARVIDSON, R. E. *ET AL.* 2009. Ground ice at the Phoenix Landing Site: stability state and origin. *Journal of Geophysical Research*, **114**, E00E07, doi: 10.1029/2009JE003417.

MILLIKEN, R. E., MUSTARD, J. F. & GOLDSBY, D. L. 2003. Viscous flow features on the surface of Mars: observations from high-resolution Mars Orbiter Camera (MOC) images. *Journal of Geophysical Research*, **108**, 5057, doi: 10.1029/2002JE002005.

MISCHNA, M. A., RICHARDSON, M. I., WILSON, R. J. & MCCLEESE, D. J. 2003. On the orbital forcing of Martian water and CO_2 cycles: a general circulation model study with simplified volatile schemes. *Journal of Geophysical Research*, **108**, 5062, doi: 10.1029/2003JE002051.

MÖHLMANN, D. T. F. 2010. Temporary liquid water in upper snow/ice sub-surfaces on Mars? *Icarus*, **207**, 140–148.

MORGAN, G. A., HEAD, J. W., FORGET, F., MADELEINE, J.-B. & SPIGA, A. 2010. Gully formation on Mars: two recent phases of formation suggested by links between morphology, slope orientation and insolation history. *Icarus*, **208**, 658–666.

MORGENSTERN, A., HAUBER, E., REISS, D., VAN GASSELT, S., GROSSE, G. & SCHIRRMEISTER, L. 2007. Deposition and degradation of a volatile-rich layer in Utopia Planitia and implications for climate history on Mars. *Journal of Geophysical Research*, **112**, E06010, doi: 10.1029/2006JE002869.

MUSTARD, J. F., COOPER, C. D. & RIFKIN, M. K. 2001. Evidence for recent climate change on Mars from the identification of youthful near-surface ground ice. *Nature*, **412**, 411–414.

MÜLLER, F. 1959. Beobachtungen über Pingos. *Meddelelser om Grønland*, **153**, 1–127.

ØDEGÅRD, R. S., ISAKSEN, K., EIKEN, T. & SOLLID, J.-L. 2003. Terrain analyses and surface velocity measurements of the Hiorthfjellet rock glacier, Svalbard. *Permafrost and Periglacial Processes*, **14**, 359–365.

PATERSON, W. S. B. 1994. *The Physics of Glaciers*, 3rd edn. Pergamon, Oxford.

PERRON, J. T., DIETRICH, W. E., HOWARD, A. D., MCKEAN, J. A. & PETTINGA, J. R. 2003. Ice-driven creep on Martian debris slopes. *Geophysical Research Letters*, **30**, 1747, doi: 10.1029/2003GL017603.

PIERCE, T. L. & CROWN, D. A. 2003. Morphologic and topographic analyses of debris aprons in the eastern Hellas region, Mars. *Icarus*, **163**, 46–65.

REID, M. E., NIELSEN, H. P. & DREISS, S. J. 1988. Hydrologic factors triggering a shallow hillslope failure. *Bulletin of the Association of Engineering Geologists*, **25**, 349–361.

REISS, D., VAN GASSELT, S., NEUKUM, G. & JAUMANN, R. 2004. Absolute dune ages and implications for the time of formation of gullies in Nirgal Vallis, Mars. *Journal of Geophysical Research*, **109**, E06007, doi: 10.1029/2004JE002251.

REISS, D., HIESINGER, H., HAUBER, E. & GWINNER, K. 2009. Regional differences in gully occurrence on Mars: a comparison between the Hale and Bond craters. *Planetary and Space Science*, **57**, 958–974.

REISS, D., ERKELING, G., BAUCH, K. E. & HIESINGER, H. 2010. Evidence for present day gully activity on the Russell crater dune field, Mars. *Geophysical Research Letters*, **37**, L06203, doi: 10.1029/2009GL042192.

SCHOLTEN, F. & GWINNER, K. 2004. Operational parallel processing in digital photogrammetry – strategy and results using different multi-line cameras. *International Archives of Photogrammetry and Remote Sensing*, **XXXIV**, 408–413.

SCHOLTEN, F., GWINNER, K. *ET AL.* & THE HRSC CO-INVESTIGATOR TEAM 2005. Mars Express HRSC data processing – methods and operational aspects. *Photogrammetric Engineering & Remote Sensing*, **71**, 1143–1152.

SCHON, S. C., HEAD, J. W. & FASSETT, C. I. 2009*a*. Unique chronostratigraphic marker in depositional fan stratigraphy on Mars: evidence for ~1.25 Ma old gully activity and surficial meltwater origin. *Geology*, **37**, 207–210.

SCHON, S. C., HEAD, J. W. & MILLIKEN, R. E. 2009*b*. A recent ice age on Mars: evidence for climate oscillations and the buried ice hypothesis revealed by regional layering in dissected mid-latitude mantling deposits. *Geophysical Research Letters*, **36**, L15202, doi: 10.1029/2009GL038554.

SCHORGHOFER, N. 2007. Dynamics of ice ages on Mars. *Nature*, **449**, 192–194.

SCHORGHOFER, N. & EDGETT, K. S. 2006. Seasonal surface frost at low latitudes on Mars. *Icarus*, **180**, 321–334.

SHAKESBY, R. A. 1997. Pronival (protalus) ramparts: a review of forms, processes, diagnostic criteria and palaeoenvironmental implications. *Progress in Physical Geography*, **21**, 394–418.

SHAKESBY, R. A., DAWSON, A. G. & MATTHEWS, J. A. 1987. Rock glaciers, protalus ramparts and related phenomena, Rondane, Norway: a continuum of large-scale talus-derived landforms. *Boreas*, **16**, 305–317.

SHUMSKII, P. A. 1964. *Principles of Structural Glaciology*. Dover, New York.

SHUSTER, D. L. & WEISS, B. P. 2005. Martian surface paleotemperatures from thermochronology of meteorites. *Science*, **309**, 597.

SMITH, P. H., TAMPPARI, L. K. ET AL. 2009. H_2O at the Phoenix landing site. *Science*, **325**, 58–61.

SMOLUCHOWSKI, R. 1968. Mars: retention of Ice. *Science*, **159**, 1348–1350.

SOARE, R. J., BURR, D. M. & WAN BUN TSEUNG, J.-M. 2005. Possible pingos and a periglacial landscape in northwest Utopia Planitia. *Icarus*, **174**, 373–382.

SODERBLOM, L. A., KREIDLER, T. J. & MASURSKY, H. 1973. Latitudinal distribution of a debris mantle on the Martian surface. *Journal of Geophysical Research*, **78**, 4117–4122.

SØRBEL, L. & TOLGENSBAKK, J. 2002. Ice-wedge polygons and solifluction in the Adventdalen area, Spitsbergen, Svalbard. *Norsk Geografisk Tidsskrift*, **56**, 62–66.

SOLLID, J. L., HOLMLUND, P., ISAKSEN, K. & HARRIS, C. 2000. Deep permafrost boreholes in western Svalbard, northern Sweden and southern Norway. *Norsk Geografisk Tidsskrift*, **54**, 186–191.

SQUYRES, S. W. 1978. Martian fretted terrain – Flow of erosional debris. *Icarus*, **34**, 600–613.

THORN, C. E. 1978. The geomorphic role of snow. *Annals of the Association of American Geographers*, **68**, 414–425.

THORN, C. E. 1988. Nivation: a geomorphic chimera. *In*: CLARK, M. J. (ed.) *Advances in Periglacial Geomorphology*. Wiley, Chichester, 3–31.

TOLGENSBAKK, J., SØRBEL, L. & HØGVARD, K. 2000. *Adventdalen, Geomorphological and Quaternary Geological Map of Svalbard, scale 1:100,000. Spitsbergen Sheet C9Q*. Norsk Polarinstitutt Temakart, Tromsø, **32**.

TOON, O. B., POLLACK, J. B., WARD, W., BURNS, J. A. & BILSKI, K. 1980. The astronomical theory of climatic change on Mars. *Icarus*, **44**, 552–607.

TOUMA, J. & WISDOM, J. 1993. The chaotic obliquity of Mars. *Science*, **259**, 1294–1297.

VAN EVERDINGEN, R. (ed.). 2005. *Multi-language Glossary of Permafrost and Related Ground-ice Terms*. National Snow and Ice Data Center/World Data Center for Glaciology, Boulder, CO. World Wide Web Address: http://nsidc.org/fgdc/glossary.

VAN GASSELT, S., REISS, D., THORPE, A. K. & NEUKUM, G. 2005. Seasonal variations of polygonal thermal contraction crack patterns in a south polar trough, Mars. *Journal of Geophysical Research*, **110**, E08002, doi: 10.1029/2004JE002385.

VAN GASSELT, S., HAUBER, E. & NEUKUM, G. 2010. Lineated valley fill at the Martian dichotomy boundary: nature and history of degradation. *Journal of Geophysical Research*, **115**, E08003, doi: 10.1029/2009JE003336.

VINCENDON, M., MUSTARD, J., FORGET, F., KRESLAVSKY, M., SPIGA, A., MURCHIE, S. & BIBRING, J.-P. 2009. Mid latitude CO_2 ice deposits analyzed with CRISM and OMEGA. *In*: *Third International Workshop on Mars Polar Energy Balance and the CO_2 Cycle, held July 21–24, 2009 in Seattle, Washington*. Lunar and Planetary Institute, Houston, TX. LPI Contribution No. **1494**, abstract 7021, pp. 61–62.

VINCENDON, M., FORGET, F. & MUSTARD, J. 2010*a*. Water ice at low to mid latitudes on Mars. *Journal of Geophysical Research*, **115**, E10001, doi: 10.1029/2010JE003584.

VINCENDON, M., MUSTARD, J., FORGET, F., KRESLAVSKY, M., SPIGA, A., MURCHIE, S. & BIBRING, J.-P. 2010*b*. Near-tropical subsurface ice on Mars. *Geophysical Research Letters*, **37**, L01202, doi: 10.1029/2009GL041426.

WALKER, A. S. 1997. *Deserts – Geology and Resources*. USGS Online Book. World Wide Web Address: http://pubs.usgs.gov/gip/deserts/.

WARD, W. R. 1973. Large-scale variations in the obliquity of Mars. *Science*, **181**, 260–262.

WEWEL, F., SCHOLTEN, F. & GWINNER, K. 2000. High Resolution Stereo Camera (HRSC) – Multispectral 3D-data acquisition and photogrammetric data processing. *Canadian Journal of Remote Sensing*, **26**, 466–474.

WILLIAMS, K. E., TOON, O. B., HELDMANN, J. L., MCKAY, C. & MELLON, M. T. 2008. Stability of mid-latitude snowpacks on Mars. *Icarus*, **196**, 565–577.

WILLIAMS, K. E., TOON, O. B., HELDMANN, J. L. & MELLON, M. T. 2009. Ancient melting of mid-latitude snowpacks on Mars as a water source for gullies. *Icarus*, **200**, 418–425.

WORSLEY, P. & GURNEY, S. D. 1996. Geomorphology and hydrogeological significance of the Holocene pingos in the Karup Valley area, Traill Island, northern east Greenland. *Journal of Quaternary Science*, **11**, 249–262.

YOSHIKAWA, K. 1993. Notes on open-system pingo ice, Adventdalen, Spitsbergen. *Permafrost and Periglacial Processes*, **4**, 327–334.

YOSHIKAWA, K. & HARADA, K. 1995. Observations on nearshore pingo growth, Adventdalen, Spitsbergen. *Permafrost and Periglacial Processes*, **6**, 361–372.

ZANETTI, M., HIESINGER, H., REISS, D., HAUBER, E. & NEUKUM, G. 2010. Distribution and evolution of scalloped terrain in the southern hemisphere, Mars. *Icarus*, **206**, 691–706.

Water ice sublimation-related landforms on Mars

N. MANGOLD

Laboratoire Planétologie et Géodynamique de Nantes, CNRS/Université Nantes, 44322 Nantes, France (e-mail: nicolas.mangold@univ-nantes.fr)

Abstract: Sublimation-related landforms are ubiquitous on Mars, especially at mid to high latitudes. This paper reviews the main landforms interpreted to form due to sublimation of subsurface ice on Mars. Pits, knobs and dissected terrains are classical landforms thought to form due to subsurface ice sublimation as observed with high-resolution imagery. Sublimation-related processes on Mars are strongly latitude dependent, with sublimation being increasingly important from high (>60°) to low latitudes (down to 25°) due to correspondingly higher mean annual temperatures. Equatorial regions (within 25° latitude) are mainly devoid of any sublimation-related landforms, reflecting an ice-free shallow subsurface. Mean temperatures and water vapour pressure strongly control the sublimation rate, but diffusion and water adsorption are fundamental and vary depending on the regolith porosity and composition, leading to variations in the theoretical depth at which water ice becomes stable. From a geomorphological point of view, this review highlights the importance of subsurface structure (fractures, layering) in the shaping of landforms and in the control of sublimation rates, in addition to usual physicochemical parameters.

Sublimation is the change of state in which a solid becomes a vapour without passing through the liquid phase. On Earth, liquid water plays such an important role that this process is minor, difficult to distinguish from snowmelt-related process and is only well identifiable in the driest periglacial environments, such as Antarctica (e.g. Marchant *et al.* 2002). On Mars, this process is a major process creating unique landforms. Indeed, the low temperatures (−50 °C on average) and the low water vapour pressure favour sublimation, limiting any liquid water phase processes at the surface. Landforms currently interpreted as having been formed by water ice sublimation on Mars are reviewed, and the results from experimental and theoretical studies required for a better understanding of this process discussed.

Observations

Early investigations

Martian cold surface temperatures and low atmospheric pressure suggest that the role of ice sublimation may be enhanced compared to Earth (Smoluchowski 1968). Sublimation-related processes are generally invoked when landforms such as closed depressions are present, not explainable by classic erosional processes. Pioneering studies using *Mariner* and *Viking* images (resolving landforms typically ≫ 100 m in size) discovered pits possibly related to volatile loss. A series of deep troughs, Cavi Angusti and Sysiphi Cavi (Cutts 1973; Ghatan *et al.* 2003), surrounding the south polar cap show a unique form in which eolian deflation and sublimation may play a role (Cutts 1973). The formation of these pits is unrelated to any volcanic or tectonic activity, as demonstrated by the lack of predominant directions in pitting and the lack of apparent caldera (Fig. 1a). Most recent studies suggest that they may be the result of the melting of deep ice embedded inside ancient polar deposits (Milkovich *et al.* 2002; Ghatan *et al.* 2003). Such melting could reflect an increased thermal gradient in the past, possibly associated with magmatic activity, causing the deepest polar layers to melt. Hence, in such cases, ice sublimation may not be the major process involved.

Other typical examples are the pitted craters, in the range of 10–100 km in diameter, usually observed at mid latitudes. These craters display strong degradation, with pits 1–3 km in depth and 1–10 km in width, which become coalescent and sometimes create a pitted ring (Fig. 1b). These pits require a high loss in volume to form, which is difficult to explain without involving either dissolution, or volatile loss, by sublimation or evaporation. Thaw of an ice-rich subsurface has been proposed at more equatorial latitudes (Costard & Kargel 1995), analogous to 'thermokarst'. On Earth, elliptical pits, often coalescent, are caused by the thaw of ice lenses (almost pure water ice formed by ice segregation in the ground), creating 'alases' – a thermokarst landform frequent in Siberia. A similar process due to sublimation may be an alternative that would not require any climatic or magmatic effect. In both cases (a dry or a wet volatile loss), the presence of water at a depth of more than 1 km in the subsurface is still debated

From: Balme, M. R., Bargery, A. S., Gallagher, C. J.,& Gupta, S. (eds) *Martian Geomorphology*. Geological Society, London, Special Publications, **356**, 133–149.
DOI: 10.1144/SP356.8 0305-8719/11/$15.00

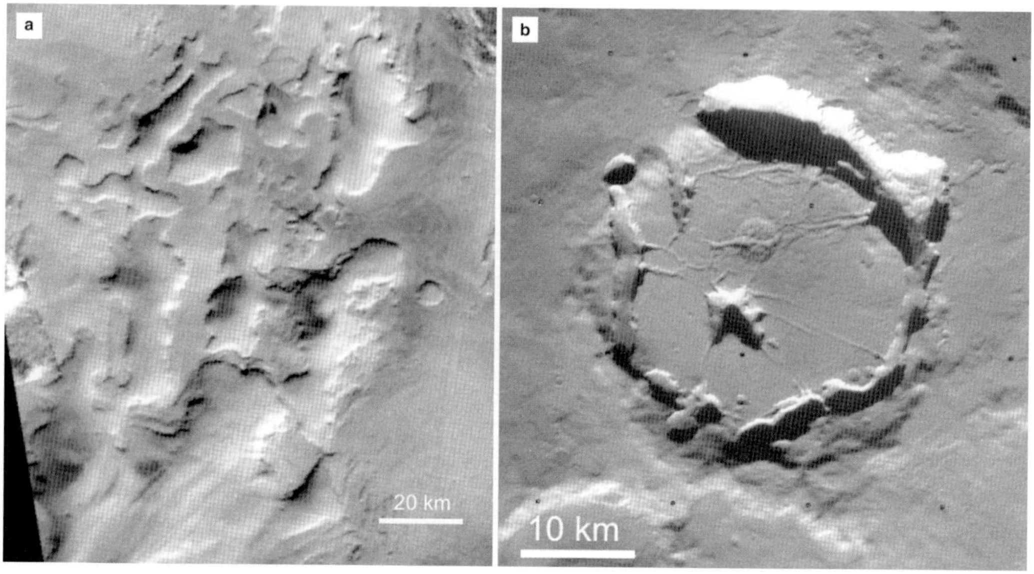

Fig. 1. (**a**) Cavi Angusti in the south polar layered deposits (*Viking* mosaic). (**b**) Pitted crater suggestive of volatile loss. *Viking* image located at 46°S, 354°W. Image credit: NASA; see prelim viii for acronym definitions.

on Mars: on one hand, craters with lobate ejecta suggest that subsurface ice exists at depths of more than 300 m, depending on latitude (e.g. Costard 1989); on the other hand, the lack of positive geophysical detection, such as by radar sounding (Farrell *et al.* 2009), maintains a degree of doubt as to its presence.

These examples show how sublimation-related landforms are difficult to identify in the absence of ice identification by geophysical sounding. However, a closer look using high-resolution imagery gives a better view of potential sublimation-related landforms.

Sublimation of water ice on polar caps

The most direct evidence of water ice sublimation on Mars comes from the observation of the polar caps. A scalloped texture is found at high resolution all over the northern residual water ice cap (Fig. 2). Ice pits probably form as the result of ablation, combining sublimation during the northern summer and wind entrainment of residual dust (Malin & Edgett 2001). The lack of small craters on the polar caps shows that this process acts quickly at the geological scale, degrading craters until they are removed completely. As the dust content of the cap is low (<20%) and the wind is strong, dust particles liberated by the ice sublimation are probably lifted into the air and do not accumulate over the residual cap. At the southern polar caps, well developed pits, typically 100 m in diameter, 10 m in depth, are interpreted as the effect of sublimation of CO_2-rich ice (e.g. Piqueux *et al.* 2003; Bibring *et al.* 2004). These landforms are specific to the southern residual cap, but, given that they form in a non-water ice material (solid CO_2), they are not described here.

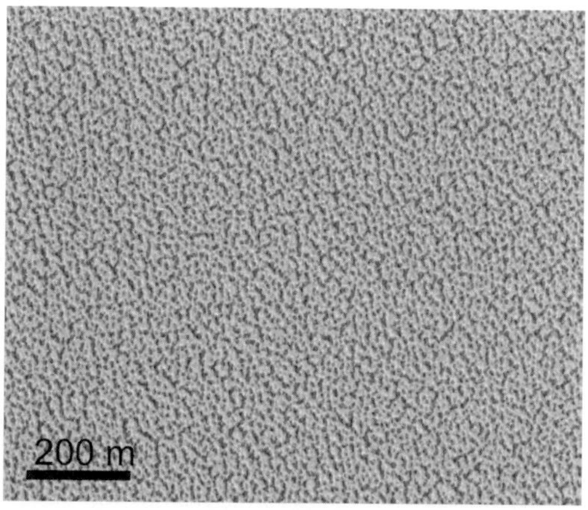

Fig. 2. Mars Orbiter Camera (MOC) image of the north polar residual ice cap (PIA09387; 85.1°N, 284.6°W: taken during the northern summer with illumination from the lower left). As water ice sublimes away a little bit each summer, dark-floored pits have formed, trapping dust and other debris. Image credit: NASA/JPL/MSSS.

Sublimation of subsurface ice at high latitudes

At latitudes poleward of 50°–55°, polygonal shapes are observed all over the planet, in both northern and southern hemispheres (Mangold 2005; Levy *et al.* 2009). Here, water ice should be close enough to the surface that thermal contraction resulting from temperature variations by seasonal cycles creates a quasi-systematic polygonal pattern through widespread regions (Fig. 3). The *Phoenix Lander* recently imaged the surface at approximately 68°N and showed that water ice is present below a dry soil (Smith *et al.* 2009). At latitudes of more than 60°, gamma-ray and neutron spectroscopy have shown the presence of excess hydrogen, reflecting either pore ice in the regolith or ice-dominant layers, under a thin desiccated regolith (Boynton *et al.* 2002; Feldman *et al.* 2002). Water ice is, therefore, close to stability at a few centimetres into the ground.

Nevertheless, a difference is seen in Mars Orbiter Camera (MOC) images between: (1) fresh polygons at latitudes of more than 70°, where cracks are thin and polygons well defined; (2) slightly degraded polygons at latitudes 60°–70°, where a degradation of cracks is visible but polygons are still well identifiable; and (3) strongly degraded polygons at latitudes roughly below 55°–60°, where most polygons are modified and sometimes difficult to identify (Mangold 2003). In this last range of latitudes, terrains are often named basket-ball terrain (Kreslavsky & Head 2002) and may represent a final stage of degradation of polygonal structures. Nevertheless, recent High Resolution Imaging Science Experiment (HiRISE) data show the presence of well-preserved small (<30 m) polygons at 45°–55° latitude, which are well preserved presumably because they are very young (Levy *et al.* 2009). Thus, the progressive change of polygon morphology from the highest latitudes to the lower can be explained by the combined role of age variability and fracturing, which amplifies the effect of sublimation and is stronger at low latitudes. Similar polygonal shapes combining thermal contraction in ground ice and sublimation are observed in Antarctica over debris-covered glaciers (Marchant *et al.* 2002). Although the surface texture is not purely related to sublimation, this latitude-dependence suggests a strong morphological change by subsurface ice sublimation.

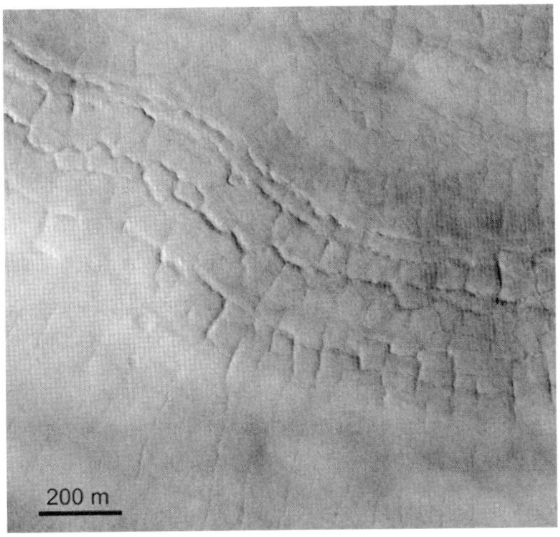

Fig. 3. HiRISE image PSP_01942-2310 of polygonal structures at 65°N. Polygons probably formed by thermal contraction are bounded by fractures. These fractures display varying widths, suggesting that sublimation widened them compared to thinner fractures in the surrounding area. Image credit: NASA/JPL/UofA.

Sublimation of subsurface ice at mid latitudes

At mid latitudes, a distinctive assemblage of small landforms (<100 m high) comprising pitted and knobby terrains with heterogeneous shapes and residual mesas is visible on MOC and HiRISE images (Figs 4 & 5). In detail, these terrains can be described as composed of three main units (Mangold *et al.* 2000*a*): (1) a smooth non-degraded unit with only few pits, which forms an homogeneous mantle of regular thickness draping the underlying topography; (2) a partially dissected unit with many knobs and a residual texture; and (3) a nearly fully dissected unit with few residual knobs and mesas. Coalescing pits on the non-degraded terrain may correspond to the first stage of degradation (CSP in Fig. 5). Cracks lacking specific orientation cross the dissected terrains and may come from the sublimation process itself, although their exact origin remains uncertain (C in Fig. 5). The thickness of the degraded unit may be in the range of metres to tens of metres at maximum (Mustard *et al.* 2001). Such features usually do not follow specific patterns, in contrast to polygonal shapes at higher latitudes. Locally, the dissection displays a regular wavelength along a given orientation, the patterns resembling eolian ripples or dunes (Fig. 4c). It may be that the upper layer consists not only of loose dust (similar to loess on Earth) but also sand-sized grains, and formed as small dunes initially stabilized by water ice degraded subsequently by sublimation. In addition, asymmetry between pole-facing slopes and equator-facing slopes is also observed (Fig. 6) in regions where the relief is sufficient to create

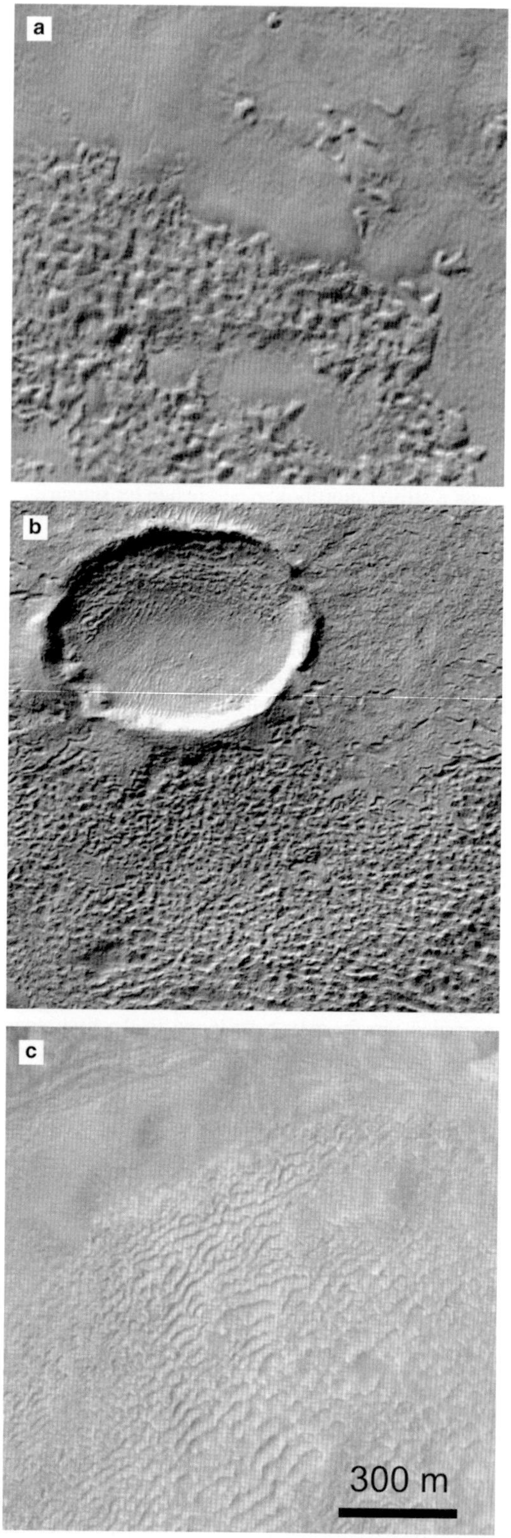

insolation differences between slopes, either during ice deposition or its later sublimation.

The development of pits suggests a volume change or differential removal of material. The occurrence of certain landforms on slopes with specific solar aspect (Fig. 6), and their correlation with middle latitudes, suggests that volatiles were emplaced within the mantling dust and that ice sublimation created the observed textures (Mangold *et al.* 2000*a*; Mustard & Cooper 2000; Mustard *et al.* 2001; Malin & Edgett 2001; Mangold 2003). Mapping of these landforms shows that they are concentrated in two approximately symmetrical latitude bands spanning 30°N–70°N and 25°S–65°S (Mustard *et al.* 2001), suggesting a correlation with processes associated with atmospheric general circulation. Fresh impact craters of any size are very rare, indicating that ages less than 1 Ma are common, with youngest ages of only several tens of thousands of years close to the pole (Mustard *et al.* 2001; Levy *et al.* 2009). On the basis of these lines of evidence, it can be concluded that this eroded mantle represents regions on Mars where an eolian layer of dust mixed with ice is experiencing erosion by processes involving water ice sublimation. The formation of this latitude-dependent mantle (LDM) is still under debate.

Mid-latitude regions on Mars have long been suspected of displaying patterns created by ice (e.g. Squyres & Carr 1986). Lobate debris aprons, lineated valley fill and concentric crater fills were observed in *Viking* images and related to viscous ice flows, that is, glacier movements (Carr & Schaber 1977; Squyres 1978; Squyres 1989). More recent laser altimeter profiles of these landforms (Mangold & Allemand 2001) and radar data (Plaut *et al.* 2009) have confirmed these early interpretations. Current interpretations involve a high proportion of ice (80% from radar data: Plaut *et al.* 2009) and formation by atmospheric precipitation (snow) as the predominant process (Head *et al.* 2005; Forget *et al.* 2006). At the time of *Viking* images, the highest resolution imagery (at 30–40 m per pixel) of the surface of these lobate aprons showed pits and closed depressions that Squyres (1989) tentatively interpreted to reflect sublimation of the glacial ice. Post-*Viking* imagery now confirms this interpretation (Mangold 2003). Indeed, a dissected mantle similar to other

Fig. 4. Close-ups of three MOC images at 45°–50° latitude south showing the latitude-dependent mantle with dissection and pitting that occurred by water ice sublimation. (**a**) MOC image FHA00982. (**b**) MOC image FHA1450. (**c**) MOC FHA00734. Image credit: NASA/JPL/MSSS.

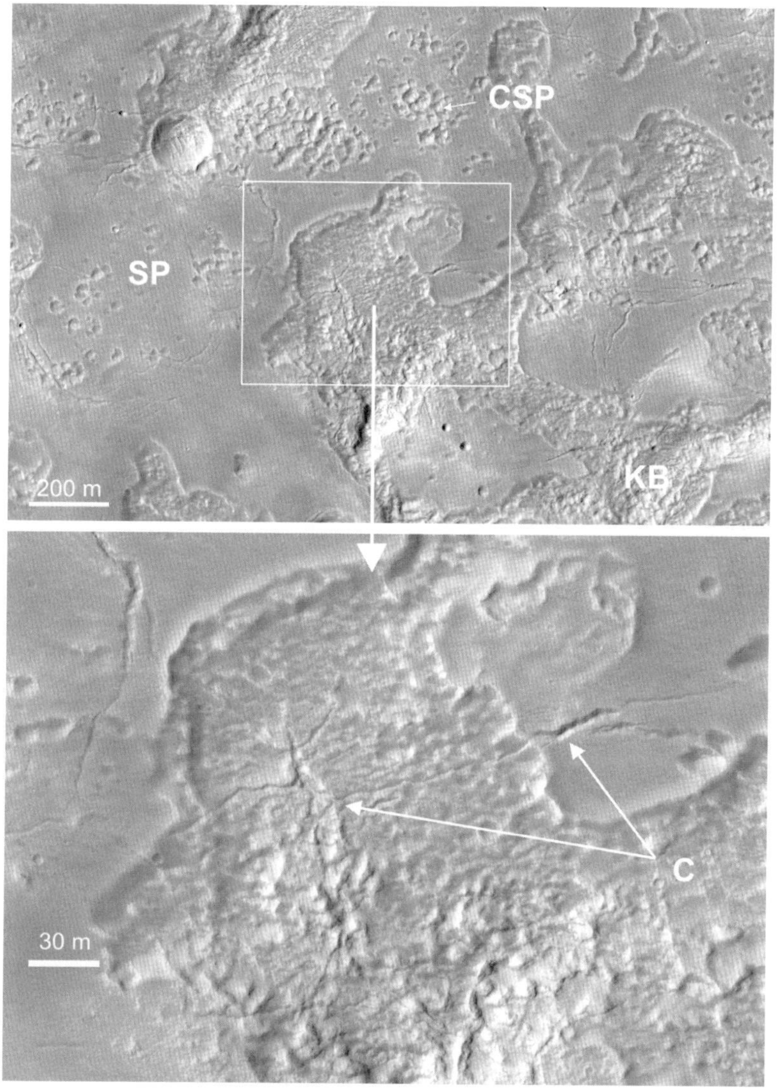

Fig. 5. Close-ups of HiRISE image 9772_2080 over the dissected mid-latitude mantle at 45°N. SP, smooth terrains with local pits; CSP, coalescing pits on smooth terrains; KB, knobby terrains after strong degradation; C, cracks. Image credit: NASA/JPL/UofA.

mid-latitude regions is observed over lobate debris aprons but pre-existing ice patterns created a strong diversity of shapes (Fig. 7). Mid-latitude glaciers exhibit ghost craters, reflecting the progressive degradation of impact craters, complex patterns named brain-like structures (Fig. 7c) or, sometimes, brain-coral terrain, and a variety of dissected terrains. These morphologies all relate to the role of ice sublimation over pre-existing landforms of various aspect and size (Malin & Edgett 2001; Mangold 2003; Levy *et al.* 2009). The presence of fractures resulting from past glacier movement, which can have a regular orthogonal shape or more curved pattern, induces preferential zones of ice sublimation (Mangold 2003). In such a process, ice near the fracture comes quickly into contact with the atmosphere and can sublime more rapidly than on a flat surface, where a residual lag would slow sublimation. This effect probably explains why the surface of viscous landforms is affected by such a variety of patterns. The fact that these landforms are sometimes very thick (>800 m in some lobate aprons) also implies that water ice is present in large amounts.

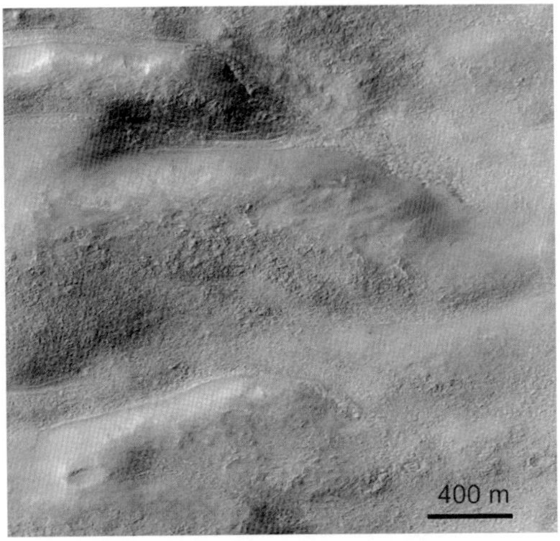

Fig. 6. MOC image (M04-01761) of alternating smooth textured and pitted terrains at 38°S. The smooth terrains are present on equator-facing slopes probably because ice was not able to condense on these warmer surfaces. Pitted textures correspond to ice-bearing regolith currently under sublimation. Image credit: NASA/JPL/MSSS.

Two mid-latitude locations in which ice sublimation landforms are common are Utopia Planitia in the northern plains, and Peneus Planum, south of Hellas Planitia in the southern hemisphere. Both regions are characterized by scalloped terrain, which is found only sparsely outside these two regions (Fig. 8). The surface of western Utopia Planitia is mostly flat, locally pitted and crossed by a polygonal pattern of troughs (Seibert & Kargel 2001). Scalloped depressions and other periglacial-like landforms such as polygons and small mounds are particularly concentrated in the western part of Utopia Planitia at approximately 45°N (Lefort *et al.* 2009). Scalloped depressions are rimless, shallow and ovoid in form, ranging from circular to elongate with morphology independent of their size. Individual scallops range from a few hundred metres to nearly 3 km wide and appear isolated or in clusters of varying density (Lefort *et al.* 2009). The main feature of a typical scallop is a pole-facing scarp inclined 15°–30°. Opposite the scarp is a gentle equator-facing slope, sometimes almost flat, but typically with a slope of 2° (Lefort *et al.* 2009). Thermal Emission Imaging System (THEMIS) maps show temperatures about 10K higher on the gentle rise than on the pole-facing scarp, consistent with the relative difference in local angles of solar incidence. The proposed process of scalloped terrain formation involves initially slight hummocks or depressions with relatively higher near-surface temperatures on their equator-facing slopes, leading in turn to enhanced sublimation of ground ice on these equator-facing slopes (Lefort *et al.* 2009). This enhanced sublimation leads to an asymmetric scallop-shaped depression with a progressively retreating gentle rise and a steeper pole-facing scarp. Over time this process deepens and extends the scallop, mainly by erosion of the equator-facing rise. This process is a plausible process involving ground ice sublimation with local heterogeneities. Similar patterns are observed in the southern hemisphere in the Peneus–Amphitrites region located at 50°–55°S, south of Hellas Planitia. They support the idea of recent degradation of a mid-latitude, ice-rich mantle as in Utopia (Zanetti *et al.* 2008; Lefort *et al.* 2009). Thus, scalloped depressions are landforms resulting from sublimation of the mantling terrains but this does not explain the lack of such scalloped terrains elsewhere if this entire latitude range is covered by an ice-rich mantle. Alternatively, the presence of pure ice lenses, or a high proportion of ice (>60%), may be necessary to explain this process. A possibility explaining this enrichment in ice is that climate models predict the presence of thick ice originating from snow accumulation in the regions surrounding both Hellas, which is close to Peneus Planum, and Utopia (Forget *et al.* 2006; Séjourné *et al.* 2009).

An alternative to a high proportion of water ice originating from precipitation is that ice lenses may have been formed in repeated freeze–thaw. The possibility that the scalloped depressions originated as ancient thermokarst lakes has been proposed by several authors (Soare *et al.* 2008). A thermokarst lake is formed by subsidence and subsequent settling of the ground resulting from the thawing of ice in the active layer above the thick permafrost. On Earth, liquid water usually fills the resulting depression, forming a lake. However, scallop floors are tilted and occur at different altitudes even when being coalescent, and no landforms involving water processes have been observed at high resolution.

In some areas of Utopia, small circular or elongated pits (Fig. 8) form chains within wide, north–south-oriented polygon troughs. These pits do not form in the east–west troughs (Lefort *et al.* 2009). They are symmetrical with respect to trough centres, range up to 20 m wide and 150 m long, occasionally coalesce, have a flat floor and are apparently shallower than the scalloped depressions. These pits also are inferred to form by sublimation. By analogy, terrestrial mudcracks form polygons similar to those formed by thermal contraction, with the exception of their smaller scale. On Earth, water evaporation from supersaturated mud can create some collapse along cracks (Fig. 9). Similar effects seem present in the Utopia

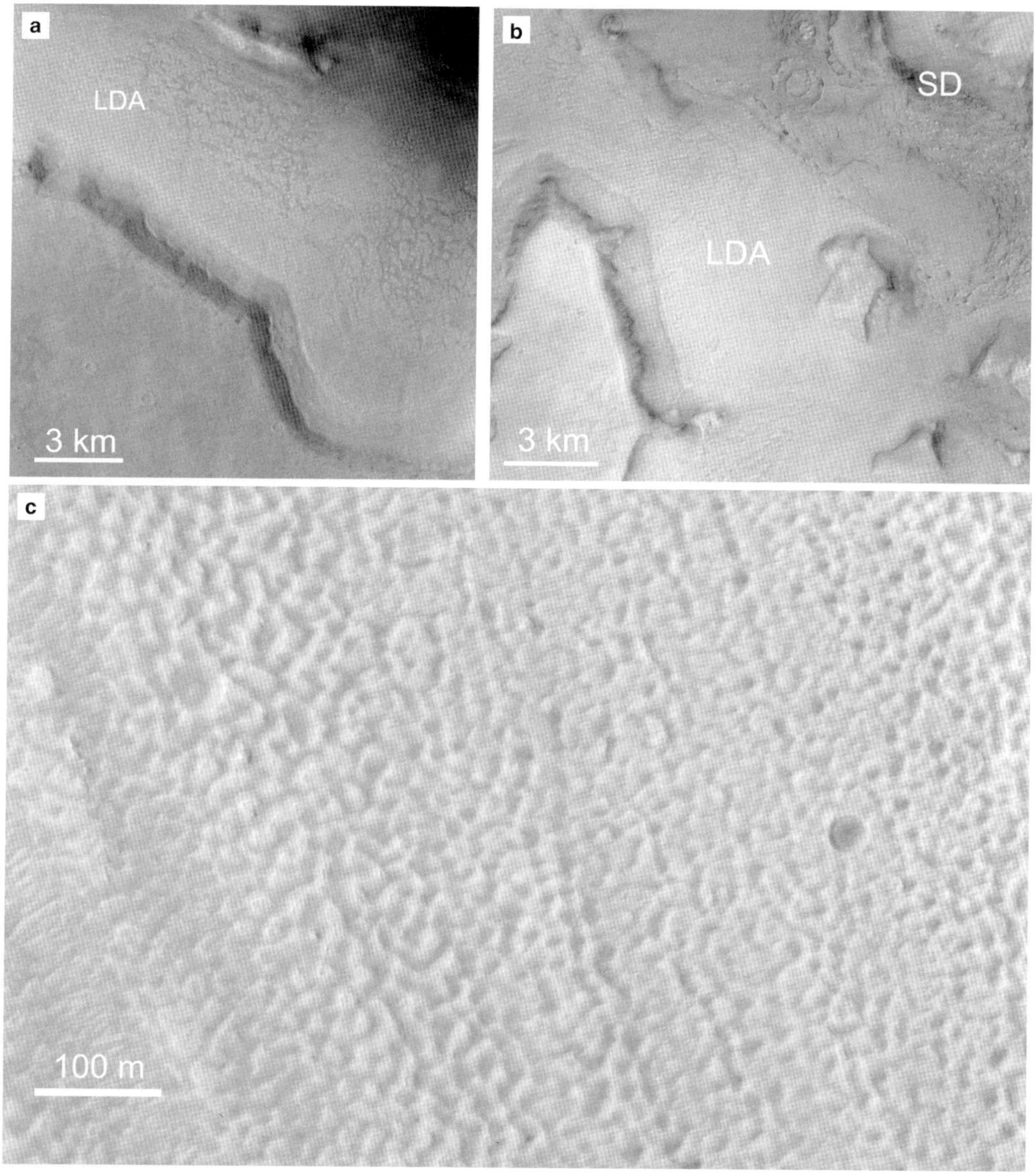

Fig. 7. (Top) Two HRSC images in Deuteronilus Mensae. LDA, lobate debris aprons; SD, strongly degraded aprons. These terrains display pits and degradation at 100 m scale. (Bottom) In the high-resolution HiRISE image (10854-1325), terrains apparently smooth at HRSC scale are strongly pitted at the 10 m scale. On the right a fresh crater not yet affected by sublimation is present. Such craters are rare, showing that degradation by sublimation is a rapid process at this latitude. Image credit: ESA/DLR/FUB, NASA/JPL/UofA.

polygons, with the difference that the collapse is controlled by surface temperature and solar aspect, and the pitting is consistent with an ice-removal process (sublimation). Scalloped landforms and pitted polygonal cracks are sublimation-related landforms for which north–south dissymmetry and collapse are the indicators of the presence of a volatile in the ground. No action of liquid water seems necessary, although freeze–thaw cycles would have helped the process if they occurred.

In summary, dissected terrains in the mid latitudes (30°–55°) display pits, knobby textures, usually without regular patterns. Regular patterns are sometimes observed for those terrains present

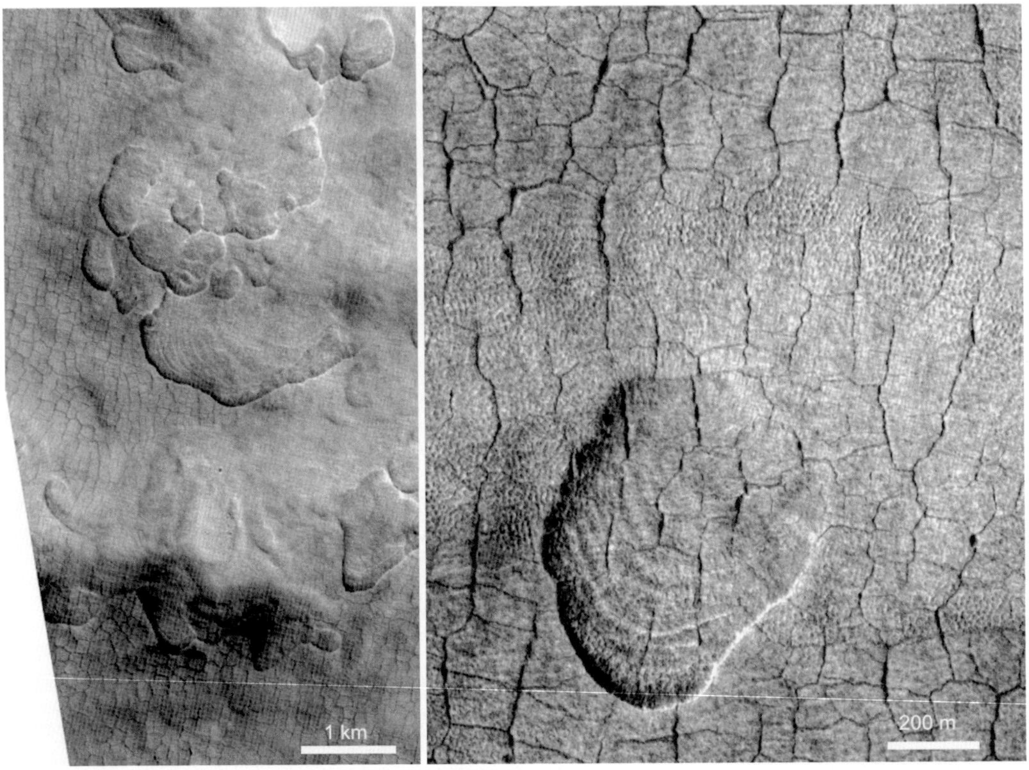

Fig. 8. Scalloped terrain in Utopia Planitia. From left to right: MOC image M04-1631 (left) and HiRISE close-ups of PSP_010034_2250 (centre). Scalloped terrains are the ovoid dissymmetric kilometre-scale depressions crossing polygonal patterns. North is to the top. Image credit: NASA/JPL/MSSS, NASA/JPL/UofA.

on viscous landforms (lobate aprons, etc.), containing eolian landforms (ripples, etc.) or presenting asymmetry resulting from aspect-controlled slope effects. High-latitude landforms generally display polygonal patterns related to the widespread presence of thermal contraction cracks. All of these textures could be named 'dry thermokarst' or 'cryokarst' by analogy to terrestrial thermokarst, in which ice melting is involved. Differential sublimation plays a strong role in the shaping of these different landforms (Fig. 10).

The role of impact craters at mid latitudes

Impact craters can provide excellent examples of sublimation effects related to different material properties. Crater ejecta on Mars is very diverse and continuous ejecta blankets are common, in contrast to the Moon, on which ejecta rays are predominant (e.g. Costard 1989). Continuous ejecta blankets are indicative not only of subsurface composition but can also provide information on morphogenesis related to sublimation. As the ejecta are continuous they can protect buried ground ice from sublimation (Meresse *et al.* 2006). This effect is spectacular for pedestal craters that are observed in the northern regions of the Utopia and Chryse Planitia (Fig. 11). Some of these pedestal craters display excess volumes with regard to the crater diameter (Meresse *et al.* 2006). Sublimation of ground ice in the northern lowland plains combined with wind deflation may be responsible for both of the apparent high ejecta volumes for the perched craters. Indeed, it is postulated that the northern plains were filled by ice-rich eolian material (Meresse *et al.* 2006) and that pedestal craters formed in this ice-rich unit by a process in which sublimation initially dissects the plain. However, because the erosion of material over the ejecta occurs at a lower rate than the removal of materials from the surrounding unit, the perched crater morphology develops (Meresse *et al.* 2006), the form being a residual. Continuous ejecta may be more cohesive and less porous than the ice-rich surface unit and pedestal craters sometimes show pits at the location of the ejecta boundary. This is indicative of enhanced sublimation at this location (Kadish *et al.* 2009), confirming the role of

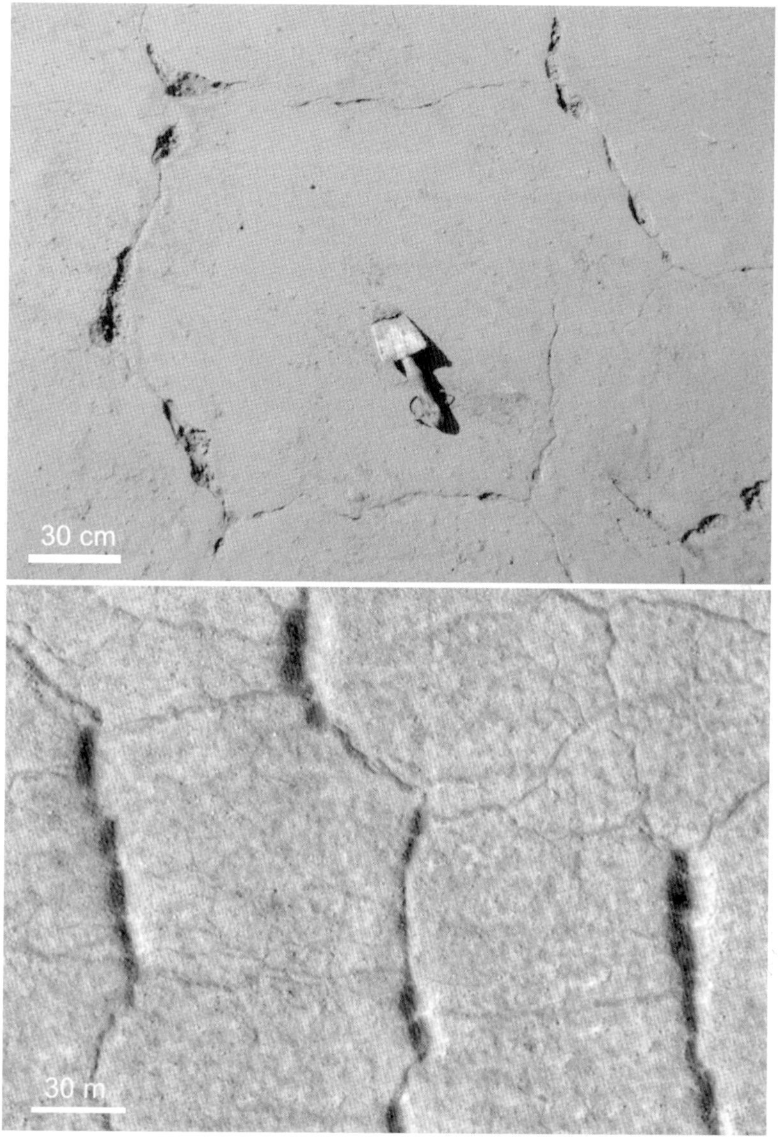

Fig. 9. Comparison between desiccation cracks with collapse at crack boundaries (outwash plain, Iceland: image by N. Mangold), and thermal contraction cracks in Utopia with sublimation pits along cracks (image PSP_010034_2250). Image credit: NASA/JPL/UofA.

differential sublimation rates in the shaping of these landforms.

Sublimation-related landforms in equatorial regions

In Martian equatorial latitudes several regions display possible sublimation-related landforms. Residual moraines of tropical glaciers are observed at the foot of the Tharsis volcanoes (e.g. Head *et al.* 2005; Fastook *et al.* 2008). The material left behind is similar to sublimation tills observed in Antarctica (Marchant *et al.* 2002). It is unlikely that any equatorial ice exists on Mars close to the surface owing to the strong sublimation rate at the equator, but deeper ice (>10 m) cannot be excluded.

A series of periglacial landforms is observed in the vicinity of Cerberus Fossae and Central Elysium Planitia (Page 2007). Polygonal patterns, only sparsely present close to the equator, are

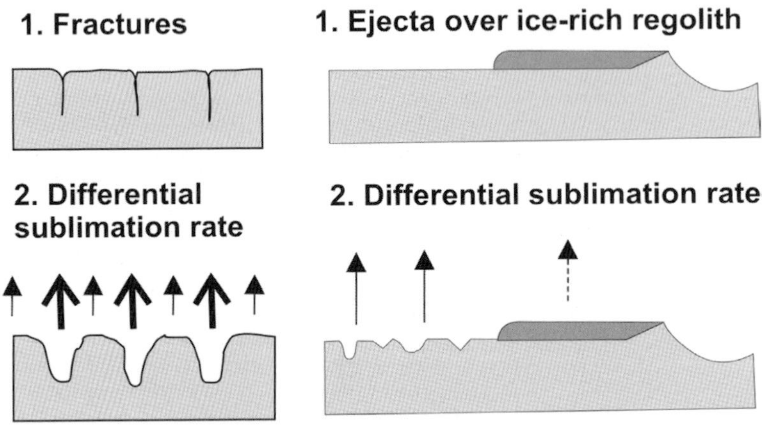

Fig. 10. Two examples of differential sublimation rates shaping two types of landforms. (Left) Cracks enhanced by sublimation widen progressively. Cracks become very degraded if subjected to long periods of sublimation or if located in regions where this process is especially efficient. (Right) Crater ejecta playing the role of a protective cap above ice-rich terrains and limiting further regolith–atmosphere exchange below the ejecta.

observed in many high-resolution images of the Cerberus plains and Marte Vallis channel, together with possible pingos and sorted circles (features formed by freeze–thaw cycles with water present: Page 2007; Balme *et al.* 2009). Local 100 m-wide depressions are observed in these polygonally patterned terrains. They may require volatile loss or local subsidence (e.g. fig. 7 in Page 2007). Explanations for these depressions include sublimation of a shallow ice-rich layer or the melting of the same layer forming alases, as observed in Ares Vallis (Costard & Kargel 1995). This region is characterized by recent lava flows, small shield volcanoes and a few outflow channels that emerge from volcanic fissures (Burr *et al.* 2002; Page 2007; Vaucher *et al.* 2009), and, in general, landforms in these regions are morphologically different from the usual pitted terrains observed at mid latitudes. Hence, they probably involved more complex interactions with local geology because of this specific context.

Equatorial regions also display enigmatic landforms that have no definitive explanation yet. Among these exists a pattern identified only in the vicinity of recent craters of several tens of kilometres in diameter. For example, Mojave Crater

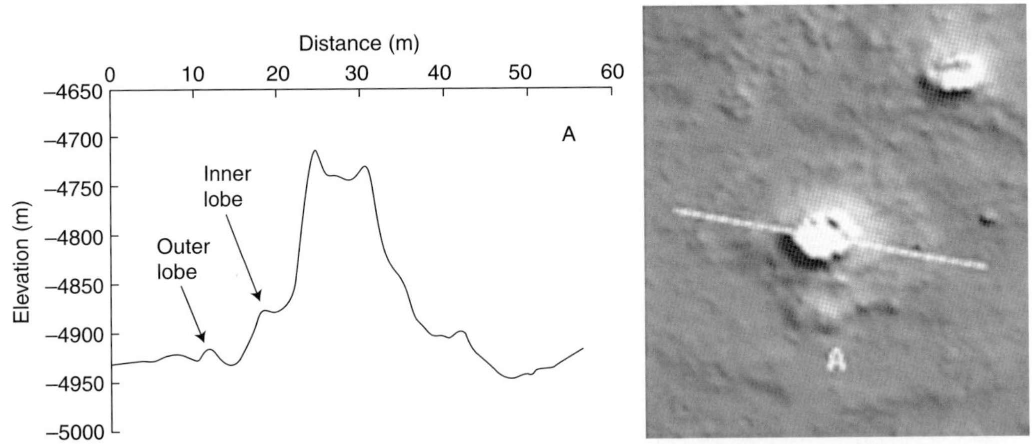

Fig. 11. Topographical cross-section of a pedestal impact crater that shows a strong inner filling and thick ejecta that cannot be explained by the volume of material excavated by the impact (from Meresse *et al.* 2006). Ejecta blankets preserve the underlying ice-rich terrains from sublimation, whereas the terrain outside ejecta blankets experience enhanced sublimation.

(Fig. 12), located 7°N, displays striking fluvial patterns suggesting that intense runoff occurred possibly as a result of the warming of the impact crater on the crust or its environmental consequences (e.g. Tornabene *et al.* 2007; Williams & Malin 2008). Some specific pitted terrains exist on the bottom of this crater, some just below debris fans (Fig. 10). These pits are very similar to the pitted textures observed at mid latitudes (Fig. 5) and therefore may involve volatile loss. Possible explanations include the devolatilization of impact melt, formed as a consequence of the crater impact (Tornabene *et al.* 2007). However, relationships with fluvial landforms, as shown in Figure 12, more probably suggest enhanced sublimation/evaporation of ice-rich/water-rich deposits that formed in after a period of intense fluvial activity. If so, these pitted landforms would then form only when the material is strongly out-of-equilibrium, as is the case for water ice at the equator and for liquid water in general on Mars.

Thus, although sublimation-driven landforms are not as common at low latitudes as they are at mid latitudes, they are observed in local environments.

The role of orbital parameters

Sublimation-driven landforms are ubiquitous in mid and high latitudes, while only locally observed in equatorial latitudes. This latitude dependence relies on the interactions between the Martian hydrological cycle and the forcing from orbital parameters. Current obliquity is 25°, but it could have reached 45° in the last 10 Ma (e.g. Laskar *et al.* 2002). Such high obliquity (45°) profoundly modulates water ice transport and deposition. During high-obliquity periods, polar caps tends to vanish whereas equatorial glaciers can build (Forget *et al.* 2006), such as around Tharsis Montes. These periods are the ones during which ice can accumulate in equatorial regions, locally forming glaciers that subsequently experience sublimation when obliquity decreases again. In addition, enhanced sublimation processes may occur in specific locations of the equatorial regions where ice is preserved in localized deposits, such as those associated with recent outflow channels. In contrast, water ice is close to stability in the mid and high latitudes, and these zones are so sensitive to small

Fig. 12. Mojave crater HiRISE image PSP9076-1880: 10 m-scale coalescing pits reflect the intense degradation of the lowest slopes of a debris fan in the top right of the image. Image credit: NASA/JPL/UofA.

obliquity that they do not require the obliquity to be as high as 45° in order to have water ice in the shallow subsurface. The mid- and high-latitude ice-rich mantles are likely to be a result of the most recent variations in obliquity (Head *et al.* 2003; Forget *et al.* 2006; Schorghofer 2007), but they are still in slight disequilibrium considering current conditions. Thus, mid and high latitudes display more sublimation-related landforms than equatorial regions as a result of the more widespread presence of shallow ice in more recent ages.

Process of sublimation: experiments and theory

Sublimation at the polar cap

When ice is exposed to vacuum at a temperature close to, but less than, the freezing point of water, the ice sublimates rapidly and leaves behind a residue of the particles it contained. The loss rate $\mathrm{d}m/\mathrm{d}t$ is given by:

$$\mathrm{d}m/\mathrm{d}t = \alpha p_{\mathrm{H_2O}}\sqrt{M/2\pi RT}$$

where α is the coefficient of sublimation, $p_{\mathrm{H_2O}}$ is the vapour pressure of the ice, M is the molecular weight of the vapour, T the temperature and R is the universal gas constant. A value for α of 0.94 ± 0.06 was empirically determined by Tschudin (1946).

At the Martian poles, water ice is close to stability. Nevertheless, sublimation during the summer season is a major process controlling the polar caps' formation and evolution (Hofstadter & Murray 1990; Skorov *et al.* 2001). Sublimation is effective and certainly explains the progressive loss of small impact craters. Nevertheless, landforms related to sublimation, such as pits, are not as well developed nor as deep as such landforms require differential sublimation, which is possible only with high solid-grain content. With only 10–20% of dust grains, the polar caps may be too clean to develop the same landforms as the regolith.

Sublimation of ground ice

Smoluchowski (1968) demonstrated that under stable climatic conditions, a relatively thin (1 m), fine-grained regolith could act as an effective diffusion barrier even on timescales of billions of years, and so claimed that ice could exist at relatively shallow depths on Mars, even in equatorial regions. The results of this pioneering study are to this day in agreement with ground truth measurements from the most recent probes.

Parameters constraining sublimation rates. On Earth, sublimation of porous ice-bearing soils has been studied by generations of arctic researchers. From experience and field observations, terrestrial researchers agree that ambient temperature has the strongest influence on the sublimation of pore ice (Yershov *et al.* 1973; Gobelman 1985; Van Dijk & Law 1995). Sublimation is most pronounced at −1 °C, and a decrease in temperature increases the bond energy of the ice surface and suppresses the rate of ice sublimation (Yershov *et al.* 1973; Huang & Aughenbaugh 1987). Relative humidity was found to be the second most influential factor (Huang & Aughenbaugh 1987). As relative humidity increases, pore-ice sublimation decreases (Gobelman 1985), but when the relative humidity is high (above 80%) there is no significant change in the sublimation rate (Aguirre-Puente & Sukwhal 1984). The effects of relative humidity are significant under Martian conditions, in which the atmospheric water vapour pressure is very low (currently an average of 12 μm of precipitable water ice). Accordingly, variation in atmospheric water vapour pressure must be considered in any model of Martian sublimation rate. At the present water vapour pressure, the frost point is at 198K. Consequently, there is a net ice loss by sublimation for all regions where the mean temperature is higher than 198K, but for the processes of sublimation/condensation for polar regions there is only a net ice loss where summer temperatures are higher than this value.

Although the ice content of frozen sediment is a factor in sublimation, Huang & Aughenbaugh (1987) indicated that it is the least effective variable when compared with temperature and relative humidity. As the ice content of the frozen sediment increases, a decreasing amount of particles are loosened from the surface by sublimation. Initially, grain size has been shown to have no major influence on sublimation rates (Wellen 1979; Johansen *et al.* 1981). Nevertheless, one difference was noted between frozen sand and silt surfaces: as sublimation of the pore ice takes place, sand continuously sloughs off the frozen sand surface, whereas silt stays in place and a desiccated layer accumulates on the frozen silt below. As a result, sublimation occurs more rapidly in sand-sized than in silt-sized material (Wellen 1979). Recently, experiments showed that grain size has a strong effect, by slowing down the sublimation process, for soils composed of very small particles (<10 μm) as proposed for Mars' regolith (e.g. Chevrier *et al.* 2008).

The physical properties of the regolith are more important than the grain size. Porosity and tortuosity (the way by which the water vapour molecules escape the soil), which control the diffusion processes that lead to sublimation and adsorption of water ice on regolith grains, have a major influence on sublimation in fine-grained regolith. Small grain sizes decrease diffusion strongly, and thus significantly reduce the sublimation rate (Smoluchowski 1968). Most recent experimental data have been used to identify the coefficients of diffusion and adsorption for given particle compositions, sizes and shapes (e.g. Chevrier *et al.* 2008). These parameters create the main variations in the sublimation rate of subsurface ice. However, for clay minerals, the theoretical diffusion (as proposed by Smoluchowski 1968) is about 3–4 times less than the coefficients determined experimentally (Chevrier *et al.* 2008). This results probably from a higher tortuosity (around 10), most probably linked to the flat shape of clay particles and a broad range of particle size, both resulting in a more complex geometry. In general, the diffusion coefficient remains in the range 10^{-4}–10^{-3} $m^2 s^{-1}$, especially since larger grain sizes induce larger pore sizes and thus result in larger diffusion coefficients (Smoluchowski 1968; Hudson *et al.* 2007; Bryson *et al.* 2008).

In summary, the distribution of ice on Mars is governed both by equilibrium thermodynamics and by kinetics. The kinetics of water transfer from the subsurface to the Martian atmosphere are largely dependent on regolith diffusion and adsorption properties, which can be studied through experiments on analogues such as basalt (Fanale & Cannon 1971; Bryson *et al.* 2008) or palagonitic soil JSC Mars-1 (Chevrier *et al.* 2007). Experimental data (Chevrier *et al.* 2008) show that adsorption significantly affects the timescales of diffusion, but not its amplitude (i.e. the diffusion coefficient does not change). Many experimental workers have neglected the interaction between the evaporating water and the surrounding porous material, assuming generally that grains are too large to have a significant effect. Adsorption strongly affects the dynamics of water vapour transport because it changes the pressure in the pores of the regolith, and will be most efficient in fine-grained regolith.

In addition to these physicochemical parameters, the ground structure should be taken into account. Fractures in glaciers or cracks in polygonal terrains are known to enhance local sublimation on Earth (Marchant *et al.* 2002), and the patterns observed on Mars over mid-latitude glaciers or polygonal terrains (Fig. 10) are examples of the efficiency of differential sublimation (Marchant *et al.* 2002; Mangold 2003, 2005). Layering, characterized by variable thickness, compaction or water-ice content, leads to differential sublimation rates between layers and can explain the presence of steps or small mesas. Differing aspect can also create differences in temperatures or water vapour pressure that lead to different rates of sublimation. Hence, the large variety of landforms created by sublimation is the result of complex interactions between regolith chemistry, mechanical properties and geological characteristics.

Finally, the majority of studies neglected the effect of wind speed on sublimation from frozen sediments because they were carried out in permafrost tunnels or cold laboratories with no mechanism for simulating the wind. Outdoor studies, on pure ice surfaces, emphasized the importance of wind as a transport agent for the water molecules that had been sublimed from the surface. Experimental data show that sublimation influences the transport of sand under subzero conditions by reducing the binding effectiveness of pore ice (Van Dijk & Law 1995). The loosened grains can then be entrained by the wind and serve to enhance movement by the abrasion of frozen particles downwind. This effect has not yet been fully considered under Martian conditions. It is likely that several landforms would not be as well developed if the residual solid particles were not entrained by wind. Pre-existing fractures and wind effects were poorly taken into account in most recent modelling, which may, therefore, underestimate the effective sublimation rate in regions where these effects occur.

Subsurface water ice distribution on Mars and the latitude-dependence of sublimation processes. The difficulty of applying a single model to the whole surface of Mars is that the regolith properties have a strong role in the kinetics of water vapour. Smoluchowski (1968) theorized that regolith layers could protect ice layers from sublimation by providing a significant barrier to the diffusion of water vapour. However, this result requires very low diffusion coefficients, which are not observed even for clay powders. The measured diffusion coefficient of 1.29×10^{-4} $m^2 s^{-1}$ indicates that water diffuses very fast in a clay regolith (Chevrier *et al.* 2008). With such a diffusion coefficient, it is unlikely that the regolith will provide a significant protection against ice sublimation. Instead, temperature remains the main factor stabilizing ice on the surface of Mars, and in most regions the presence of ice is possible only if the temperature is low enough. Although actual sublimation rates on Mars may be significantly larger where wind speeds exceed a few metres per second, this conclusion is generally valid to shallow depths in unconsolidated regolith (<10 cm). Packing of the regolith at greater depths strongly reduces the porosity and eventually decreases the diffusion rates of water vapour (Clifford & Hillel 1983). This explains the preservation of ice below several tens of centimetres of regolith at Martian mid latitudes.

Mellon & Jakosky (1993) developed a model of the thermal and diffusive stability of ground ice in the Martian permafrost. In this model, the diurnal and seasonal subsurface thermal oscillations drove water vapour diffusion in exchange with the atmosphere. Phase partitioning between vapour, ice and adsorbed water was maintained, and ice was allowed to condense where stability conditions were met. Ice condensation results showed that an ice table forms, such that ice-free regolith blankets densely ice-cemented soil. The ice table generally represents a depth where ground ice is stable with respect to sublimation on annual or longer timescales.

At latitudes of more than 60°, gamma-ray and neutron spectroscopy have indicated the presence of ice below an ice-free regolith only a few centimetres thick (Feldman *et al.* 2002). This has been locally confirmed at the *Phoenix Landing Site*, where ice was exposed at depths of less than 10 cm (Smith *et al.* 2009). These findings are good first-order indications of the predictive capacity of ice-stability models and are consistent with the landforms observed. The presence of widespread polygonal patterns indicates the occurrence of ice below 1 m, which is approximately the depth at which the annual thermal wave can propagate and create thermal stress (Mangold 2005). At these latitudes the sublimation occurs in the first decimetres of soils, which explains the progressive degradation of polygonal patterns. Similar polygons are more and more degraded as the equator is approached, as a consequence of an enhanced sublimation (Mangold 2005).

At lower latitudes, calculations suggest that water-ice stability occurs at less than 1 m as far as 50° latitude (Mellon *et al.* 2004). Ice-rich material is expected to vary with latitude and exceeds about 1 m at around 45° (e.g. Mellon & Jakosky 1993; Schorghofer & Aharonson 2005). Direct evidence for ice was not found by spectrometers, although there is an ongoing debate concerning the occurrence of local water (either as ground ice, adsorbed water or hydrated minerals) in the ground, mainly in the Arabia region (Feldman *et al.* 2002). Recently, new impacts in the Martian mid latitudes have exposed near-surface ice that was observed to slowly fade over timescales of months (Dundas & Byrne 2010). Models suggest that over 1 mm of sublimation occurred in the period during which the ice was observed to fade. The persistence of visible ice through such sublimation suggests that the ice is relatively pure, rather than pore filling (Dundas & Byrne 2010). Water-ice stability models based on this new discovery show that the ice excavated is buried beneath a 15–50 cm-thick dry layer at 45° of latitude. These results suggest that the ice observed by the neutron spectrometer of the *Mars Odyssey* is likely to be sequestered in the first 10–20 cm, rather than 1 m, or that maps based on their results should include water ice down to 45° latitude. At these latitudes polygonal patterns are still apparent in high-resolution HiRISE imagery (Levy *et al.* 2009). However, their presence using MOC image resolution was much more limited (Mangold 2005), suggesting this deeper ice at mid latitudes may limit the size of the polygons with respect to those at higher latitudes.

Mid-latitude glaciers observed at 40–50° latitude display metre-scale pitted landforms, suggesting that the glacier ice is superimposed by more than 1 m of desiccated material. At a depth of 10 m, the mean temperature is no longer dependent on the annual variation in atmospheric temperature. This suggests that the ice is still present in mid-latitude glaciers not only because it is protected by a sublimation lag, but also because its temperature is close to stability at this latitude. Consequently, the ice could have accumulated during periods of higher obliquity (Chevrier *et al.* 2008). Diffusion coefficients were used to determine ice lifetimes below different regolith thicknesses (Bryson *et al.* 2008). The results indicate that a 1 m layer of ice below 2 m of fine-grained basaltic regolith could still remain from the last large obliquity change, 0.4 Ma ago, at 195K. This could explain the presence of mid-latitude concentrations of subsurface ice.

The mid-latitude mantle displaying pits and dissected layers (Fig. 4) is a common morphology typical of latitudes ranging from 30° to 55° in which landform patterns are weakly controlled by polygonal cracks. Mid-latitude ice-rich mantles may be degraded by sublimation of ice present below the depth at which thermal contraction is an efficient process; that is, below the seasonal thermal-wave propagation. This depth is not well established, but from all of these results it seems that ice more than 50 cm deep does not create significant cracking leading to polygons, but can still lead to the formation of sublimation-related landforms. As a consequence, such sublimation-related landforms are usually irregular, with the exception of lobate debris-apron surfaces on which regular patterns reflect pre-existing fractures in the glacial landforms.

Neutron spectroscopy has indicated the presence of a significant amount of hydrogen in equatorial latitudes (below 25°–30° latitude north and south) that would translate to 10–12% of water ice, assuming that hydrogen was present only in water molecules. However, it is unknown whether this hydrogen is present as bounded in minerals, as water adsorbed onto grains or as water ice (Feldman *et al.* 2002). Studies involving modified craters show that eolian dust as thick as 50 m covers a large part of this region (Mangold *et al.*

2009). As no sublimation-related features are apparent in this region, hydrogen found here by neutron spectroscopy may be unrelated to water ice. In these equatorial regions, some models predict that water ice may be present below a desiccated layer more than 100 m thick (Clifford 1993). The problem is that the timescale over which sublimation can occur is huge, that is, more than 1 billion (10^9) years, and it is difficult to take this timescale into account in models because processes such as tectonics or volcanism may play a role in increasing sublimation rates or releasing volatiles locally. So, models based on a shallow homogeneous regolith are probably not relevant. The thin sublimation tills observed over equatorial glaciers (e.g. Fastook *et al.* 2008) show that sublimation here was active at a sufficiently high rate that significant quantities of water ice have been lost since the last major episode of climatic change. Possible cold traps of equatorial ice are still a topic of research, however.

Conclusion

Sublimation-related landforms are ubiquitous on Mars. Radar sounders or neutron spectrometers have indicated the existence of subsurface ice, but its presence is difficult to verify, and its distribution, density and evolution over the recent past remain topics of ongoing investigation. Therefore, the most difficult problem in understanding the genesis of sublimation-related landforms is determining the distribution of subsurface water ice at the time of morphogenesis. At present, this involves considerable speculation. Nevertheless, this review of accepted sublimation-related landforms on Mars shows that pits, knobs and dissected terrains are classical landforms reflecting ground ice sublimation. These processes on Mars are strongly latitude dependent, with sublimation being increasingly important from high ($>60°$) to low latitudes (down to 25°) due to correspondingly higher mean annual temperatures that ultimately peak above the current frost point. Equatorial regions (within 25° latitude) display recent sublimation-related landforms in a few locations only, suggesting a broadly ice-free regolith.

While mean temperatures and water vapour pressure strongly control the sublimation rate, sublimation in the ground depends on diffusion and water adsorption, which are both very dependent on regolith porosity and composition. Our poor knowledge of these parameters is the limiting factor in the accuracy of sublimation-rate models. From a geomorphological point of view, sublimation-related landforms are often shaped by differential sublimation rates owing to the presence of fractures, such as those formed by thermal contraction. This review highlights the importance of regolith structure (fractures, layering), in addition to the usual physicochemical parameters, in the shaping of landforms, as well as in the controlling sublimation rates. Further laboratory data and *in situ* measurements will be required to better understand the variety of sublimation-related landforms on Mars.

References

Aguirre-Puente, J. & Sukwhal, R. N. 1984. Sublimation of ice in frozen dispersed media. *In*: Lunardini, V. J. (ed.) *Proceedings of the Third International Offshore Mechanics and Arctic Engineering Symposium*, Volume 3. American Society of Mechanical Engineers, 38–44.

Balme, M. R., Gallagher, C. J., Page, D. P., Murray, J. B. & Muller, J. P. 2009. Sorted stone circles in Elysium Planitia, Mars: implications for recent martian climate. *Icarus*, **200**, 30–38.

Bibring, J.-P., Langevin, Y. *et al.* 2004. Perennial water ice identified around the Mars South pole. *Nature*, **428**, 627–630.

Boynton, W. V., Feldman, W. C. *et al.* 2002. Distribution of hydrogen in the nearsurface of Mars: evidence for subsurface ice deposits. *Science*, **297**, 81–85.

Bryson, K., Chevrier, V. & Sears, D. W. G. 2008. Ice on Mars: an experimental investigation into the increased stability of ice on Mars by a surface covering of fine-grained basaltic regolith. *Icarus*, **196**, 459–476.

Burr, D. M., Grier, J., McEwen, A. & Sakimoto, S. 2002. Recent aqueous floods from the Cerberus Fossae, Mars. *Geophysical Research Letters*, **29**, 1–13, doi: 10.1029/2001GL01335.

Carr, M. H. & Schaber, G. G. 1977. Martian permafrost features. *Journal of Geophysical Research*, **82**, 4039–4054.

Chevrier, V., Sears, D. W. G. *et al.* 2007. The sublimation rate of ice under simulated Mars conditions and the effect of layers of mock regolith JSC Mars-1. *Geophysical Research Letters*, **34**, L02203, doi: 10.1029/2006GL028401.

Chevrier, V., Ostrowski, D. R. & Sears, D. W. G. 2008. Experimental study of the sublimation of ice through an unconsolidated clay layer: implications for the stability of ice on Mars and the possible diurnal variations in atmospheric water. *Icarus*, **196**, 459–476.

Clifford, S. M. 1993. A model for the hydrologic and climatic behavior of water on Mars. *Journal of Geophysical Research*, **98**, 10 973–11 016.

Clifford, S. M. & Hillel, D. 1983. The stability of ground ice in the equatorial region of Mars. *Journal of Geophysical Research*, **88**, 2456–2474.

Costard, F. & Kargel, J. 1995. Outwash plains and thermokarst on Mars. *Icarus*, **114**, 93–112.

Costard, F. M. 1989. The spatial distribution of volatiles in the Martian hydrolithosphere. *Earth, Moon, and Planets*, **45**, 265–290.

Cutts, J. A. 1973. Wind erosion in the Martian polar regions. *Journal of Geophysical Research*, **78**, 4211–4221.

Dundas, C. M. & Byrne, S. 2010. Modeling sublimation of ice exposed by recent impacts in the Martian mid-latitudes. *Icarus*, **206**, 716–728.

FANALE, F. P. & CANNON, W. A. 1971. Adsorption on the martian regolith. *Nature*, **230**, 502–504.

FARRELL, W. M., PLAUT, J. J., CUMMER, S. A., GURNETT, D. A., PICARDI, G., WATTERS, T. A. & SAFAENILI, A. 2009. Is the Martian radar table hidden from radar view? *Journal of Geophysical Research Letters*, **36**, L15206, doi: 10.1029/2009GL038945.

FASTOOK, J. L., HEAD, J. W., MERCHANT, D. R. & FORGET, F. 2008. Tropical mountain glaciers on Mars: altitude-dependence of ice accumulation, accumulation conditions, formation times, glacier dynamics, and implications for planetary spin-axis/orbital history. *Icarus*, **198**, 305–317.

FELDMAN, W. C., BOYNTON, W. V. *ET AL.* 2002. Global distribution of neutrons from Mars: results from Mars Odyssey. *Science*, **297**, 75–78.

FELDMAN, W. C., PRETTYMAN, T. H. *ET AL.* 2004. Global distribution of near-surface hydrogen on Mars. *Journal of Geophysical Research*, **109**, E09006, doi: 10.1029/2003JE002160.

FORGET, F., HABERLE, R. M., MONTMESSIN, F., LEVRARD, B. & HEAD, J. W. 2006. Formation of glaciers on Mars by atmospheric precipitation at high obliquity. *Science*, **311**, 368–371, doi: 10.1126/science.1120335.

GHATAN, G. J., HEAD, J. W., III. & PRATT, S. 2003. Cavi Angusti, Mars: characterization and assessment of possible formation mechanisms. *Journal of Geophysical Research*, **108**, 5045, doi: 10.1029/2002JE001972.

GOBELMAN, S. 1985. *Sublimation From Reconstituted Frozen Silt*. MSc thesis, University of Alaska, Fairbanks.

HEAD, J. W., MUSTARD, J. F., KRESLAVSKY, M. A., MILLIKEN, R. E. & MARCHANT, D. R. 2003. Recent ice ages on Mars. *Nature*, **426**, 797–802.

HEAD, J. W., NEUKUM, G. *ET AL.* 2005. Tropical to mid-latitude snow and ice accumulation, flow and glaciation on Mars. *Nature*, **434**, 346–351.

HOFSTADTER, M. D. & MURRAY, B. C. 1990. Ice sublimation and rheology: implications for the Martian polar layered deposits. *Icarus*, **84**, 352–361.

HUANG, S. L. & AUGHENBAUGH, N. B. 1987. Sublimation of pore ice in frozen silt. *Journal of Cold Regions Engineering*, **1**, 171–181.

HUDSON, T. L., AHARONSON, O., SCHORGHOFER, N., FARMER, C. B., HECHT, M. H. & BRIDGES, N. T. 2007. Water vapor diffusion in Mars subsurface environments. *Journal of Geophysical Research*, **112**, E05016, doi: 10.1029/2006JE002815.

JOHANSEN, N. I., CHALICH, P. C. & WELLEN, E. W. 1981. *Sublimation and its control in the CRREL Permafrost Tunnel*. Cold Regions Research and Engineering Laboratory Special Report, **81–8**.

KADISH, S. J., HEAD, J. W., BARLOW, N. G. & MARCHANT, D. R. 2009. Martian pedestal craters: marginal sublimation pits implicate a climate-related formation mechanism. *Geophysical Research Letters*, **35**, L16104, doi: 10.1029/2008GL034990.

KRESLAVSKY, M. A. & HEAD, J. W., III. 2002. Mars: nature and evolution of young latitude-dependent water-ice-rich mantle. *Geophysical Research Letters*, **29**, 1719, doi: 10.1029/2002GL015392.

LASKAR, J., LEVRARD, B. & MUSTARD, J. F. 2002. Orbital forcing of the martian polar layered deposits. *Nature*, **419**, 375–378.

LEFORT, A., RUSSELL, P. S., THOMAS, N., MCEWEN, A. S., DUNDAS, C. M. & KIRK, R. L. 2009. Observations of eriglacial landforms in Utopia Planitia with the High Resolution Imaging Science Experiment (HiRISE). *Journal of Geophysical Research*, **114**, E04005, doi: 10.1029/2008JE003264.

LEVY, J. S., HEAD, J. W. & MARCHANT, D. R. 2009. Thermal contraction crack polygons on Mars: classification, distribution, and climate implications from HiRISE observations. *Journal of Geophysical Research*, **114**, E01007, doi: 10.1029/2008JE003273.

MALIN, M. C. & EDGETT, K. S. 2001. Mars Global Surveyor Mars Orbiter Camera: interplanetary cruise through primary mission. *Journal of Geophysical Research*, **106**, 23 429–23 570.

MANGOLD, N. 2003. Geomorphic analysis of lobate debris aprons on Mars at MOC scale: evidence for ice sublimation initiated by fractures. *Journal of Geophysical Research – Planets*, **108**, 2–1.

MANGOLD, N. 2005. High latitude patterned ground on Mars: classification, distribution and climatic control. *Icarus*, **174**, 336–359.

MANGOLD, N. & ALLEMAND, P. 2001. Topographic analysis of features related to ice on Mars. *Journal of Geophysical Research Letters*, **28**, 407–411.

MANGOLD, N., COSTARD, F. & PEULVAST, J.-P. 2000*a*. Thermokarstic degradation of lobate debris aprons and fretted channels. *In*: *Proceedings of the Second International Mars Polar Science and Exploration Conference, Reykjavik, Iceland, August 21–25, 2000*. Lunar and Planetary Institute, Houston, TX, Abstract 4032.

MANGOLD, N., COSTARD, F. & PEULVAST, J.-P. 2000*b*. Thermokarstic degradation of the Martian surface. *In*: *Proceedings of the Second International Mars Polar Science and Exploration Conference, Reykjavik, Iceland, August 21–25, 2000*. Lunar and Planetary Institute, Houston, TX, Abstract 4052.

MANGOLD, N., ANSAN, V., MASSON, Ph. & VINCENDON, C. 2009. Estimate of aeolian dust thickness in Arabia Terra, Mars: implications of a thick mantle (>20 m) for hydrogen detection. *Géomorphologie*, **1/2009**, 23–32.

MARCHANT, D. R., LEWIS, A. R. *ET AL.* 2002. Formation of patterned ground and sublimation till over Miocene glacier ice in Beacon Valley, southern Victoria Land, Antarctica. *GSA Bulletin*, **114**, 718–730.

MELLON, M. T. & JAKOSKY, B. M. 1993. Geographic variations in the thermal and diffusive stability of ground ice on Mars. *Journal of Geophysical Research – Planets*, **98**, 3345–3364.

MELLON, M. T., FELDMAN, W. C. & PRETTYMAN, T. H. 2004. The presence and stability of ground ice in the southern hemisphere of Mars. *Icarus*, **169**, 324–340.

MERESSE, S., COSTARD, F., MANGOLD, N., BARATOUX, D. & BOYCE, J. M. 2006. Martian perched craters and large ejecta volume: evidence of episodes of deflation of the northern lowlands. *Meteoritics and Planetary Science*, **41**, 1423–1690.

MILKOVICH, S. M., HEAD, J. W., III & PRATT, S. 2002. Meltback of Hesperian-aged ice-rich deposits near the south pole of Mars: evidence for drainage channels and lakes. *Journal of Geophysical Research*, **107**, 5043, doi: 10.1029/2001JE001802.

Mustard, J. F. & Cooper, C. D. 2000. Global distribution of dissected duricrust on Mars. *In*: *Proceedings of the 31st Annual Lunar and Planetary Science Conference, March 13–17, 2000, League City, Texas*. Lunar and Planetary Institute, Houston, TX, Abstract 1168.

Mustard, J. F., Cooper, C. D. & Rifkin, M. K. 2001. Evidence for recent climate change on Mars from the identification of youthful near-surface ground ice. *Nature*, **412**, 411–413.

Page, D. P. 2007. Recent low-latitude freeze–thaw on Mars. *Icarus*, **189**, 83–117.

Piqueux, S., Byrne, S. & Richardson, M. I. 2003. Sublimation of Mars's southern seasonal CO_2 ice cap and the formation of spiders. *Journal of Geophysical Research*, **108**, 5084, doi: 10.1029/2002JE002007.

Plaut, J. J., Safaeinili, A. *et al.* 2009. Radar evidence for ice in lobate debris aprons in the mid-northern latitudes of Mars. *Geophysical Research Letters*, **36**, L2203, doi: 10.1029/2008GL036379.

Schorghofer, N. 2007. Dynamics of ice ages on Mars. *Nature*, **449**, 192–195.

Schorghofer, N. & Aharonson, O. 2005. Stability and exchange of subsurface ice on Mars. *Journal of Geophysical Research*, **110**, E05003, doi: 10.1029/2004JE002350.

Seibert, N. M. & Kargel, J. S. 2001. Small-scale Martian polygonal terrain: Implications for liquid surface water. *Geophysical Research Letters*, **28**, 899–902.

Séjourné, A., Costard, F., Gargani, J., Marmo, C., Forget, F., Madeleine, J.-B. & Soare, R. J. 2009. Periglacial processes in Utopia Planitia, evolution of Scalloped Terrains: new insights from Hirise observations. *In*: *Proceedings of the 41st Annual Lunar and Planetary Science Conference, March 23–27, 2009, The Woodlands, Texas*. Lunar and Planetary Institute, Houston, TX, Abstract 1733.

Skorov, Y. V., Markiewicz, W. J., Basilevsky, A. T. & Keller, H. U. 2001. Stability of water ice under a porous nonvolatile layer: implications to the south polar layered deposits of Mars. *Planetary and Space Science*, **49**, 59–63.

Smith, P. H., Tamppari, L. K. *et al.* 2009. H_2O at the Phoenix Landing Site. *Science*, **325**, 58–61.

Smoluchowski, R. 1968. Mars: retention of ice. *Science*, **159**, 1348–1350.

Soare, R. J., Osinski, G. R. & Roehm, C. L. 2008. Thermokarst lakes and ponds on Mars in the very recent (late Amazonian) past. *Earth and Planetary Science Letters*, **272**, 382–393.

Squyres, S. W. 1978. Martian fretted terrain: flow of erosional debris. *Icarus*, **34**, 600–613.

Squyres, S. W. 1989. Urey prize lecture: water on Mars. *Icarus*, **79**, 229–288.

Squyres, S. W. & Carr, M. H. 1986. Geomorphic evidence for the distribution of ground ice on Mars. *Science*, **231**, 249–252.

Tornabene, L. L., McEwen, A. S. *et al.* 2007. Impact melting and the role of subsurface volatiles: implications for the formation of valley networks and phyllosilicate-rich lithologies on early Mars. *In*: *Seventh International Conference on Mars, July 9–13, 2007, Pasadena, California*. Lunar and Planetary Institute, Houston, TX, Abstract 3288.

Tschudin, K. 1946. Rate of vaporization of ice. *Helvetica Physica Acta*, **19**, 91–102.

Van Dijk, D. & Law, J. 1995. Sublimation and aeolian sand movement from a frozen surface: experimental results from Presqu'ile Beach, Ontario. *Geomorphology*, **11**, 177–187.

Vaucher, J., Baratoux, D., Mangold, N., Pinet, P., Kurita, K. & Grégoire, M. 2009. The volcanic history of Central Elysium Planitia: implications for martian magmatism. *Icarus*, **204**, 418–442, doi: 10.1016/j.icarus.2009.06.032.

Wellen, E. W. 1979. *Sublimation of Ice from Permafrost Silt at the CRREL Tunnel*. MSc thesis, University of Alaska, Fairbanks.

Williams, R. M. E. & Malin, M. C. 2008. Sub-kilometer fans in Mojave Crater, Mars. *Icarus*, **198**, 365–383.

Yershov, E. D., Gurov, V. V. & Dostovalov, B. N. 1973. Ice sublimation in fine-grained soils of various cryogenous structure during their interaction with an air current. *In*: Sanger, F. (ed.) *Proceedings of the Second International Permafrost Conference, July 13–28, 1973*. National Academy of Sciences, Washington, DC, 330–332.

Zanetti, M., Hiesinger, H., Reiss, D., Hauber, E. & Neukum, G. 2008. Scalloped depressions in Malea Planum, southern Hellas Basin, Mars. *In*: *Proceedings of the 39th Annual Lunar and Planetary Science Conference, March 10–14, 2008, League City, Texas*. Lunar and Planetary Institute, Houston, TX, Abstract 1682.

Identifying Martian gully evolution

A. H. ASTON[1]*, S. J. CONWAY[2] & M. R. BALME[2,3]

[1]*Department of Earth Sciences, University College London, Gower Street, London WC1E 6BT, UK*

[2]*Department of Earth & Environmental Sciences, CEPSAR, Open University, Walton Hall, Milton Keynes MK7 6AA, UK*

[3]*Planetary Science Institute, Suite 106, 1700 East Fort Lowell Road, Tucson, AZ 85719, USA*

**Corresponding author (e-mail: anselm.aston.09@ucl.ac.uk)*

Abstract: Martian gullies are small-scale, geologically recent features characterized by the alcove-channel-apron morphology associated with flows with a component of liquid water. Theories advanced to explain Martian gully formation include groundwater processes and melting of near-surface ice due to climate variation. Gullies are often associated with 'mantling terrain' that drapes topography at mid to high latitudes and which has been proposed to be ice-rich.

We have morphologically classified Martian gullies into four groupings according to whether they form solely within the mantle (Type A), erode into 'bedrock' (Type B), and by how well developed they appear (1 or 2). Orientation, length, geological setting and latitude were also recorded, as well as whether more than one generation of gullies formed on a given slope (labelled 'reactivated').

About 25% of gullies form solely within the mantle; these are generally shorter than gullies that erode bedrock and the morphologically simplest gullies (A1) are the shortest. We present latitude and orientation trends for the most recent episode of gully formation. We suggest that this recent activity is probably controlled by either deposition of ice-rich material or degradation of pre-existing ice-rich material.

Martian gullies are small hillslope features with fluvial-like form that comprise a distinctive alcove-channel-apron morphology (e.g. Fig. 1). They are amongst the youngest features on the Martian surface (Schon *et al.* 2009), and are suggestive of geologically recent activity of water. As such, they have been the object of considerable interest since their discovery by Malin & Edgett (2000). They offer the possibility not only of constraining the presence and activity of water in the Martian surface and near subsurface, but of reconstructing Martian climate variations and proposed recent ice ages.

The distribution and orientation of gullies have been comprehensively surveyed (e.g. Malin & Edgett 2000; Costard *et al.* 2002; Heldmann & Mellon 2004; Balme *et al.* 2006; Dickson *et al.* 2007; Heldmann *et al.* 2007; Kneissl *et al.* 2010), and several key constraints on their distribution have emerged. They are found principally in mid latitudes, between 30° and 60°, with occasional higher-latitude clusters such as in the south polar pits, and they exhibit a regional distribution that has not yet been fully explained. Gullies are more common in the southern hemisphere than the northern (Dickson & Head 2009), and show a preference for certain orientations. In the southern hemisphere, the orientation is predominantly polewards, although analyses by latitude have shown that this becomes less marked at higher latitudes (Costard *et al.* 2002; Balme *et al.* 2006; Dickson *et al.* 2007). Surveys of the northern hemisphere gullies do not agree as to the orientations of gullies broken down by latitude (Bridges & Lackner 2006; Heldmann *et al.* 2007; Kneissl *et al.* 2010). However, the most recent survey (Kneissl *et al.* 2010), which included both a greater number of images and images from two different cameras (High Resolution Stereo Camera and Mars Orbiter Camera), found that the distribution of gully orientations was the same as in the southern hemisphere.

In terms of geological context, gullies are most often found on the inner rims of craters, with smaller numbers in valleys (e.g. Dao and Harmakhis) and on isolated prominences variously described as hills, knobs, buttes or mesas. A few have been observed on the outer rims and central peaks of craters, and a substantial number on dunes (although not all of these exhibit the classic alcove-channel-apron morphology).

From: Balme, M. R., Bargery, A. S., Gallagher, C. J. & Gupta, S. (eds) *Martian Geomorphology*. Geological Society, London, Special Publications, **356**, 151–169.
DOI: 10.1144/SP356.9 0305-8719/11/$15.00

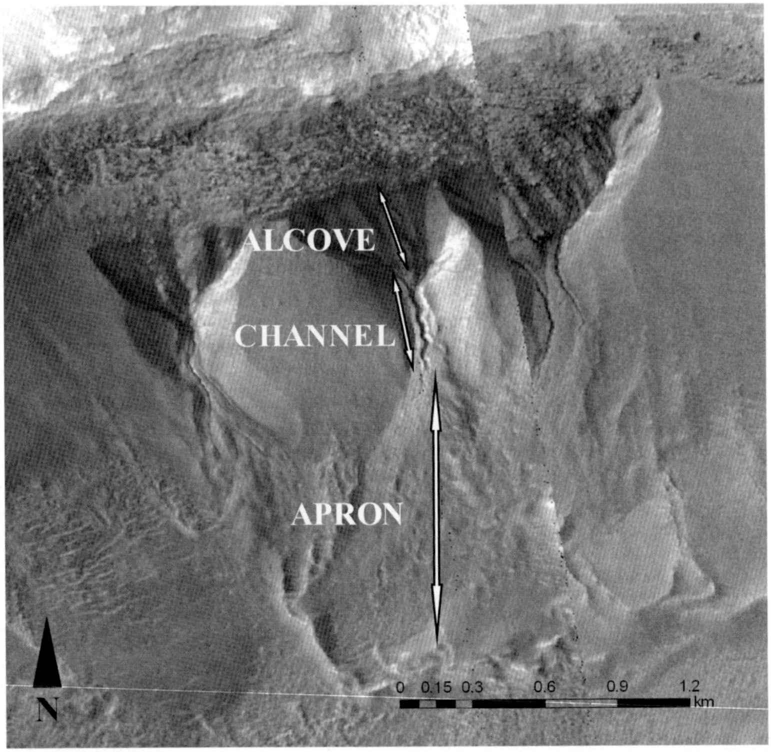

Fig. 1. A Martian gully showing the classic alcove-channel-apron morphology. This particular example would be defined by Malin & Edgett (2000) as 'abbreviated'; 42.0°S, 195.5°E. Image credit: NASA/JPL/MSSS; see prelim viii for acronym definitions.

Formation process

The process(es) by which Martian gullies formed have been a topic of some contention, and there have been several suggestions as to both the erosive agent and the mechanism of formation. The initial discovery paper (Malin & Edgett 2000) interpreted the gullies as evidence of liquid water. Alternatives offered include carbon dioxide flows (Musselwhite *et al.* 2001; Hoffman 2002), dry mass wasting (Treiman 2003) and brines (Knauth & Burt 2002; Chevrier & Altheide 2008). Liquid carbon dioxide as the agent has been refuted based on stability considerations (Stewart & Nimmo 2002; Heldmann & Mellon 2004), while the morphology, particularly the sinuosity of many gullies, argues against the dry mass-wasting hypothesis (e.g. Levy *et al.* 2009*a*; Conway *et al.* 2011).

The suggestion of brines as gully-forming fluids represents an attempt to explain activity of gullies under current surface conditions. At current surface temperatures and pressures, liquid water is not stable on the Martian surface, although models suggest that short-duration flow could remain stable over average gully channel lengths (Hecht 2002; Heldmann *et al.* 2005). However, Mars experiences extreme changes in obliquity, with axial tilt having varied between 14° and 48° over the last 10 Ma (Laskar *et al.* 2002). This has two effects that are relevant to gullies. First, permanent sunlight on the polar caps releases both CO_2 and water vapour, which increase the atmospheric pressure, and allows the water to be deposited as ice elsewhere on Mars. Secondly, on higher-latitude pole-facing slopes, sudden insolation in springtime allows the ground to warm above freezing and melt to depths of approximately 0.5 m (Costard *et al.* 2002), depending on the conductivity (Head *et al.* 2003).

The latter theory suggests that during high-obliquity excursions these processes redistribute water from the polar caps to be deposited at latitudes above 30° in the form of an ice-rich mantling material. This is observed intermittently at latitudes below 60° and continuously at higher latitudes (Mustard *et al.* 2001; Milliken *et al.* 2003). A dust lag that forms at the beginning of low-obliquity periods subsequently protects this mantle from sublimation, and provides a reservoir of ice on the surface. Liquid water is thus a plausible gully-forming agent during high-obliquity periods, and the debate on gully formation has centred on mechanisms for its release. The mechanisms proposed fall into two broad groups: one based on subsurface processes; the other on atmospheric processes.

The subsurface theories, initially proposed by Malin & Edgett (2000) who noted that gully alcoves occur within the uppermost few hundred metres of any given slope, suggest that gullies represent the release of groundwater. In this model, shallow liquid aquifers, possibly maintained by geothermal heat, exist in the near-subsurface. Icy plugs prevent their release where they intersect slopes (Mellon & Phillips 2001). Changing conditions at times of obliquity excursions can melt these plugs and release the water to form gullies. Layers of impermeable rock may act as aquicludes, trapping the water at particular depths. In a deep-aquifer variant of this theory, dykes transport water up from deep aquifers to be released at the slopes (Gaidos 2001). The 'abbreviated' alcoves of Malin & Edgett (2000), which are capped by a single stratum of rock often extending across several alcoves or a whole slope, have been invoked as morphological support for aquifer theories (Heldmann & Mellon 2004). Some support for this model was provided by the observation of an apparent pattern of regional drainage linked with gullies in the Gorgonum Basin (Marquez *et al.* 2005). While these theories are compatible with models of subsurface heating (Hartmann 2001; Heldmann & Mellon 2004) and with the 'widened' and 'abbreviated' alcove morphologies observed by Malin & Edgett (2000), they cannot explain the occurrence of gullies on isolated prominences such as hills, mesas, dunes and crater central peaks where aquifers are geologically implausible (Balme *et al.* 2006; Dickson & Head 2009; Kneissl *et al.* 2010).

Atmospheric theories, by contrast, emphasize melting of ice in the ground (<1 m) or near-surface during favourable climatic periods. Costard *et al.* (2002) proposed that water from a humid atmosphere at high obliquity concentrates in the subsurface at mid latitudes and is released by debris flows when the carbon dioxide frost cap sublimes in spring. An alternative proposed by Christensen (Christensen 2003) is that melting at the base of the icy mantle slowly carves gullies within the mantling material, which become exposed as the mantle is gradually removed. These theories depend on favourable results from ice-melting models, and have difficulty explaining the variation in orientation of gullies with latitude (Heldmann *et al.* 2007).

New models have convinced some adherents of the subsurface theories that formation of gullies by seasonal snowmelt is a plausible formation mechanism in both hemispheres (e.g. Williams *et al.* 2009), which may indicate movement towards a consensus of atmospheric formation. Overall, the patterns of gully orientation, their occurrence on isolated hills and crater central peaks, and the recognition of obliquity changes as a plausible controlling factor suggest that most gullies on Mars formed by the melting of near-surface ice and/or snow (Costard *et al.* 2002; Christensen 2003; Head *et al.* 2008; Dickson & Head 2009). Exactly where, when and how this occurred remains to be determined.

Morphology

The classic gully morphology (Malin & Edgett 2000) is an alcove-channel-apron structure (Fig. 1), with a channel that is widest and deepest at the base of the alcove, and which becomes shallower and narrower downslope. While the alcove-apron structure is similar to that produced on Earth by dry mass wasting, the channels are distinctive and an indicator of liquid flow. A descriptive classification scheme based on alcove shape was proposed by Malin & Edgett (2000), who identified three types of alcoves: 'lengthened', 'widened' and 'abbreviated'–plus a fourth type, 'occupied'–which were filled with material. Heldmann *et al.* (2007) identified a further type of alcove, dubbed 'eroded'. A second morphological study (Bleamaster & Crown 2005) noted the associations of gullies with mantle. Following the Christensen (2003) theory, they proposed an evolutionary morphological sequence on the walls of Dao and Harmakhis Valles, in which gullies are incised into mantled walls, and evolve into classic gully morphology by removal of the mantle.

However, since the latter paper, morphological classification has not been widely used. Instead, aspects such as height and length of gullies have been studied, as have contexts and orientations (e.g. Heldmann & Mellon 2004; Balme *et al.* 2006; Bridges & Lackner 2006; Dickson *et al.* 2007; Heldmann *et al.* 2007; Kneissl *et al.* 2010). In general, all gullies have been treated as morphologically identical for the purposes of regional, hemispherical or planetary surveys. This paper takes a morphological approach first, identifying a distinction between two different gully types, and then tests the hypothesis that certain morphologies represent stages in gully evolution. Gullies classified according to this scheme are also analysed according to context, orientation, latitude and length.

Classification scheme

Two types of gullies were identified inductively from an initial morphological examination of Martian gully images. Each type appears to possess distinct morphological features that are not present in the other; however, as a longitudinal study of gullies is impossible, these represent only one possible classification scheme based principally on the relationships of gullies to mantling material and bedrock. Criteria are shown in Table 1.

Table 1. *Gully classification criteria*

	A (cuts mantle only)	B (cuts mantle and bedrock)
1 (immature)	No separate alcove	Does not reach slope line
2 (mature)	V-shaped alcove	Reaches slope line and cuts backwards

Type A

This type (Fig. 2, Table 1) appears to exist entirely within the mantling material, with negligible effect on the underlying bedrock. Type A1 gullies are simple vertical slits incised into the mantle,

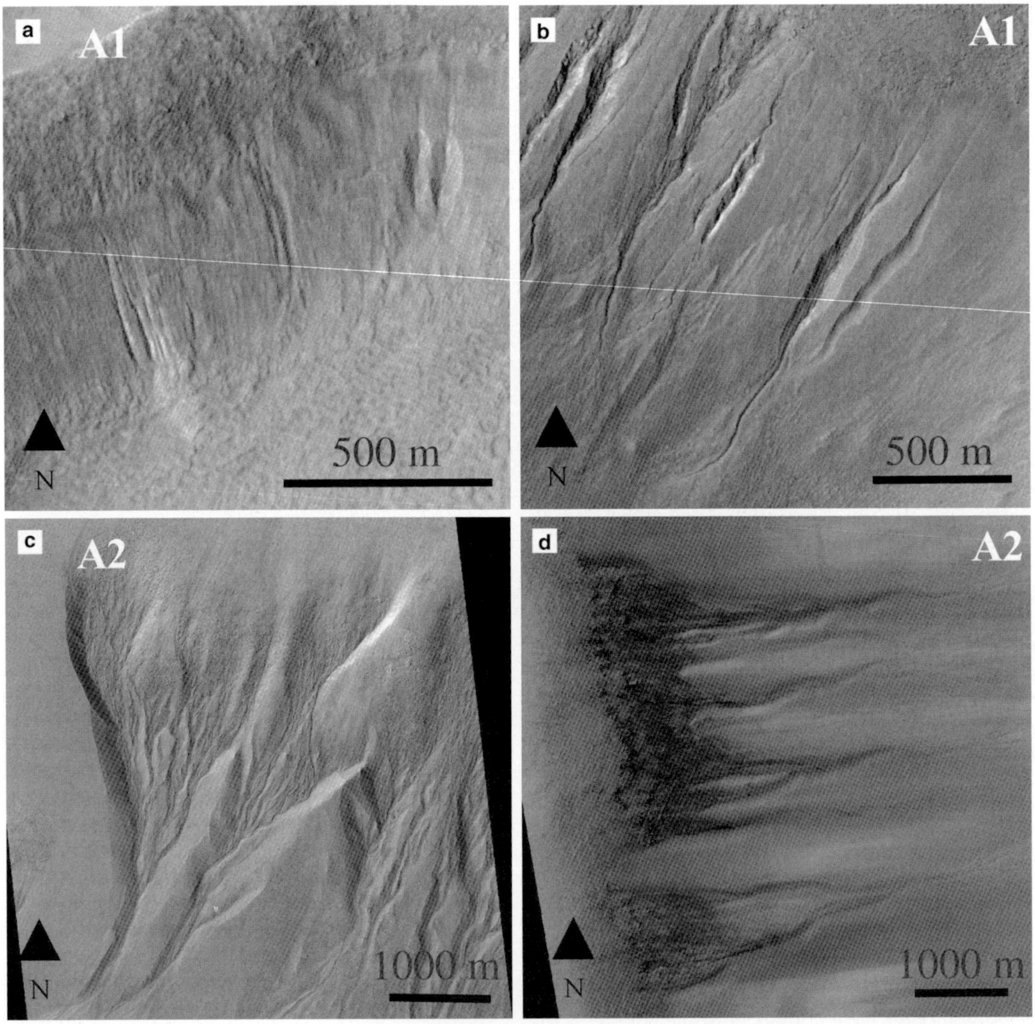

Fig. 2. Type A gullies. (**a**) Part of HiRISE image PSP_003596_1435; 36.2°S, 198.3°E. (**b**) Part of HiRISE image PSP_002884_1395; 40.4°S, 196.9°E. (**c**) Part of HiRISE image PSP_003170_1330; 46.6°S, 309.1°E. (**d**) Part of HiRISE image PSP_005160_1150; 64.8°S, 344.6°E. Image credit: NASA/JPL/UofA.

and sometimes display thin vertical lineaments (inferred to be subtle troughs) leading from the top of the alcove to a higher point on the cliff (Fig. 2a, b). The classic alcove-channel-apron morphology is not observed. The channel may widen at the top, but the alcove essentially consists of a deep channel leading to an apron.

Type A gullies erode the mantling material above the tip of the channel/alcove until none remains, at which point a more open V-shaped alcove is formed, leading into a deeply incised channel, resulting in an Type A2 gully (Fig. 2c, d). Essentially, mantling material is stripped away from the bedrock to make the alcove, but no removal of the bedrock takes place.

Type B

This type of gully (Fig. 3, Table 1) affects both mantling material and bedrock. A number of different morphologies are observed but which, for this study, we classify together. The 'immature' Type B1 gullies (Fig. 3a, b) do not reach the top of the hillslope and include the 'abbreviated' alcoves identified by Malin & Edgett that appear capped by rock strata (Fig. 1). These abbreviated gullies are

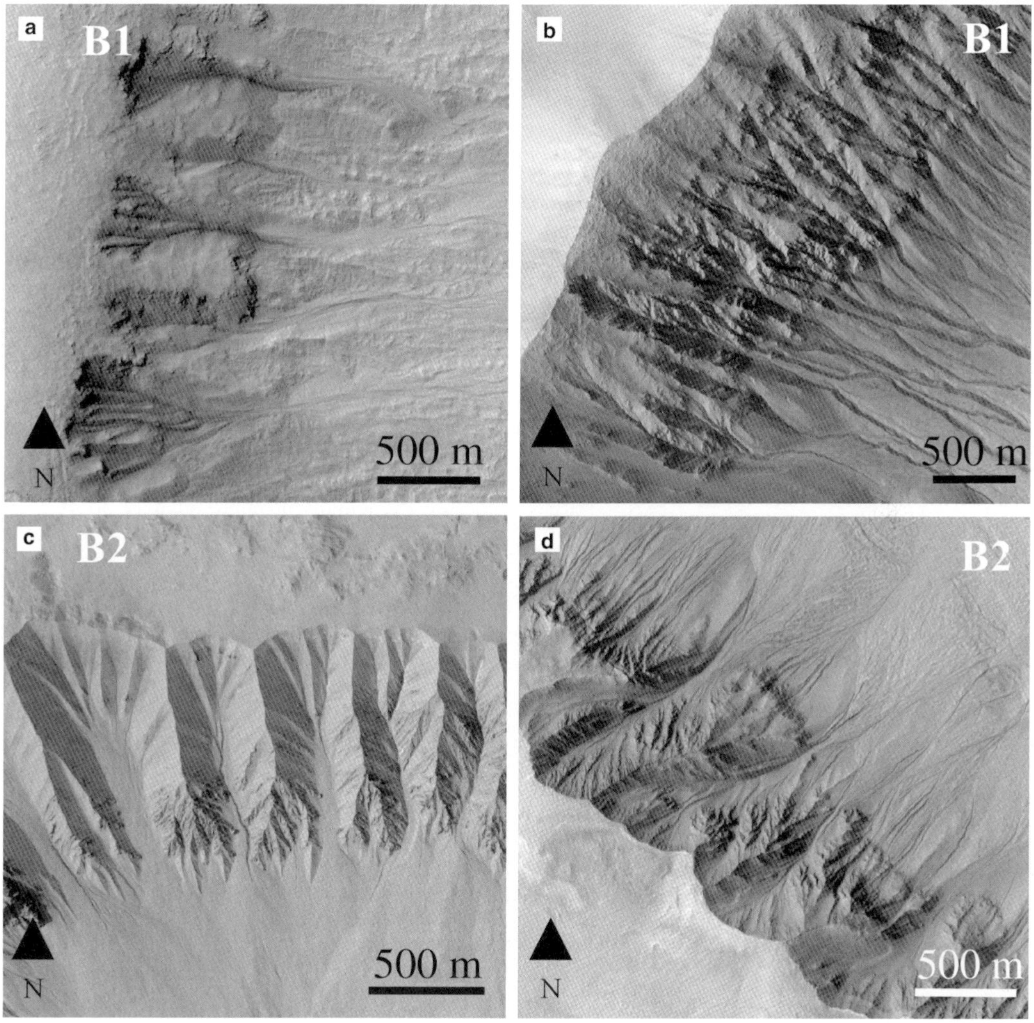

Fig. 3. Type B gullies. (**a**) Part of HiRISE image ESP_014427_1340; 45.9°S, 45.7°E. (**b**) Part of HiRISE image PSP_003583_1425; 37.1°S, 191.9°E. (**c**) Part of HiRISE image PSP_004060_1040; 35.7°S, 129.4°E. (**d**) Part of HiRISE image PSP_003675_1375; −42.3°S, 201.8°E. Image credit: NASA/JPL/UofA.

by no means a majority, and alcoves are observed that break through the capping layers. This suggests that the 'abbreviated' gullies noted by Malin & Edgett (2000) eventually grow to such an extent that the capping strata collapse, and the gully extends progressively back and upslope. However, a number of gullies in this category show no sign of ever having possessed such a capping layer, so this is unlikely to be the only evolutionary path followed by these gullies. 'Mature' B-type gullies are those that have reached the top of the cliff and begun cutting backwards into the slope line, creating a characteristic scalloped pattern in plane-view images (Fig. 3c, d). Gullies that have begun such horizontal erosion are classified as Type B2.

'Reactivated' gullies

A second variable, 'reactivation', was encountered during construction of the classification scheme, with relevance to the 'occupied' alcoves of Malin & Edgett (2000). We observed that while some gullies are cut into fresh mantling material (Fig. 4a), the alcoves of others show mantling material overlying eroded bedrock (e.g. Fig. 4b, c). The shape of the covered alcove can be seen clearly beneath the mantle, but the bedrock is fully or partially obscured by the mantle. This implies either: (1) that the bedrock has been eroded whilst underneath the mantling material, which subsequently collapses into the cavity (i.e. model of Christensen 2003); or (2) that these alcoves represent multiple stages of erosion.

This latter theory is supported by the presence of small, immature gullies (e.g. Type A1 or B1) within alcove-channel-apron systems much too large to have been formed by those gullies themselves (Fig. 4d). Thus, a second morphological classification was incorporated into the study to test these theories: gullies showing evidence of previous cycles of activity were labelled 'reactivated'. Gullies that do not show such evidence (e.g. Fig. 4a) were tagged 'single active phase'. The aim of recording these data was to explore the possibility of multiple generations of gully formation. In the absence of suitable age-dating of large numbers of individual gullies (challenging, given the time-consuming nature of crater counting many such landforms), comparing the distribution and orientation of reactivated gullies with the wider population provides a qualitative means to explore gully formation over time.

Data and methods

The classification was applied to a sample of gullies imaged by the Mars Orbiter Camera (MOC) on the *Mars Global Surveyor* during mission phases M03–R09 (July 1999–September 2003) at a resolution of 1.5 – 12.5 m per pixel. Data collection was performed using ArcGIS 9.2 software. The dataset used was that assembled by Balme *et al.* (2006), which comprises a database of images from the southern hemisphere, each of which contains at least one gully. The high southern latitude gullies were omitted due to the sparse number of gullies and their unique geological setting (within specific polar pits). The dataset for this study thus covers the region 30°–60°S (the region in which the vast majority of gullies are found).

Each MOC image or series of overlapping images showed one or more gullied slopes. These were divided into eight sections by orientation (north, NE, east, etc.). For each section, the following data were recorded:

- total number of gullies;
- number of each type of gully (A1, A2, B1, B2);
- number that appeared reactivated and which appeared to have a single active phase;
- number that could not be classified (due to deep shadow or the edge of the image);
- context: one of six categories (inner crater rim; outer crater rim; hills/knobs; valleys; crater central peak; and 'other');
- latitude.

Sampling involved working through the catalogue of Balme *et al.* (2006), who recorded simple statistics on the orientation and aspect for slope sections rather than individual gullies, and entering the above data into the geographical information system (GIS). A subsidiary survey was made to measure the lengths of approximately 50 of each type of gully (A1, A2, B1, B2) for analysis, for a total of 200 measurements. Both HiRISE and MOC images were used in this part of the study. To ensure a broad geographical coverage, no more than four gully lengths were measured on any given slope section, and no more than two of each type on any given slope section.

Results

Survey

The main survey classified 1626 gullies in 200 slope sections, broadly distributed across the southern hemisphere (Fig. 5) in a pattern similar to that found by larger surveys, such as that of Dickson *et al.* (2007). Of the 1626 gullies, the alcoves of 122 (7.5%) were not visible due either to deep shadow (a handful of steep pole-facing slopes) or because they lay beyond the edges of MOC images, and thus were classified as 'not known'. Neither type nor age could be

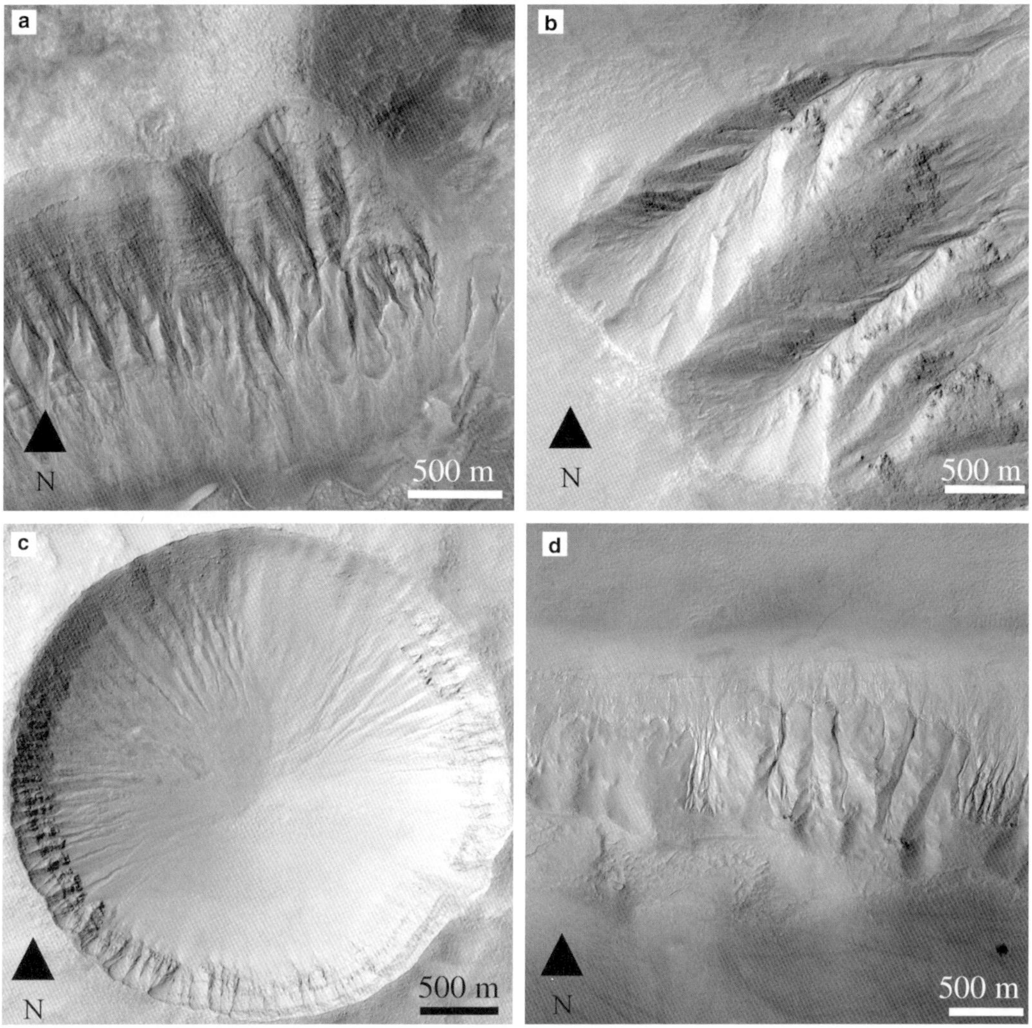

Fig. 4. Reactivated and single active phase gullies. (**a**) Single active phase: these gullies are eroding fresh bedrock, with no evidence of earlier activity. Part of HiRISE image PSP_004176_1405; 39.4°S, 202.7°E. (**b**) Reactivated: these alcoves were cut by an earlier phase of activity and subsequently covered by mantling material, now being removed by fresh gully activity. Part of HiRISE image ESP_014301_1270; 52.9°S, 245.8°E. (**c**) Crater showing single active phase A-Type gullies top left, and reactivated B-Type top right and bottom left. Alcoves at the bottom have not been reactivated. Part of HiRISE image ESP_014355_1380; 41.5°S, 210.6°E. (**d**) Gullies on the right and centre left are single active phase; gullies between them are reactivated. Note that gully alcoves are at the bottom of the image and the slope is downwards to the top of the image. Part of HiRISE image ESP_012603_1300; 49.5°S, 163.0°E. Image credit: NASA/JPL/UofA.

determined–information was only available for orientation and context. As MOC images are longer in a north–south than an east–west direction, this affected principally east- or west-facing gullies.

Latitudes ranged from 28° to 65°S–the full range of known gully latitudes excluding the south polar pits–and were initially sorted by 10° latitude bands. However, only six gullies were found in the <30°S band, and only 25 in the 60°–65°S band. These numbers are too small for meaningful statistics, and so these bands were amalgamated into their neighbours to create <40°S and >50°S bands, respectively. The majority of gullies were found between 30°S and 39°S, with 60% of the total gullies being within these latitudes.

The distribution of orientations (Fig. 6a) matches previous studies of the whole gully population (Balme *et al.* 2006), with a strong overall poleward

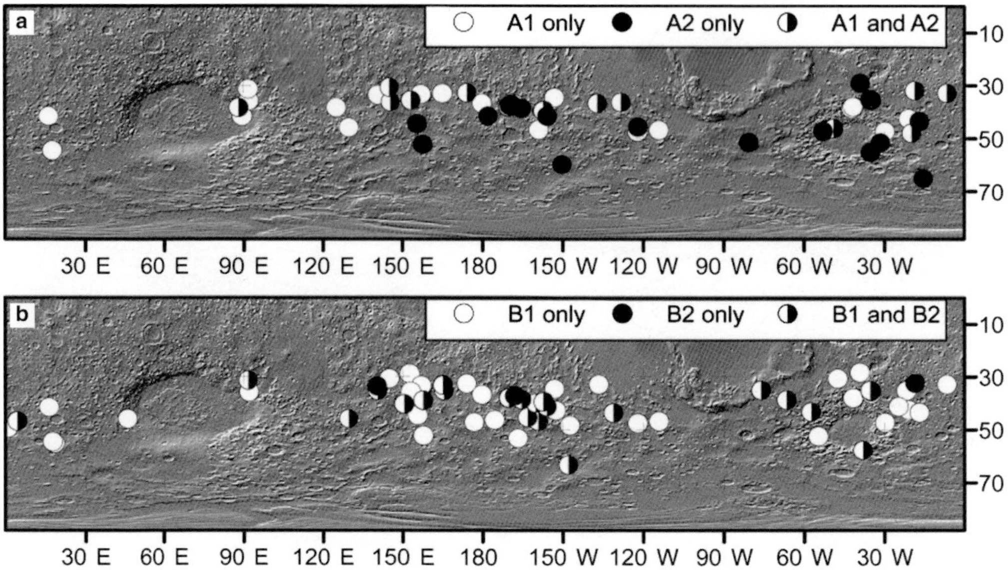

Fig. 5. Map showing the geographical distribution of the slopes surveyed by morphological types: (**a**) Type A gullies and (**b**) Type B gullies.

preferred orientation; although this study found a slightly higher proportion facing eastward. The orientations were also analysed for different latitude bands (Fig. 6b–d), and here the pattern differs to some extent from the Balme *et al.* (2006) study in which a significant proportion of gullies remained pole-oriented at higher latitudes, even as increasing numbers began to appear with other orientations. No poleward orientation was observed for the 40°–49°S band, and only a weak poleward orientation in the ≥50°S band, although the latter may be an artefact of the small population sampled.

Classification

Figure 7 shows that the majority of identifiable gullies (73%) were Type B, with B1 being much the most common (60% of all identifiable gullies); a result expected as it covered the largest range of morphologies. The proportions of the remaining three gully types were all in the range 10–15% (Fig. 7). If considered by latitude (Fig. 8), the overall proportions are similar in the <40 and 40°–49°S bands, with a change only in the ≥50°S band where 90% of identified gullies were B1 or A2. The most obvious and important trends are the steady increase in A2 proportion, and the steady decrease in A1 proportion with higher latitude (Fig. 8).

When broken down by orientation (Fig. 9), the 'immature' gully types (A1 and B1) showed similar distributions to each other, with a strong poleward orientation. Type A1 showed a westward bias, and Type B1 a slight eastward bias. However, the 'mature' gullies (A2 and B2) were much more evenly distributed, with no southward orientation bias, and lacked sizable populations only in the northwestward (A2 and B2) and northward (A2) orientations. Type A1 gullies appear to be pole-facing at all latitudes; however, A2 gullies are mainly west, south and SW facing at low latitude but east and NE facing at higher latitude (Fig. 10).

The distribution of types also differs between contexts (Fig. 11). The overall picture is dominated by the inner crater rim gullies (67% of total). Valley contexts show a similar distribution, although with a lower proportion of A-type gullies. However, the hills context is dominated by Type A (64%), with no B2 gullies. This latter result is a consequence of the definition of B2 gullies as eroding backwards into the slope line: knobs and hills rarely, if ever, have a clear slope line to erode, unlike craters and valleys. The impact crater central peaks context is entirely devoid of Type A1 gullies, which cannot be explained simply as an artefact of the classification scheme. However, it should be noted that: (1) there is only a small number (55) of central peaks gullies clustered at a very few locations; and (2) all of the central peak gullies occurred at latitudes higher than 40° S, where there are few A1 gullies anyway (about 5% of total gullies at this latitude; Fig. 9), so this result is not statistically significant.

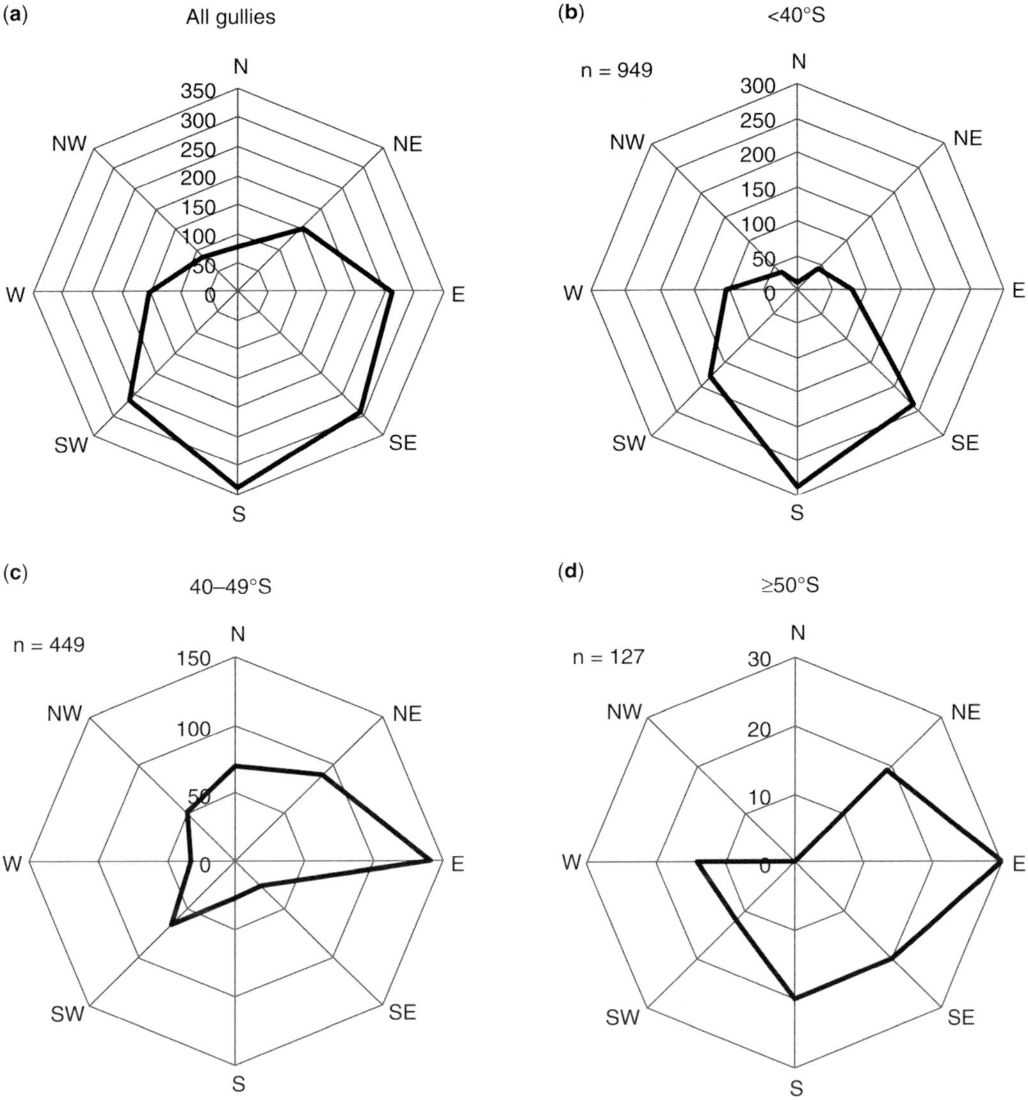

Fig. 6. Orientation of classified gullies: (**a**) all gullies, (**b**) gullies <40°S, (**c**) gullies 40°–49°S and (**d**) gullies ≥50°S.

Reactivated gullies

Of the 1504 identifiable gullies, 33% were classified as 'reactivated'. This proportion was broadly similar for the >40 and 40°–49° S latitude bands, but in the ≥50°S latitude band it decreased sharply to 13% (Fig. 12). In addition, the orientations of reactivated gullies appear different to the general population (compare Figs 13a & 6a). Overall, reactivated gullies have little orientation trend but, when the results are broken down by latitude (Fig. 13b–d), a more complex picture emerges. The poleward and westward orientation of reactivated gullies is clear for low latitudes, but there is also a trend towards north- and east-facing gullies at mid latitudes. There are too few data to determine any trend at higher latitudes.

Lengths

A total of 123 gullies were measured in MOC plus a further 65 from HiRISE. A total of 188 gullies were thus analysed (*c.* 50 in each category), ranging in length from 270 m to 15.4 km (the second largest, however, was 8.3 km). The results (Fig. 14) show that quartile values for A2 and B2 were higher

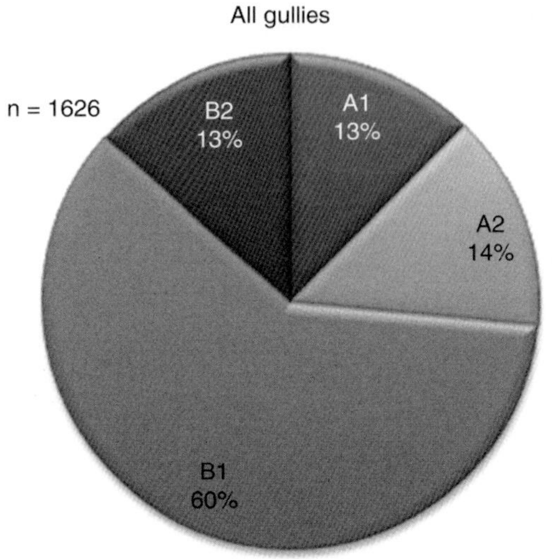

Fig. 7. Breakdown of classified gullies by morphological type.

than for A1 and B1. Although the longest B-type gully (4177 m) was a B1, mean and quartile values for B2 were higher, and twice as many B2 as B1 gullies exceeded 2000 m. There was generally no clear trend in length with latitude, although A2 gullies seem to have greater lengths at higher latitudes (Fig. 15). All of the six longest gullies measured were A2, and four of these occur south of 60°S.

Discussion

Classification

While a morphological approach involves an element of subjective assessment, it is capable of revealing information that cannot be found using solely quantitative surveys. Gullied slopes have been very comprehensively surveyed and quantified on a planetary scale, providing firm benchmarks for complementary studies that do not treat all gullies as equal. The quantitative analysis of morphologically classified data can both reveal potential second-order variations in the gully distribution and highlight gully characteristics that merit further investigation.

One example of particular interest to studies linking gullies with climate–and thus obliquity cycles–is the population of gullies that have formed most recently. One way of determining a population of 'young' gullies would be to determine the size–frequency distribution of impact craters on all documented gullies and to take a sample of the youngest. Such an approach is impractical, given the lack of very-high-resolution data, the relative paucity of craters on such young deposits and the obliteration of craters by mass wasting. However, by identifying a distinct morphological subset of gullies that exist wholly within the mantle–Type A gullies–we have found a plausible 'young' population that are limited in their degree of erosion, and seem to have formed since the mantle was emplaced. In addition, we have identified a subset of gullies that display reactivation or 'gullies within gullies' that might also indicate a youthful population. We explore in detail, below, the distribution, setting and orientation of these populations.

Gully evolution

The length data (Fig. 14) are consistent with the idea that the morphological differences between gullies represent evolutionary stages: A1 gullies are generally shorter than A2, and B1 gullies shorter than B2. B-Type gullies in general are also longer than A-type. That gullies do not form instantaneously, and instead evolve over time, has been noted by several authors: Dickson & Head (2009) and Schon *et al.* (2009) described gullies with multiple debris aprons and with interleaved deposits or channels that indicate multiple flows within the same gully. We suggest that our length data reinforce this view that gullies evolve over time and that there is, from Type A1 to Type A2 at least, an obvious progression in morphology and length.

Type A gullies

The A-type gullies represent about a quarter of the total population. This is an important result which demonstrates that many gullies do not erode 'bedrock' and exist only in a surficial layer, inferred to be ice-rich (Mustard *et al.* 2001; Milliken *et al.* 2003; Vincendon *et al.*2010). This reinforces the link between ice-rich material and the gully-forming process.

The immature A1 gullies are much more common at low latitude, and few exist at high latitude. This trend is reversed for the more developed A2 gullies, which are more numerous nearer the poles. Type A2 is also the only subset to show a trend of length with latitude: A2 gullies are longer at higher latitudes than at lower latitudes (Fig. 15). Nearly all A1 gullies are pole-facing (Figs 9 & 10) in contrast to A2 gullies, which are only generally pole-facing at lower latitudes (Fig. 10). This suggests that, of all gully types, the formation of A1 gullies is most strongly controlled by patterns of insolation. That A1 gullies are more common at low latitude, but A2 are more common at high latitude, suggests either that the recent gully-forming process has acted to a greater extent at high latitude

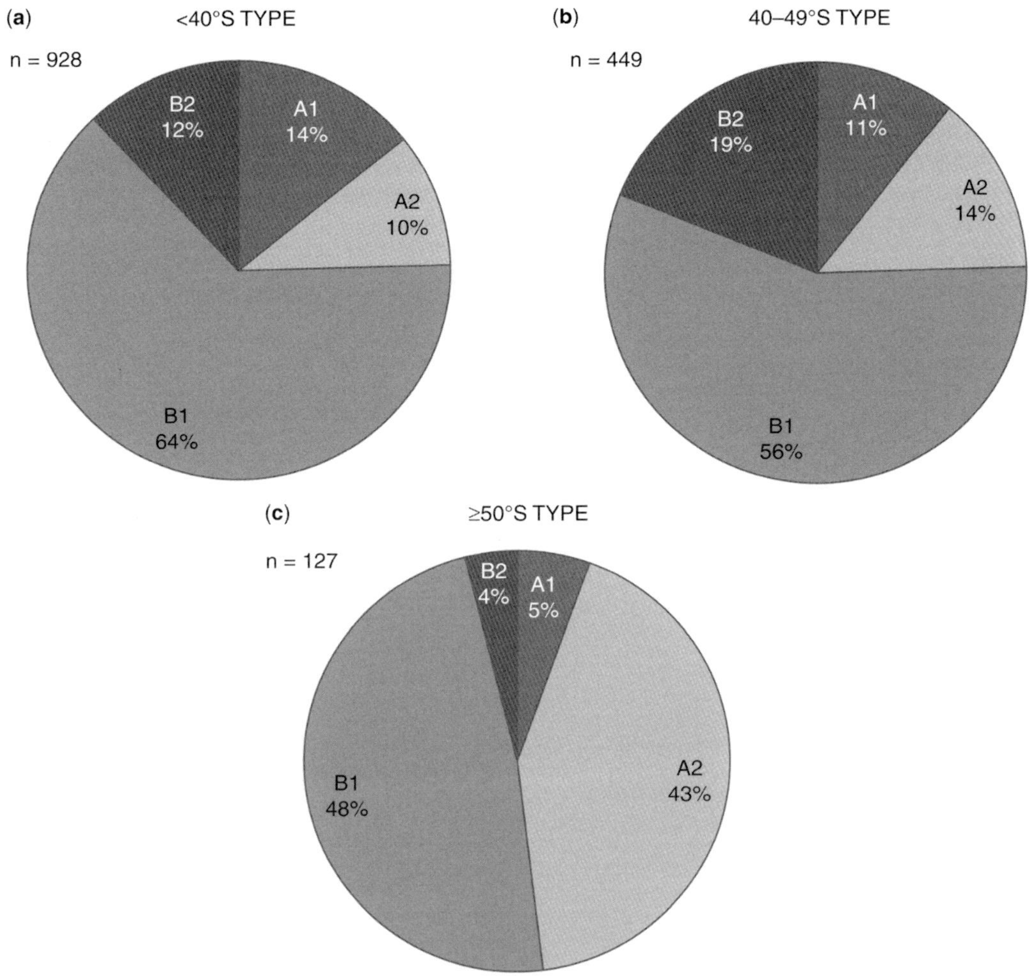

Fig. 8. Breakdown of classified gullies by morphological type and latitude: (**a**) gullies <40°S, (**b**) gullies 40°–49°S and (**c**) gullies ≥50°S.

than low, perhaps due to thicker mantle, or that the low-latitude A1 gullies are the most recent of the latest episode of gully formation that has not yet begun at higher latitude. Furthermore, that A1 gullies are so commonly pole-facing suggests one of two things: (i) conditions at low latitude are only just suitable for gully formation on pole-facing slopes, whereas at higher latitude conditions were such that gullies evolved further and formed on slopes of nearly any aspect; or (ii) only at low latitude and on pole-facing slopes are conditions suitable for the latest gullies to have begun to form. That the higher latitude A2 gullies are more 'mature' is further strengthened by the observation that A2 gullies are longer at higher latitude, although this could also be a function of the greater thickness of the mantle at higher latitudes.

Reactivated gullies

As a whole, the population of reactivated gullies (Fig. 13) shows no clear orientation trend. They are more numerous at lower latitudes, where they show a similar orientation pattern to the A1 gullies (the pronounced NE-facing trend at 40°–50°S is an intriguing exception, but based on a small number of slope sections). The contrasting population of single active phase gullies has the same orientation and latitude trend as the overall gully population described by this and other studies. We

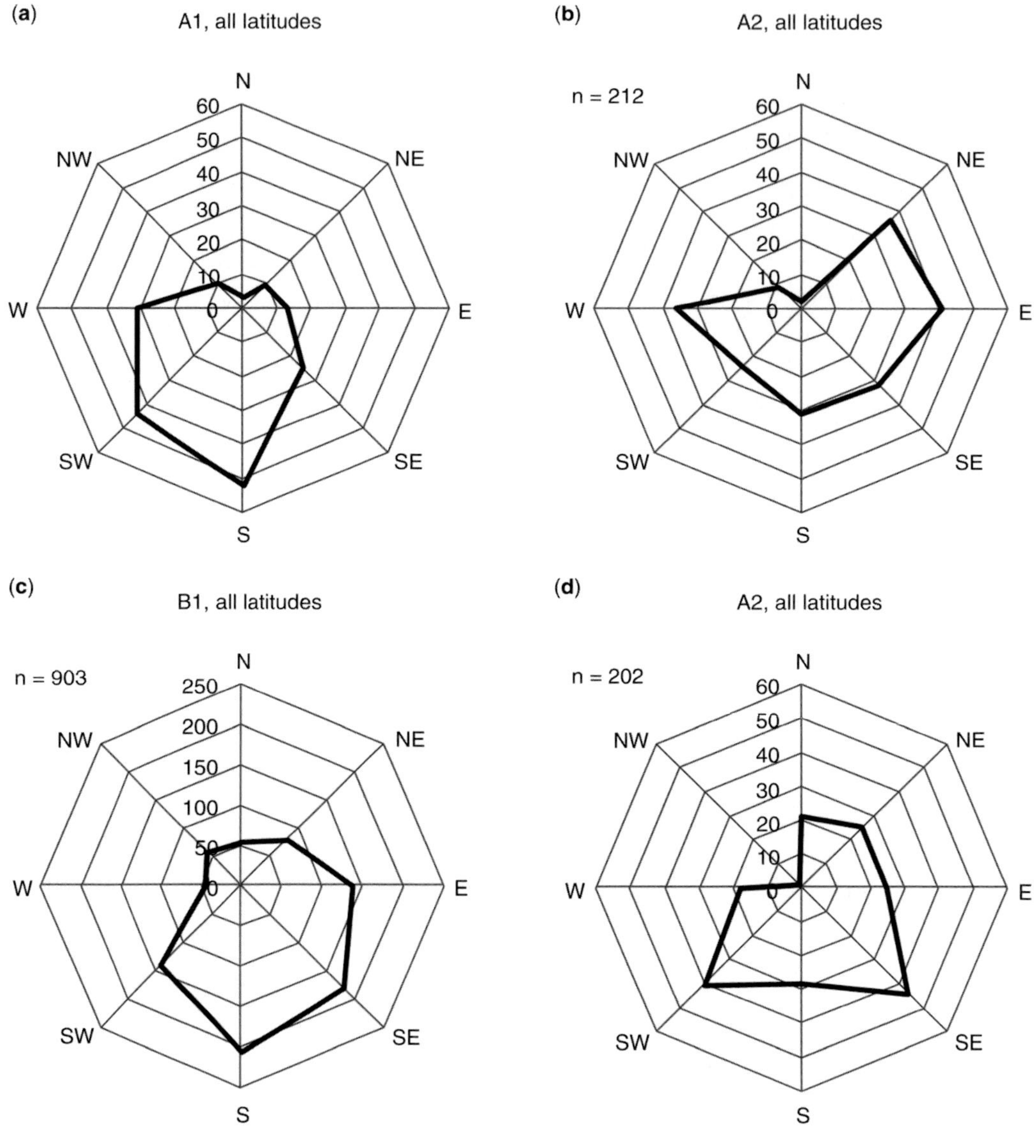

Fig. 9. Orientation of morphological type across all latitudes: (**a**) Type A1, (**b**) Type A2, (**c**) Type B1 and (**d**) Type B2.

suggest that reactivated gullies represent locations where recent gully formation is promoted due to latitude, slope or local conditions. If this is the case, then our results provide tentative support for the idea that low-latitude, pole-facing slopes have favoured gully formation most recently.

'Recent' gullies

Adding the observations of reactivated gullies to the discussion of the orientation and latitudinal distribution of A1 gullies suggests that the current population of low-latitude A1 gullies could represent the onset of a recent phase of gully formation, and that the majority of the A2 gullies, especially at high latitude, are the product of an earlier gully-forming episode. However, the evidence is still insufficient to rule out the alternative theory that the A1 gullies represent an aborted phase of a gully-formation episode that became fully developed in the south, perhaps benefiting from thicker mantle there. Interestingly, B1 gullies are also generally pole facing compared to B2, and are less common at higher latitudes. While the case for B1 gullies

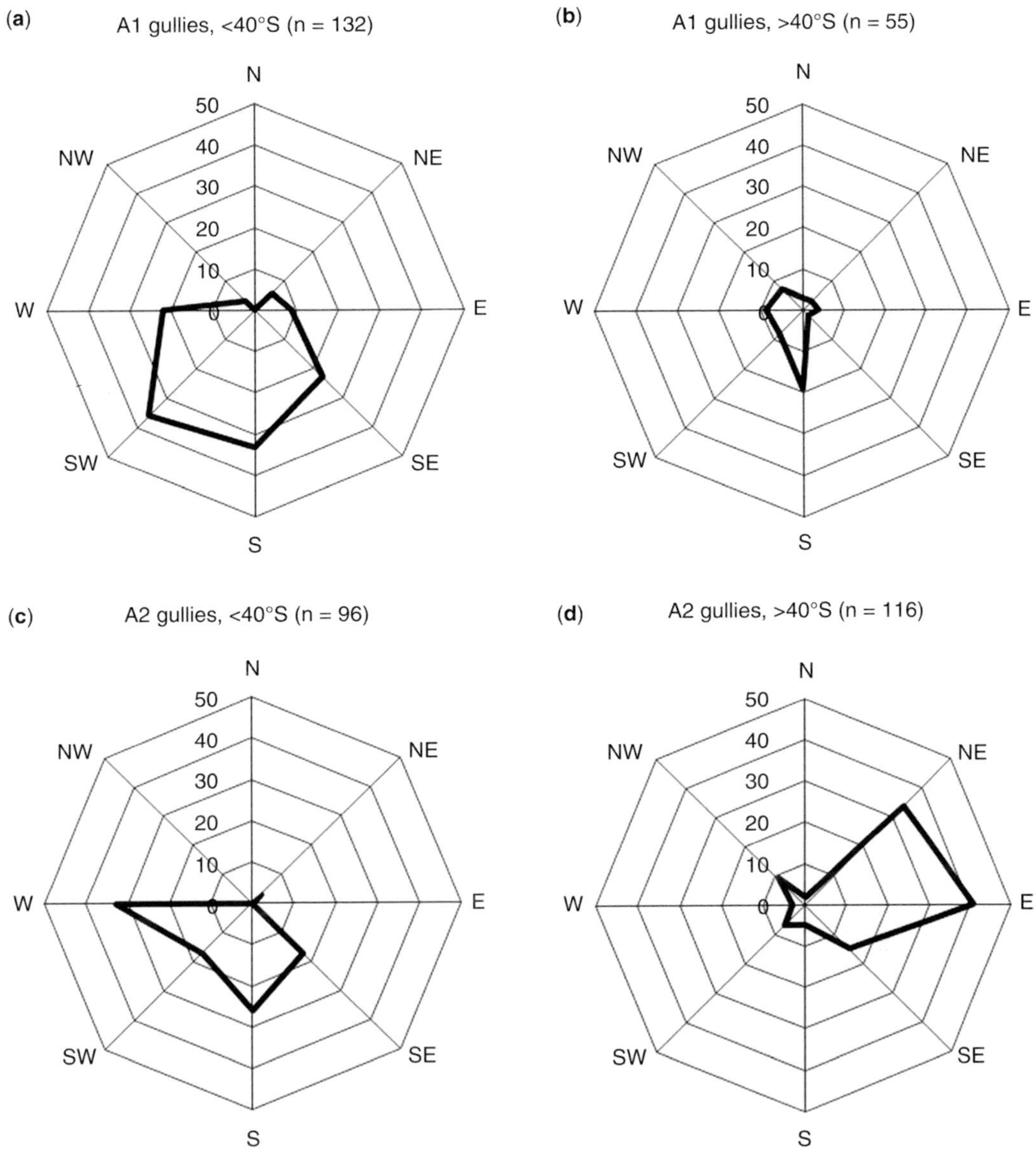

Fig. 10. Orientation of Type A gullies by latitude: (**a**) Type A1 <40°S, (**b**) Type A1 >40°S, (**c**) Type A2 <40°S and (**d**) Type A2 >40°S.

being immature versions of B2 gullies is weaker than the case for A1 gullies evolving to A2, this might also be taken as supporting evidence that low-latitude, pole-facing gullies reflect a recent episode of gully formation.

In the context of the Costard *et al.* (2002) obliquity-driven climate model, the orientation preference for these new gullies suggests that the latest phase of gully activity could have occurred less than 1 Ma ago (the last period of high obliquity of >35°). Under such obliquity conditions, pole-facing slopes at mid latitudes receive the greatest insolation (Kreslavsky *et al.* 2008). However, the model of Costard *et al.* (2002) was based on likely locations for melting, and if melting is actually more pervasive (e.g. Hecht 2002) then the accumulation of near-surface ice instead becomes the limiting control.

Under moderate and high obliquities, pole-facing slopes are the locations of preferential ice

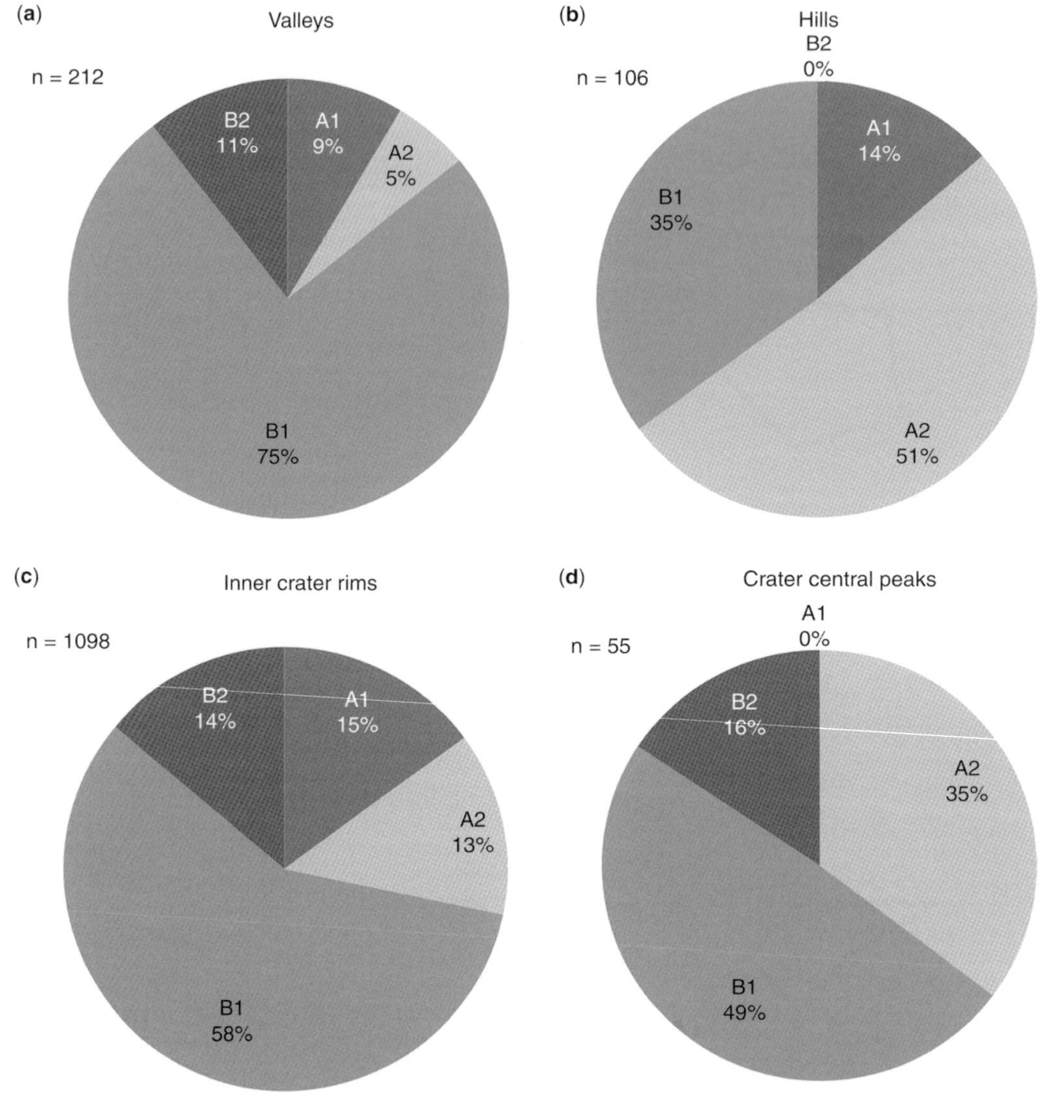

Fig. 11. Breakdown of classified gullies by morphological type and topographic context: (**a**) valleys, (**b**) hills, (**c**) inner crater rims and (**d**) crater central peaks.

and CO_2 deposition. However, sufficient ice needs to be deposited on a seasonal timescale to allow for the creation of enough meltwater to form gullies (even considering the smaller quantities of water required for debris flow than overland flow: Iverson 1997), and, for this to occur, higher levels of precipitable water are required in the atmosphere. Interestingly, predictions from General Circulation Models of the climate suggest this could have been the case as recently as 100 000 years (100 ka) before present (e.g. Mischna *et al.* 2003).

The question of whether pole-facing A1 gullies at low latitude are a young population or an aborted phase could be resolved in a number of ways. First, crater counting of the debris aprons of the large A2 gullies at high latitudes could be compared with crater counts of low-latitude A1 gullies. Such a study is probably the best way to link gully morphology, setting and orientation with age, although accounting for impact crater retention on poorly consolidated mass-wasting deposits and working out how accumulation of craters on steep

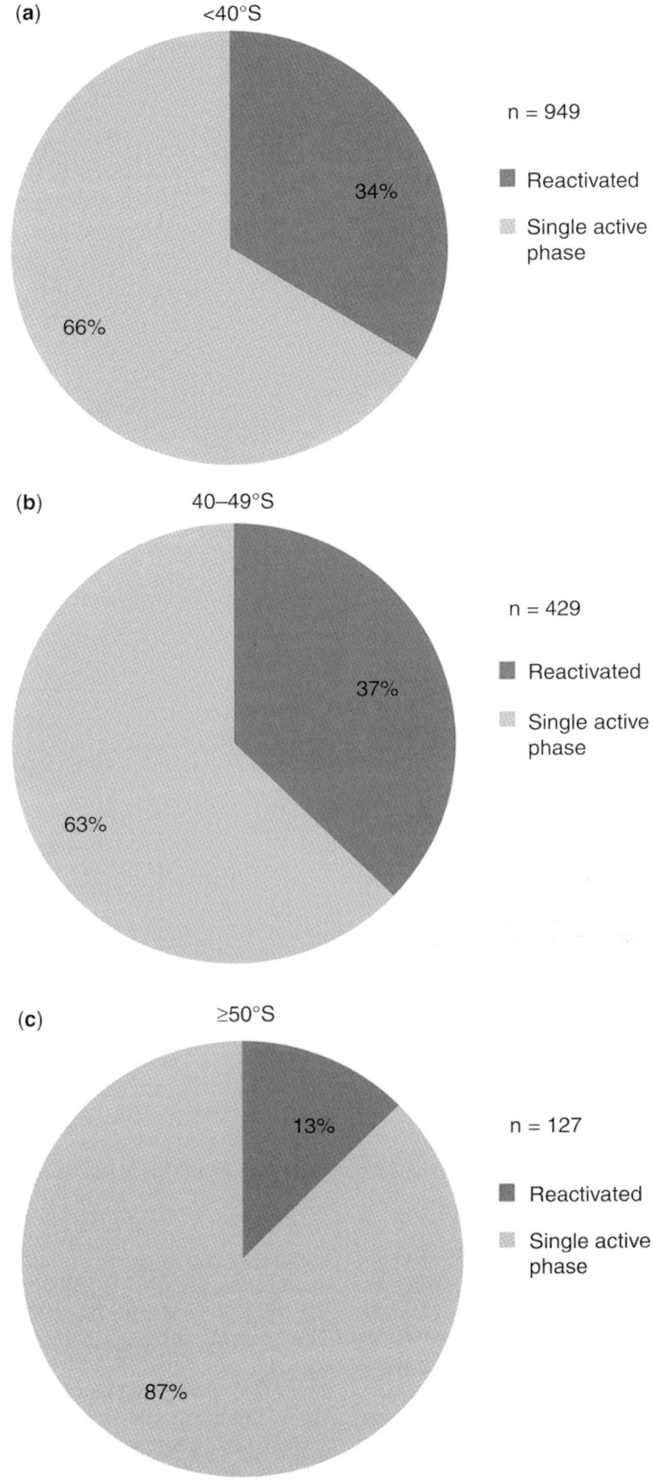

Fig. 12. Proportion of reactivated gullies by latitude: (**a**) gullies $<40°$S, (**b**) gullies 40°–49°S and (**c**) gullies $\geq 50°$S.

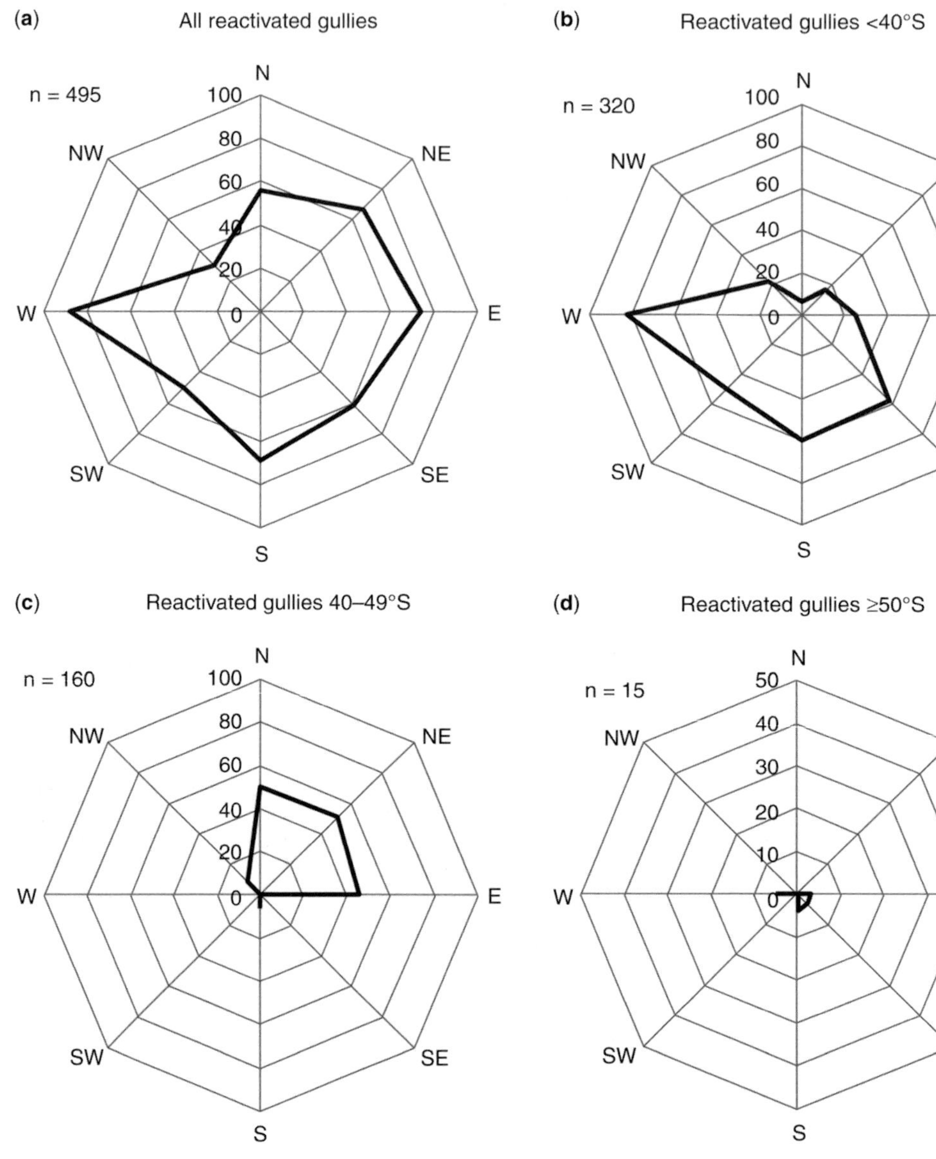

Fig. 13. Orientation of reactivated gullies: (**a**) all latitudes, (**b**) <40°S, (**c**) 40°–49°S and (**d**) ≥50°S.

slopes compares with flat surface makes this a challenging problem. Nevertheless, if these problems can be solved, the quantitative ages derived could determine where in the obliquity history these gullies formed and discriminate whether it is insolation or ice accumulation that controls gully distribution. However, it should also be noted that, for suitable crater count statistics to be obtained, appropriate HiRISE images would need to be available and a statistically significant population of gullies of both types would be required. A second possibility might be to classify gullies by degradation state, comparing gullies that appear morphologically pristine with those that appear degraded. This could include determining cross-cutting relationships between gully terminal deposits and small-scale polygonal fractures, found to be particularly common at mid–high latitude (e.g. Levy *et al.* 2009*b*; Gallagher & Balme 2011). Although this method provides relative ages, as opposed to the crater-counting technique, it could be more amenable to study.

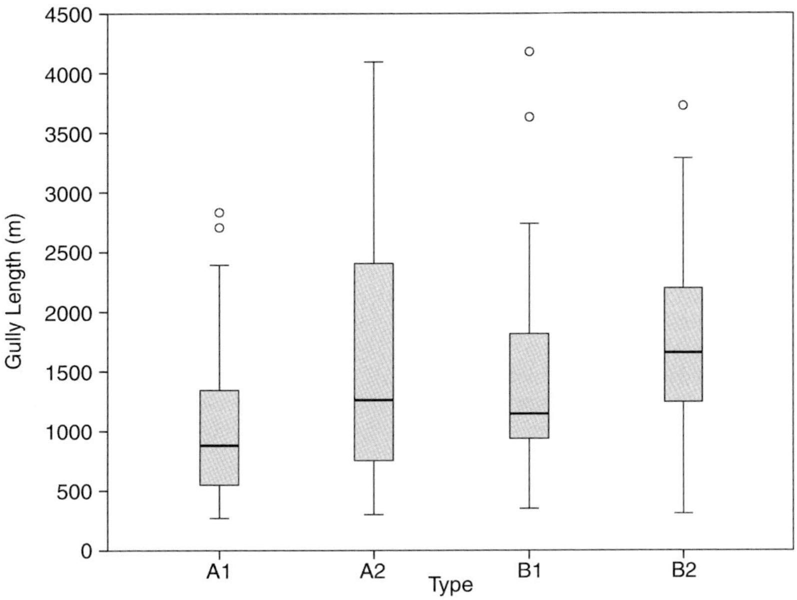

Fig. 14. Boxplot of gully lengths for each morphological type. The boxes represent the first and the third quartiles of the distribution, with the black bar marking the median. The narrow bars mark the maximum and minimum of the distribution, with the circle symbols representing 'mild' outliers (between 1.5 and 3 interquartile ranges beyond the bars). For clarity, the five mild outliers and two extreme outliers in Type A2 at more than 4500 m have been omitted.

Geographical context

The data showing gully distribution by context (Fig. 11) show that there are a larger proportion of A-type gullies on slopes on positive topography (hills) than on slopes in negative topography (craters or valleys). This may be due to the steeper uphill sections in negative topographies, which promote backwards incision and thus the formation of B-type gullies.

The central peak setting appears to be an exception to the trends mentioned above, as the

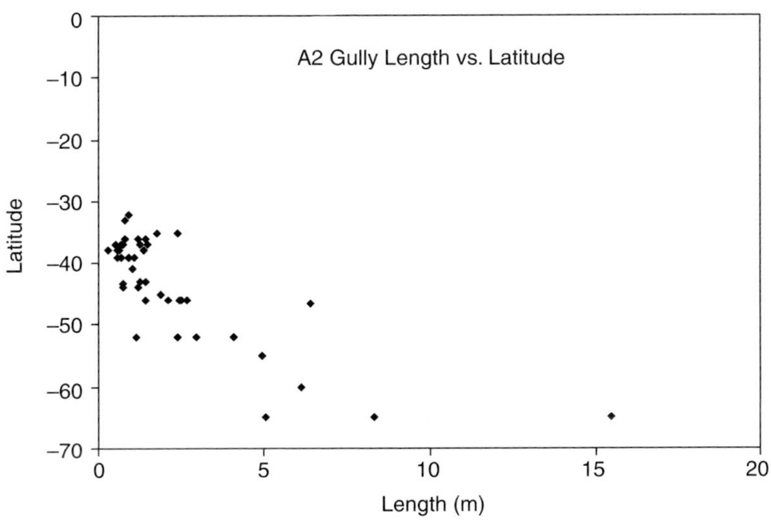

Fig. 15. Type A2 gullies: plot of length against latitude. There appears to be a slight increase in length with latitude.

distribution bears little resemblance to the hills settings, in particular the complete absence of A1 gullies. However, as all of the central peak gullies occur above 40°S where A1 gullies are rare, the absence of this type is more likely to be a reflection of latitudinal trends.

Conclusion

Our data suggest that the extent to which gullies have developed varies at different latitudes, and that A-type gullies in the mantle represent the most recent phase of gully formation. The study has demonstrated the importance of analysing the relationship of gullies and mantle on a gully-by-gully basis, and the key role of morphological examination in understanding gullies. We suggest that about 25% of gullies form only within the 'mantling deposits' and do not erode bedrock or the hillslope material beneath the mantle. This reinforces the link between ice-rich mantle and gully formation.

In general, our data and observations imply that individual gullies evolve over time. Our measurements of lengths of different morphological types support our inference that morphologically simple A1-type gullies evolve into larger, more complex A2-type gullies. The availability of very-high-resolution HiRISE image data means that these suggestions can be tested using size–frequency statistics of very small impact craters on A-type gullies.

This work was funded jointly by the Open University Centre for Earth, Planetary, Space and Astronomical Research (CEPSAR), the UK Natural Environment Research Council (NERC), and the UK Science and Technology Facilities Council (STFC). The helpful suggestions and comments of two anonymous reviewers significantly improved the paper.

References

BALME, M. R., MANGOLD, N *ET AL.* 2006. Orientation and distribution of recent Gullies in the southern Hemisphere of Mars: observations from HRSC/MEX and MOC/MGS data. *Journal of Geophysical Research–Planets*, **111**, E05001, doi: 10.1029/2005JE002607.

BLEAMASTER, L. F. & CROWN, D. A. 2005. Mantle and Gully associations along the walls of Dao and Harmakhis Valles, Mars. *Geophysical Research Letters*, **32**, L20203.

BRIDGES, N. T. & LACKNER, C. N. 2006. Northern hemisphere Martian Gullies and Mantled Terrain: implications for near-Surface water Migration in Mars' recent past. *Journal of Geophysical Research–Planets*, **111**, E09014.

CHEVRIER, V. F. & ALTHEIDE, T. S. 2008. Low temperature aqueous ferric sulfate solutions on the surface of Mars. *Geophysical Research Letters*, **35**, L22101.

CHRISTENSEN, P. R. 2003. Formation of recent Martian Gullies through melting of extensive water-rich snow deposits. *Nature*, **422**, 45–48.

CONWAY, S. J., BALME, M. R., MURRAY, J. B., TOWNER, M. C., OKUBO, C. H. & GRINDROD, P. M. 2011. The indication of Martian gully formation processes by slope–area analysis. *In*: BALME, M. R., BARGERY, A. S., GALLAGHER, C. J. & GUPTA, S. (eds) *Martian Geomorphology*. Geological Society, London, Special Publications, **356**, 171–201.

COSTARD, F., FORGET, F., MANGOLD, N. & PEULVAST, J. 2002. Formation of recent Martian debris flows by melting of near-surface ground ice at high obliquity. *Science*, **295**, 110–113.

DICKSON, J. L. & HEAD, J. W. 2009. The formation and evolution of youthful gullies on Mars: Gullies as the late-stage phase of Mars' most recent ice age. *Icarus*, **204**, 63–86.

DICKSON, J. L., HEAD, J. W. & KRESLAVSKY, M. 2007. Martian gullies in the southern mid-latitudes of Mars: evidence for climate-controlled formation of young fluvial features based upon local and global topography. *Icarus*, **188**, 315–323.

GAIDOS, E. J. 2001. Cryovolcanism and the recent flow of liquid water on Mars. *Icarus*, **153**, 218–223.

GALLAGHER, C. J. & BALME, M. R. 2011. Landforms indicative of ground-ice thaw in the northern high latitudes of Mars. *In*: BALME, M. R., BARGERY, A. S., GALLAGHER, C. J. & GUPTA, S. (eds) *Martian Geomorphology*. Geological Society, London, Special Publications, **356**, 87–110.

HARTMANN, W. K. 2001. Martian seeps and their relation to youthful geothermal activity. *Space Science Reviews*, **96**, 405–410.

HEAD, J. W., MUSTARD, J. F., KRESLAVSKY, M. A., MILLIKEN, R. E. & MARCHANT, D. R. 2003. Recent ice ages on Mars. *Nature*, **426**, 797–802.

HEAD, J. W., MARCHANT, D. R. & KRESLAVSKY, M. A. 2008. Formation of gullies on Mars: link to recent climate history and insolation microenvironments implicate surface water flow origin. *Proceedings of the National Academy of Sciences*, **105**, 13 258–13 263.

HECHT, M. H. 2002. Metastability of liquid water on Mars. *Icarus*, **156**, 373–386.

HELDMANN, J. L. & MELLON, M. T. 2004. Observations of Martian Gullies and constraints on potential formation mechanisms. *Icarus*, **168**, 285–304.

HELDMANN, J. L., TOON, O. B., POLLARD, W. H., MELLON, M. T., PITLICK, J., MCKAY, C. P. & ANDERSEN, D. T. 2005. Formation of Martian Gullies by the action of liquid water flowing under current Martian environmental conditions. *Journal of Geophysical Research–Planets*, **110**, E05004.

HELDMANN, J. L., CARLSSON, E., JOHANSSON, H., MELLON, M. T. & TOON, O. B. 2007. Observations of Martian gullies and constraints on potential formation mechanisms: II. The Northern Hemisphere. *Icarus*, **188**, 324–344.

HOFFMAN, N. 2002. Active polar gullies on Mars and the role of carbon dioxide. *Astrobiology*, **2**, 313–323.

IVERSON, R. M. 1997. The physics of debris flows. *Reviews of Geophysics*, **35**, 245–296.

KNAUTH, L. P. & BURT, D. M. 2002. Eutectic brines on Mars: origin and possible relation to young seepage features. *Icarus*, **158**, 267–271.

KNEISSL, T., REISS, D., VAN GASSELT, S. & NEUKUM, G. 2010. Distribution and orientation of northern-hemisphere gullies on Mars from the evaluation of HRSC and MOC-NA data. *Earth and Planetary Science Letters*, **294**, 357–367.

KRESLAVSKY, M. A., HEAD, J. W. & MARCHANT, D. R. 2008. Periods of active permafrost layer formation during the geological history of Mars: implications for circum-polar and mid-latitude surface processes. *Planetary and Space Science*, **56**, 289–302.

LASKAR, J., LEVRARD, B. & MUSTARD, J. F. 2002. Orbital forcing of the Martian polar layered deposits. *Nature*, **419**, 375–377.

LEVY, J., HEAD, J., DICKSON, J., FASSETT, C., MORGAN, G. & SCHON, S. 2009*a*. Identification of gully debris flow deposits in Protonilus Mensae, Mars: characterization of a water-bearing, energetic gully-forming process. *Earth and Planetary Science Letters*, **294**, 368–377.

LEVY, J., HEAD, J., MARCHANT, D., DICKSON, J. & MORGAN, G. 2009*b*. Geologically recent gully–polygon relationships on Mars: insights from the Antarctic dry valleys on the roles of permafrost, microclimates, and water sources for surface flow. *Icarus*, **201**, 113–126.

MALIN, M. C. & EDGETT, K. S. 2000. Evidence for recent groundwater seepage and surface runoff on Mars. *Science*, **288**, 2330.

MARQUEZ, A., DE PABLO, M. A., OYARZUN, R. & VIEDMA, C. 2005. Evidence of gully formation by regional groundwater flow in the Gorgonum–Newton Region (Mars). *Icarus*, **179**, 398–414.

MELLON, M. T. & PHILLIPS, R. J. 2001. Recent gullies on Mars and the source of liquid water. *Journal of Geophysical Research–Planets*, **106**, 23 165–23 179.

MILLIKEN, R., MUSTARD, J. & GOLDSBY, D. 2003. Viscous flow features on the surface of Mars: observations from high-resolution Mars Orbiter Camera (MOC) images. *Journal of Geophysical Research–Planets*, **108**, 5057.

MISCHNA, M. A., RICHARDSON, M. I., WILSON, R. J. & MCCLEESE, D. J. 2003. On the orbital forcing of Martian water and CO_2 cycles: a General Circulation Model study with simplified volatile schemes. *Journal of Geophysical Research-Planets*, **108**, 5062.

MUSSELWHITE, D. S., SWINDLE, T. D. & LUNINE, J. I. 2001. Liquid CO_2 breakout and the formation of recent small gullies on Mars. *Geophysical Research Letters*, **28**, 1283–1285.

MUSTARD, J. F., COOPER, C. D. & RIFKIN, M. K. 2001. Evidence for recent climate change on Mars from the identification of youthful near-surface ground ice. *Nature*, **412**, 411–414.

REISS, D., VAN GASSELT, S., NEUKUM, G. & JAUMANN, R. 2004 Absolute dune ages and Implications for the time of formation of gullies in Nirgal Vallis, Mars. *In*: *Proceedings of the 35th Annual Lunar and Planetary Science Conference, March 15–19, 2004, League City, Texas*. Lunar and Planetary Institute, Houston, TX, Abstract 1639.

SCHON, S. C., HEAD, J. W. & FASSETT, C. I. 2009. Unique chronostratigraphic marker in depositional fan stratigraphy on Mars: evidence for ca 1.25 Ma gully activity and surficial meltwater origin. *Geology*, **37**, 207–210.

STEWART, S. T. & NIMMO, F. 2002. Surface runoff features on Mars: testing the carbon dioxide formation hypothesis. *Journal of Geophysical Research–Planets*, **107**, 1–7.

TREIMAN, A. H. 2003. Geologic settings of Martian gullies: implications for their origins. *Journal of Geophysical Research–Planets*, **108**, 8031.

VINCENDON, M., MUSTARD, J. F., FORGET, F., KRESLAVSKY, M., SPIGA, A., MURCHIE, S. & BIBRING, J.-P. 2010. Near-tropical subsurface ice on Mars. *Geophysical Research Letters*, **37,** L01202.

WILLIAMS, K., TOON, O., HELDMANN, J. & MELLON, M. 2009. Ancient melting of mid-latitude snowpacks on Mars as a water source for gullies. *Icarus*, **200**, 418–425.

The indication of Martian gully formation processes by slope–area analysis

SUSAN J. CONWAY[1,5]*, MATTHEW R. BALME[1], JOHN B. MURRAY[1], MARTIN C. TOWNER[2], CHRIS H. OKUBO[3] & PETER M. GRINDROD[4]

[1]*Earth and Environmental Sciences, Open University, Walton Hall, Milton Keynes MK7 6AA, UK*

[2]*Department of Earth Science and Engineering, Impacts and Astromaterials Research Centre, Imperial College, London SW7 2AZ, UK*

[3]*Astrogeology Science Center, US Geological Survey, 2255 North Gemini Drive, Flagstaff, AZ 86001, USA*

[4]*Department of Earth Sciences, University College London, Gower Street, London WC1E 6BT, UK*

[5]*Present address: Laboratoire de planétologie et géodynamique, CNRS UMR 6112, Université de Nantes, 2 rue de la Houssinière, BP 92208, 44322 Nantes cedex, France*

**Corresponding author (e-mail: susan.conway@univ-nantes.fr)*

Abstract: The formation process of recent gullies on Mars is currently under debate. This study aims to discriminate between the proposed formation processes – pure water flow, debris flow and dry mass wasting – through the application of geomorphological indices commonly used in terrestrial geomorphology. High-resolution digital elevation models (DEMs) of Earth and Mars were used to evaluate the drainage characteristics of small slope sections. Data from Earth were used to validate the hillslope, debris-flow and alluvial process domains previously found for large fluvial catchments on Earth, and these domains were applied to gullied and ungullied slopes on Mars. In accordance with other studies, our results indicate that debris flow is one of the main processes forming the Martian gullies that were being examined. The source of the water is predominantly distributed surface melting, not an underground aquifer. Evidence is also presented indicating that other processes may have shaped Martian crater slopes, such as ice-assisted creep and solifluction, in agreement with the proposed recent Martian glacial and periglacial climate. Our results suggest that, within impact craters, different processes are acting on differently oriented slopes, but further work is needed to investigate the potential link between these observations and changes in Martian climate.

Martian 'gully' landforms were first described by Malin & Edgett (2000), and were defined as features that have an alcove, channel and debris apron with the general appearance of gullies carved by water. Within this definition gullies have a wide range of morphologies (Fig. 1) and are found in abundance on steep slopes at mid latitudes in both hemispheres on Mars (e.g. Heldmann & Mellon 2004; Heldmann *et al.* 2007). They are interpreted to be geologically young features because of their pristine appearance and the paucity of superposed impact craters. Recent work has suggested that some gullies have been active in the last 3–1.25 Ma (Reiss *et al.* 2004; Schon *et al.* 2009). Malin *et al.* (2006) observed new, high-albedo, dendritic deposits (named light-toned deposits) located along the paths of some gullies, and which formed between subsequent images taken by the Mars Orbiter Camera (MOC). These light-toned deposits have been attributed to either dry mass wasting (Pelletier *et al.* 2008; Kolb *et al.* 2010), or debris flow (Heldmann *et al.* 2010), involving up to 50% water (Iverson 1997). However, the origins of these deposits are still under debate, and it is not clear whether they are related to the formation processes of the gullies or are formed by a secondary process.

The formation process for Martian gullies in general is also still under debate. Three main candidates exist: (1) aquifer outflow; (2) surface melting; or (3) dry granular flow. In the aquifer model, the water is either released from a near-surface confined aquifer (Malin & Edgett 2000; Heldmann *et al.*

From: Balme, M. R., Bargery, A. S., Gallagher, C. J. & Gupta, S. (eds) *Martian Geomorphology*. Geological Society, London, Special Publications, **356**, 171–201.
DOI: 10.1144/SP356.10 0305-8719/11/$15.00

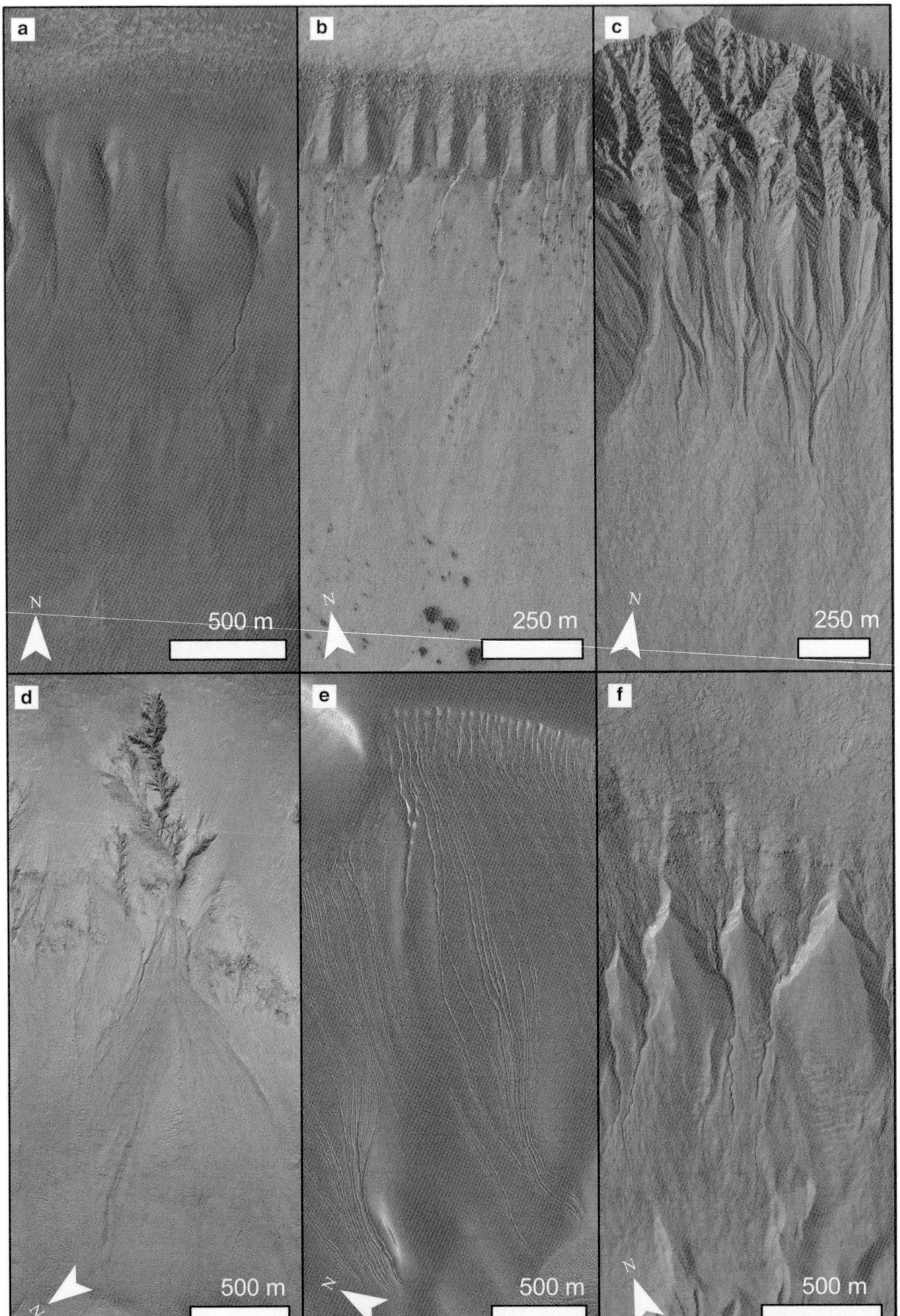

Fig. 1.

2005) or brought up from depth by cryovolcanic processes (Gaidos 2001). The main criticism of the aquifer-based models is their failure to explain the location of some gullies on isolated hills, impact crater central peaks, mesas and sand dunes. Melting of near-surface ground ice or surface ice has been proposed for the formation of gullies under recent obliquity excursions (Costard *et al.* 2002). There is growing support for this model, with the most compelling arguments being: (1) the majority of gullies exists at mid latitudes; (2) the dominance of pole-facing gullies (Balme *et al.* 2006; Dickson *et al.* 2007; Kneissl *et al.* 2010); and (3) observations of coincidence with sites of seasonal surface-ice accumulation (Dickson & Head 2009). Granular flow has been suggested as either unassisted (Treiman 2003; Shinbrot *et al.* 2004) or carbon- dioxide-assisted flow (Musselwhite *et al.* 2001). The main criticism of the granular flow model is that it fails to replicate some commonly observed features of gullies; in particular, channel sinuosity, and complex tributary and distributary systems (McEwen *et al.* 2007).

There is also debate about the type of fluid involved: pure water or brine. Whilst pure water is not stable under the current surface environment on Mars, it can persist in a metastable form (Hecht 2002), although its flow behaviour may be substantially different to water on Earth (Conway *et al.* 2011). Brines are a likely product of water sourced from underground and, moreover, the presence of some common geological compounds can substantially depress the freezing point of water (e.g. Chevrier & Altheide 2008). Brines are less likely in a surface melting scenario because water ice condensed from the atmosphere will have had less opportunity to dissolve salts than an underground water body. Both pure water and brine can support very high concentrations of entrained sediment, and form a flow commonly termed a 'debris flow'. Debris flow is an attractive candidate process for forming gullies because large amounts of erosion and deposition can be brought about with only a 10–50% water content (Iverson 1997). Several authors have proposed debris flow as a potential gully-forming mechanism on Mars due to the supply of loose sediment combined with the steep slopes on which gullies are found (e.g. Malin & Edgett 2000; Balme *et al.* 2006). The inclusion of debris might also limit evaporation and freezing of the water within the flow. Debris flows on Earth are commonly triggered by sudden and intense or prolonged rainfall (e.g. Ben David-Novak *et al.* 2004; Decaulne & Sæmundsson 2007; Godt & Coe 2007; Crosta & Frattini 2008; Morton *et al.* 2008), which is not a possible mechanism on Mars under recent climate. However, debris flows can also be triggered by snowmelt or melting permafrost (Harris & Gustafson 1993; Decaulne *et al.* 2005). As noted by Lanza *et al.* (2010), infiltration rates on Mars are likely to exceed the low discharge rates produced by a surface melting source. Hence, overland flow is unlikely unless there is a shallow impermeable barrier, such as near-surface permafrost, or frozen layer formed at the base of the water flow on contact with a cold substrate (Conway *et al.* 2011). The dominance of infiltration satisfies the conditions for triggering debris flow, sediment saturation and elevated pore pressures. The lack of vegetation and the associated lower cohesion of the Martian soil, compared to Earth, potentially means that debris flows can be triggered on much lower slope gradients than they are on Earth.

Gullies formed by dilute-water flow and debris flow on Earth can be visually very similar to each other, and the basic structure of gullies can be formed by dry granular flow (Mangeney *et al.* 2007). In many geomorphological problems, convergence of visual form means that using images alone can make it very difficult to determine process. The ongoing debate regarding the formation mechanisms of gullies on Mars is a prime example of this. For example, some workers have dismissed debris flow as a mechanism for forming Martian gullies because they have not observed the levees that are one of the diagnostic features of debris flow (e.g. Innes 1983). However, the ability to identify levees depends on viewing geometry and sun angle; metre-sized levees are often not visible on 25 cm per pixel air photographs of Earth. It is also possible that a combination of the lower gravity and different sediment type on Mars means that the levees might be small compared to those on Earth.

The amount of water required to carve channels, and to transport and deposit sediment, differs

Fig. 1. (*Continued*) HiRISE images of a variety of gullies on Mars. Image credits: NASA/JPL/UofA. (**a**) Gullies on the wall of a small impact crater within Kaiser Crater, site KC in this study, image number: PSP_003418_1335, at 18.8°E, 54.3°S. (**b**) Gullies within a polar pit, image number: PSP_003498_1090 at 1.6°E, 70.6°S. (**c**) Gullies on the wall of Galap Crater, near Sirenum Fossae, image number: PSP_003939_1420, at 192.9°E, 37.7°S. (**d**) Gullies on the wall of Wirtz Crater, a large impact crater to the east of Argyre basin, image number: PSP_002457_1310, at 335.3°E, 48.2°S. (**e**) Gullies on the slip face of dunes in Russell Crater, located in Noachis Terra, image number: PSP_001440_1255, at 12.9°E, 54.2°S. (**f**) Gullies on the wall of an impact crater to the west of Newton Crater in Terra Sirenum, image number: PSP_005930_1395, at 196.8°E, 40.3°S.

substantially between debris flow, water or brine flow (termed 'alluvial' throughout the rest of this paper) and granular flow. Determining the amount of water available at the Martian surface is important for questions of Martian climate, hydrology and the study of potential Martian habitats. Hence, an accurate determination of active processes is needed that in turn can constrain the quantity of fluid required to form gullies. Quantitative geomorphological study can provide the tools to discriminate between these three processes. The recent availability of high-resolution digital elevation models (DEMs) of Mars has opened up the possibility of using quantitative geomorphic methods that have, until now, been restricted to analysing landscapes on Earth. By taking well-developed slope–area analyses and other geomorphic process indicators for the Earth and applying them to Mars, this study aims to give insights into both the processes that formed the gullies on Mars and the source of any water involved.

We used three geomorphic tools commonly applied in terrestrial geomorphology to identify active processes forming gullies on Mars: slope–area plots (Fig. 2a), cumulative area distribution (CAD) plots (Fig. 2b) and wetness index maps. These analytical techniques are described in more detail in the following sections. They are usually used to assess active processes within catchment areas and other larger-scale landscape analyses. To

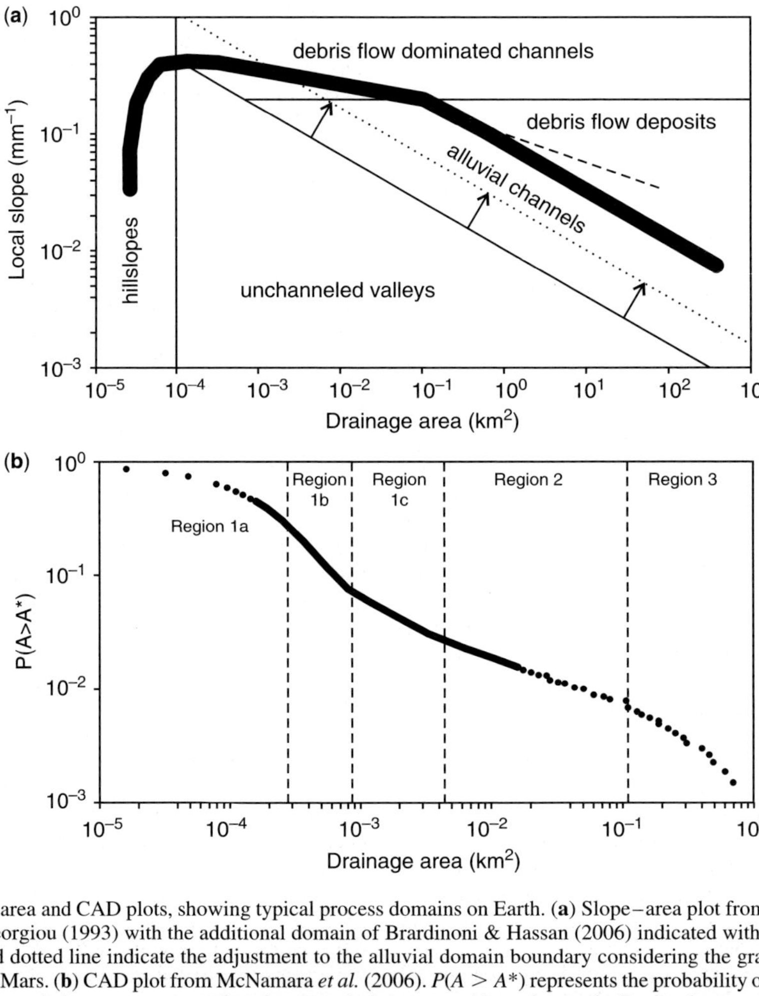

Fig. 2. Slope–area and CAD plots, showing typical process domains on Earth. (**a**) Slope–area plot from Montgomery & Foufoula-Georgiou (1993) with the additional domain of Brardinoni & Hassan (2006) indicated with a dashed line. The arrows and dotted line indicate the adjustment to the alluvial domain boundary considering the gravitational acceleration of Mars. (**b**) CAD plot from McNamara *et al.* (2006). $P(A > A^*)$ represents the probability of a point in the landscape having a drainage area greater than the given drainage area, A^*, on the x-axis. Region 1a represents hillslopes that diverge and do not gather drainage. Region 1b represents hillslopes with convergent topography. Region 1c represents pore-pressure- triggered landsliding or debris flow. Region 2 represents incision or channel formation. Region 3 has large steps where large tributaries join the channel.

test whether they are equally applicable to smaller areas, we first applied them to five study sites on Earth at an equivalent scale to gullies on Mars. Recently deglaciated areas were preferred as these have: (1) a geologically short and well-defined slope development history (i.e. since deglaciation); and (2) a glacial trough-valley slope-profile that strongly resembles that of fresh impact craters (compare the relationships in Brook *et al.* 2008 and Garvin *et al.* 1999). However, suitable quality data could not be found for the alluvial end-member process in glacial environments, so two desert study sites were also included.

When we were satisfied that different geomorphic processes could be discriminated on Earth using slope–area plots, CAD plots and wetness index maps, we applied these analyses to slopes containing gullies on Mars.

Method

Slope–area and CAD methods

The so-called 'stream power law' was first proposed by Hack (1957) and has been widely used to investigate landscape evolution on Earth (e.g. Kirkby *et al.* 2003; Stock & Dietrich 2003). It is based on the detachment and transport limited rate of bedrock erosion, otherwise known as the shear-stress incision model, which is stated as follows:

$$S = kA^{-\theta} \quad (1)$$

where S is local slope, A is upslope drainage area, k is a process-related constant, which is different for detachment and transport cases, and θ is the concavity index, which is process dependent. It has also been noted that if the drainage area is plotted against the local slope for drainage basins then process domains can be defined in log–log plots, as shown in Figure 2a (after Montgomery & Foufoula-Georgiou 1993). These process domains were initially schematic, based on few data, but have been supported by later work (e.g. Whipple & Tucker 1999; Snyder *et al.* 2000; Kobor & Roering 2004; Marchi *et al.* 2008). Brardinoni & Hassan (2006) added an additional domain in which systems, dominated by debris-flow deposition, occupy that part of the alluvial domain of Montgomery & Foufoula-Georgiou (1993) that is located towards higher drainage areas and steeper slopes (Fig. 2a). This domain was proposed from field observations in glacially modified area and has since been supported by additional observations by Mao *et al.* (2009) in a different geomorphic setting. Process information can be obtained both from the position of the data points relative to defined domains on this slope–area plot and from the trend of the data within these domains; for example, whether the data points plot in a concave, convex, upwards-trending or downwards-trending curve (Tucker & Bras 1998). The general trend for an alluvial system is shown in Figure 2a, which passes through several process domains. The data for such plots are generally derived from DEMs or topographical maps. The slope and contributing area data are either extracted from the channel only or from the whole drainage basin, depending on the focus of the study. In Figure 2a these data are taken from every pixel contained within the catchment of the whole fluvial system (encompassing valley hillslopes, tributaries, main channels and estuary system) sampled at a single point in time.

Cumulative area distribution (CAD) is the probability distribution of points in the landscape with a drainage area greater than any particular area, A^*. The log–log plot of $P(A > A^*)$ against A^* gives information on the processes acting within a catchment (Perera & Willgoose 1998; McNamara *et al.* 2006). Interpretation of this index varies, but generally it is split into three areas: (1) in small drainage areas the plot usually evolves from convex to concave, and represents diffusive erosion; (2) intermediate drainage areas are linear in a log–log plot and this is thought to represent incision (i.e. channel formation); and (3) in large drainage areas there are small steps where major tributaries join the channel (Fig. 2b). McNamara *et al.* (2006) split domain (1) into three subdomains (Fig. 2b): (1a) a convex section, representing hillslopes that diverge and do not gather drainage; (1b) linear and steep section in a log–log plot, indicating hillslopes with convergent topography; and (1c) a concave section, which they suggest is a reach dominated by pore- pressure-triggered landsliding (including debris flows, which are triggered by this mechanism).

The stream-power law (Equation 1), and the process interpretations in slope–area and CAD plots of Montgomery & Foufoula-Georgiou (1993), Tucker & Bras (1998) and Brardinoni & Hassan (2006), are based on empirical hydraulic geometry functions that are predicated on, and developed for, studies of large fluvial systems with channel morphology well adjusted to perennial discharge. It could, therefore, be argued that these systems are unlike the hillslope systems in this study. Hence, we have tested these interpretive analysis techniques on small gully systems on Earth where we know the active processes in order to demonstrate that they can still be valid. It is, of course, necessary to bear in mind that there is always some uncertainty in inferring process from landscape form, in part due to the intrinsic variability and complexity of natural systems but also due to the effects of vegetation, tectonics, climate and, perhaps, human interaction with the landscape. However, on Mars the surface processes are likely to be simpler, with little chance of factors, such as rain, vegetation or human action confounding the process domains, so these indices should provide an important addition to the 'visual' morphology when inferring process from form.

Application of the slope–area method to Mars

The reduced gravitational acceleration of Mars shifts the slope–area boundary of the alluvial slope–area domain vertically (dotted line in Fig. 2a). This means that the unchanneled domain extends to higher slopes for a given drainage area for Mars (extending into the alluvial and debris-flow domains for Earth); however, the hillslope domain is unaffected. Considering the fact that gullies on Mars do not have large tributary-channel networks, it seems unlikely that this domain would be well developed. Appendix 1 gives details of the calculations performed to account for the gravitational acceleration of Mars. The relative gradients and curvatures of the trends described by the alluvial data in slope–area plots are unaffected by the reduced gravity. We have not been able to revise the position of the domain added by Brardinoni & Hassan (2006) as a function of gravitational acceleration because this domain was added empirically, based on field observations.

The slope threshold for dry mass wasting or landsliding in loose material is the same as on Earth (Moore & Jakosky 1989; Peters *et al.* 2008). The slope thresholds for pore-pressure failure are also unaffected by the difference in gravitational acceleration. Hence, there would be no change to these process domains or trends for either dry mass-wasting or pore-pressure-triggered processes such as debris flow.

We note that on Earth vegetation cover, soil type and geology can have profound impacts on the slope values in a landscape for a given drainage area (Yetemen *et al.* 2010), but we would expect only variations in soil type and geology to affect the data on Mars. Despite these differences in surface properties, basins with similar processes on Earth show a similar pattern or trend of data, but displaced vertically in slope–area plots (Yetemen *et al.* 2010).

Datasets and the generation of DEMs

Slope–area analysis is only possible with high-quality elevation data, preferably at a resolution better than 10 m per pixel or 1:25 000 map scale (Montgomery & Foufoula-Georgiou 1993; Tarolli & Fontana 2009). For each of the terrestrial sites 1 m-resolution DEMs were derived from airborne laser altimeter (LiDAR) data. These were then resampled to 5 m resolution to match the Mars data, as described later. Table 1 lists the data sources for the study sites on Earth. The DEM for NW Iceland was produced from the raw LiDAR point data collected by the UK's Natural Environment Research Council's Airborne Research and Survey Facility in 2007 using techniques described by Conway *et al.* (2010) and correcting for between-track shifts using methods developed by Akca (2007*a*, *b*).

For Mars we used four 1 m-resolution DEMs produced using stereo photogrammetry from 25 cm per pixel High Resolution Science Imaging Experiment (HiRISE) images. The DEMs for sites PC, GC, KC and TS were produced by the authors from publicly released HiRISE images using methods described by Kirk *et al.* (2008). Significant metre-scale random noise present in the DEMs of sites GC, KC and TS had a detrimental effect on preliminary slope–area analyses. Hence, all of the DEMs were resampled to 5 m per pixel before the reanalysis was performed.

The precision of elevation values in the DEMs used here can be estimated based on viewing geometry and pixel scale. For the DEM of site PC, the attendant image pair PSP_004060_1440 (0.255 m per pixel) and PSP_005550_1440 (0.266 m per pixel) have a 12.6° stereoscopic convergence angle. Assuming 1/5 pixel matching error and using a pixel scale of 0.266 m per pixel from the more oblique image, the vertical precision is estimated to be approximately 0.24 m (cf. Kirk *et al.* 2008). DEMs for sites GC, KC and TS have a similar magnitude of vertical precision. The pixel matching error is influenced by signal-to-noise ratio, scene contrast and differences in illumination between images. Pattern noise can also be introduced by the automatic terrain extraction algorithm, especially in areas of low correlation. Manual editing is necessary to correct spurious topography in areas of poor correlation (e.g. smooth, low contrast slopes and along shadows).

Finally, a synthetic crater was constructed to test whether the results from the Mars study sites in general reflected the process or, instead, were a result of the geometry imposed by the impact crater setting (all the Mars study areas were on the inner walls of bowl-shaped depressions, but none of the ones on Earth). A 10 km-diameter synthetic crater was created by applying a smooth parabolic radial profile that was derived by fitting curves through ungullied radial profiles of the craters in sites PC and GC. Metre-scale 'pink' (also called '1/f') noise was added to simulate a natural rough surface (Jack 2000).

Table 1. *Summary of the data for the study sites on Earth*

Site	Location	Date flown	Data source	Approximate precipitation (mm a^{-1})	Landscape type	Latitude	Longitude	Average elevation (m)	Relief (m)
A	San Jacinto Fault (SJF Segment 3) – Santa Rosa Mountains	mid 2005	NCALM B4 Project	150	Desert	33° 25′ 58.55″ N	116° 28′ 57.55″ W	597	677
B	Death Valley California	28/02/2005	NCALM	<85	Desert	39° 38′ 01.77″ N	105° 49′ 13.88″ W	3664	1345
C	St. Elias, Alaska	02-15/9/2005	NCALM	2000	Periglacial	60° 18′ 18.59″ N	144° 32′ 14.98″ W	490	831
D	Front Range, Colorado	30/09/2005	NCALM	600	Periglacial	37° 04′ 28.50″ N	117° 26′ 37.60″ W	258	854
E	Westfjords, Iceland	05/08/2007	ARSF	700	Periglacial	66° 04′ 13.20″ N	023° 07′ 14.19″ W	271	807

Average elevation is given relative to datum; for A–D this is NAD 1983 and for site E this is WGS 1984, in both cases the difference between the datum and sea level is approximately 60 m.
Abbreviations: NCALM, National Center for Airborne Laser Mapping supported by the USA's National Science Foundation; ARSF, Airborne Research and Survey Facility supported by the UK Natural Environment Research Council.

Derivation of drainage area and local slope

Representative slope sections were chosen in each DEM (Figs 3 & 4). For Earth, these were chosen to represent end-member and intermediate-process domains, including dry mass wasting, debris flow and alluvial processes. On Mars, some areas were chosen that covered the complete slope on which gullies are found, whilst others covered a single gully system or ungullied slope for comparison. Slope sections always included the drainage divide at the top and extended downslope as far as the visible signs of the distal extent of the gully (or slope) deposits. Where possible, lines delineating drainage basins were followed to define the lateral extent of slope sections, but on poorly incised hillslopes this was not always possible and the lateral extent was defined as a straight line. For site KC, on Mars, we chose different configurations of slope sections to test the sensitivity of our analyses to the exact method used to delineate the slope sections. Careful delineation of slope sections is necessary for two reasons. First, because the larger the sample area, the more processes are included within it, and the more difficult the results will be to interpret. Secondly, if parts of the slope that are integral to the process to be identified are omitted, then the process signal will not be complete.

The slope and the flow directions of each pixel in each DEM were determined using a 'Dinf' algorithm. This algorithm gives flow directions in any direction, rather than only towards one of the eight neighbouring pixels (Tarboton *et al.* 1991). This has been shown to produce better results from slope–area analysis because it gives a more accurate approximation of the real path of flow through the landscape (Borga *et al.* 2004). For each pixel, the accumulation of flow was calculated from the flow directions by summing the number of pixels located upstream and then multiplying by the pixel area. These analyses were performed using the TauDEM extension for ArcGIS, based on the algorithms developed by Tarboton (1997). For each DEM the 'wetness index' was also calculated. This is the natural logarithm of the ratio of contributing area to slope. It provides information on the potential connectivity of the landscape drainage and the potential ability of the surrounding landscape to route drainage (Woods & Sivapalan 1997). However, in the case of Earth and particularly in the case of Mars, this index should not be interpreted literally as implying that the terrain is 'wet'. In our study, it is used as a visual aid to interpret the spatial variability of the slope–area plot. For example, highly permeable talus slopes on Earth are essentially dry, but they may have a moderate–high wetness index. However, we would expect a talus slope on Earth to show a characteristic spatial pattern of wetness index, indicative of dry mass-wasting processes. All the DEMs underwent the same processing steps.

We extracted the drainage area and slope for every pixel within the chosen slope sections. To simplify the representation of these data we calculated the mean slope for 0.05-wide logarithmic bins of drainage area, and then constructed the slope–area and CAD plots. Binning data in this way make the trends in slope–area and CAD plots clearer, and is a commonly used display technique (e.g. Snyder *et al.* 2000).

In addition, for one site on Mars (site KC) we visually identified the initiation sites of the gullies on orthorectified HiRISE images. The initiation points for the gullies were defined as the furthest upstream extents defined by a distinct cut or scarp (Fig. 5a). For each of these locations we extracted the slope and drainage area for the underlying pixel. This analysis was not performed for site PC because edge contamination and noise made it impractical. The analysis was also omitted for site GC because the gullies start at the top of the slope, and so would, by definition, occur in the lowest drainage areas.

Study areas

Earth

All of the study sites on Earth are located in the northern hemisphere and most are within continental USA. Table 1 provides a summary of the sites and Figure 3 shows the setting of the areas studied.

Site SJ – San Jacinto, California. This site is located in California along a splay of the San Andreas Fault, called the San Jacinto Fault. This area is a desert with little rainfall (*c.* 150 mm, annual average recorded by a NOAA weather station in nearby Borrego Springs), which has undergone rapid recent uplift caused by the fault system. The landscape has a well-developed ephemeral gully network with large alluvial fans. From the study of the 1 m LiDAR data and aerial images, we infer the processes forming these fans to be sheet flow rather than debris flow, based on the lack of levees and lobate terminal deposits. The vegetation is sparse, consisting of small scrub bushes. The underlying geology of the study area is mainly granite, schist and gneiss, with minor outcrops of Quaternary older-fan deposits (Moyle 1982). For our analyses we used three study areas that contained small complete gully systems, including sources, channels and debris aprons, but avoided large fan systems and debris aprons from neighbouring

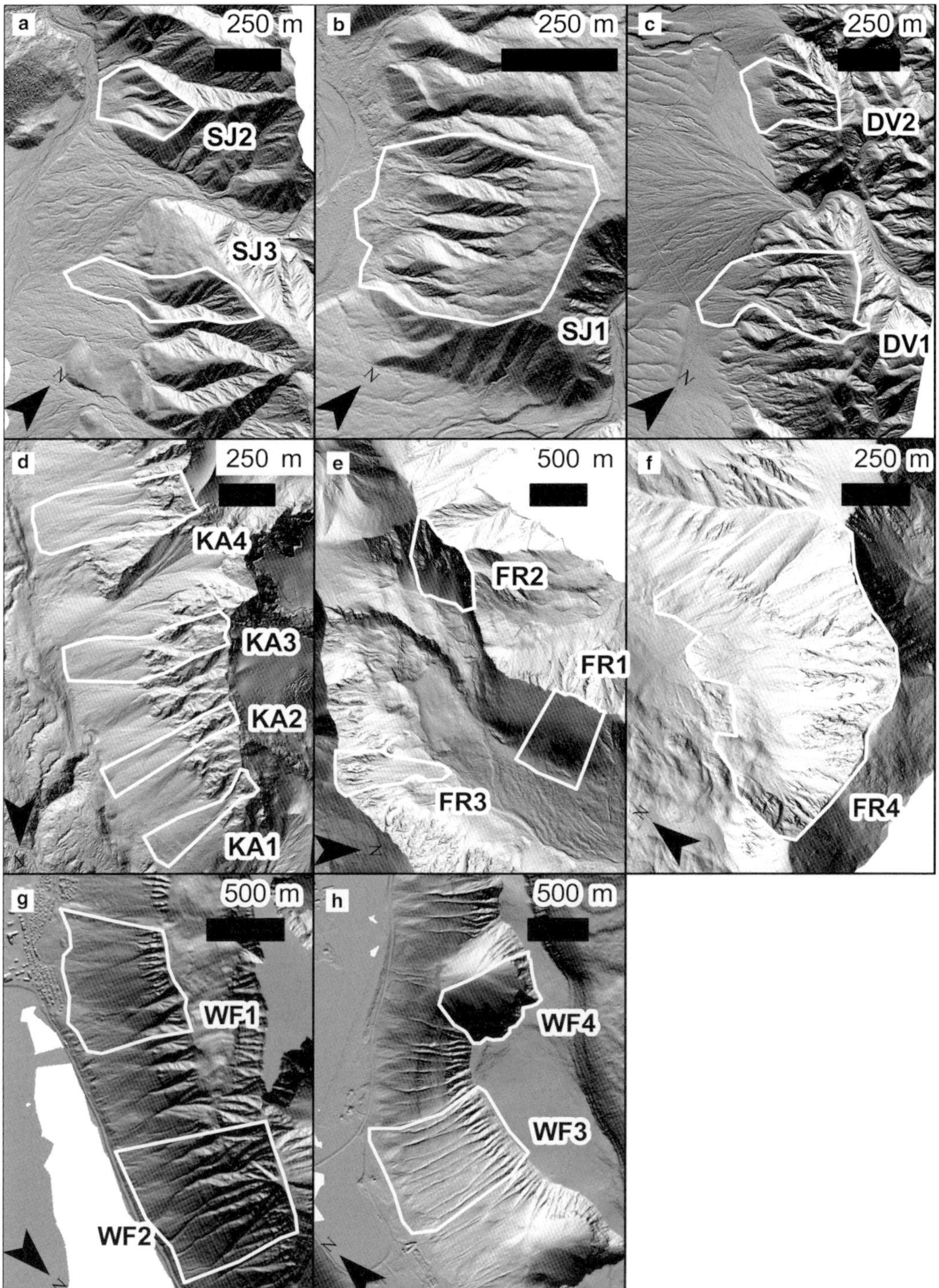

Fig. 3. Hillshade representations made from DEMs of the study site locations on Earth. Areas included in this study are outlined and labelled in the figure. (**a**) & (**b**) Site SJ, San Jacinto, California. (**c**) Site DV, Death Valley, California. (**d**) Site KA, St Elias Mountains, Alaska. (**e**) & (**f**) Site FR, Front Range, Colorado. (**g**) & (**h**) Site WF, Westfjords, NW Iceland.

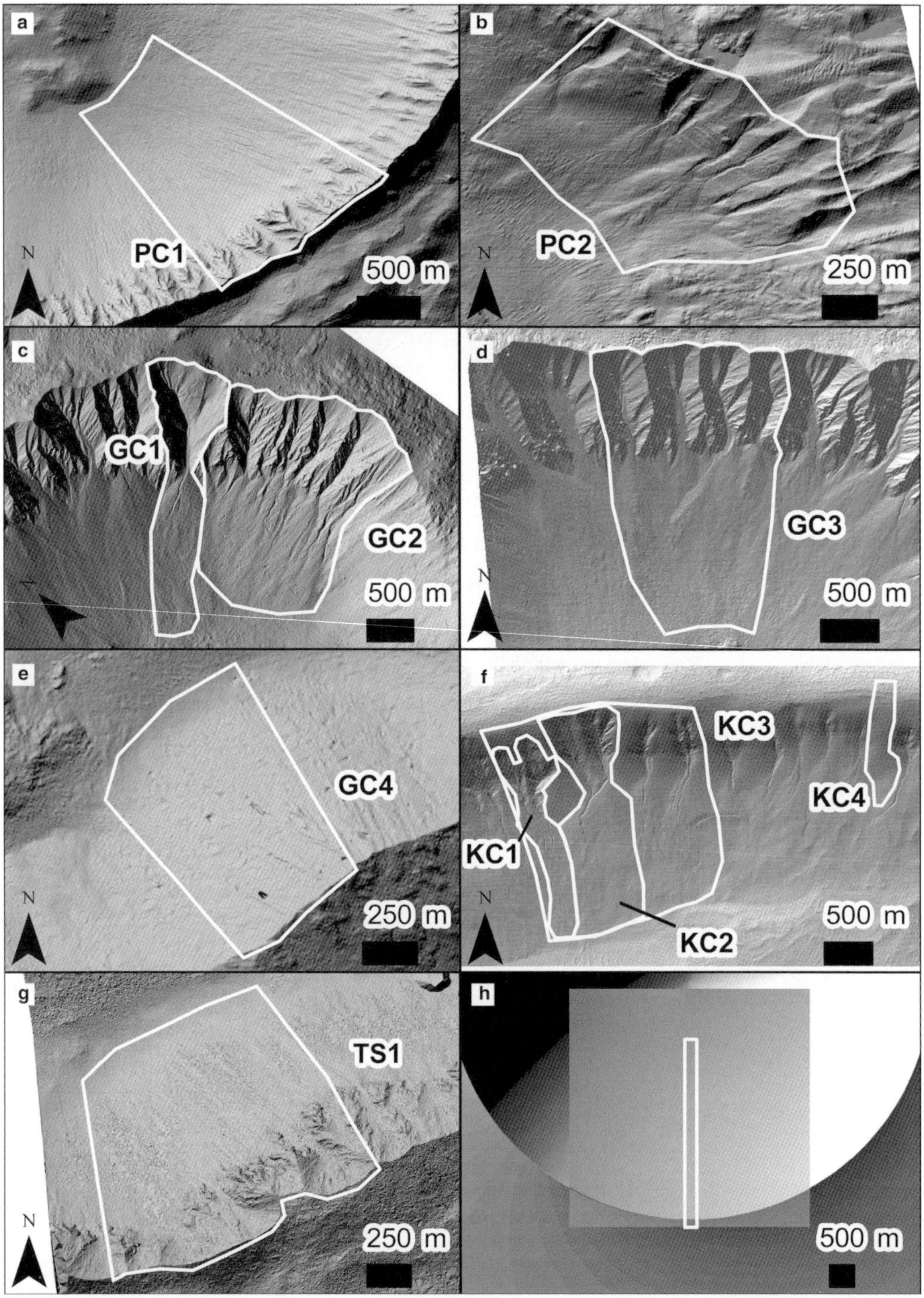

Fig. 4. Hillshade representations made from DEMs of the study site locations on Mars. Areas included in this study are outlined and labelled in the figure. (**a**) &(**b**) Site PC, Penticton Crater in Eastern Hellas. (**c**)–(**e**) Site GC, Gasa Crater in Terra Cimmeria.(**f**) Site KC, a crater inside Kaiser Crater in Noachis Terra. (**g**) Site TS, a crater in Terra Sirenum. (**h**) The 10 km-diameter synthetic crater, in which the square area is where the pink noise has been applied.

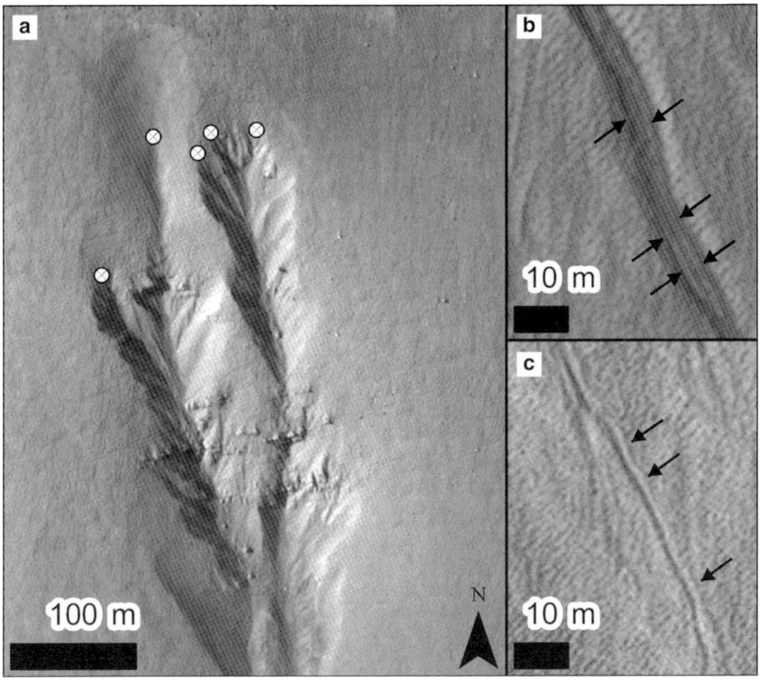

Fig. 5. Close-up views of gullies in Kaiser Crater (site KC), subset of HiRISE image PSP_003418_1335. Image credits: NASA/JPL/UofA. (**a**) Examples of gullyheads identified for individual slope–area analysis, marked by circles containing white crosses. (**b**) Levees interior to a channel, arrows point to levees within the channel on each side. (**c**) Single leveed channel, arrows point to the more obvious levee on the right, but there is an indication that there is another on the left.

systems (Fig. 3a, b: study areas SJ1, SJ2 and SJ3). Owing to the small size of the fans in area SJ1, it is difficult to entirely rule out debris flow as a potential process in forming these alluvial fans.

Site DV – Death Valley, California. This site is located a few kilometres NE of Ubehebe Volcano, in Death Valley, California. This is a desert area that has well-developed ephemeral gully networks with large alluvial fans. There is little precipitation in this area, although the nearby mountains receive as much as 85 mm of rain per year (Crippen 1979) and rare large storms can do much geomorphic work. Debris flows are found on the fans in the area (e.g. Blair 1999, 2000), but the primary process active in the gullies is alluvial transport (Crippen 1979). We inspected the 1 m LiDAR data for presence of levees and depositional lobes on the fans, and found no evidence of these. However, without direct field observations, the fact that debris flows do not act on these fans remains an assumption. The bedrock consists of Palaeozoic sedimentary rocks (Workman *et al.* 2002). We chose two study areas (Fig. 3c: study areas DV1 and DV2) with gully systems that were not affected by neighbouring alluvial fans or gully systems and so only receive local rainfall levels.

Site KA – St Elias Mountains, Alaska. This site is located east of the abandoned town of Katalla close to the recently deglaciated mountain range of St Elias, near the coast of Alaska and on the border with Yukon, Canada. The area has been unglaciated for approximately the last 10 ka (10^4 years) (Sirkin & Tuthill 1987) and receives very high precipitation, which falls as snow on the upper slopes and rain on the lower. Our study area overlies Tertiary volcanic materials. The slope scarp was generated by the active Ragged Mountain Fault (Miller 1961). The area was neither snow covered nor tree covered at the time of survey, and the slopes are composed of steep bedrock cliffs that lead directly into large talus aprons. Debris-flow tracks are apparent across this talus slope, especially in study areas KA3 and KA4, and might also have occurred in study area KA3 (Fig. 3d). Study area KA1 shows no evidence of debris-flow processes (Fig. 3d).

Site FR – Front Range, Colorado. This site is located in the mountainous eastern side of the continental divide. The area was deglaciated around 14–12 ka before present (Godt & Coe 2007) and the landscape is dominated by glacially carved valleys. This area has experienced recent debris flows (Coe *et al.* 2002; Godt & Coe 2007) and has no permanent snowpack. Our study slopes, located above the tree line, are dominated by Precambrian biotitic gneiss and quartz monzonite, scattered Tertiary intrusions, and by various surface deposits, all of which host debris flows (Godt & Coe 2007). The head wall and side walls of the cirques have large rockfall talus deposits, which have also experienced recent debris flows. These slopes have little or no vegetation. Three of our study areas (Fig. 3e, f: study areas FR2–FR4) include debris flows located on talus. By way of contrast, we also examined a partially vegetated slope (study area FR1) that is unchanneled and which we infer to be dominated by creep processes (Fig. 3e).

Site WF – Westfjords, Iceland. The site is located in NW Iceland and is dominated by fjords and glacially carved valleys. The last glacial retreat occurred approximately 10 ka before present (Norðdalh 1990). The valley walls have many active debris flows (Conway *et al.* 2010), and on the slopes above Ísafjörður (Fig. 3g: study area WF1) they occur in most years (Decaulne *et al.* 2005). The site has a maritime climate, so has high levels of both snow and rainfall, but does not have permanent ice or snow patches. The site is underlain by Miocene basalts, although the debris flows occur most often in glacial till. From this site we chose a study area above the town of Ísafjörður that has very active debris flows (Fig. 3g: study area WF1), two study areas with fewer active debris flows and more alluvial processes (Fig. 3g, h: study areas WF2 and WF3), and one study area dominated by rockfall and rockslide processes, although there are some debris-flow tracks visible in the field (Fig. 3h: study area WF4). All of these study areas have patchy vegetation, but no trees.

Mars

All of the gullies that we studied on Mars were located on the inner walls of craters in the southern hemisphere (Table 2). Slopes both with and without gullies were analysed for comparison. Sites PC, GC and KC were analysed by Lanza *et al.* (2010) because all of the sites showed visual evidence of debris flows.

Site PC – Penticton Crater in Eastern Hellas. This site contains the very recent, light-toned deposits observed by Malin *et al.* (2006) and interpreted by them to be a recent 'gully-forming' event. These flows were later suggested by Pelletier *et al.* (2008) to be produced by dry granular flow or, possibly, also debris flow. This slope does not have any well-defined channels. We used two study areas within the approximately 7.5 km-diameter crater for our slope–area analyses, shown in Figure 4a, b. Study area PC1 is located over the equator-facing, light-toned deposits (Fig. 4a) and study area PC2 on the west-facing crater wall, which contains small gullies (Fig. 4b). These gullies appear to be incised into 'mantle deposits' (Mustard *et al.* 2001). The mantle is hypothesized to be the remnants of a previously extensive volatile-rich deposit (e.g. Mangold 2005). This crater is very asymmetric, with the east and north rims being subdued in terms of elevation (the rim is nearly absent on the east side) whilst the southern rim is abrupt and steep.

Site GC – Gasa Crater in Terra Cimmeria. This approximately 7 km-wide crater, shown in Figure 4c, d, has well-developed alcoves or indentations into the rim of the crater. Gully channels are most obvious on the west-facing to pole-facing slopes (Fig. 4c, d), and the equator-facing slope lacks these well-defined alcoves and channels (Fig. 4e). We chose sections on the pole- (study areas GC1 and GC2), west- (study area GC3) and equator-facing (study area GC4) slopes. This

Table 2. *Summary of the data for the study sites on Mars*

Site	HiRISE image pair	Latitude	Longitude	Average elevation (m)	Relief (m)
F	PSP_001714_1415 PSP_001846_1415	−38.4°	96.8°	−2648	1124
G	PSP_004060_1440 PSP_005550_1440	−35.7°	129.4°	300	1205
H	PSP_003418_1335 PSP_003708_1335	−46.1°	18.8°	595	687
J	PSP_003674_1425 PSP_005942_1425	−37.4°	229.0°	1904	961

Average elevation is given relative to the Mars datum, as defined from the MOLA dataset. The average elevation has been estimated from the MOLA dataset and relief from the HiRISE DEMs.

crater is located within a larger crater, which also has gullies on its west- to pole-facing slopes. There is no evidence of mantle deposits being present anywhere within this crater.

Site KC – crater inside Kaiser Crater in Noachis Terra. The study crater, approximately 12 km across, is located within the larger Kaiser Crater, which not only has gullies down its own rim, but also gullies on the dunes within it (Bourke 2005). Gullies in this crater have alcoves at various positions on the slope, which converge to form well-defined tributary networks. Lateral levees bound some of the channels (Fig. 5b, c). This slope has the subdued appearance often attributed to the presence of volatile-rich mantle deposits (Mustard *et al.* 2001). We chose study areas that encompass the drainage area of two gullies (study area KC2), a single gully (study area KC1) and also the slope section as a whole (study area KC3), all of which are shown in Figure 4f. We chose study area KC4, an area of the slope not affected by gullies, for comparison (Fig. 4f).

Site TS – crater in Terra Sirenum. This approximately 7 km-diameter crater is located to the south of Pickering Crater in Terra Sirenum and contains pole-facing gullies. We analysed an equator-facing slope (Fig. 4g: study area TS1) that has no evidence of channels but contains an apparently well-developed talus apron. There is no evidence of mantle deposits being present on this slope.

Results

Earth

Initially we chose two study areas with talus and with active creep. The slope–area analysis results for these are shown in Figure 6a. The study areas with well-developed talus (WF4 and KA1) show the following pattern on log–log plots: (1) with a small drainage area, the curves are initially flat; (2) there is then a linear decrease in slope with increasing drainage area; and (3) the curve then becomes horizontal again in a higher drainage area with a lower slope value. Talus slopes that have a mixture of processes (e.g. KA2) show a curve that drops off linearly in log–log plots then flattens in higher drainage areas.

The CAD plot (Fig. 7a) provides additional information: the talus-dominated study areas have a very smooth convex shape. The gradient of the curve is low until the drainage area is approximately 0.001 km^2, after which the curve drops sharply and continues to steepen with increasing drainage area.

The soil-creep diffusive process study area (FR1 in Fig. 6a) shows a distinctive signature in slope–area plots: (1) the curve is initially horizontal to gently downwards sloping; (2) between drainage areas of 0.0001 and 0.001 km^2 the slope increases linearly with increasing drainage area; and (3) there is then a marked slope turnover at which the curve switches to decreasing slope with increasing drainage area. The soil-creep diffusive process study area resembles the talus slopes in CAD plots (FR1, Fig. 7a).

Figures 6b and 7b show the debris-flow study areas that are influenced by talus processes, and Figures 6c and 7c show those that are more influenced by alluvial processes. Generally, in slope–area plots, debris flow produces a curve that drops off linearly in log–log plots, flattening off before finally dropping away steeply. The difference between the talus study areas (e.g. KA2, Fig. 6a) and the debris-flow study areas influenced by talus (Fig. 6b) is subtle in some cases. In a similar way, the difference between the debris-flow areas influenced by talus processes (Fig. 6b) and those influenced by alluvial processes (Fig. 6c) is also subtle. Without field information it would be difficult to differentiate talus-dominated and debris-flow-dominated slopes reliably in slope–area plots (e.g. compare Fig. 6a, KA2, and 6b). However, in CAD plots, it is possible to differentiate between the two process types. The debris-flow-dominated study areas (Fig. 7b, c) show the following pattern: (1) the curve drops away from the horizontal slowly (but faster than the talus slopes) in small drainage areas; (2) the curve then either dips down linearly or follows a flattened convex path; and (3) in high drainage areas the curve drops away sharply with increasing drainage area.

Study areas modified by ephemeral water flow have distinct signatures in slope–area plots (Fig. 6d) and in CAD plots (Fig. 7d). In slope–area plots they show a shallow linearly decreasing trend in small drainage areas, which gets steeper in higher drainage areas and drops into the alluvial domain. The CAD plot drops away from the horizontal slowly and then dips down linearly (or even with a concave profile) until the tail of the curve drops sharply off in the highest drainage areas.

Synthetic crater

The slope–area and CAD plots for the synthetic crater are easily differentiated from the process study areas that we have examined on Earth. In slope–area plots, the synthetic crater produces a hump-backed curve (Fig. 8d): in small drainage areas the curve rises steeply, then levels off and drops in high drainage areas. In appearance, as expected, the curve is closest to study area FR1,

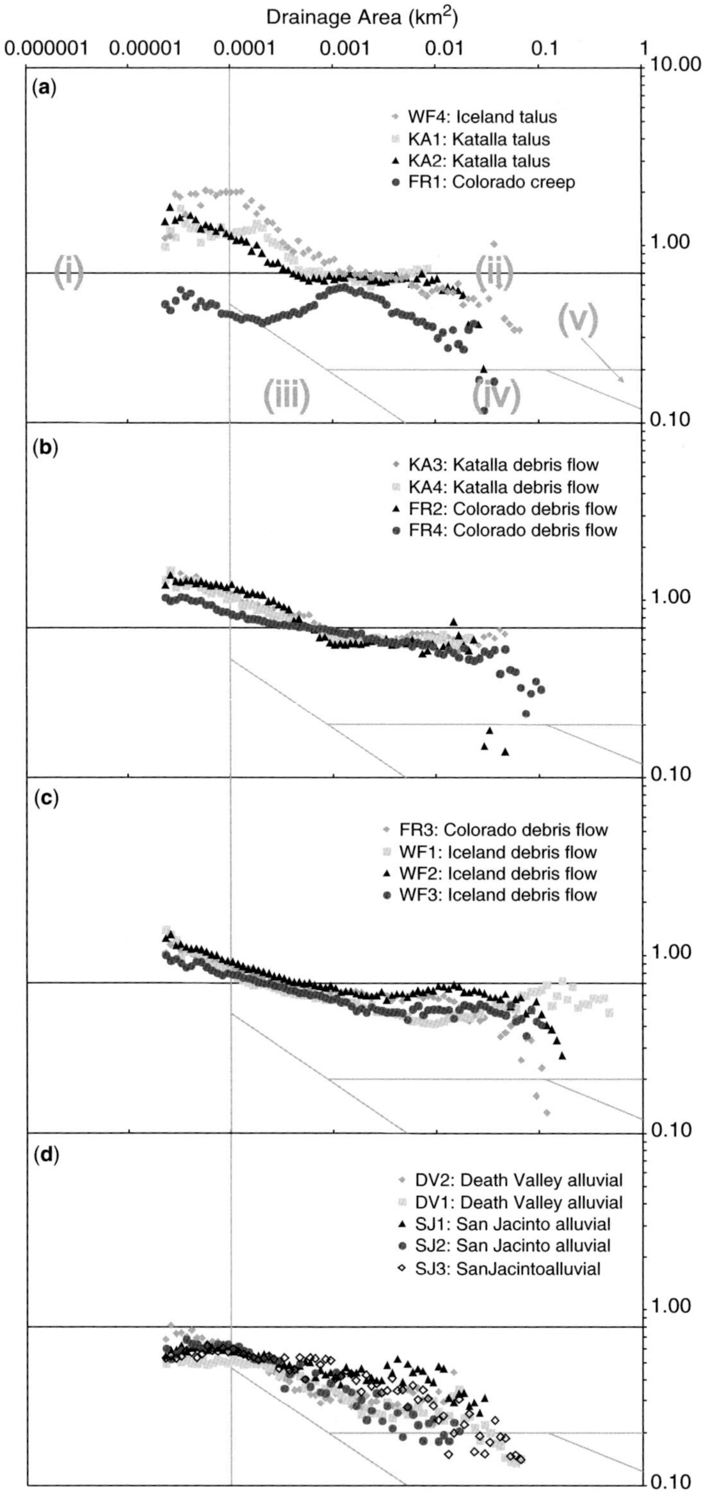
Drainage Area (km^2)
0.000001
0.00001
0.0001
0.001
0.01
0.1
1
10.00
1.00
0.10
Local slope (mm^{-1})
(a)
WF4: Iceland talus
KA1: Katalla talus
KA2: Katalla talus
FR1: Colorado creep
(i)
(ii)
(iii)
(iv)
(v)
(b)
KA3: Katalla debris flow
KA4: Katalla debris flow
FR2: Colorado debris flow
FR4: Colorado debris flow
(c)
FR3: Colorado debris flow
WF1: Iceland debris flow
WF2: Iceland debris flow
WF3: Iceland debris flow
(d)
DV2: Death Valley alluvial
DV1: Death Valley alluvial
SJ1: San Jacinto alluvial
SJ2: San Jacinto alluvial
SJ3: SanJacintoalluvial

Fig. 6.

the area dominated by diffusive creep (Fig. 6a). In CAD plots (Fig. 9d), the line follows a smooth convex arc, similar to that shown by talus on Earth, except without a break in gradient.

Mars

The slope–area plots for sites PC and GC (Penticton Crater and Gasa Crater inner slopes, respectively) closely resemble each other (Fig. 8a, b). The resulting curve can be divided into three zones: (1) a short initial increase in slope with increasing drainage area, followed by a slope turnover in very small drainage areas; (2) a linear or slightly concave decreasing slope trend with increasing drainage area that continues for most of the plot; and (3) finally, in the largest drainage areas, there is a steep decrease in slope with increasing drainage area. For study area PC1 there is a distinct and linear decline in slope with drainage area, whereas for study areas PC2, GC1, GC2 and GC3 this section is slightly concave. The drop-off in the highest drainage areas occurs in lower absolute drainage area values than for site GC. In the CAD plot, study areas PC1 and GC4 have a smooth convex form, whereas study areas PC2, GC1 and GC2 all have a nearly linear, flattened section in intermediate drainage areas (Fig. 9a, b). Study area GC3 lies close to PC1, GC1 and GC2 but without any sign of flattening.

The slope–area plots for gullies in study areas KC1, KC2 and KC3 (Fig. 8c) can be split into three sections as follows: (1) in small drainage areas the curve is subhorizontal with a subtle upward trend. This trend is more apparent for the data from individual gullies than the data obtained from the whole slope section and is somewhat variable between gully systems. (2) In intermediate drainage areas there is a transitional zone, occurring at different drainage areas for each gully system, in which slope drops off markedly with drainage area. (3) In higher drainage areas there is a gently declining relationship between slope and drainage area, which is the same for all the gully systems.

The ungullied study area (KC4) is also shown in Figure 8c. This study area has a hump-back shape, resembling that seen for the synthetic crater. The hump occurs across the same slope values as the transition zone (2) for the gullied slopes. In CAD plots (Fig. 9c), study areas KC2 and KC3 have a flattened section in intermediate drainage areas, followed by a steepening decrease in higher drainage areas. The study area without gullies (KC4) has a curve that is convex and initially declines slowly, before dropping off steeply. Study area KC1 has a less flattened profile than study areas KC2 or KC3, and it seems to be a mixture between slope types typified by gullied study areas KC2 or KC3 and ungullied study area KC4.

In slope–area plots, study area TS1, an ungullied slope, shows a slope–area turnover in small drainage areas, followed by a decreasing and slightly concave trend in slope with drainage area (Fig. 8d). There is a slight upturn in the highest drainage areas, but this is likely to be an artefact caused by only a few data points being used to calculate the mean slope in these bins. In CAD plots (Fig. 9d), study area TS1 has a very smooth convex curve.

The slope and drainage area of the gully-head initiation points were recorded for site KC. These data are displayed on Figure 8c. Interestingly, the locations of the gully heads cluster around the range of drainage areas of the transitional section in the slope–area plot, but are located at higher slope values.

Wetness index on Earth and Mars

The spatial distribution of the slope–area data is most easily visualized using a wetness index map. Maps of wetness index are presented for Earth (Fig. 10) and for Mars (Fig. 11). The alluvial study areas in Earth sites SJ and DV show very low overall wetness indices – only the channels have a significant wetness index (Fig. 10a–c). Debris-flow study areas are slightly more complex (Fig. 10d–h): the slopes generally have a moderate wetness index, but there are localized paths along which the wetness index is higher. Site WF (Fig. 10g, h) is the best example of this pattern, but it is also the area with the highest influence of overland flow. For site KA (Fig. 10d) this signature is poorly developed, but this site has been influenced by talus processes. The creep-dominated study area, FR1, has moderate wetness index throughout

Fig. 6. (*Continued*) Slope–area plots for the study areas on Earth. Marked with solid grey lines are the domains of Montgomery & Foufoula-Georgiou (1993) and Brardinoni & Hassan (2006), as shown in Figure 2a. Labels are included in (a), but omitted for clarity in the other plots and are as follows: (i) hillslopes domain; (ii) debris-flow dominated channels; (iii) unchanneled valleys; (iv) alluvial channels; and (v) debris-flow deposition domain. The horizontal dotted line represents the threshold for unconsolidated dry mass wasting at 0.7 gradient, which is equivalent to a 35° slope. (**a**) Plots for those areas dominated by talus and creep processes. (**b**) Plots for those areas dominated by debris flow, with some influence from talus processes. (**c**) Plots for those areas dominated by debris flow, with influence from alluvial processes. (**d**) Plots for those areas dominated by ephemeral water flow or alluvial processes.

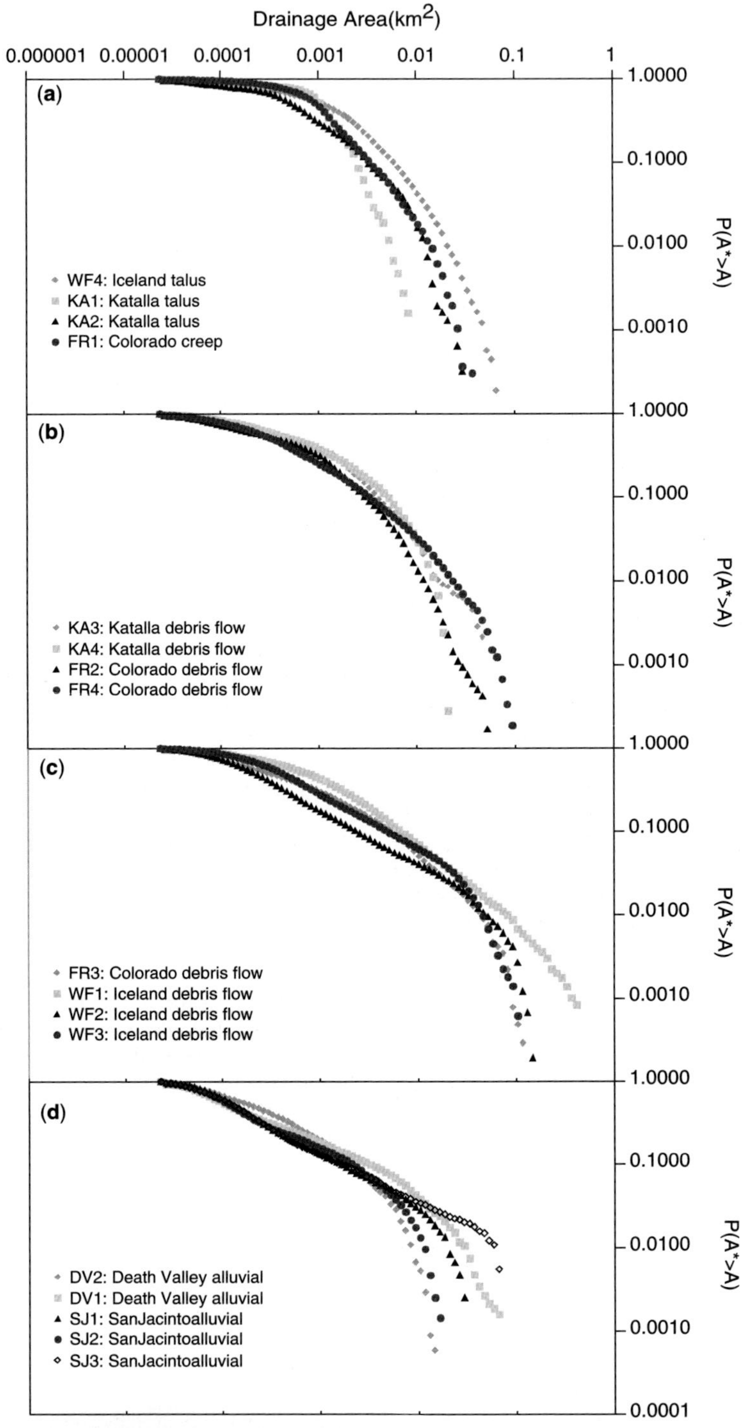

Fig. 7. Cumulative area distribution plots for the study areas on Earth. (**a**) Plots for those areas dominated by talus and creep processes. (**b**) Plots for those areas dominated by debris flow, with some influence from talus processes. (**c**) Plots for those areas dominated by debris flow, with influence from alluvial processes. (**d**) Plots for those areas dominated by ephemeral water flow or alluvial processes.

(Fig. 10e). The talus study areas KA1, KA2 (Fig. 10d) and WF4 (Fig. 10h) show lobe-like areas of low wetness index, with widening streaks of higher wetness index in between.

On Mars, study area PC1 (Fig. 11a) and the synthetic crater (Fig. 11h) have similar wetness index maps: the slope generally increases in wetness index going downhill and there are quasi-linear streaks of higher wetness index that increase in value going downslope. Study area PC2 (Fig. 11b) has an overall low wetness index, apart from concentrated lines of high wetness index within the gully alcoves that spread and become more diffuse in the debris aprons. A similar overall pattern is shown for study areas GC1, GC2 and GC3 (Fig. 11c, d), but the ridges around the alcoves have a very low wetness index. Study area GC2, in particular (Fig. 11c), shows very concentrated, slightly sinuous, high wetness index lines on its debris apron. However, this part of the DEM contains significant noise, making it hard to judge whether this is simply an artefact. Study areas GC4 (Fig. 11e) and TS1 (Fig. 11g) have similar wetness index maps: there is a low wetness index at the crest of the slope and where bedrock is exposed, and the wetness index generally increases downslope, but this trend is superposed with diffuse linear streaks of higher relative wetness index. Site KC (Fig. 11f) generally has a moderate wetness index, with the alcoves and channels of the gullies showing focused high wetness index values. The gullies are flanked by a much lower wetness index, with the debris aprons generally having a high wetness index and a diffuse downslope streaking.

Discussion

Comparison of Earth data to previously published slope–area process domains

There are two interlinked methods of determining slope processes from slope–area plots:

- the data points fall within domains in the plots, which have been found both theoretically and empirically to relate to particular processes;
- the data points exhibit trends and gradients that provide information on active processes.

We compared our data from Earth to the slope–area process domains of Montgomery & Foufoula-Georgiou (1993) and the additional domain added by Brardinoni & Hassan (2006), shown as solid lines in Figure 6. The data from our creep, talus and debris-flow analyses fall into the debris-flow domain of Montgomery & Foufoula-Georgiou (1993). However, some of our debris-flow data drop into the alluvial domain in the highest drainage areas. Because they are small systems with limited drainage areas, however, only a few points fall within the alluvial domain. Some of our data approach the additional domain added by Brardinoni & Hassan (2006), but do not extend towards sufficiently high drainage areas (or low drainage areas) to enter it (Fig. 6b, c). Our data from the alluvial systems (Fig. 6d) fall into both the debris-flow and the alluvial domains. They start to trend downwards in slope–area plots in lower drainage areas than our debris-flow systems.

Tucker & Bras (1998) simulated the effects of different dominant processes on slope–area plots and we now compare their model results to the patterns in slope–area plots shown by our data. Our talus systems (Fig. 6a) closely fit their model of a landscape dominated by landsliding (which includes the process of debris flow). In slope–area plots our talus data have an initial flat section in small drainage areas, which represents the slope threshold for the rock-wall failure and so differs between localities. In higher drainage areas the curves are again flat, representing the failure threshold of loose talus, which is consistent for all areas at approximately 0.7 gradient, equivalent to a slope of approximately 35°. This is an approximate mean slope angle for talus slopes on Earth (Chandler 1973; Selby 1993) and is shown by a dotted horizontal line in Figures 6 and 8. Between these two horizontal sections there is a transition where the dominance shifts from rock-wall failure to unconsolidated talus failure.

Within the framework of Tucker & Bras (1998), the pattern shown by the debris-flow slopes on Earth (Fig. 6b, c) is most consistent with the transition from unsaturated landsliding (dry mass wasting of both talus and rock wall) to pore-pressure-triggered landsliding (which we interpret to also include debris flow) in a landscape dominated by landsliding. The presence of processes with a slope-failure threshold cause data in slope–area plots to fall along horizontal lines. Hence, as the process dominance changes from rock-wall failure (highest threshold) to unsaturated landsliding (intermediate threshold) to saturated landsliding (lowest threshold), the curve declines and levels off at the slope value of the saturated landslide threshold in that particular area. As each physical locality has its own saturation threshold, this horizontal section occurs at different slope values for different localities but is always located below the dry stability line at 0.7.

In slope–area plots our data from alluvial systems on Earth (Fig. 6d) show a simple decline of slope with drainage area, possibly steepening in higher drainage areas. The data are scattered in drainage areas of more than 0.001 km^2 owing to the limitations of the small sizes of the gully systems

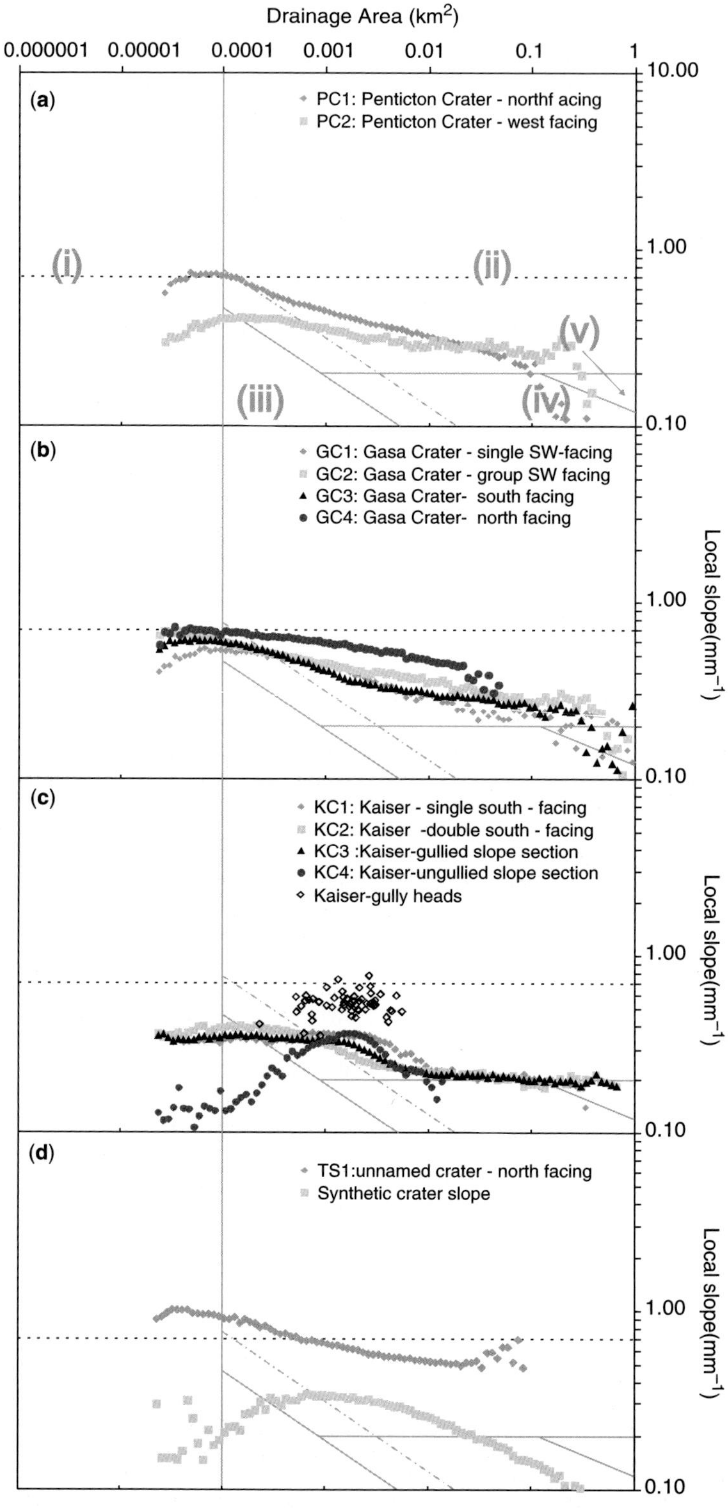
Drainage Area (km2)
0.000001 0.00001 0.0001 0.001 0.01 0.1 1
10.00
1.00
0.10
Local slope(mm-1)
(a)
PC1: Penticton Crater - northf acing
PC2: Penticton Crater - west facing
(i)
(ii)
(iii)
(iv)
(v)
(b)
GC1: Gasa Crater - single SW-facing
GC2: Gasa Crater - group SW facing
GC3: Gasa Crater- south facing
GC4: Gasa Crater- north facing
(c)
KC1: Kaiser - single south - facing
KC2: Kaiser -double south - facing
KC3 :Kaiser-gullied slope section
KC4: Kaiser-ungullied slope section
Kaiser-gully heads
(d)
TS1:unnamed crater - north facing
Synthetic crater slope

Fig. 8.

available. This means a relatively small number of pixels were used to generate each point, leading to random scatter. However, even taking into account the scatter, the data are below the slope threshold for dry slope failure at 0.7 gradient, which suggests a gradual transition from pore-pressure-dominated landsliding to fluvial processes.

The main feature of our creep-dominated hillslope data (FR1, Fig. 6a) is a turnover from increasing slope with drainage area to decreasing slope with drainage area. One of the alluvial systems in site SJ (study area SJ3) shows a weak slope turnover in the lowest drainage areas, but none of the other plots show this feature. The slope–area turnover is shown in Figure 2 and is generally expected to occur in slope–area plots (e.g. Tucker & Bras 1998). It usually occurs in, or close to, the 'hillslope' domain of Montgomery & Foufoula-Georgiou (1993). The turnover represents a transition from convex slopes dominated by diffusive processes (which include soil creep often modified by plant roots and other biota) to concave slopes dominated by advective, or alluvial, processes. Within the diffusive processes domain in slope–area plots, slope increases with drainage area. The most likely reason that most of our data do not show this turnover is that the slopes we studied lack stable vegetation (Dietrich & Perron 2006; Marchi *et al.* 2008). Another potential contributing factor is that the bedrock and colluvium in our study areas are not naturally cohesive; for example, clay-rich rocks can exhibit convex creep-dominated slopes in unvegetated badlands on Earth.

The pattern of data in slope–area plots shown by our alluvial systems and by some of our debris-flow systems (slow decline in small drainage areas followed by a steep decline in higher drainage areas) has been shown from numerous remote-sensing and field studies to mark the transition from the colluvial (including debris flow) regime to that of a fully fluvial regime (e.g. Lague & Davy 2003; Stock & Dietrich 2003, 2006). Some have described the transition as a separate linear portion of the plot between the colluvial and the fluvial (Lague & Davy 2003), and some as a gradual curved transition (Stock & Dietrich 2003). However, both are consistent with Tucker & Bras' (1998) transition from pore-pressure-triggered landsliding into a fully fluvial system. Our plots do not show a well-developed alluvial regime, but this is owing to the use of high-resolution data of very small areas rather than large, well-developed fluvial catchments.

In summary, our terrestrial data are consistent with published slope–area process domains, and provide reassurance that the method is applicable and that the Mars data can be used to infer process in a similar way. The caveat to this is that the environmental differences between Earth and Mars, as detailed in the introduction to this paper, must be considered when comparing terrestrial process domains to data from Mars. Furthermore, improved process discrimination can be made by considering CAD profiles in addition to slope–area analysis.

Comparison of Earth data to published CAD process domains

Comparison of all of our CAD plots for Earth (Fig. 7) to the published process domains for CAD (Fig. 2) reveals that our data do not generally follow the cited trends. This is possibly because we are studying small areas rather than large catchments. However, the shape of the curve outlined by our data in CAD plots does allow process discrimination and does follow some of the framework outlined by McNamara *et al.* (2006). Specifically, region 1 on Figure 2 has three subregions whose shapes can be recognized in our datasets. The talus data (Fig. 7a) and synthetic crater (Fig. 9d) are both convex in their CAD plots, resembling most closely region 1a of McNamara *et al.* (2006). They describe this region as 'composed primarily of divergent topography characteristic of convex hillslopes' (p. 153) and thus do not gather drainage. Our alluvial data and some of our debris-flow data show a flattening of the CAD plot curve in the middle region, giving a steep linear section corresponding to either region 1b or region 2 (Fig. 2b), which McNamara *et al.* (2006) describe as slopes that are convergent (1b) or channel forming (2). Two debris flows (WF2 and WF3 in Fig. 7c) show a concave section, which would correspond to region 1c of

Fig. 8. (*Continued*) Slope–area plots for the study areas on Mars. Marked with solid grey lines are the domains of Montgomery & Foufoula-Georgiou (1993) and Brardinoni & Hassan (2006), as shown in Figure 2a. Labels are included in (a), but omitted for clarity in the other plots and are as follows: (i) hillslopes domain; (ii) debris-flow-dominated channels; (iii) unchanneled valleys; (iv) alluvial channels; and (v) debris-flow deposition domain. The horizontal dotted line represents the threshold for unconsolidated dry mass wasting at 0.7 gradient, which is equivalent to a 35° slope. The dash–dot line represents the adjustment of the alluvial domain when taking into account Mars' gravitational acceleration. (**a**) Plots for site PC, Penticton Crater in Eastern Hellas. (**b**) Plots for site GC, Gasa Crater in Terra Cimmeria. (**c**) Plots for site KC, a crater inside Kaiser Crater in Noachis Terra. (**d**) Plots for site TS, a crater in Terra Sirenum and the 10 km-diameter synthetic crater.

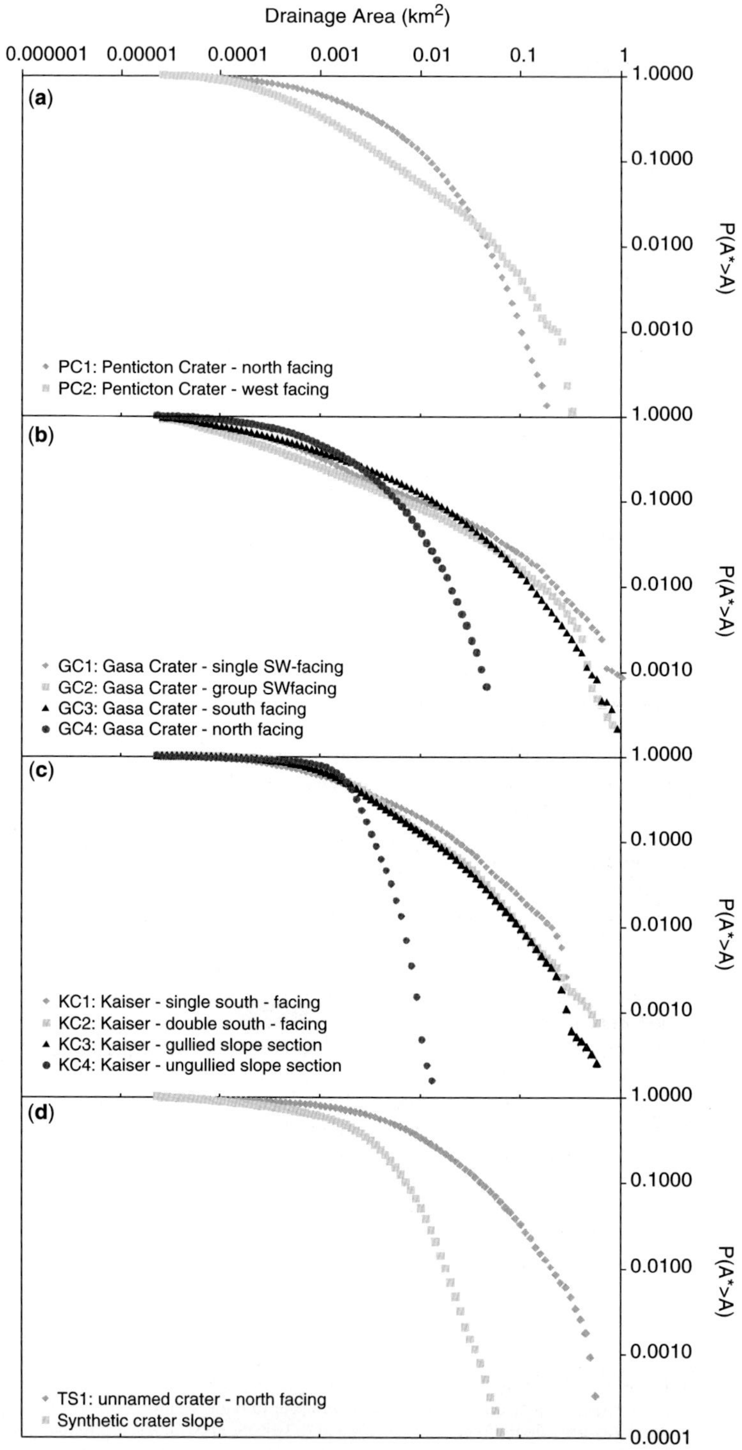

Fig. 9. Cumulative area distribution plots for the study areas on Mars. (**a**) Plots for site PC, Penticton Crater in Eastern Hellas. (**b**) Plots for site GC, Gasa Crater in Terra Cimmeria. (**c**) Plots for site KC, a crater inside Kaiser Crater in Noachis Terra. (**d**) Plots for site TS, a crater in Terra Sirenum and the 10 km-diameter synthetic crater.

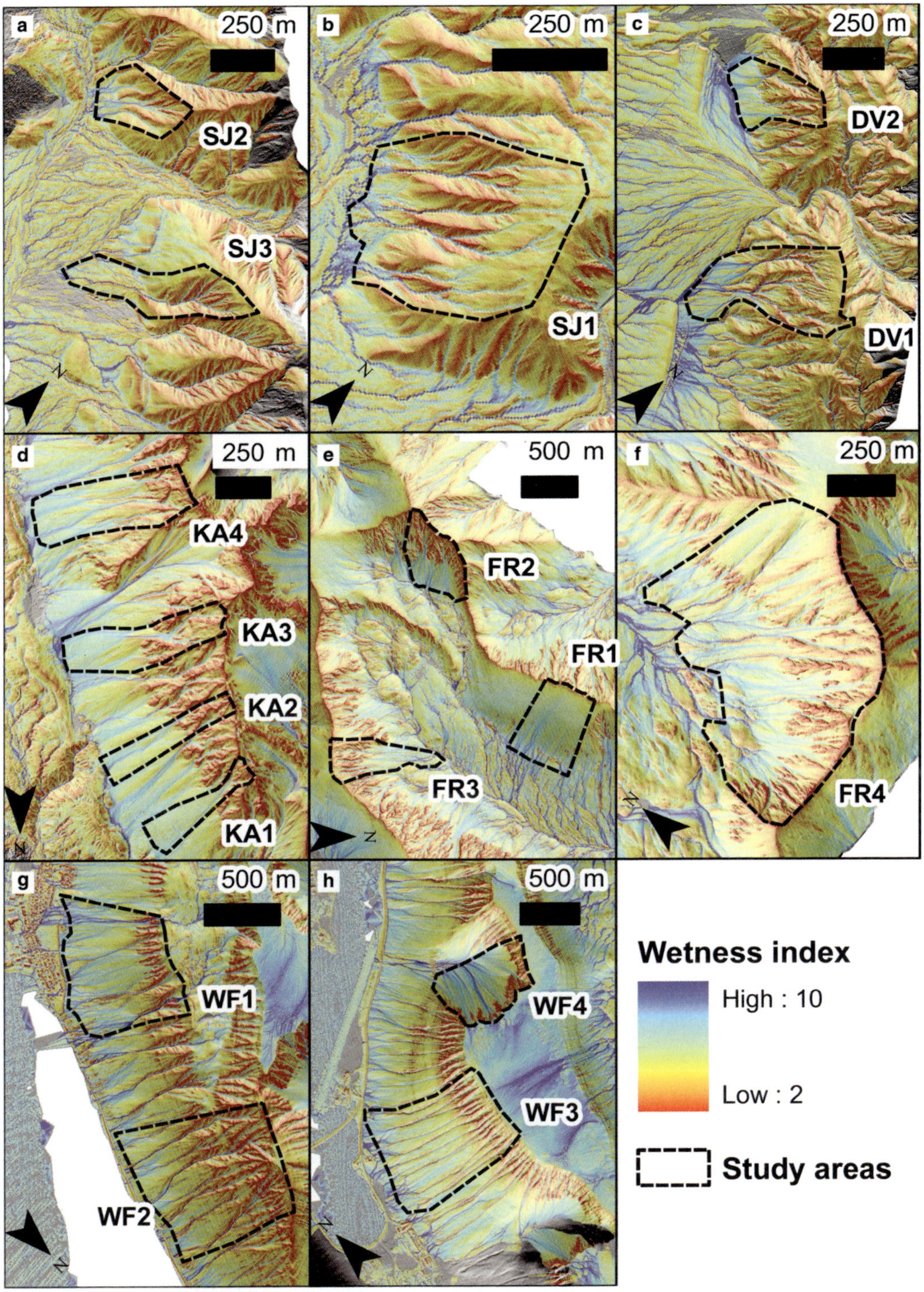

Fig. 10. Wetness index maps made from DEMs of the study site locations on Earth. Areas included in this study are outlined and labelled in the figure. Wetness index values are represented by the same colours in Figure 11 to allow direct comparison. (**a**) & (**b**) Site SJ, San Jacinto, California. (**c**) Site DV, Death Valley, California. (**d**) Site KA, St Elias Mountains, Alaska. (**e**) & (**f**) Site FR, Front Range, Colorado. (**g**) & (**h**) Site WF, Westfjords, NW Iceland.

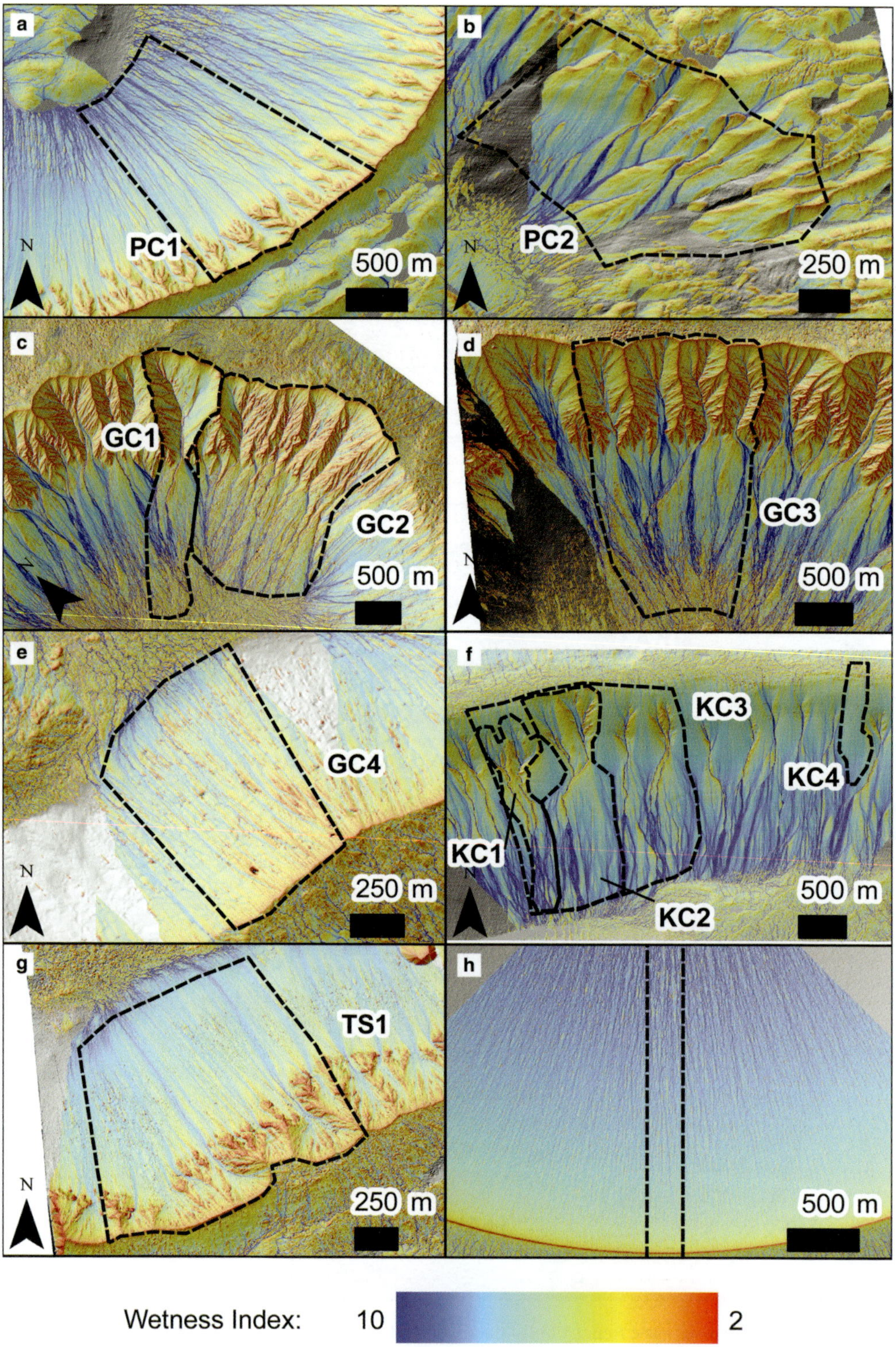
a
PC1
500 m
N
b
PC2
250 m
N
c
GC1
GC2
500 m
d
GC3
500 m
N
e
GC4
250 m
N
f
KC3
KC4
KC1
KC2
500 m
g
TS1
250 m
N
h
500 m
Wetness Index:
10
2

Fig. 11.

McNamara *et al.* (2006) and that they attribute to pore-pressure-triggered landsliding or debris flow.

The similarity of talus and debris flow in slope–area plots can be attributed to their similarly linear long profiles. However, the two processes produce different patterns in CAD plots because talus slopes tend to disperse drainage, but debris-flow slopes tend to have convergent drainage. This can also be seen in the wetness index plots (Fig. 10).

This difference of behaviour in CAD and wetness index plots, in addition to the information from the slope–area plots, shows that we can detect slopes dominated by alluvial, debris flow and dry mass wasting on the basis of these parameters, even for small catchments such as individual gullies or debris-flow tracks. However, it should be noted here that these analyses have been performed on relatively few sample sites on Earth and some of the differences are subtle. Future work has to include extending this analysis to a greater number of test sites on Earth to verify that this kind of process discrimination is robust. Using these initial results we continue and apply these methods of process discrimination to Mars.

Process domains for gullies on Mars

In slope–area plots all of the Mars slope sections, except study area TS1, fall below the slope threshold for dry mass wasting (dotted line in the plots in Fig. 8). This means that talus-like dry mass wasting is not a dominant process in these areas. However, study area TS1, visually similar to talus on Earth, is not only above the slope threshold for dry mass wasting, but also bears a signature similar to talus on Earth in the combination of its slope–area plot, CAD plot and wetness index map.

Within the process domains of Montgomery & Foufoula-Georgiou (1993), the majority of the Mars data lie within the debris-flow domain, with some data located in the debris-flow deposition domain added by Brardinoni & Hassan (2006) and a few in the alluvial domain. The difference in gravity between Earth and Mars requires an upwards slope adjustment to the alluvial channel's domain boundary (see Fig. 2a) in slope–area plots (Appendix 1), but does not change the gradient of the line. This is marked by the dash–dot line on the plots in Figure 8. This shift places more data in the unchanneled domain, but does not place any additional data into the alluvial or debris-flow domains. This distribution, in itself, does not provide very detailed information on the formation mechanisms for gullies. However, by combining slope–area trends, CAD plots and wetness index maps, we can make more-detailed assessments. We examine each of the study areas on Mars in turn and then discuss the overall implications for the gully formation processes.

Synthetic crater. The pattern in slope–area plots of the interior of impact craters is, in part, a result of the inherent shape of the crater slope, which in turn is due to the impact process and the modification that occurs immediately afterwards. The slope of a fresh impact crater is concave and exponentially shaped in profile (Garvin *et al.* 1999). Thus, in slope–area plots it resembles a well-developed alluvial system on Earth (e.g. Hack 1957). This reinforces the uncertainty in inferring a unique process from slope form. In CAD plots, however, the synthetic crater data show a similar pattern to that of talus slopes on Earth, indicating that at short length-scales this type of slope cannot channelize flow on its own. This interpretation is supported by the wetness index plot (Fig. 11), which shows a slowly coalescing flow rather than discrete areas of fluid concentration.

Site PC – Penticton Crater in Eastern Hellas. In slope–area plots the slope turnover is well expressed for both study areas in site PC (Fig. 8a). This suggests a strong diffusive or creep influence on both slopes. Study areas PC1 and PC2 both resemble either poorly developed talus or debris flow in slope–area plots. In the CAD plot (Fig. 9a), however, study area PC2 has the distinctive profile associated with debris flow, whereas study area PC1 more closely resembles talus. Talus processes can only be active in study area PC1 in small drainage areas, where it lies on the dry mass-wasting threshold in slope–area plots. Hence, the shape of the CAD curve must be explained by another process, which has a slope threshold but does not concentrate drainage. This unknown process must be pore pressure triggered as it is below the slope for dry mass wasting. In addition, the wetness index plot reveals that study areas PC1 and PC2 are very different. Study area PC1 has a similar wetness index map to the synthetic crater (Fig. 11h), whereas study area PC2 resembles debris-flow areas on Earth (e.g.

Fig. 11. (*Continued*) Wetness index maps made from DEMs of the study site locations on Mars. Areas included in this study are outlined and labelled in the figure. Wetness index values are represented by the same colours in Figure 10 to allow direct comparison. (**a**) & (**b**) Site PC, Penticton Crater in Eastern Hellas. (**c**)–(**e**) Site GC, Gasa Crater in Terra Cimmeria. (**f**) Site KC, a crater inside Kaiser Crater in Noachis Terra. (**g**) Site TS, a crater in Terra Sirenum. (**h**) The 10 km-diameter synthetic crater.

Fig. 10f), with a strongly concentrated high wetness index within alcoves and channels, becoming more diffuse downslope on the debris aprons.

The combined evidence suggests that the west-facing slope, which contains small gullies, has been modified by debris flow, whereas the equator-facing slope is more similar to dry mass-wasting deposits. This agrees with the interpretation of Pelletier *et al.* (2008), who, using numerical modelling, concluded that the new bright-toned deposits on this slope were more similar in form to deposits of dry granular flows than debris flows.

Site GC – Gasa Crater in Terra Cimmeria. In the slope–area plot for site GC (Fig. 8b), the slope turnover occurs in very small drainage areas (one or two pixels) and is thus partly abbreviated. This suggests that creep has not strongly influenced this site. This interpretation is supported by the observation that the gully heads originate at the very top of the slope. Study areas GC1, GC2 and GC3 resemble either poorly developed talus on Earth (study area KA2, Fig. 6a) or debris flows on Earth (Fig. 6b, c) in slope–area plots. However, in CAD plots (Fig. 9b), they have a flattened mid-section, resembling debris-flow systems on Earth. Their wetness index plots (Fig. 11c, d) have strong similarities with debris-flow systems on Earth (e.g. Fig. 10g); showing flow concentration in the alcove and channel, with more diffuse flow on the debris apron. Study area GC2 (Fig. 11c) shows a similar pattern of wetness index to the alluvial systems on Earth, with focused flow throughout.

In slope–area plots (Fig. 8b), study area GC4 has a flatter profile than study areas GC1, GC2 and GC3. The drop in slope in high drainage areas in GC4 is probably an artefact of the low number of pixels included in the slope calculations in the last five to 10 points. In the CAD plot (Fig. 9b), study area GC4 has a similar shape to talus systems on Earth (Fig. 7a). The talus interpretation for GC4 is supported by additional evidence: (1) there is no evidence for channels (Fig. 4e); (2) the wetness index plot (Fig. 11e) is similar to talus slopes on Earth; and (3) part of the slope–area curve lies on the threshold for dry mass wasting (Fig. 8b). The dip of the slope–area curve away from the threshold for dry mass wasting suggests that another process with a lower slope threshold is acting either without having an effect on the CAD plot or with the same CAD plot as talus. We hypothesize that this may be the same unknown process as noted in study area PC1.

The combined evidence suggests that the pole- and east-facing slopes of the crater have been affected by debris-flow processes, and the equator-facing slope by mass wasting and an unknown process.

Site KC – crater inside Kaiser Crater in Noachis Terra. Our ungullied study area (KC4) shows patterns in slope–area (Fig. 8c) and CAD plots (Fig. 9c) very similar to the synthetic crater and creep slopes on Earth. The difference between this study area and the gullied study areas (KC1–KC3) is presumably a result of the process of gully formation. Study areas KC1–KC3 do not have slope–area plots (Fig. 8c) that fit easily within the framework established so far. However, if we refer to the modelling work of Tucker & Bras (1998), then the patterns in slope–area plots can be explained. In small drainage areas our curves for study areas with gullies have a horizontal or slightly positive trend compared to our ungullied study area, which has a definite positive trend. This suggests the weak influence of diffusive processes (which generate a positive relationship in slope–area plots) combined with slope-threshold processes (which tend to produce horizontal trends). As all of the data are below the dry mass-wasting threshold, this threshold process is likely to be a pore-pressure-triggered process, such as debris flow. In intermediate drainage areas, there is a transitional region that occurs in a similar drainage area to the slope turnover in the ungullied section. In high drainage areas, the gullied study areas show a slightly decreasing subhorizontal trend as opposed to the ungullied study area, which has a well-defined decrease in slope with drainage area. This also can be attributed to a pore-pressure-triggered threshold process but at a lower slope threshold than the previous process. In CAD plots (Fig. 9c), study areas KC1–KC3 are consistent with debris-flow processes. The wetness index plots for these study areas (Fig. 11f) are similar to terrestrial debris-flow study areas that have been influenced by alluvial processes (e.g. site WF, Fig. 10g, h). This suggests that the first pore-pressure threshold in slope–area plots is the result of debris flow and the second lower one is due to an unknown process, which again could be the same process affecting sites PC and GC.

In slope–area plots, the gully heads on this slope (Fig. 8c) coincide with the drainage area of the slope turnover in study area KC4 and the transitional study areas of KC1–KC3. This coincident relationship matches the observations made by many authors who have studied gullies on Earth (e.g. Hancock & Evans 2006). Our channel heads lie mainly in the domain attributed to 'pore-pressure landsliding channel initiation' processes, but some also lie in the 'unchanneled' domain (McNamara *et al.* 2006). Notably, the gully heads occur below the dry mass-wasting threshold, again suggesting that these Martian gullies are initiated by a pore-pressure threshold process. The gully heads occur on slope gradients of 0.55 similar to those described by Lanza *et al.* (2010), but in drainage areas are an order of magnitude lower. This is possibly due to the different approach used by Lanza *et al.* (2010) to measure the contributing area, and possible differences in their interpretation of the location of channel initiation. The co-occurrence of the gully heads with the slope turnover in slope–area plots suggests that the gullies are a result of whole-slope drainage, as previously found by Lanza *et al.* (2010), either at the surface or shallow subsurface. Our work provides additional evidence to support the conclusions of Lanza *et al.* (2010) that these gullies originate from a distributed source and, hence, supports the surface melting model for Martian gully formation, rather than an aquifer source model. Further, this observation provides additional evidence that a threshold process, probably debris flow, is forming these gullies, as previously suggested by Lanza *et al.* (2010).

From the combination of the slope–area, CAD and wetness index plots, we infer that the gullies in this crater are produced by debris flow and were initiated by surface, or near-subsurface, flow of water. Creep and an unknown process were likely to have been the dominant processes on the ungullied crater slopes. This is consistent with the setting of these gullies within the ice-rich mantle deposits, which is likely to be susceptible to melting and provides a distributed source of water for the gullies.

Site TS – crater in Terra Sirenum. Unlike the other areas we have studied on Mars, parts of the slope–area data for study area TS1 in lower drainage areas (Fig. 8d) are above the threshold slope for dry mass wasting. This is an indication that rock-strength-limited dry mass wasting is occurring in the upper parts of the slope. In CAD plots (Fig. 9d), this study area has the classic shape of a talus or creep slope. However, the slope–area trend shown by study area TS1 is very different from that of the synthetic crater (Fig. 8d), which we assume to have been similar to the starting point for study area TS1. This assumption carries the implication that the slope in study area TS1 has evolved over time from concave to linear in profile. Study area TS1 shows a very similar trend in slope–area plots as study area GC4 (Fig. 8b), but originates above the 0.7 slope threshold. As discussed previously for study area GC4, in the framework of Tucker & Bras (1998) such a pattern is likely to reflect a gradual transition from the dominance of a dry mass-wasting threshold in lower drainage areas to the dominance of a pore-pressure-triggered slope threshold due to an unknown process in higher drainage areas. However, in the case of TS1, this signal not only includes dry mass wasting of non-cohesive material, but rock-wall mass wasting as well. The wetness index map shows that the slope does not concentrate drainage, except for some diffuse linear areas, again resembling talus slopes on Earth. The combination of the slope–area plot, CAD plot and wetness index map suggests a dominantly dry mass-wasting evolution of this slope, which fits well with the visual observations.

Solifluction on slopes on Mars. In many of the Mars study sites we have inferred an unknown process that is responsible for a second, lower-slope pore-pressure-triggered threshold in the slope–area plots. However, this process seems to produce slopes that yield a CAD plot that is similar to talus on Earth, that is, it does not concentrate drainage. As suggested by Tucker & Bras (1998), another threshold process that would produce a similar response in slope–area plots to pore-pressure-induced landsliding is solifluction. Solifluction in frozen landscapes comprises the combined action of gelifluction and frost creep, and describes the slow downslope movement of water-saturated debris or soils. Solifluction requires freezing and thawing to generate elevated pore pressures and occurs at lower slope angles than pore-pressure-induced failure, which can trigger landslides and debris flow (Harris *et al.* 2008). This process is consistent with the recently observations of periglacial landform assemblages on Mars (Balme & Gallagher 2009; Balme *et al.* 2009; Soare & Osinski 2009).

Implications for the formation process of Martian gullies

Dietrich & Perron (2006) suggested that the lack of biotic processes on Mars would promote erosion by rilling and gullying and stripping of the fine surface materials, given a suitable water source. This would lead to a slope–area plot that lacked a distinct slope turnover, similar to the slope–area plots seen in the Death Valley data (our site DV – Fig. 6d). However, inspecting the trends in the slope–area plots for the Mars systems in Figure 8, one of the most apparent differences from Earth is the presence of this slope turnover. This indicates that creep is a more dominant process on Martian hillslopes than on those we studied on Earth; contradictory to the predictions made by Dietrich & Perron (2006). The creep signal in most published slope–area plots on Earth is induced predominantly by biota, hence on Mars the creep must be facilitated using a different mechanism. Perron *et al.* (2003) observed, using Mars Orbiter Laser Altimeter (MOLA) data, that slopes on Mars have average gradients well below 35°

and suggested that ice-driven creep is the cause. Other potential creep mechanisms include frost heave and shrink–swell in clays and hydrated salts, both of which produce creep on unvegetated and unbioturbated slopes on Earth. These mechanisms, however, would require widespread and relatively large amounts of liquid water, which is considered unlikely under current or geologically recent Martian climate. Hence, we believe that ice-driven creep provides the best explanation for the signals seen in our slope–area data from Mars. In accordance with their results, most of the slopes we studied on Mars also have average gradients well below 35°, with the exception of slope TS1, whose average gradients are partially above 35°.

Virtually every gully that we have studied on Mars has the distinct signal of debris flow as the dominant gully-forming process. Lanza *et al.* (2010) also found visual and morphometric evidence of debris flows in these areas. The notable exception is area PC1, the slope containing the new light-toned deposits. However, this area does not include gullies of a normal form (Fig. 1) as they lack well-defined alcoves and channels. Examination of a far greater number of DEMs containing gullies would be needed to confirm debris flow as the main gully-forming process on Mars. However, if this is the principal mechanism, it raises the following hypotheses and predictions for the formation of gullies on Mars.

(1) The high sediment concentrations and low infiltration rates could protect the water from evaporation.
(2) The energy released by grain interactions within the flow could retard freezing.
(3) Basal freezing (Conway *et al.* 2011) or a permafrost layer could facilitate the saturation of the sediment that is required to generate the high pore-water pressures to trigger debris flow.
(4) Expected depositional features include levees and lobes.
(5) Expected erosional features include discrete slip scars.

Points (1)–(3) are hard to observe or test, but the erosional and depositional features can be detected in the high-resolution HiRISE images. Failure scars have been noted by other authors (Dickson & Head 2009) from HiRISE images and are present within our study areas. Depositional lobes have also been noted by other authors (Levy *et al.* 2009; Lanza *et al.* 2010). Visual observations have been made of debris-flow levees (Lanza *et al.* 2010), but DEMs from HiRISE are not yet of sufficient quality to reliably resolve debris-flow levees. High-quality DEMs would allow the estimation of individual flow volumes (Conway *et al.* 2010), which could be used to constrain models of gully formation. This should be a priority for future work, as it would allow more accurate estimates of the amounts of water associated with the formation of gully landforms.

A debris flow, once triggered, results in more erosion and deposition with less water than pure water flow. This means that high discharges, invoked by other workers (Heldmann *et al.* 2005; Hart *et al.* 2009), are not required to form Martian gullies. Modelling has shown that surface melting produces only small amounts of liquid water (Williams *et al.* 2009). This has been one of the major criticisms of the surface melting model. However, if gullies are formed mainly by debris flow, points (1) and (2) above indicate that relatively small amounts of water are needed.

Implications for the water source of Martian gullies

The observed relationship in slope–area plots between the slope turnover and the location of gully heads at site KC on Mars is an important observation, and indicates that the transition from concave to convex topography is closely linked to gully formation. This would not be expected in an aquifer system, as channel formation would be controlled predominantly by the location of aquifer bodies rather than the shape of the landscape (Fetter 2001). Our work indicates that a widely distributed source of surface or shallow subsurface flow at site KC would be the most satisfactory explanation, in support of the conclusions of Lanza *et al.* (2010). Because our data do not show a definite trend in slope–area plots, this indicates that the channels originate from shallow subsurface flow (Hattanji *et al.* 2006; Jaeger *et al.* 2007; Imaizumi *et al.* 2010) or, more probably, surface flow in a soil-poor landscape (Larsen *et al.* 2006). A potential source for this near-surface water is the mantle deposits, which have been observed on both this slope and at site PC2, that have been linked to gully formation by other authors (Christensen 2003; Aston *et al.* 2011; van Gasselt *et al.* 2011).

The development of equally spaced incised alcoves at site GC can be attributed to either geological controls (e.g. faulting) or landscape self-organization from an interlinked debris-flow–alluvial system (Perron *et al.* 2009). We argue against a structural control because there is a lack of these organized alcoves on the equator-facing slope. Hence, considering that we conclude debris flow to be the dominant gully-forming process on this crater slope, it would seem most likely that these self-organized alcoves are a result of this process. This kind of self-organization requires a landscape that responds to a distributed water source as on Earth rather than an aquifer source.

Kreslavsky & Head (2003) and Kreslavsky *et al.* (2008) found that pole-facing slopes between 40° and 50° latitude in both hemispheres were systematically gentler than equator-facing slopes. They suggest that this is due to insolation asymmetry and melting of ice on pole-facing slopes during periods of high obliquity, similar to the model proposed by Costard *et al.* (2002) for gully formation on pole-facing slopes. Our study sites also show this asymmetry: pole-facing slopes are longer, and have a greater variety of slope angles and are more concave; whereas the equator-facing slopes are shorter and have a more uniform distribution of slopes and are more linear. There is a marked difference in geomorphological process between crater walls with different aspects in the two craters that we studied (sites PC and GC). The observed asymmetry of process and form supports the model of a climatic influence on gully formation and general slope development of the craters. However, many more sites would have to be studied to verify this for gullies in general.

Conclusions

We have shown the potential of applying quantitative geomorphological analysis techniques commonly used on Earth for discriminating between different active processes on Mars. Specifically, we have validated the use of slope–area plots, cumulative area distribution (CAD) plots and wetness index maps on small slope sections of less than 1 km$^{2.}$ We have shown that pure-water (alluvial) flow-, debris-flow- and dry mass-wasting-dominated slopes can be satisfactory discriminated on Earth. By applying these techniques to four areas of Mars containing recent gullies we have inferred that debris flow is the dominant gully-forming process. However, we have also inferred that, as on Earth, gully formation on Mars is a complex process: slopes on Mars are likely to have been affected by a variety of processes that lead to a mixture of signals from our geomorphological analyses. Despite this, we have not found the distinctive geomorphological fingerprint of pure-water flow on slopes that host gullies. Its absence, however, does not prove the absence of the process. Our results are consistent with the possibility that ice-driven creep and solifluction are, or have recently been, active in modifying crater slopes on Mars.

From the location of gully heads within the landscape, and by studying the form of alcoves, it is apparent that at least two of the sites examined contain gullies that have been formed from a widely distributed source of water. This is most easily explained by a surface melting source for the water. The model of Costard *et al.* (2002) provides a mechanism by which the cause of this melting was increased insolation during past high-obliquity excursions. Our preliminary observations of an asymmetry in process and form around the impact craters provides additional support for this model, but we cannot rule out surface melting during present day or during other epochs.

Our geomorphological evidence for debris flow as an active process in forming gullies is reinforced by visual observations. Debris flow as a process leaves distinct geomorphological features, such as failure scars and lobate deposits, which have been observed both here and in previous studies (Dickson & Head 2009; Levy *et al.* 2009; Lanza *et al.* 2010). Unfortunately, the topographical data on Mars are not yet sufficient for the discrimination of these features and flanking levees in DEMs, which would allow accurate estimations of individual flow volumes and thus an estimation of the volumes of water needed to form the gullies (Conway *et al.* 2010).

Thanks go to G. Meyer and one anonymous reviewer for their constructive comments that greatly improved this manuscript. This work would not have been possible without a postgraduate studentship grant from the UK Natural Environment Research Council (NERC). We thank the NERC ARSF for obtaining the LiDAR data on which part of this paper relies. We thank the UK NASA RPIF-3D Facility at UCL for enabling the production of one of the HiRISE DEMs. Additional funding was awarded to S.J. Conway by Earth and Space Awards, the Geological Society's W.G. Fearnsides Award, The Dudley Stamp Fund and the British Society for Geomorphology's postgraduate funds. P.M. Grindrod is funded by an STFC Aurora Fellowship (ST/F011830/1). Thanks to J. Yearsley for the creation of spatialPattern script to create pink noise in MatLab.

Appendix 1

The derivation of the shear-stress erosion model relies on the assumption that erosion rate (E) is a power law of bed shear stress (τ_b):

$$E = k\tau_b^a \tag{2}$$

where k and a are positive constants. Following Whipple & Tucker (1999) and Snyder (2000), we use the assumptions of conservation of mass (water) and steady uniform flow to obtain the following expression of basal shear stress:

$$\tau_b = \rho C_f^{1/3}\left(\frac{gSQ}{W}\right)^{2/3} \tag{3}$$

where ρ is the density of water, C_f is a dimensionless friction factor, g is the acceleration due to gravity, S is the local channel slope, Q is the stream discharge and W is the stream width. We then include a relationship for basin hydrology and hydraulic geometry given by:

$$Q = k_Q A^c \quad (4)$$

$$W = k_W Q^b \quad (5)$$

where k_Q and k_W are constants, A is the drainage area, and b and c are positive dimensionless constants. Combining Equations (2)–(5) leads to:

$$E = k_E A^m S^n \quad (6)$$

where:

$$k_E = k_b k_W^{-2a/3} k_Q^{2a(1-b)/3} \rho^a g^{2a/3} \quad (7)$$

$$m = (2ac/3)(1-b) \quad (8)$$

$$n = 2a/3. \quad (9)$$

Given this, if we now define a constant k_M for Mars, based on the assumption that gravitation acceleration is approximately one third that on Earth:

$$k_M = (1/3)^{2a/3} \quad k_E = (1/3)^n k_E \quad (10)$$

To derive Equation (1), in the main text, we have to include the expectation that, over long timescales, uplift rate (U) and erosion rate compete to change the landscape elevation (z):

$$\frac{\partial z}{\partial t} = U - E = U - k_E A^m S^n \quad (11)$$

where t is a given time-step. Now if we assume that the system is in equilibrium in which erosion is balanced by uplift rate, $\partial z/\partial t = 0$, then:

$$S = (U/k_E)^{1/n} A^{-m/n}. \quad (12)$$

Comparing this to Equation (1), in the main text, we have:

$$k = (U/k_E)^{1/n} \quad (13)$$

$$\theta = -m/n. \quad (14)$$

If we then include our k_M constant (Equation 7) for Mars in Equation (9) we get:

$$S = (U/k_M)^{1/n} A^{-m/n} \quad (15)$$

$$S = 3(U/k_E)^{1/n} A^{-m/n}. \quad (16)$$

Thus, for a given drainage area on Earth we would expect the slope on Mars to be three times smaller or in log–log terms:

$$\log S = \log 3 + 1/n \log(U/k_E) - m/n \log A. \quad (17)$$

References

AKCA, D. 2007*a*. *Least Squares 3D Surface Matching*. PhD thesis, Eidgenössische Technische Hochschule, Zürich.

AKCA, D. 2007*b*. Matching of 3D surfaces and their intensities. *ISPRS Journal of Photogrammetry and Remote Sensing*, **62**, 112–121.

ASTON, A. H., CONWAY, S. J. & BALME, M. R. 2011. Identifying Martian gully evolution. *In*: BALME, M. R., BARGERY, A. S., GALLAGHER, C. J. & GUPTA, S. (eds) *Martian Geomorphology*. Geological Society, London, Special Publications, **356**, 151–169.

BALME, M., MANGOLD, N. *ET AL.* 2006. Orientation and distribution of recent gullies in the southern hemisphere of mars: observations from High Resolution Stereo Camera/Mars Express (HRSC/MEX) and Mars Orbiter Camera/Mars Global Surveyor (MOC/MGS) data. *Journal of Geophysical Research – Planets*, **111**, E05001, doi: 10.1029/2005JE002607.

BALME, M. R. & GALLAGHER, C. 2009. An equatorial periglacial landscape on Mars. *Earth and Planetary Science Letters*, **285**, 1–15.

BALME, M. R., GALLAGHER, C. J., PAGE, D. P., MURRAY, J. B. & MULLER, J. P. 2009. Sorted stone circles in Elysium Planitia, Mars: implications for recent martian climate. *Icarus*, **200**, 30–38.

BEN DAVID-NOVAK, H., MORIN, E. & ENZEL, Y. 2004. Modern extreme storms and the rainfall thresholds for initiating debris flows on the hyperarid western escarpment of the Dead Sea, Israel. *Geological Society of America Bulletin*, **116**, 718–728.

BLAIR, T. C. 1999. Cause of dominance by sheetflood vs. debris-flow processes on two adjoining alluvial fans, Death Valley, California. *Sedimentology*, **46**, 1015–1028.

BLAIR, T. C. 2000. Sedimentology and progressive tectonic unconformities of the sheetflood-dominated Hell's Gate alluvial fan, Death Valley, California. *Sedimentary Geology*, **132**, 233–262.

BORGA, M., TONELLI, F. & SELLERONI, J. 2004. A physically based model of the effects of forest roads on slope stability. *Water Resources Research*, **40**, 1–9.

BOURKE, M. C. 2005. Alluvial fans on dunes in Kaiser crater suggest niveo-aeolian and denivation processes on Mars. *In*: *Proceedings of the 36th Annual Lunar and Planetary Science Conference, March 14–18, 2005, League City, Texas*. Lunar and Planetary Institute, Houston, TX, Abstract 2373.

BRARDINONI, F. & HASSAN, M. A. 2006. Glacial erosion, evolution of river long profiles, and the organization of process domains in mountain drainage basins of coastal British Columbia. *Journal of Geophysical Research – Earth Surface*, **111**, F01013, doi: 10.1029/2005JF000358.

BROOK, M. S., KIRKBRIDE, M. P. & BROCK, B. W. 2008. Temporal constraints on glacial valley cross-profile evolution: two thumb range, central Southern Alps, New Zealand. *Geomorphology*, **97**, 24–34.

CHANDLER, R. J. 1973. The inclination of talus, Artic Talus terraces, and other slopes composed of granular materials. *Journal of Geology*, **81**, 1–14.

Chevrier, V. & Altheide, T. S. 2008. Low temperature aqueous ferric sulfate solutions on the surface of Mars. *Geophysical Research Letters*, **35**, L22101.

Christensen, P. R. 2003. Formation of recent martian gullies through melting of extensive water-rich snow deposits. *Nature*, **422**, 45–48.

Coe, J. A., Godt, J. W. & Henceroth, A. J. 2002. *Debris flows Along the Interstate 70 Corridor, Floyd Hill to the Arapahoe Basin Ski Area, Central Colorado – A Field Trip Guidebook.* United States Geological Survey Open-File Report, **02-398**.

Conway, S. J., Decaulne, A., Balme, M. R., Murray, J. B. & Towner, M. C. 2010. A new approach to estimating hazard posed by debris flows in the Westfjords of Iceland. *Geomorphology*, **114**, 556–572.

Conway, S. J., Balme, M. R., Lamb, M. P., Towner, M. C. & Murray, J. B. 2011. Enhanced runout and erosion by overland flow under subfreezing and low pressure conditions: experiments and application to Mars. *Icarus*, **211**, 443–457.

Costard, F., Forget, F., Mangold, N. & Peulvast, J. P. 2002. Formation of recent Martian debris flows by melting of near-surface ground ice at high obliquity. *Science*, **295**, 110–113.

Crippen, J. R. 1979. *Potential Hazards from Floodflows and Debris Movement in the Furnace Creek Area, Death Valley National Monument, California–Nevada.* United States Geological Survey Open-File Report, **79–991**.

Crosta, G. B. & Frattini, P. 2008. Rainfall-induced landslides and debris flows. *Hydrological Processes*, **22**, 473–477.

Decaulne, A. & Sæmundsson, Þ. 2007. Spatial and temporal diversity for debris-flow meteorological control in subarctic oceanic periglacial environments in Iceland. *Earth Surface Processes and Landforms*, **32**, 1971–1983.

Decaulne, A., Sæmundsson, Þ. & Pétursson, O. 2005. Debris flow triggered by rapid snowmelt: a case study in the Gleiðarhjalli area, northwestern Iceland. *Geografiska Annaler. Series A, Physical Geography*, **87A**, 487–500.

Dickson, J. L. & Head, J. W. 2009. The formation and evolution of youthful gullies on Mars: gullies as the late-stage phase of Mars' most recent ice age. *Icarus*, **204**, 63–86.

Dickson, J. L., Head, J. W. & Kreslavsky, M. 2007. Martian gullies in the southern mid-latitudes of Mars: evidence for climate-controlled formation of young fluvial features based upon local and global topography. *Icarus*, **188**, 315–323.

Dietrich, W. E. & Perron, J. T. 2006. The search for a topographic signature of life. *Nature*, **439**, 411–418.

Fetter, C. W. 2001. *Applied Hydrogeology*. Prentice-Hall, Englewood Cliffs, NJ.

Gaidos, E. J. 2001. Cryovolcanism and the recent flow of liquid water on Mars. *Icarus*, **153**, 218–223.

Garvin, J. B., Sakimoto, S. E. H., Schnetzler, C. & Frawley, J. J. 1999. Global geometric properties of Martian impact craters: a preliminary assessment using Mars Orbiter Laser Altimeter (MOLA). *In*: *Fifth International Conference on Mars, July 19–24, 1999, Pasadena, California*. Lunar and Planetary Institute, Houston, TX, Abstract 6163.

Godt, J. W. & Coe, J. A. 2007. Alpine debris flows triggered by a 28 July 1999 thunderstorm in the central front range, Colorado. *Geomorphology*, **84**, 80–97.

Hack, J. T. 1957. Studies of longitudinal stream profiles in Virginia and Maryland. *United States Geological Survey Professional Paper*, **294–B**, 45–97.

Hancock, G. R. & Evans, K. G. 2006. Channel head location and characteristics using digital elevation models. *Earth Surface Processes and Landforms*, **31**, 809–824.

Harris, C., Smith, J. S., Davies, M. C. R. & Rea, B. 2008. An investigation of periglacial slope stability in relation to soil properties based on physical modelling in the geotechnical centrifuge. *Geomorphology*, **93**, 437–459.

Harris, S. A. & Gustafson, C. A. 1993. Debris flow characteristics in an area of continuous permafrost, St. Elias Range, Yukon Territory. *Zeitschrift für Geomorphologie*, **37**, 41–56.

Hart, S. D., Gulick, V. C., Parsons, R. A. & Barnhart, C. J. 2009. Gully slopes and discharges on Lyot Crater's Central Peak. *In*: *Proceedings of the 40th Annual Lunar and Planetary Science Conference, March 23–27, 2009, The Woodlands, Texas*. Lunar and Planetary Institute, Houston, TX, Abstract 2349.

Hattanji, T., Onda, Y. & Matsukura, Y. 2006. Thresholds for bed load transport and channel initiation in a chert area in Ashio Mountains, Japan: an empirical approach from hydrogeomorphic observations. *Journal of Geophysical Research – Earth Surface*, **111**, F02022, doi: 10.1029/2004JF000206.

Hecht, M. H. 2002. Metastability of liquid water on Mars. *Icarus*, **156**, 373–386.

Heldmann, J. L. & Mellon, M. T. 2004. Observations of martian gullies and constraints on potential formation mechanisms. *Icarus*, **168**, 285–304.

Heldmann, J. L., Toon, O. B., Pollard, W. H., Mellon, M. T., Pitlick, J., McKay, C. P. & Andersen, D. T. 2005. Formation of Martian gullies by the action of liquid water flowing under current Martian environmental conditions. *Journal of Geophysical Research – Planets*, **110**, E05004, doi: 10.1029/2004JE002261.

Heldmann, J. L., Carlsson, E., Johansson, H., Mellon, M. T. & Toon, O. B. 2007. Observations of martian gullies and constraints on potential formation mechanisms II. The northern hemisphere. *Icarus*, **188**, 324–344.

Heldmann, J. L., Conley, C., Brown, A. J., Fletcher, L., Bishop, J. L. & McKay, C. P. 2010. Possible liquid water origin for Atacama desert mudflow and recent gully deposits on Mars. *Icarus*, **206**, 685–690.

Imaizumi, F., Hattanji, T. & Hayakawa, Y. S. 2010. Channel initiation by surface and subsurface flows in a steep catchment of the Akaishi Mountains, Japan. *Geomorphology*, **115**, 32–42.

Innes, J. L. 1983. Debris flows. *Progress in Physical Geography*, **7**, 469–501.

Iverson, R. M. 1997. The physics of debris flows. *Reviews of Geophysics*, **35**, 245–296.

Jack, J. L. 2000. Red-shifts and red herrings in geographical ecology. *Ecography*, **23**, 101–113.

Jaeger, K. L., Montgomery, D. R. & Bolton, S. M. 2007. Channel and perennial flow initiation in headwater streams: management implications of variability in source-area size. *Environmental Management*, **40**, 775–786.

KIRK, R. L., HOWINGTON-KRAUS, E. *ET AL.* 2008. Ultrahigh resolution topographic mapping of Mars with MRO HiRISE stereo images: meter-scale slopes of candidate Phoenix landing sites. *Journal of Geophysical Research – Planets*, **113**, E00A24, doi: 10.1029/2007JE003000.

KIRKBY, M. J., BULL, L. J., POESEN, J., NACHTERGAELE, J. & VANDEKERCKHOVE, L. 2003. Observed and modelled distributions of channel and gully heads – with examples from SE Spain and Belgium. *Catena*, **50**, 415–434.

KNEISSL, T., REISS, D., VAN GASSELT, S. & NEUKUM, G. 2010. Distribution and orientation of northern-hemisphere gullies on Mars from the evaluation of HRSC and MOC-NA data. *Earth and Planetary Science Letters*, **294**, 357–367.

KOBOR, J. S. & ROERING, J. J. 2004. Systematic variation of bedrock channel gradients in the central Oregon Coast Range: implications for rock uplift and shallow landsliding. *Geomorphology*, **62**, 239–256.

KOLB, K. J., PELLETIER, J. D. & MCEWEN, A. S. 2010. Modeling the formation of bright slope deposits associated with gullies in Hale Crater, Mars: implications for recent liquid water. *Icarus*, **205**, 113–137.

KRESLAVSKY, M. A. & HEAD, J. W. 2003. North–south topographic slope asymmetry on Mars: evidence for insolation-related erosion at high obliquity. *Geophysical Research Letters*, **30**, 1715, doi: 10.1029/2003GL017795.

KRESLAVSKY, M. A., HEAD, J. W. & MARCHANT, D. R. 2008. Periods of active permafrost layer formation during the geological history of Mars: implications for circum-polar and mid-latitude surface processes. *Planetary and Space Science*, **56**, 289–302.

LAGUE, D. & DAVY, P. 2003. Constraints on the long-term colluvial erosion law by analyzing slope–area relationships at various uplift rates in the Siwaliks Hills (Nepal). *Journal of Geophysical Research – Solid Earth*, **108**, 2129, doi: 10.1029/2002JB001893.

LANZA, N. L., MEYER, G. A., OKUBO, C. H., NEWSOM, H. E. & WIENS, R. C. 2010. Evidence for debris flow gully formation initiated by shallow subsurface water on Mars. *Icarus*, **205**, 103–112.

LARSEN, I. J., PEDERSON, J. L. & SCHMIDT, J. C. 2006. Geologic versus wildfire controls on hillslope processes and debris flow initiation in the Green River canyons of Dinosaur National Monument. *Geomorphology*, **81**, 114–127.

LEVY, J. S., HEAD, J. W., DICKSON, J. L., FASSETT, C. I., MORGAN, G. A. & SCHON, S. C. 2009. Identification of gully debris flow deposits in Protonilus Mensae, Mars: characterization of a water-bearing, energetic gully-forming process. *Earth and Planetary Science Letters*, **294**, 368–377.

MALIN, M. C. & EDGETT, K. S. 2000. Evidence for recent groundwater seepage and surface runoff on Mars. *Science*, **288**, 2330–2335.

MALIN, M. C., EDGETT, K. S., POSIOLOVA, L. V., MCCOLLEY, S. M. & DOBREA, E. Z. N. 2006. Present-day impact cratering rate and contemporary gully activity on Mars. *Science*, **314**, 1573–1577.

MANGENEY, A., BOUCHUT, F., THOMAS, N., VILOTTE, J. P. & BRISTEAU, M. O. 2007. Numerical modeling of self-channeling granular flows and of their levee-channel deposits. *Journal of Geophysical Research – Earth Surface*, **112**, F02017, doi: 10.1029/2006JF000469.

MANGOLD, N. 2005. High latitude patterned grounds on Mars: classification, distribution and climatic control. *Icarus*, **174**, 336–359.

MAO, L., CAVALLI, M., COMITI, F., MARCHI, L., LENZI, M. A. & ARATTANO, M. 2009. Sediment transfer processes in two Alpine catchments of contrasting morphological settings. *Journal of Hydrology*, **364**, 88–98.

MARCHI, L., DALLA FONTANA, G., CAVALLI, M. & TAGLIAVINI, F. 2008. Rocky headwaters in the Dolomites, Italy: field observations and topographic analysis. *Arctic Antarctic and Alpine Research*, **40**, 685–694.

MCEWEN, A. S., HANSEN, C. J. *ET AL.* 2007. A closer look at water-related geologic activity on Mars. *Science*, **317**, 1706–1709.

MCNAMARA, J. P., ZIEGLER, A. D., WOOD, S. H. & VOGLER, J. B. 2006. Channel head locations with respect to geomorphologic thresholds derived from a digital elevation model: a case study in northern Thailand. *Forest Ecology and Management*, **224**, 147–156.

MILLER, D. J. 1961. *Geology of the Katalla District, Gulf of Alaska Tertiary Province, Alaska.* United States Geological Survey Open-File Report, **206**.

MONTGOMERY, D. R. & FOUFOULA-GEORGIOU, E. 1993. Channel network source representation using digital elevation models. *Water Resources Research*, **29**, 3925–3934.

MOORE, H. J. & JAKOSKY, B. M. 1989. Viking landing sites, remote-sensing observations, and physical properties of Martian surface materials. *Icarus*, **81**, 164–184.

MORTON, D. M., ALVAREZ, R. M., RUPPERT, K. R. & GOFORTH, B. 2008. Contrasting rainfall generated debris flows from adjacent watersheds at Forest Falls, southern California, USA. *Geomorphology*, **96**, 322–338.

MOYLE, W. R. 1982. *Water Resources of Borrego Valley and Vicinity, California: Phase 1 – Definition of Geologic and Hydrologic Characteristics of a basin.* United States Geological Survey Open-File Report, **82–855**.

MUSSELWHITE, D. S., SWINDLE, T. D. & LUNINE, J. I. 2001. Liquid CO_2 breakout and the formation of recent small gullies on Mars. *Geophysical Research Letters*, **28**, 1283–1285.

MUSTARD, J. F., COOPER, C. D. & RIFKIN, M. K. 2001. Evidence for recent climate change on Mars from the identification of youthful near-surface ground ice. *Nature*, **412**, 411–414.

NORÐDALH, H. 1990. Late Weichselian and early Holocene deglaciation history of Iceland. *Jökull*, **40**, 27–50.

PELLETIER, J. D., KOLB, K. J., MCEWEN, A. S. & KIRK, R. L. 2008. Recent bright gully deposits on Mars: Wet or dry flow? *Geology*, **36**, 211–214.

PERERA, H. & WILLGOOSE, G. 1998. A physical explanation of the cumulative area distribution curve. *Water Resources Research*, **34**, 1335–1343.

PERRON, J. T., DIETRICH, W. E., HOWARD, A. D., MCKEAN, J. A. & PETTINGA, J. R. 2003. Ice-driven creep on Martian debris slopes. *Geophysical Research Letters*, **30**, 1747, doi: 10.1029/2003GL017603.

PERRON, J. T., KIRCHNER, J. W. & DIETRICH, W. E. 2009. Formation of evenly spaced ridges and valleys. *Nature*, **460**, 502–505.

PETERS, G. H., ABBEY, W. *ET AL.* 2008. Mojave Mars simulant–Characterization of a new geologic Mars analog. *Icarus*, **197**, 470–479.

REISS, D., VAN GASSELT, S., NEUKUM, G. & JAUMANN, R. 2004. Absolute dune ages and implications for the time of formation of gullies in Nirgal Vallis, Mars. *Journal of Geophysical Research – Planets*, **109**, E06007, doi: 10.1029/2004JE002251.

SCHON, S. C., HEAD, J. W. & FASSETT, C. I. 2009. Unique chronostratigraphic marker in depositional fan stratigraphy on Mars: evidence for ca. 1.25 Ma gully activity and surficial meltwater origin. *Geology*, **37**, 207–210.

SELBY, M. J. 1993. *Hillslope Materials and Processes*. Oxford University Press, Oxford.

SHINBROT, T., DUONG, N. H., KWAN, L. & ALVAREZ, M. M. 2004. Dry granular flows can generate surface features resembling those seen in Martian gullies. *Proceedings of the National Academy of Sciences of the United States of America*, **101**, 8542–8546.

SIRKIN, L. & TUTHILL, S. J. 1987. Late Pleistocene and Holocene deglaciation and environments of the southern Chugach Mountains, Alaska. *Geological Society of America Bulletin*, **99**, 376–384.

SNYDER, N. P., WHIPPLE, K. X., TUCKER, G. E. & MERRITTS, D. J. 2000. Landscape response to tectonic forcing: Digital elevation model analysis of stream profiles in the Mendocino triple junction region, northern California. *Bulletin of the Geological Society of America*, **112**, 1250–1263.

SOARE, R. J. & OSINSKI, G. R. 2009. Stratigraphical evidence of late Amazonian periglaciation and glaciation in the Astapus Colles region of Mars. *Icarus*, **202**, 17–21.

STOCK, J. & DIETRICH, W. E. 2003. Valley incision by debris flows: evidence of a topographic signature. *Water Resources Research*, **39**, 1093, doi: 10.1029/2001WR001057.

STOCK, J. D. & DIETRICH, W. E. 2006. Erosion of steepland valleys by debris flows. *Bulletin of the Geological Society of America*, **118**, 1125–1148.

TARBOTON, D. G. 1997. A new method for the determination of flow directions and upslope areas in grid digital elevation models. *Water Resources Research*, **33**, 309–319.

TARBOTON, D. G., BRAS, R. L. & RODRIGUEZ-ITURBE, I. 1991. On the extraction of channel networks from digital elevation data. *Hydrological Processes*, **5**, 81–100.

TAROLLI, P. & FONTANA, G. D. 2009. Hillslope-to-valley transition morphology: new opportunities from high resolution DTMs. *Geomorphology*, **113**, 47–56.

TREIMAN, A. H. 2003. Geologic settings of Martian gullies: implications for their origins. *Journal of Geophysical Research – Planets*, **108**, 8031, doi: 10.1029/2002JE001900.

TUCKER, G. E. & BRAS, R. L. 1998. Hillslope processes, drainage density, and landscape morphology. *Water Resources Research*, **34**, 2751–2764.

VAN GASSELT, S., HAUBER, E., ROSSI, A.-P., DUMKE, A., OROSEI, R. & NEUKUM, G. 2011. Periglacial geomorphology and landscape evolution of the Tempe Terra region, Mars. *In*: BALME, M. R., BARGERY, A. S., GALLAGHER, C. J. & GUPTA, S. (eds) *Martian Geomorphology*. Geological Society, London, Special Publications, **356**, 43–67.

WHIPPLE, K. X. & TUCKER, G. E. 1999. Dynamics of the stream-power river incision model: implications for height limits of mountain ranges, landscape response timescales, and research needs. *Journal of Geophysical Research – Solid Earth*, **104**, 17 661–17 674.

WILLIAMS, K. E., TOON, O. B., HELDMANN, J. L. & MELLON, M. T. 2009. Ancient melting of mid-latitude snowpacks on Mars as a water source for gullies. *Icarus*, **200**, 418–425.

WOODS, R. A. & SIVAPALAN, M. 1997. A connection between topographically driven runoff generation and channel network structure. *Water Resources Research*, **33**, 2939–2950.

WORKMAN, J. B., MENGES, C. M. *ET AL.* 2002. *Geologic Map of the Death Valley Ground-water Model Area, Nevada and California*. United States Geological Survey Miscellaneous Field Studies Map, 2381-A.

YETEMEN, O., ISTANBULLUOGLU, E. & VIVONI, E. R. 2010. The implications of geology, soils, and vegetation on landscape morphology: inferences from semi-arid basins with complex vegetation patterns in Central New Mexico, USA. *Geomorphology*, **116**, 246–263.

Fill and spill in Lethe Vallis: a recent flood-routing system in Elysium Planitia, Mars

M. R. BALME[1,2]*, C. J. GALLAGHER[3], S. GUPTA[4] & J. B. MURRAY[1]

[1]*Department of Earth and Environmental Sciences, Open University, Walton Hall, Milton Keynes MK7 6AA, UK*

[2]*Planetary Science Institute, Suite 106, 1700 East Fort Lowell Road, Tucson, AZ 85719, USA*

[3]*UCD School of Geography, Planning and Environmental Policy, Newman Building, University College Dublin, Belfield, Dublin 4, Ireland*

[4]*Department of Earth Sciences and Engineering, Impacts and Astromaterials Research Centre, South Kensington Campus, Imperial College, London SW7 2AZ, UK*

**Corresponding author (e-mail: m.r.balme@open.ac.uk)*

Abstract: Lethe Vallis is an approximately 230 km-long and 1.5 km-wide channel connecting several shallow basins in the Elysium Planitia region of Mars. It sits within a distinctive morphological unit defined by a platy-ridged-polygonized texture. We have documented the geomorphology of the system, and constructed topographical long profiles of the channel thalweg and the contacts of the platy-ridged-polygonized material.

The Lethe thalweg is shallow (with a slope of about 0.0001) but contains steeper sections that match the locations of observed cataract systems. The contact profiles suggest that the small basins linked by Lethe progressively ponded and over-spilled as the system developed, the cataracts being associated with this over-spill. Other landforms observed in the system include streamlined islands, anastomosing distributary systems, fluvial hanging channels and terraces on the channel margins. There are also possible dunes and/or antidunes within the channel. These all point to catastrophic fluvial flooding. Estimates of formative discharge are of the order of 1×10^4–5×10^4 m^3 s^{-1}, similar to the discharge of the Mississippi River.

We infer that Lethe Vallis formed as a fluvial 'fill and spill' catastrophic flood system. This demonstrates that the main Western Elysium Basin, the upstream source of Lethe Vallis, contained a substantial transient lake.

Lethe Vallis is a discontinuous erosional channel that connects two basins in the equatorial Elysium Planitia region of Mars. Elysium Planitia (Fig. 1) lies topographically below and to the north of the 'dichotomy boundary', the broad scarp that marks the transition between Mars' older, more heavily cratered southern highlands and the younger, low-lying northern plains. To the north of Elysium Planitia is the Elysium volcanic rise, comprising three large volcanoes that include Elysium Mons: the fifth tallest volcano on Mars.

Elysium Planitia has been the focus of considerable interest in Martian geomorphology because it contains a complex of interconnected basins and channels that provide evidence for the most geologically recent, large-scale flood events on Mars (Burr *et al.* 2002). The primary channel, Athabasca Vallis, is over 10 km wide, contains kilometre-scale streamlined islands and extends more than 300 km from its point of origin: a series of kilometre-wide fractures called the Cerberus Fossae, and which themselves extend for over 1000 km to the SE of the Elysium rise. The floods that carved Athabasca Vallis could have occurred as recently as 2–8 Ma (Burr *et al.* 2002). Athabasca Vallis debouches into the Western Elysium Basin, a large (*c.* 150 000 km^2), extremely flat-floored basin that also appears to be a young surface (3–7 Ma: Murray *et al.* 2005). The basin is covered by distinctive 'platy-ridged-polygonized' terrain (Fig. 2) characterized by 'plates' of low relative albedo material, rough at the metre-scale and commonly containing metre-sized clasts. The plates often appear 'ridged', by curvilinear, positive relief linear features in which metre-scale clasts can also sometimes be identified. Between the plates are the polygonized areas, much smoother at metre-scale and of relatively higher albedo. They are defined by their gently undulating surfaces that form a network of low polygonal mounds and troughs. These mounds can be as small as a few metres across or as large as 10–20 m. Networks of

From: Balme, M. R., Bargery, A. S., Gallagher, C. J. & Gupta, S. (eds) *Martian Geomorphology*. Geological Society, London, Special Publications, **356**, 203–227.
DOI: 10.1144/SP356.11 0305-8719/11/$15.00

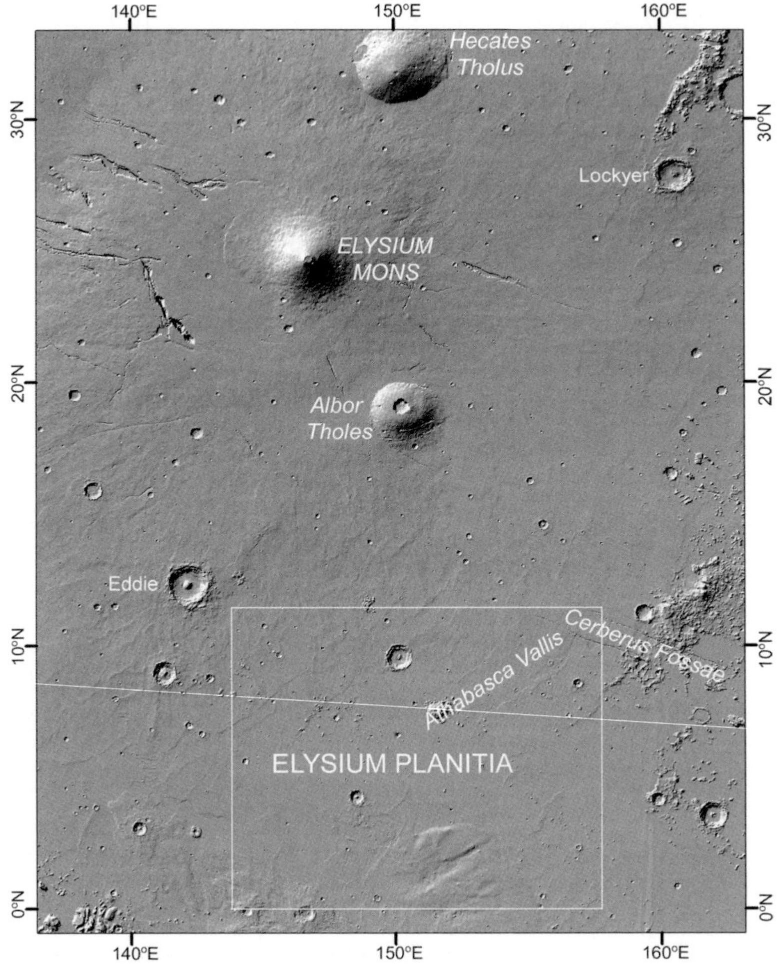

Fig. 1. Regional setting showing Elysium volcanic rise and Elysium Planitia. The white box shows the extent of Figure 3. Background image MOLA hillshade. North is up in this and all following images. Image credit: NASA/JPL/MOLA Science team; see prelim viii for acronym definitions.

smaller mounds often form within larger, 'master' networks. This combination of surfaces and structure forms a distinctive terrain type that consistently superposes the older, degraded terrains of the Elysium plains. Such material, with some (but never inconsistent) variations in morphology, is found throughout the Elysium Basin complex and within the erosional channels that link them.

The platy-ridged terrain has been likened to terrestrial pack ice (Brackenridge 1993; Rice *et al.* 2002; Murray *et al.* 2005). Several authors have, therefore, suggested that these deposits represent a debris-covered frozen palaeolake or sea (Scott *et al.* 1991; Brackenridge 1993; Murray *et al.* 2005) that formed as a result of catastrophic fluvial floods. Others have suggested that the surface seen today is a palimpsest landscape representing periglacial modification of the once ice-rich terrain (Balme *et al.* 2010). A radically different interpretation is that the platy-ridged-polygonized deposits represent flood lavas that occurred after the fluvial episode (Plescia 1990; McEwen *et al.* 1998; Hartmann & Berman 2000; Keszthelyi *et al.* 2000; Plescia 2003; Keszthelyi *et al.* 2004*a*, *b*; Jaeger *et al.* 2007; Vaucher *et al.* 2009*a*), and that any original fluvial deposits have long since been covered by a thin veneer of lava. In this interpretation, both the platy and polygonized surfaces represent primary volcanic deposits rather than secondary modification of an ice-rich surface. A more speculative volcanic interpretation is that the entire complex was formed over a period of only a

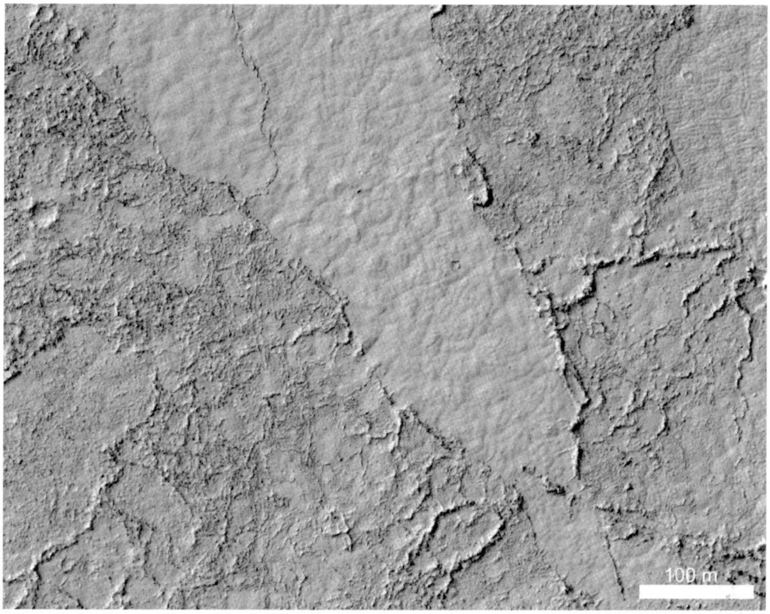

Fig. 2. Platy-ridged-polygonized terrain. The rougher, lower albedo material is characterized as 'platy', and the smoother, higher albedo hummocky material as 'polygonized'. The ridges in the lower part of the image demonstrate the last part of the description. The plates can often be reassembled like a jigsaw, and demonstrate relative movement. The polygonization occurs at several scales, with the domed patterns having a wavelength from a few metres to about 50 m. Part of HiRISE image PSP_009056_1840. Image credit: NASA/JPL/UofA.

few weeks by fluid, turbulent lavas that both eroded and then solidified the channels to form the platy-ridge-polygonal terrain that is seen today (Jaeger *et al.* 2010).

Irrespective of whether the platy-ridged-polygonized surface represents icy, periglacial or volcanic deposits, it is confined within a well-defined morphological contact (Fig. 3) that defines a 'highstand' mark and which (at least in the main Western Elysium Basin) closely follows an equipotential surface (Balme *et al.* 2010). Outside the contact, the terrains are more heavily impact-cratered and, at least to the north, are morphologically identical to the subdued volcanic landscape associated with the Elysium volcanic rise. Here we describe the morphology of Lethe Vallis and investigate what it can tell us about the evolution of the wider region. The geographical locations of the many figures described in this chapter are shown in the context figure at the end of this chapter.

In general, the topographical data presented here are derived from the NASA Mars Orbiter Laser Altimeter (MOLA) instrument (Smith *et al.* 2001) and are presented in the form of height above (or in this region, below) the Mars areoid. The MOLA data use the IAU2000 reference system, and the areoid is defined by the Goddard Mars potential model GMM3 (mgm1025) evaluated to degree and order 50 (Neumann *et al.* 2003). Other datasets used were obtained from the NASA Planetary Data System (PDS; http://pds.nasa.gov/) archive, processed and map-projected using ISIS (http://isis.astrogeology.usgs.gov/) software.

Overview: linked basins in Elysium Planitia

The extents of the platy-ridged-polygonized terrain in the Western Elysium Planitia region of Mars (Balme *et al.* 2010) were mapped using HRSC (High Resolution Stereo Camera; 12 m per pixel), THEMIS (Thermal Emission Imaging Spectrometer; 18–100 m per pixel), MOCNA (Mars Orbiter Camera Narrow Angle; 2–10 m per pixel), CTX (*Mars Global Surveyor* Context Imager; 6 m per pixel) and HiRISE (*Mars Global Surveyor* High Resolution Imaging Science Experiment; 0.25 m per pixel) data. Figure 3 demonstrates that the main Western Elysium Basin is linked to the various subsidiary basins by over-spill channels. The topographically highest sub-basin (sub-basin 1) is fed both from an over-spill in a broad channel south of Athabasca Vallis and also from a narrow channel

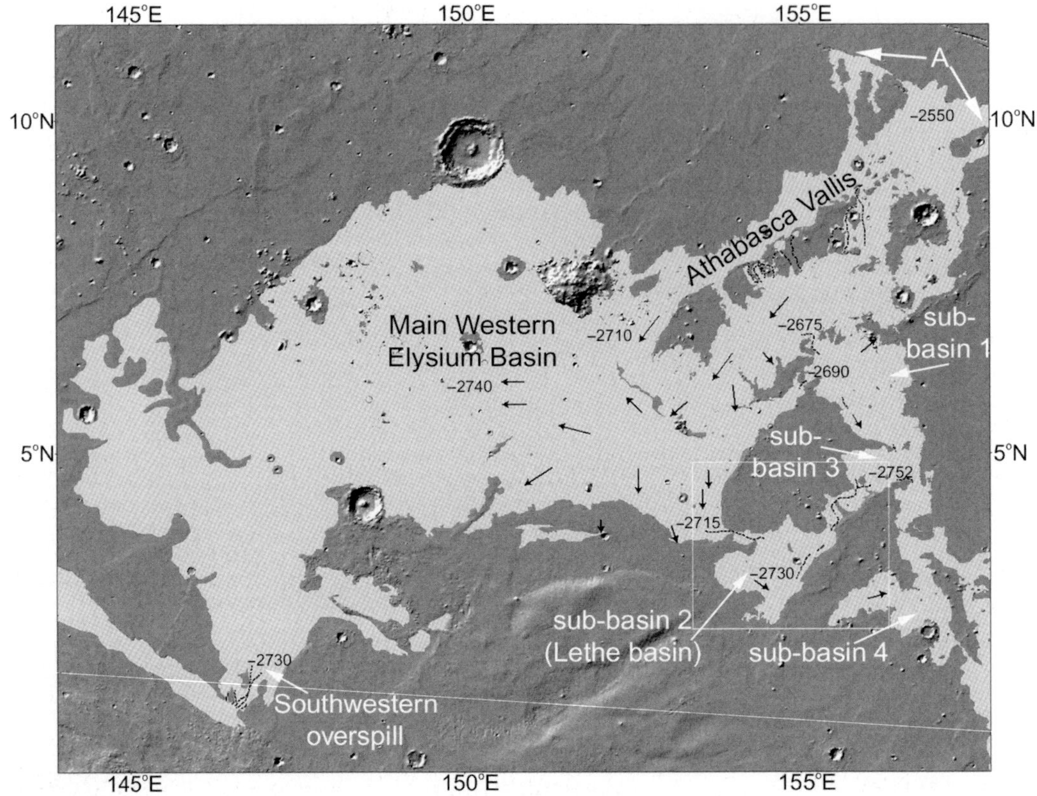

Fig. 3. Mapped contact of platy-ridged-polygonized terrain in Elysium Planitia. The black arrows indicate the inferred flow directions in the platy material. Small black figures show locations of topographical measurements. The white box marks the extent of Figures 4 & 5 and also the context figure at the end of the chapter. The source of the entire flow is interpreted to be the Cerberus Fossae fracture system designated 'A'.

leading from the eastern edge of the main Western Elysium Basin. Lethe Vallis originates at the SE margin of the main Western Elysium Basin, flows through sub-basin 2 (which we hereafter refer to as the Lethe Basin) and then debouches into sub-basin 3. Sub-basin 1 also drains into sub-basin 3 via a short, erosional channel system. Sub-basin 3 then drains into sub-basin 4. The elevation of various features in the system obtained from MOLA are shown in Figure 3. The spatial and topographical relationships between the basins suggest that sub-basin 3 was filled before the return-flow from the main Western Elysium Basin entered it via Lethe Vallis.

Lethe Vallis: description

Lethe Vallis is approximately 230 km in length. The path of the main channel, and the mapped highstand contact either side of the main channel, is shown in Figure 4 and the topography within the contact in Figure 5. The inlet to Lethe Vallis is topographically higher (*c.* −2715 m with respect to Mars datum) than both the lowest point in the main Western Elysium Basin (*c.* −2740 m) and the SW over-spill channel (*c.* −2730 m). Thus, Lethe Vallis must have become abandoned when the level in the main Western Elysium Basin dropped as it was drained to the SW. The inlet to Lethe Vallis is about 35–40 m higher than the debouchment into sub-basin 3, equating to a mean slope of less than 0.0002. The long profile shown in Figure 6 was constructed using a geographical information system (GIS) in which individual MOLA points within 100 m of the thalweg of the main channel (as delineated by the path shown by the solid white line in Fig. 4) were selected. The same method was used to extract profiles for the north and south highstand contacts (Fig. 7). The highstand contact profiles are very similar in terms of absolute topography and are symmetrical about the channel-floor

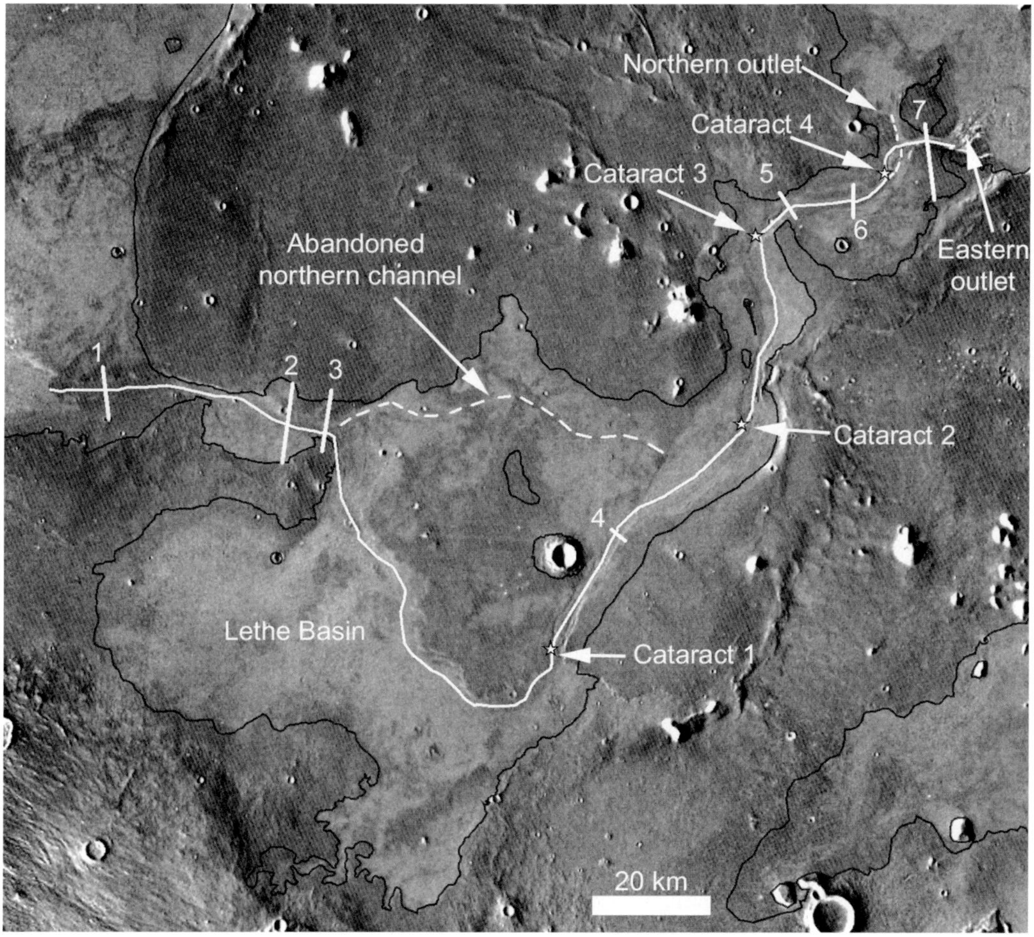

Fig. 4. Lethe Vallis system overview. The solid white line represents the most obvious flow route. Dashed white lines represent relict channels. The transverse white lines show the position of topographical cross-profiles. The thin dark lines show the mapped contact between the platy-ridged-polygonized terrain and the older, volcanic background material. Positions of the headward alcoves of cataracts are marked with stars. The background image is a mosaic of THEMIS daytime thermal infrared images. Image credit: NASA/JPL/ASU.

profile for a given distance downstream. This confirms that the mapped contact is, indeed, a high-stand mark of a fluid. In plan view, the mapped high-stand contact divides the system into four reaches (which we identify as zones 1–4 in Fig. 5), each defined by a narrowing of the contact to form a constriction. Topographical profiles were constructed using MOLA data (Table 1). These profiles can be used to analyse channel shape but it should be noted that, because the along-track point-to-point spacing of MOLA data is about 300 m and the channel width is only about 1 km, few details of the shape other than the maximum depth can be determined. Next, we describe the various reaches of Lethe Vallis system, from its source in Western Elysium Basin to its termination in sub-basin 3.

The source region of Lethe Vallis

Lethe begins as a shallow narrowing valley south of a lobate scarp (probably the edge of an ancient lava flow associated with the Elysium volcanic rise) at the SE edge of the Western Elysium Basin. Amphitheatre-shaped alcoves in this scarp (white arrows, Fig. 8) suggest headward erosion, and it is likely that this scarp once hosted a series of hydraulic steps or rapids. Figure 8 also shows how the rafts of darker material have rotated out from the main deposit of platy material, presumably as a result of flow over the scarp, and southwards into the Lethe Vallis channel. That the darker, platy material was once buoyant and mobile, and was degraded by flow from underneath it, is reinforced by the

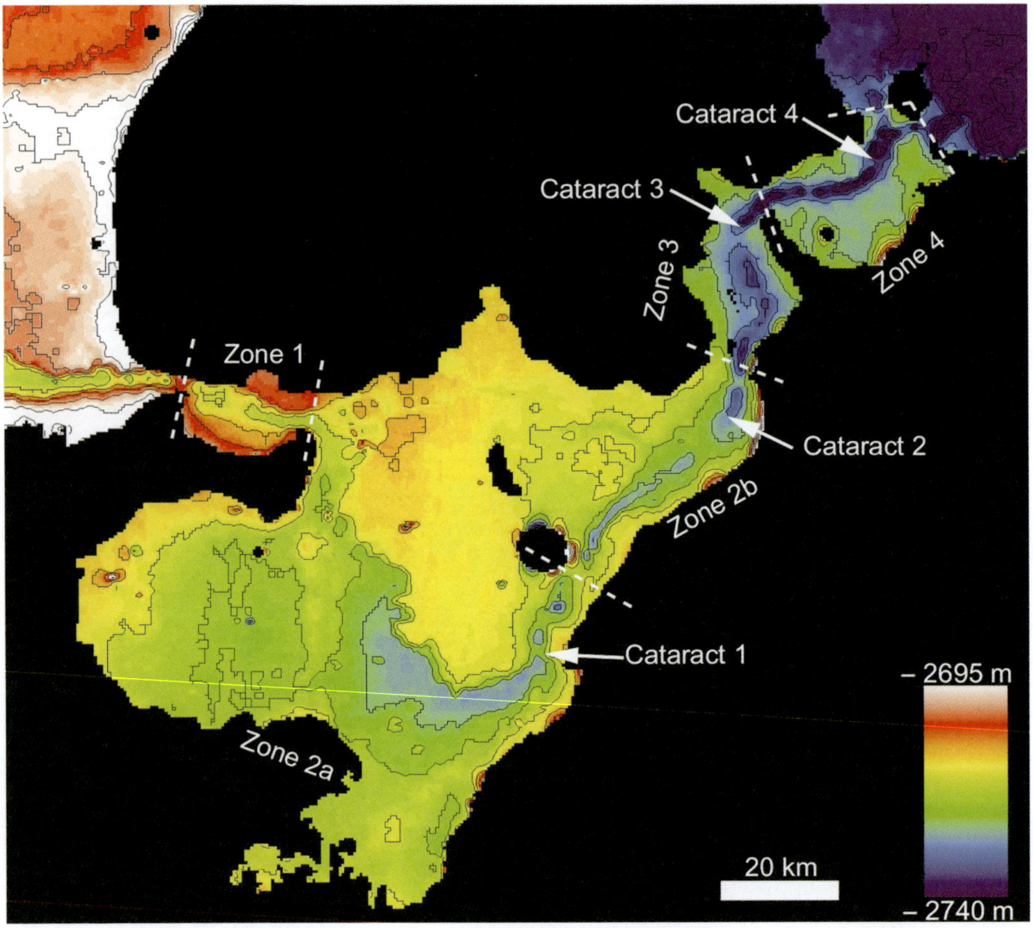

Fig. 5. Gridded MOLA topographical data within the mapped contact. The Lethe system is split into four zones, shown by dashed white lines, each marked by constriction in the mapped contacts. The 'beaded' discontinuities (especially in zone 4) in the channel floor are due to interpolation artefacts in the gridded data.

'arching' developed above the channel (dark arrows in Fig. 8), similar in form to sea ice trapped in narrow channels on Earth (Sodhi 1977).

For the first 25 km or so of its length, the channel-floor climbs uphill by a few metres. This occurs as the inlet to Lethe Vallis narrows significantly; the point at which the gradient reverses is where the channel is narrowest. This suggests that there was flow convergence and acceleration through this narrow strait. The margins of the first 10–20 km of the channel are about 15 m higher than the floor, and the mapped highstand contact of the deposits is about 10 m higher still. This demonstrates that there was (at least temporarily) a sufficiently high fluid level to drive flow through the shallowing and narrowing inlet to Lethe Vallis.

Zone 1

Zone 1 is defined by a channel reach of about 20 km in length incised into a basin approximately 15 km wide. The highstand is about 15 m higher than the channel floor. The channel floor is approximately horizontal, but the contacts drop in elevation by about 5 m over the 20 km length of the channel.

Zone 2

Zone 2 is the largest geomorphic zone of the flow-routing system and consists of a broad basin about 70 km across. At the transition from zone 1 to zone 2, about 45 km from the source, the channel bifurcates: one branch of the channel loops south around the edge of a lobate scarp (probably

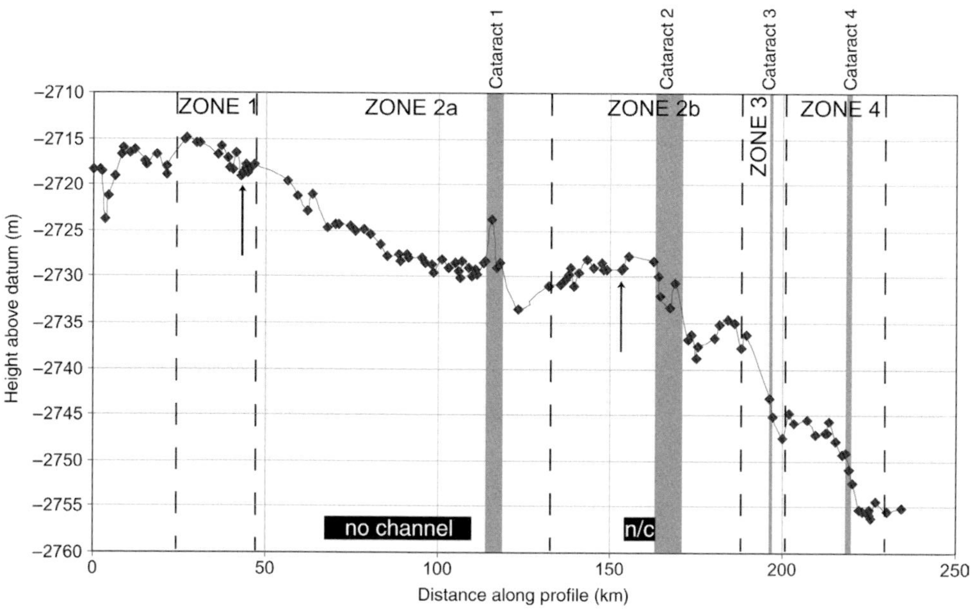

Fig. 6. Lethe Vallis long profile. Data extracted from individual MOLA points, not gridded data. Arrows represent the point at which the northern abandoned channel section leaves (first arrow) and then returns (second arrow) to the main channel. Dark bars at the bottom of the plot represent sections of the profile where there is no clear morphological signature of an erosional channel. The vertical dark bars represent the position of cataract systems. The zones marked by vertical dashed lines are defined in Figure 5.

another ancient lava flow), but the other cuts east across this topographical obstacle. The eastern arm is nearly completely infilled with platy-ridged-polygonized material and is topographically higher than the southern, suggesting that it became relict while the southern branch was still active. The southern branch of the channel is poorly defined between about 70 and 110 km from the source, and forms a broad valley rather than a channel.

This broad valley terminates at the first of four relict cataract or dry waterfall systems in Lethe Vallis (Fig. 9). These comprise amphitheatre-shaped alcoves eroded back into the upstream channel reach and form the head of one or more incised interior channels. Streamlined islands occur within the interior channel reach downstream of the cataract system and appear to be erosional bedrock remnants.

Slightly further downstream, the eastern branch of the first bifurcation returns to the main channel at about 150 km from the source. There is no obvious hanging channel at the intersection. About 10 km further downstream there is a second, very-well-defined cataract system (Fig. 10), and about 10 km further downstream a marked constriction in the highstand contact marks the beginning of zone 3. Both cataracts mark a break in slope in the thalweg (Fig. 6).

Cataract 1, within the basin, suggests that zone 2 can be divided into two subsidiary regions (a and b). This is reinforced by the presence of topographical obstacles (an impact crater and a small outlier of unflooded material) at the point in the channel where the inner channels downstream of cataract 1 terminate. This suggests that these obstacles represent what was once a topographical divide, that cataract 1 represents headward erosion of these inner channels into zone 2a, and that the cataract system has migrated upstream by more than 10 km. This is supported by the topography: in zone 2a the contact is 15–20 m higher than the channel floor and is approximately horizontal (except at the proximal part of the zone). The channel floor has a shallow, convex slope and drops by about 10 m over 70 km or so. Zone 2b is dissimilar to zone 2a: the channel floor is approximately horizontal to gently sloping, and appears to have a slight convex 'hump.' The highstand contacts drop in elevation by about 5–10 m in zone 2b.

Zone 3

Zone 3 is a basin about 15 km wide containing an approximately 30 km reach of channel. The transition from zone 3 to zone 4 is defined by a constriction of the highstand and by the presence of a

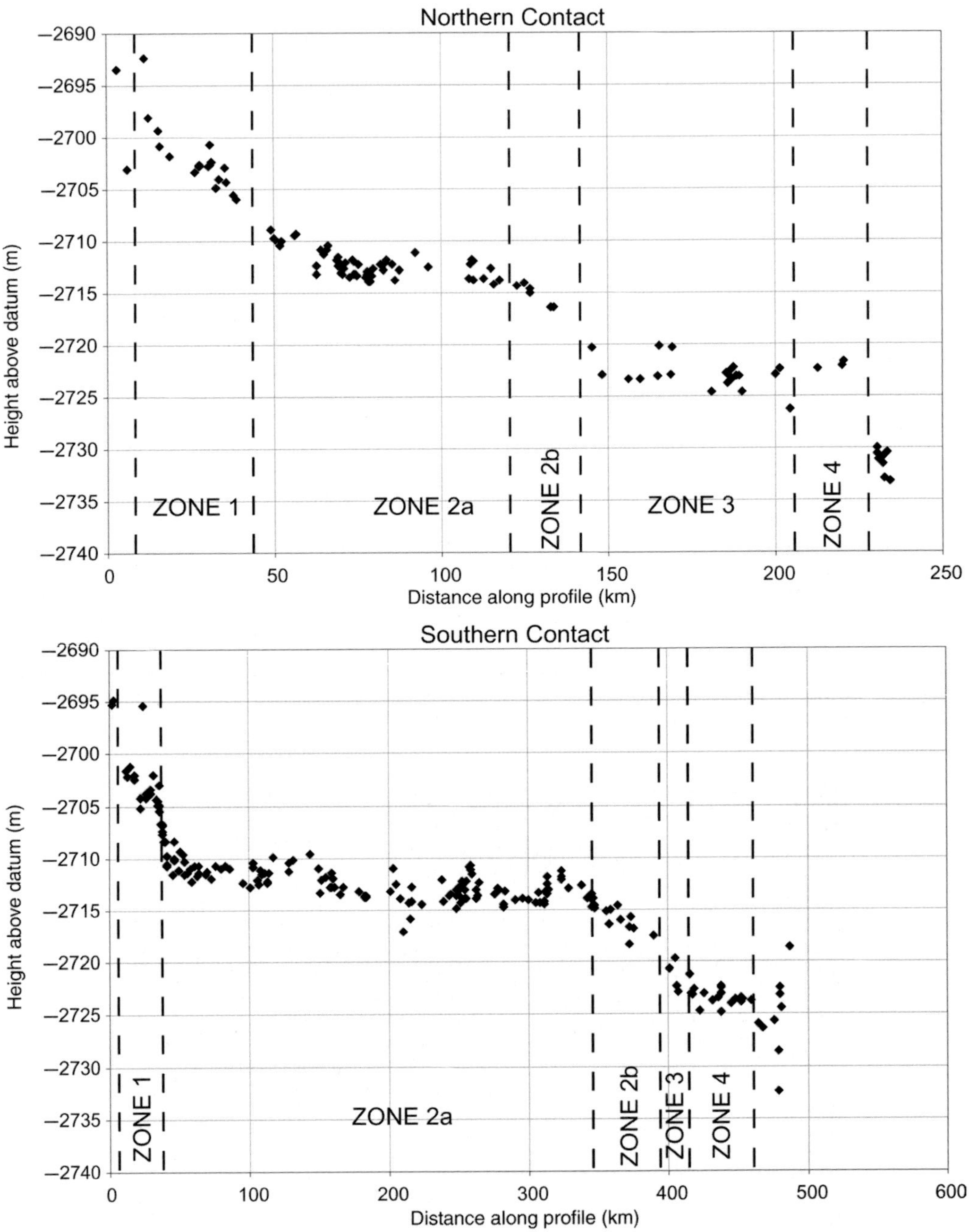

Fig. 7. Topographical profile of mapped highstand contacts, plotted in an easterly direction (i.e. from upstream to downstream). Again, these profiles are extracted from individual MOLA point, not gridded data.

streamlined island around a small (*c.* 750 m diameter) impact crater. This streamlined island marks a short bifurcation of the flow, the southern branch of which contains a small cataract system (Fig. 11a) at about 195 km from the start of Lethe Vallis. This slope discontinuity marks the transition from flow around both sides of the obstacle to flow along only one side, a flow diversion. The channel floor drops in elevation by about 10 m across the zone, but the highstand contact is approximately level.

Table 1. *Channel cross-sections. Note that the channel width refers to the measured width of the erosional channel, not the profile length. See Figure 4 for the locations of profiles*

Profile	Channel width (km)	Channel depth (m)
1	2.9	11
2	1.6	2.5
3	3.3	11
4	1.1	5
5	1.8	10
6	1.7	11
7	1.0	10

Zone 4

Zone 4, again defined by a constriction in the highstand contact, covers a slightly larger area and longer reach than zone 3 but is similar in many ways. There is, again, a small cataract system at 220 km from the source (cataract 4; Fig. 11b) on one side of a streamlined island that represents a bifurcation in flow. Immediately downstream of the streamlined island below cataract 4, the channel bifurcates once again, with branches heading north and east. The northern branch of the channel clearly became abandoned before the flooding ceased, as there is now a hanging channel incised into the bedrock flank of the deeper, eastern-flowing channel (Fig. 12). Each branch terminates only a few kilometres further downstream in similar distributary networks of islands and channels as they enter sub-basin 3 (Figs 12 & 13).

The channel floor drops about 10 m along its reach within zone 4, but, like zone 3, the highstand contact is nearly level. The break in slope in zone 4 is associated with cataract 4. The surface above cataract 4 appears to be at a similar elevation to the floor of the northward abandoned channel, suggesting that the upstream migration of the knickpoint that formed cataract 4 was responsible for beheading the northern terminal branch of Lethe shown in Figure 12.

The highstand contacts in zones 3 and 4 are approximately horizontal and at the same elevation. The channel floor, however, drops steeply across the approximately 10 m steps associated with the cataracts. In sub-basin 3, where Lethe terminates, the channel floor is about 25 m below the southern contact, which itself is at about −2730 to −2735 m. This is about the same elevation as the floor of Lethe Vallis just below cataract 2, suggesting that if basin 3 was fluid-filled when Lethe Vallis first debouched into it, then the Lethe channel in zones 3 and 4 was completely inundated at this time.

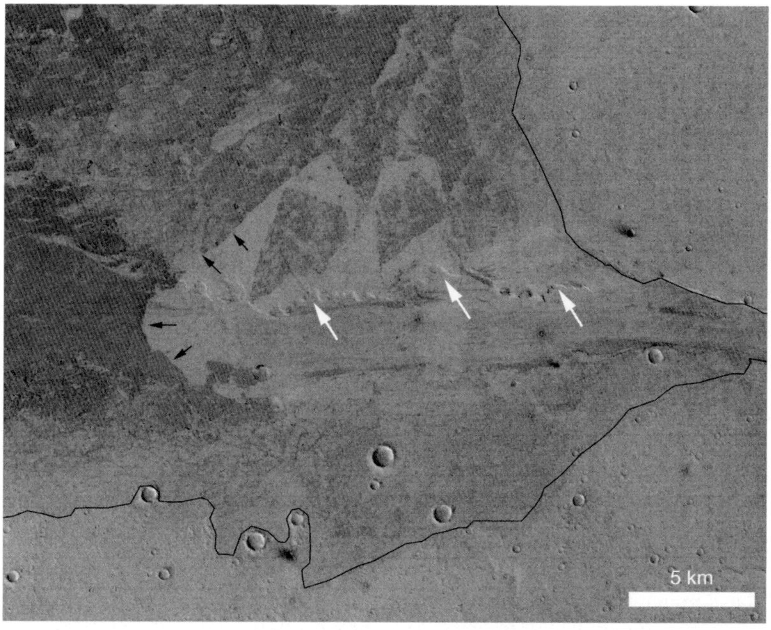

Fig. 8. Mouth of Lethe. Note the amphitheatre-shaped incisions into the northern scarp (white arrows) and the lower albedo 'plate' overlying the channel edge (black arrows), and showing 'arching' that indicates headward erosion of the plate by flow underneath. Part of HRSC nadir image from orbit h2121. Image credit: ESA/DLR/FUB.

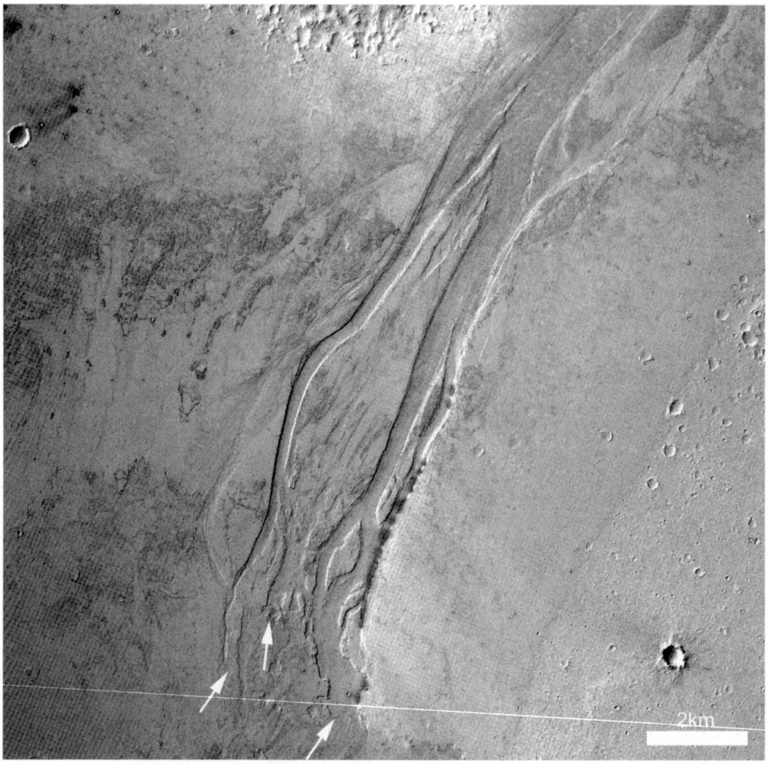

Fig. 9. Cataract 1. Flow is from bottom of the image to the top. White arrows indicate the upstream position of the amphitheatre-shaped head scarps that front interior channels. Note the streamlined islands within the cataract system. Part of CTX image P21_009201_1834. Image credit: NASA/JPL/MSSS.

Twin terminations of Lethe Vallis in sub-basin 3

The twin terminations of Lethe Vallis where it enters sub-basin 3 are distinguished by a series of distributary channels divided by streamlined islands (Figs 12 & 13). The islands appear to have been heavily modified: they do not have a smooth shape, in contrast to mid-channel islands with streamlined tails behind obstacles (Figs 10 & 11a), and instead appear ragged. They are steep sided, with little evidence for slumping, so appear to be composed of competent, cliff-bearing material. In full-resolution HiRISE images, horizontal parting or layering can be seen in the steep sides of the islands. There is no evidence for dipping beds.

Some islands show deeply incised divide crossings, whilst others are cut by shallow surface channels (Fig. 14). Many of the divide crossings are at a high angle to the dominant flow direction, some even are perpendicular to the flow, and in one case both ends of a divide-crossing channel point in what is now the downstream direction. There is a faint polygonization on the surface of some of the islands, and many are covered by the same rubbly surface texture seen on the channel floors and inferred to have once floated on the liquid within the channel. The islands are generally flat topped and are thus streamlined only in plan view and not in cross section. These observations demonstrate that they were either nearly, but not fully, submerged, or that, if fully submerged, the flow was insufficiently powerful or long lasting to create a three-dimensionally streamlined obstacle.

Because it was supplied by an inlet topographically higher (and closer to source; Fig. 3) than the inlet to Lethe Vallis, sub-basin 3 is interpreted to have already contained fluid before the return flow from the main Western Elysium Basin joined it. Interestingly, a similar, but smaller, distributary system occurs within sub-basin 1, where another return flow from the Western Elysium Basin re-enters this sub-basin.

Lethe Vallis: long profile

The Lethe Vallis thalweg long profile and highstand long profiles are shown in Figures 6 and 7,

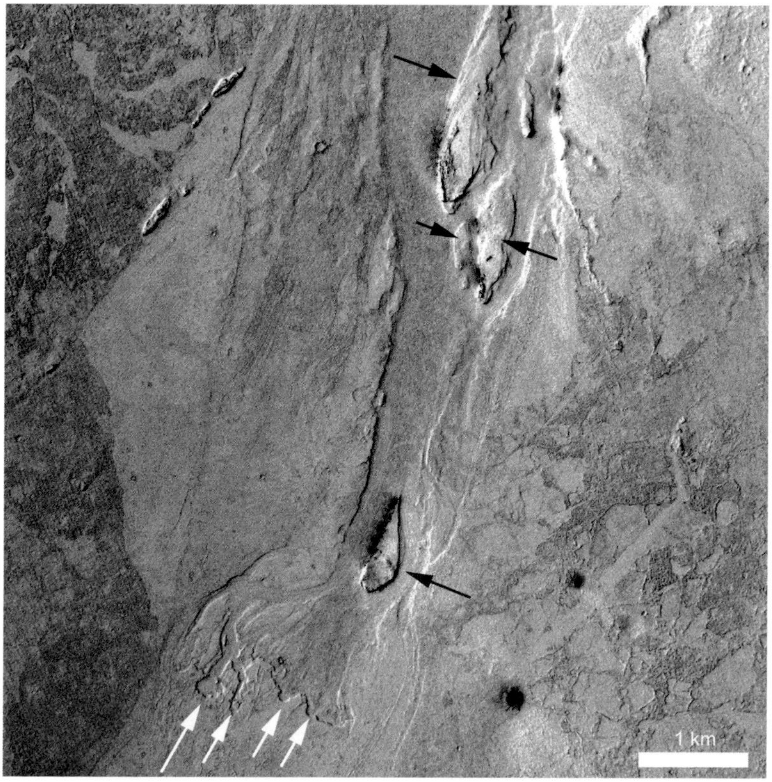

Fig. 10. Cataract 2. Flow from the bottom of the image to the top. Note the multiple amphitheatre-shaped head scarps at the bottom of the image (white arrows) and the multiple streamlined islands (balck arrows). Platy-ridged terrain textures can be seen on both sides of the channel. Part of CTX image P02_002015_1858. Image credit: NASA/JPL/MSSS.

respectively. Between the highest point in the channel at 25 km from source and the first cataract, the channel-floor topography follows a shallow concave profile. Between the first and second cataracts the profile is approximately horizontal. After the second cataract the profile steepens and topography decreases in convex 'steps' associated with the position of the cataracts. In addition to their concave as opposed to stepped profiles, there is also a marked difference in the absolute gradient between the reaches of the channel upstream of cataract 2 (a drop of about 0.1 m km^{-1}) and the channel downstream of this feature (about 0.3 m km^{-1}). Interestingly, although cataracts 1 and 2 are large landforms, each about 1 km wide, they each represent a drop in channel floor level of less than 5 m, whereas the smaller cataracts 3 and 4 represent a drop in channel-floor elevation of about 10 m.

Landforms related to flooding

Steep banks, terraced channel margins (Fig. 15) and streamlined islands (e.g. Figs 9–11) are common within Lethe Vallis. The streamlined islands, in particular, appear to be residual bedrock macroforms as they contain large impact craters (e.g. Fig. 11a), and frequently have topographical surface elevations and surface textures consistent with terrain outside the channel. Together with fluvial hanging channels (Figs 12 & 15) that indicate locations where previous flow paths have become relict due to downcutting in the main channel, such landforms demonstrate the erosional power of the flow that formed the channel.

Terraces are generally narrow and discontinuous over more than a few kilometres and are often superposed by platy-ridged-polygonized material. Their small size and relative geological youth means that insufficient crater-count data can be gathered to determine whether the terraces were created by multiple flood events widely separated in time (in contrast to, e.g. Warner *et al.* 2009). The presence of similar superposing platy-ridged-polygonized material on both the terraces and the channel floor might be taken to indicate that the terraces represent differential erosion of a layered or horizontally

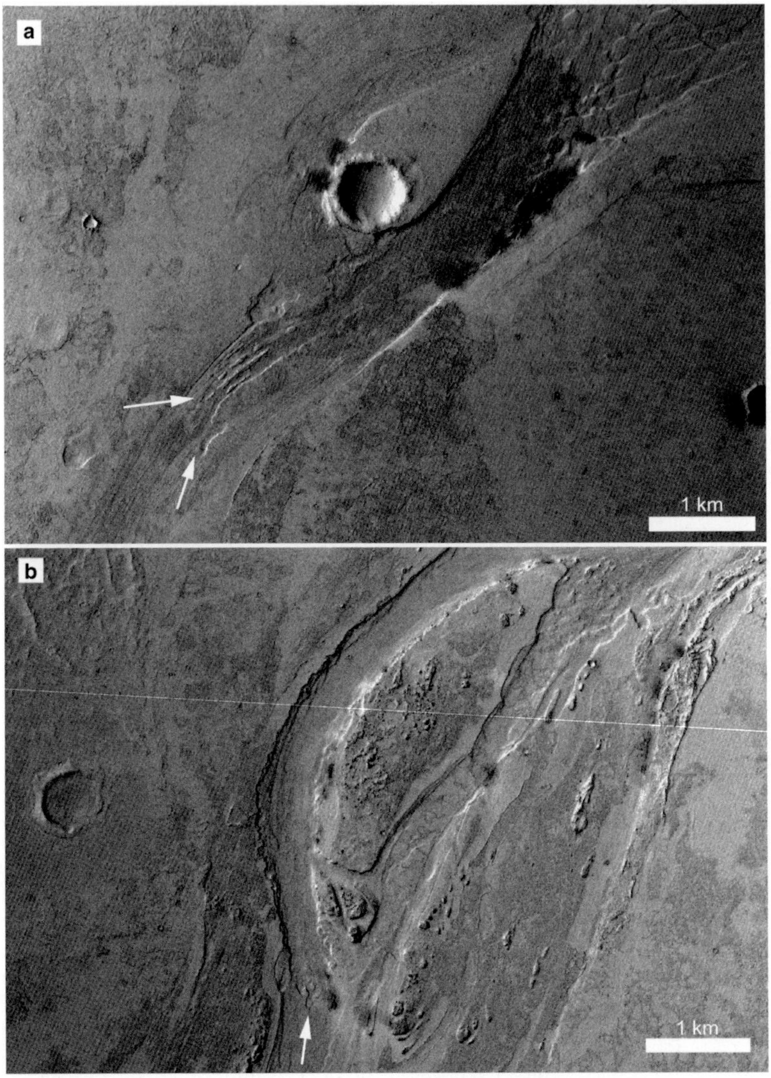

Fig. 11. Cataracts 3 (**a**) and 4 (**b**). Flow is from the bottom left to the top right in both cases. White arrows indicate head scarps. Both examples occur at the upstream edge of large streamlined islands, and indicate flow abandonment of one side of the island and promoted erosion on the other. Note the crescentic forms in the upper right of (a) that are also shown in Figure 17. Parts of CTX images P02_002015_1858 (a) and P16_007118_1845 (b). Image credits: NASA/JPL/MSSS.

structured pre-existing substrate, and that they represent structural benches or bedrock terraces (i.e. strathes) produced by changes in the base level during a single flow event rather than multiple events. Paired terraces are seen (e.g. Fig. 15) indicating channel narrowing with time and implying that, ultimately, downcutting locally dominated lateral erosion. This is particularly evident in the distal parts of Lethe Vallis.

Cataracts 1 and 2 are defined by a series of multiple amphitheatre-shaped head scarps that head long (5 km in the case of cataract 2 and nearly 10 km for cataract 1) interior channels. Downstream of the cataracts the interior channels contain streamlined islands or locally anabranching channels. The inner channel morphology is indicative of upstream knickpoint migration, and it is likely that the cataracts indicate regions of enhanced erosion.

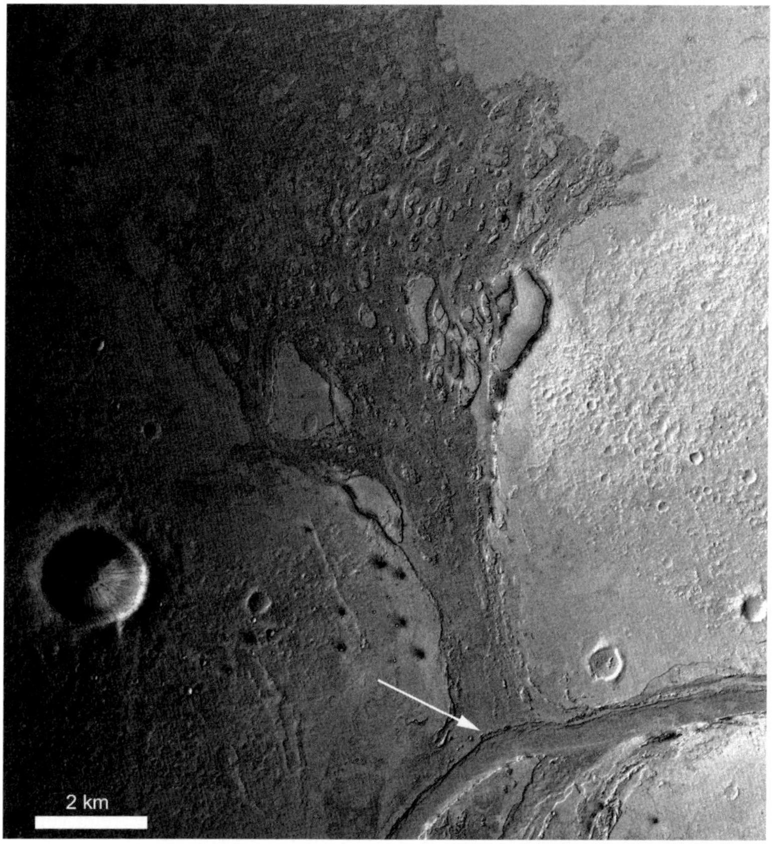

Fig. 12. Fluvial hanging valley and northern outlet of Lethe Vallis. The northern outlet was abandoned while the eastern outlet was still active, forming the hanging valley indicated by the white arrow. The northern termination is marked by a series of anastomosed channels and streamlined islands. Part of CTX image P16_007118_1845. Image credit: NASA/JPL/MSSS.

The floor of Lethe Vallis hosts all of the surface types typical of the wider Elysium Complex: rubbly and ridged material (sometimes forming fractured plates), polygonally patterned intra-plates, and sinuous ridges superposing both rubbly and polygonal terrains. Along its entire course, Lethe Vallis appears at least partially infilled by these deposits, and there are few landforms indicative of flood erosion or deposition preserved within the channel itself. Moreover, variations of the same surface textures occur outside the channel and extend to the mapped highstand contact, which is sometimes 15–20 m higher than the edge of the channel.

There are some landforms within the channel, however, that seem not to have been completely overprinted by these textures. These include transverse-crescentic forms, chevron-like or rhomboid forms, and longitudinal lineations and grooves. The crescentic and rhomboid forms occur in sets downstream of the cataracts, and are of the order of 100 m in length. Examples can be seen in Figures 9a, 16 and 17. Although no suitable topographical data exist, the shapes of these forms can be estimated from patterns of lighting and shadow. Both types have gently sloping upstream faces, but steeper downstream faces. Between some of the crescentic forms are box-shaped channels or troughs. The crescentic forms have a morphology similar to fluvial duneforms formed as a result of subcritical flow (Froude number < 1: Carling *et al.* 2009*a*). Very similar landforms have previously been identified on Mars in Athabasca Vallis (Burr *et al.* 2004), the major flood channel in this regional system. The rhomboid forms are, perhaps, more similar to bedforms created by standing-wave action within floods (Morton 1978). We therefore speculate that they are fluvial bedforms similar to antidunes that formed in trans-critical flow

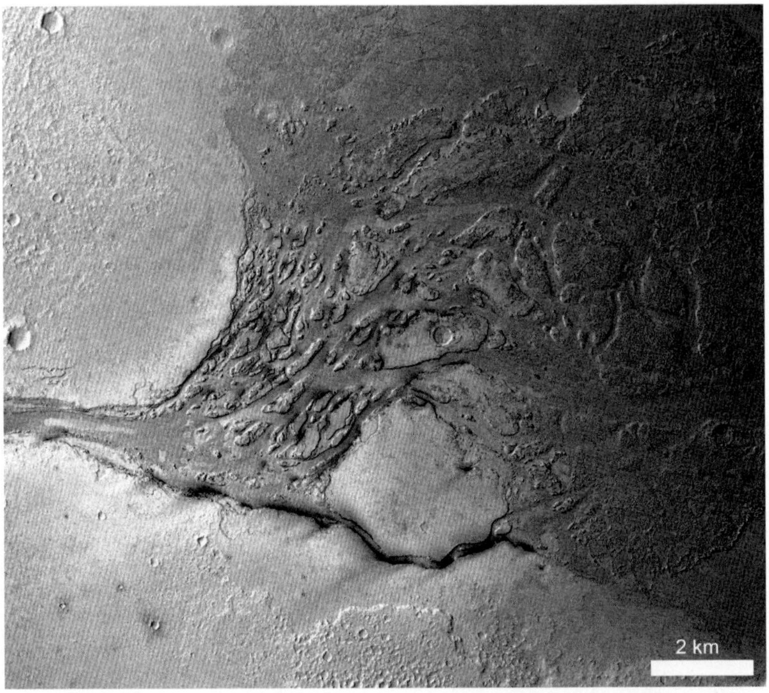

Fig. 13. Eastern outlet of Lethe Vallis, showing anastomosed channels and streamlines islands. Part of CTX image P16_007118_1845. Image credit: NASA/JPL/MSSS.

(Woodford 1935; Carling *et al.* 2009*a*) in which the Froude number is between 0.7 and 1.3.

The rhomboid forms are superimposed by lineations parallel to the direction of flow (Fig. 18), but because there are only relatively low-resolution images available of the crescentic forms no further details can be seen. The lineations consist of furrows and ridges, about 1 m across, which

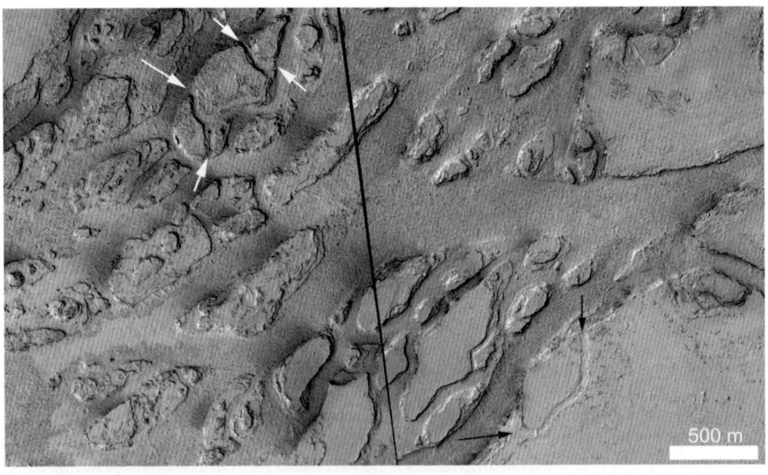

Fig. 14. Divide crossings and abandoned channels in the distributary system. Divide crossing channels can be shallowly (black arrows) or deeply (white arrows) incised. Note the large angles that the white-arrowed channels make to the dominant flow direction (bottom left to top right). The image covers the central portion of Figure 13, and comprises parts of HiRISE images PSP_004072_1845 and PSP_007262_1845. Image credits: NASA/JPL/UofA.

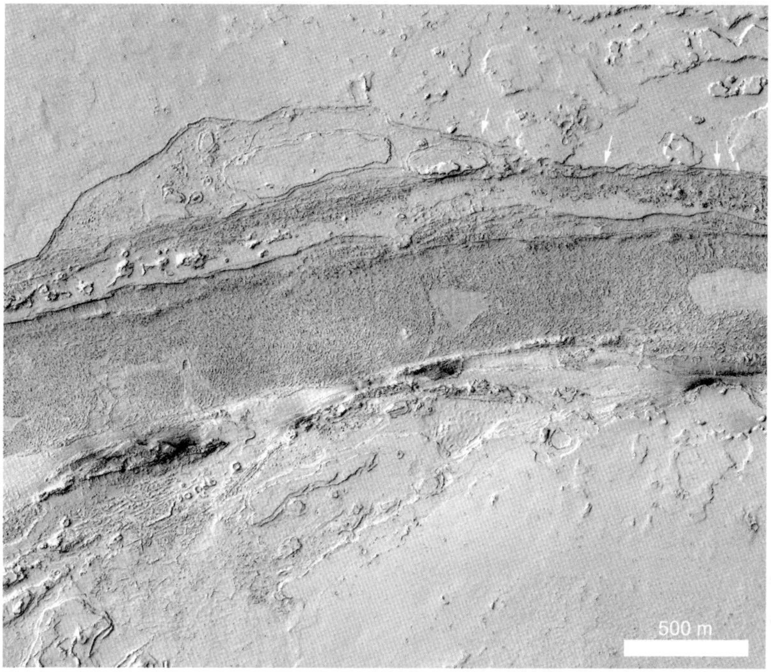

Fig. 15. Distal part of Lethe Vallis, just west of the eastern outlet. Flow is from left to right. Paired terraces can be seen on either side of the channel. Platy-ridged textures are seen both on the channel floor and on the terraces. A shallow fluvial hanging valley is marked by the white arrows. Part of HiRISE image PSP_007263_1845. Image credit: NASA/JPL/UofA.

generally fade out on the steeper, downslope sides of the chevron forms. Where the lineations occur with the polygonized texture, they appear to superpose the polygon centre but fade out within the troughs (Fig. 18). This implies either that the lineations deformed only the high polygon centres or that the polygons formed beneath the lineations and erased them at the polygon margins. We suggest that these linear forms are scour or tool marks caused by physical deformation of the underlying bedform by material carried within, or floating on top of, a shallow flow and that gouged the bed as it moved.

In summary, landforms consistent with catastrophic flood erosion are common within Lethe Vallis. However, there is generally a lack of channel-floor landforms, which probably reflects the extensive fill by platy-ridged-polygonized material (rubbly-textured material overlying the crescentic dune-like forms provides a good example of this: Fig. 17). The fill does not appear to reflect a later episode of shallow flooding, however, as rubbly material is seen on top of erosional islands, on terraces, in abandoned channels and all the way up to the highstand contact, but does not consistently overlie channel landforms. For example, most of the cataract head scarps have pristine morphology and are not draped by rubbly material. In addition, some parts of the channel appear to be incised into material with this platy-ridged-polygonized texture (Fig. 19): the distinctive surface texture is subdued or removed within the channel but pristine immediately outside the channel. This suggests that the channel was either formed by multiple episodes of flooding, was then choked by the same fluid that formed it, with later flow continuing over grounded 'plates' or that there was preferential modification of the platy-ridged texture within the channel.

Analysis

Discharge

Given suitable measurements of width, slope and channel depth, it is possible to estimate the formative flow velocity and discharge responsible for the geomorphology of Lethe Vallis. To obtain these measurements we have again used MOLA laser altimeter data. Although the MOLA data are precise in the vertical direction, individual point measurements are the result of averaging over 'spots' of about 150 m in diameter. Thus, height

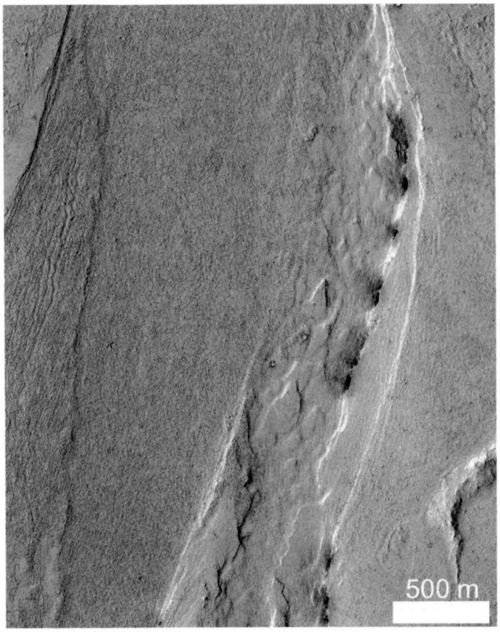

Fig. 16. Chevron or rhomboid-shaped channel-floor landforms in Lethe Vallis. These features, visible on the right-hand side of the channel (flow is from bottom to top), appear to be superposed by platy-ridged textures on the central channel floor. They occur just downstream of cataract 2 (this image overlaps the top right-hand corner of Fig. 10). Part of HiRISE image PSP_0010335_1840. Image credit: NASA/JPL/UofA.

variations within the spot footprint can cause anomalies when considering small variations in topography. Furthermore, although the along-track (approximately north–south) MOLA point spacing is about 300 m, tracks can be separated from each other by up to several kilometres. Hence, reliable channel cross-sections can only be obtained from those reaches of Lethe Vallis that run approximately east–west. Seven such cross-sections were obtained (Table 1). The channel width was defined from higher-resolution visual images, whilst the channel depth was measured using MOLA data.

Assuming that the fluid that eroded Lethe Vallis was a Newtonian fluid with minimal viscosity, the Darcy–Weisbach equation can be used to estimate the mean flow speed as described by Wilson *et al.* (2009):

$$u = [(8gRS)/f_c]^{1/2} \quad (1)$$

where g is gravitational acceleration (3.72 m s^{-1}), R is the hydraulic radius of the channel (width multiplied by depth divided by width plus depth),

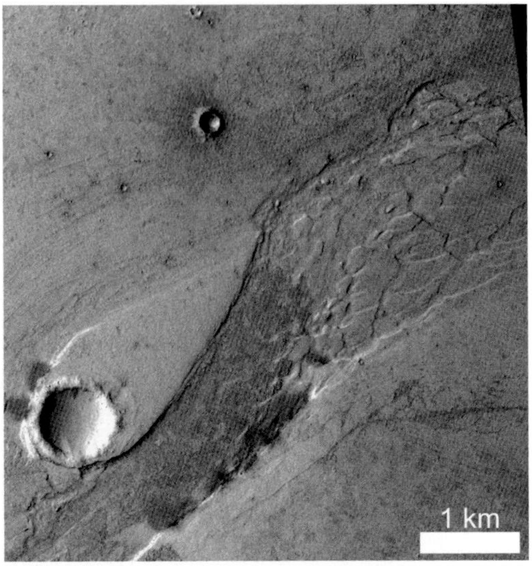

Fig. 17. Crescentic channel-floor landforms. Note the overlying, low albedo platy-ridged textures. These landforms occur downstream of cataract 3. This image overlaps the top right-hand corner of Figure 11a. Part of MOC NA image S1101466. Image credit: NASA/JPL/MSSS.

S is sine of the channel bed slope and f_c the dimensionless friction factor. As the depth of Lethe Vallis is commonly about 1% of its width, R can be replaced by d (depth in metres) in this and following

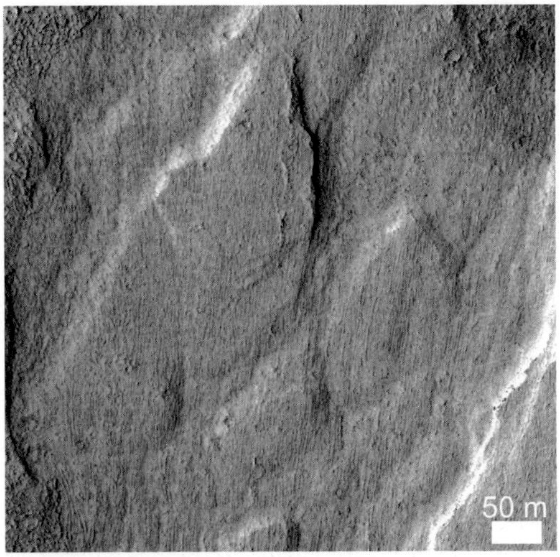

Fig. 18. Close-up view of chevron/rhomboid landforms shown in Figure 16. Flow-parallel grooves and furrows can be seen, as well as a subtle polygonization. This image covers the central part of Figure 16. Part of HiRISE image PSP_0010335_1840. Image credit: NASA/JPL/UofA.

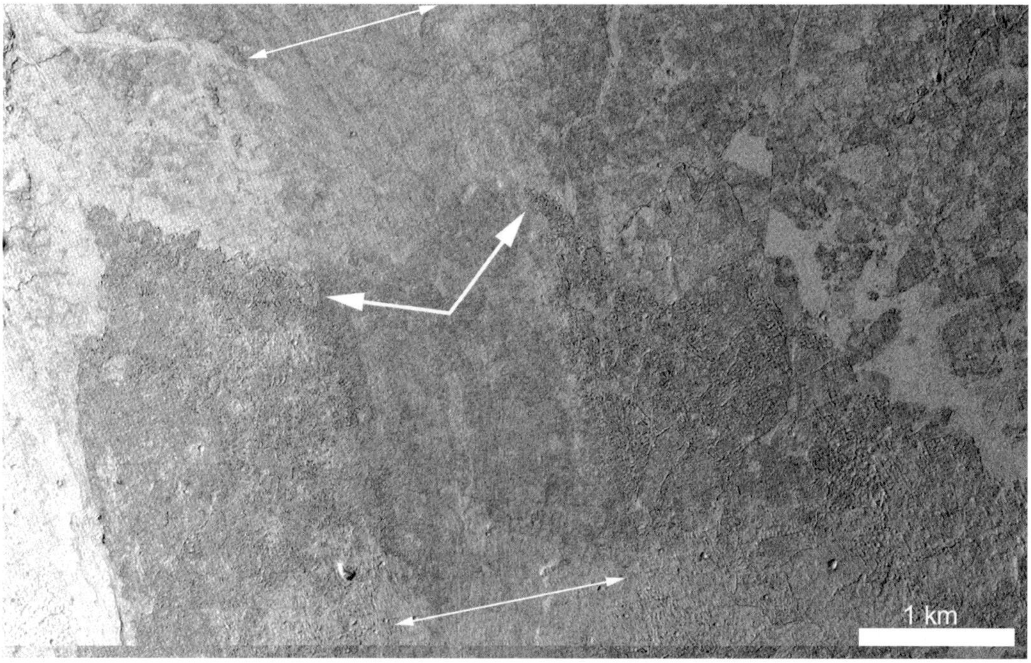

Fig. 19. Superposition relationships within Lethe Vallis. Flow is from bottom to top. The Lethe channel (width shown by straight, double-headed arrows) cross-cuts the platy texture, which can be seen to be subdued within the channel, but rougher outside. This is particularly obvious where the margins of a 'plate' become blurred where it crosses the channel (bold double arrow). This shows that the plate was mobile when the basin was deeply filled, but grounded as the fluid level dropped, and was later modified by further continued flow. Part of CTX image P16_007118_1845. Image credit: NASA/JPL/MSSS.

equations. The dimensionless friction factor describes the relationship between the mean velocity of the flow and the shear velocity of the flow, and is a factor of the depth, width and roughness of the channel. At this point, several further assumptions must be made. First, as we have no knowledge of the depth to which Lethe Vallis is infilled with post-flow deposits, nor whether the interior erosional channel was ever bankfull, we must assume a depth of flow. We have chosen a depth of 10 m as this seems to be a characteristic depth for five of the seven profiles.

Secondly, we have no knowledge of the roughness of the original channel floor nor whether roughness elements were fixed or mobile. The friction factor must, therefore, be estimated. Some authors (e.g. Wilson *et al.* 2004) have used estimates for channel roughness based on the size distribution of clasts seen by the *Viking* and *Mars Pathfinder* (which landed in outflow-channel distal deposits), whereas others have used models tested on terrestrial analogues to infer roughness element size (Kleinhans 2005). Given that Lethe is small in comparison to many Martian flood channels, we use the formulation for f_c based on a bed of gravel-dominated clasts:

$$f_c = 8[5.75 \log_{10}[(d_m^{0.314} R^{0.686})/D_{84}] + 2.8822]^{-2} \tag{2}$$

where d_m is maximum channel depth and D_{84} the 84th percentile clast size. Because d_m is much less than the channel width, and thus R is equivalent to d_m, Equation (2) becomes:

$$f_c = 8[5.75 \log_{10}[d_m/D_{84}] + 2.8822]^{-2}. \tag{3}$$

A reasonable range of values for D_{84} is 10–50 cm (Wilson *et al.* 2004). If D_{84} is 10 cm then f_c is about 0.04. If D_{84} is 50 cm then f_c is about 0.075. The gradient of the channel floor is about 0.0001 overall, and about 0.0003 for the steeper second half of the channel. Thus, mean flow speeds within Lethe are estimated from Equation (1) to have been between 1 and 1.75 m s^{-1}. Given Lethe's channel width of 1–3 km and a depth estimate of 10 m, this equates to a discharge of

1×10^4–5×10^4 $m^3 s^{-1}$, similar to the discharge of the Mississippi River on Earth (Baker 2001). As noted above, the Lethe Vallis channel appears to be infilled. If the original channel was three times deeper (30 m) then the flow rates would have been of the order of 1.8–3 m s^{-1} and the formative discharge of the order of 10^5 $m^3 s^{-1}$.

Formative duration and erosional power

While these discharge values are necessarily only an order of magnitude estimate, they can be used to calculate an approximate formative time over which the morphology we have described was produced. This can also serve as a minimum time over which the system was active. The Lethe Basin (zone 2) has an area of about 4800 km^2, and the difference between the elevation of the high-stand contact and the deepest point on the basin is about 20 m. Thus, a conservative estimate for the maximum amount of liquid contained within the Lethe Basin is approximately 50 km^3 or 5×10^{10} m^3. Dividing this value by the calculated formative discharge of Lethe Vallis – as there is only one outlet from the Lethe Basin – gives a drainage time of 10–50 days. Hence, the morphology now evident was formed within this time frame. If, as is almost certainly the case, there was prolonged, post-formation flow through the Lethe Basin from the topographically higher main Western Elysium Basin, the calculated formative discharge represents peak, rather than average, flow and Lethe Vallis would have been active for many times longer than the formative period. However, the dominant fluvial morphology of Lethe probably represents a short catastrophic event that lasted only days or weeks.

The stream power per unit area can be estimated from channel dimensions, flow density, gravity, slope and reconstructed flow velocities (Rhoads 1987). Estimates for stream power in Lethe are of the order of 10–50 W m^{-2}. This is much lower than values derived for catastrophic floods on Earth (Benito 1997; Baker 2009). It should be noted that, although the mean slope of the Lethe Vallis thalweg is extremely low and that local variations in slope are only of a factor of 10, combined with local variations in channel depth of a factor of 2 or 3 would increase the stream power by at least one order of magnitude – bringing it in line with the lower end of stream-power calculations associated with streamlined hill formation on Earth (500–2000 W m^{-2}; Baker 2009). The low gradient (and, hence, low stream power) could account for the lack of incision throughout the majority of zone 2a in the Lethe Basin, which might represent a reach in which there was little to no erosion.

Discussion

Terminal distributary systems

Together with the channel cataracts, the twin distributary systems at the terminus of Lethe Vallis are among the most spectacular and distinctive features in the region. Although the deeply incised distributary networks must reflect an increase in erosivity, this inference does not lead to an unequivocal morphogenetic interpretation of the distributary substrate. Two hypotheses are plausible. One hypothesis is that these are channels incised into alluvial deposits, with incision driven by base-level changes. Secondly, they may represent bedrock erosional remnants similar to the streamlined islands within the channel. This determination is important in the context of understanding the evolution of sub-basin 3 and its relationship to fluvial processes in Lethe Vallis.

Kehew *et al.* (2009) described stages of channel development in megafloods: 'Initial stages of erosion carve a wide shallow tract, commonly with anastomosing channels. Further erosion leads to channel deepening with more organized flow resulting in longitudinal grooves. A central, large, deep inner channel results from coalescence and capture of lateral flow'. Anastomosing patterns are characteristic of megaflooding, in which the rapidly increasing discharge spills across divides, and creates a network of dividing and rejoining channels (Baker 2009). Given that zone 4 (and possibly zone 3) clearly represent what was once a rapidly filling basin that over-spilled first to the north, and then to the east, the formation of these anastomosing networks as 'classic' megaflood landforms seems likely. The inference that the islands are composed of competent material, the visibility of horizontal layers within the island walls, and the observation that the island surfaces are topographically and morphologically similar to the surfaces outside of the basin seem to confirm their formation as remnant islands.

However, the fluvial hanging channel that indicates beheading of the outflow to the north by outflow to the east runs counter to this conclusion. Once the divide to the north of zone 4 was overtopped, why did erosion not proceed to continue to deepen and entrench this flow path into sub-basin 3? Another clue that something other than simple erosion occurred comes from the observation that the traces of complex anastomosing patterns are not seen at the margins of zone 1, 2a, 2b or 3. Finally, if they are catastrophic erosional forms, these anastomosing channel patterns probably formed when local discharge was greatest, and this is most likely when the zone 3 and zone 4 basins over-spilled. This begs the question: why are there two such distributary networks at the termination of Lethe Vallis? Does this require the northern branch to have been temporarily dammed to allow a second episode of fill and spill to occur? It appears that this type of network might not represent simple erosional outflow alone. Further, given the inference that sub-basin 3 was fluid-filled prior to the return flow along Lethe Vallis, are these types of networks restricted to flood discharges into already filled basins? It is clear that sub-basin 3 itself over-topped and drained to the SE, so this perhaps suggests that base-level changes within sub-basin 3 were also involved in creating the incised network seen today.

The highstand contact within sub-basin 3 is at an elevation of about −2736 m: below the height of the top surfaces of the islands in the upstream part of both Lethe's northern and eastern distributary networks. This means that there is a drop of more than 10 m in the contact elevation in the transition between zone 4 and sub-basin 3. The eastern outflow channel floor is lower than the sub-basin 3 contact (Fig. 6), but the northern outflow channel floor is not. Thus, if there was liquid within sub-basin 3 when the Lethe return flow breached zone 4, this liquid would have created a higher base level than if Lethe had discharged into an empty basin.

To accommodate these observations we propose the following model. Flow from Lethe Vallis created a temporary filled basin in zones 3 and 4, which then over-spilled via the northern outflow branch. We suggest that over-spill to the east occurred soon after and that the two branches were active simultaneously for at least a short period. This was probably because the northern outflow channel could not contain the discharge from Lethe and the eastern outflow channel was incised by the excess discharge. Temporary damming within the northern channel might also provide a mechanism. We suggest that, for an as-yet unknown reason, the eastern outflow soon dominated the northern. As discharge dropped, the system became less erosive and the fluid level in sub-basin 3 rose, possibly inundating the distal part of Lethe Vallis once again but not completely covering the streamlined islands in the distributary systems. However, sub-basin 3 eventually over-spilled and drained, and the drop in base level led to increased erosivity within Lethe Vallis, especially in the eastern branch. The steep sides of the streamlined islands reflect this later stage of downcutting. Knickpoint retreat as a result of the base-level drop deepened the eastern branch at the expense of the northern, beheading the northern valley completely as the knickpoint migrated upstream.

Finally, there is a hint that the dominant channel within the eastern distributary system continues some way into sub-basin 3. Whether this represents an evolution of the formative stage of the flooding or later incision by continued flow through Lethe is unknown. It is also unknown where material eroded from the Lethe channel has been deposited. Although there seems to be some evidence for deposition within the Lethe system, we can only assume that the majority of this material now occupies one of the downstream basins or has been transported even further east into Elysium Planitia.

Evolution of the basin-channel sequence

The overall topography, planform geomorphology and landform assemblage clearly demonstrate that Lethe Vallis evolved as a flood channel. The horizontal margins and large ratio of wetted perimeter to channel length of Lethe Basin (or at least zone 2a) suggest that this part of the system was temporarily a filled basin. Overall, the planform of the highstand and stepped topography of the long profile suggest that the Lethe Vallis system represents a series of basins that formed by filling, over-spilling and draining.

The shallow, concave profile of the upper reaches of Lethe Vallis and the morphologically poorly defined channel within zone 2a, suggests a low-energy flow. The profile appears similar to the graded profile of a terrestrial equilibrium river and, perhaps, indicates a continuous flow over a protracted period of time. We note that the initial spill event here probably occurred within the now-abandoned northern channel and that flow around the southern part of the Lethe Basin was more likely to be quiescent before it too over-spilled, about 10 km downstream of what is now cataract 1. We suggest that zone 2a represents the most likely reach of Lethe Vallis to be incised into an alluvial, rather than bedrock, substrate.

In contrast, the convex steps of the lower reaches instead point to rejuvenations of flow and changes in base level. The choke-points within the system that delineate the different geomorphic zones, and their association with the downstream terminations of inner channels forming below cataracts, strongly suggests that Lethe Vallis links basins that were once fluid-filled. Furthermore, this morphology suggests that they filled and over-spilled as a result of the continued inflow from the main Western Elysium Basin. The hanging channels and abandoned northern termination of Lethe demonstrate that the flow path changed over time. We propose the following evolutionary scheme to describe how the system developed, with Figure 20 showing how the flood highstand might have evolved.

- As the main Western Elysium Basin filled (Fig. 20a), fluid levels overtopped a topographical divide in the SE, which quickly became a

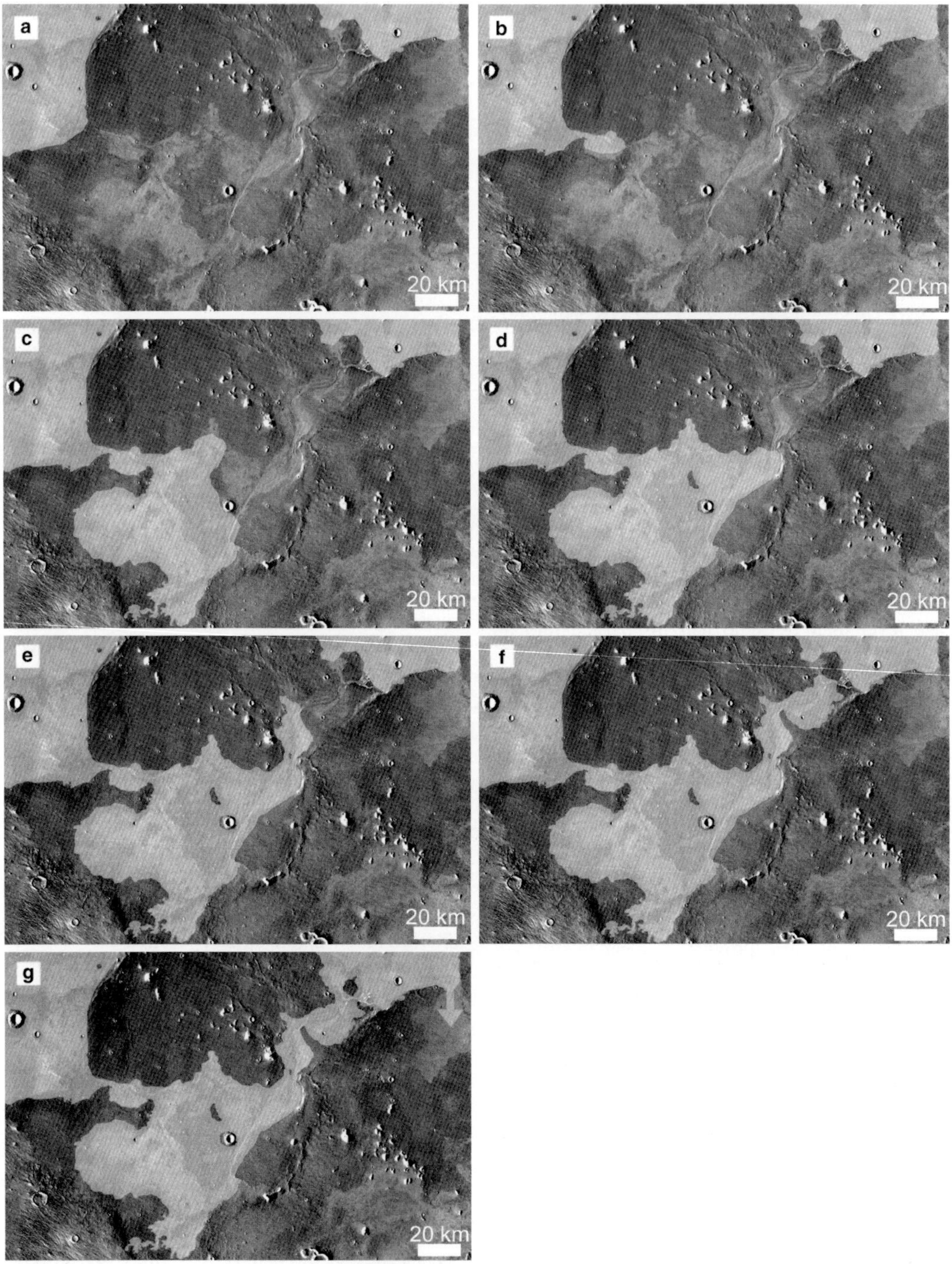

Fig. 20. Fill and spill sequence for the Lethe system. The white shaded area shows how the highstand of the floods evolved over time. Note that we assume here that sub-basin 3, to the NE, was filled before Lethe formed. We do not attempt to show how the actual water level changed with time because this cannot be reliably inferred from the existing topography. The background image is a mosaic of THEMIS daytime thermal infrared images. Image credit: NASA/JPL/ASU.

spillway into zone 1, upstream of what is now the Lethe Basin.

- The small area of zone 1 filled quickly (Fig. 20b) and over-spilled to the east into zone 2. From this point onwards there was a continuous (although not necessarily uniform) supply of liquid from the main Western Elysium Basin. The sloping profiles of both the north and south contacts suggest that zone 1 was not a long-lived standing body of quiescent fluid, but was initially a basin-wide, bankfull flow. Overflow and outflow, with retreat to the current erosional channel, occurred after the first influx.
- The large spatial extent of zone 2 (*c.* 4800 km^2) meant that it filled slowly and became a large reservoir of fluid containing several tens of cubic kilometres of liquid when full. The east-trending abandoned branch of Lethe Vallis (shown by the dashed line in Fig. 4) initially connected the western parts of the zone to the eastern part. The nearly horizontal profile of the contacts in zone 2 (especially zone 2a) demonstrate that this part of the system was a standing body of liquid that was in gravitational equilibrium. Zone 2a filled before zone 2b (Fig. 20c) and there was initially a topographical divide between zone 2a and 2b that is now marked by a large impact crater and a small outlier of older material (just west of profile 4 in Fig. 4). The abandoned northern channel in zone 2a marks the original path of flow between zone 2a and zone 2b. Thus, zone 2b may have been in the process of filling while zone 2a was mainly a standing body of liquid.
- Zone 2a over-spilled about 5–10 km downstream of the current position of cataract 1 (Fig. 20d). The cataract migrated headwards following the breach until it reached its current position. As zone 2b filled and in turn over-spilled into zone 3 (Fig. 20e), the northern branch of Lethe within zone 2 became abandoned as the liquid level slowly dropped and the flow was instead routed to the south of the topographical bulge that now occupies the centre of the Lethe Basin. It is likely that the upstream part of zone 2b ceased to be a fluid-filled basin at this point, and that it was instead dominated by an erosional channel, but the lower part of zone 2b remained fluid-filled. The initial breach was a few kilometres downstream of cataract 2, which, as for cataract 1, migrated headward as the flow continued.
- Zone 3 filled and over-spilled into zone 4 (Fig. 20f). Zones 3 and 4 may, instead, represent a single basin, given that their highstand contacts are topographically similar.
- The horizontal topographical profiles of their contacts suggest that zones 3 and 4 were standing bodies of liquid. However, continued inflow from Lethe meant that these basins soon over-spilled into sub-basin 3 through the northern outlet. We infer that sub-basin 3 was liquid-filled because its northern inlets were topographically higher and closer to source than the ingress from Lethe Vallis.
- As described above, Lethe further added to the supply of liquid to sub-basin 3 leading to an increase in fluid level in the sub-basin. Continued inflow from Lethe meant that zone 4 over-spilled at a second point, forming the eastern outlet from Lethe Vallis into sub-basin 3.
- The increased liquid supply from Lethe caused sub-basin 3 to over-spill catastrophically to the SE (Fig. 20e), leading to a drop in base level. This fall in base level triggered a complete channel avulsion in Lethe Vallis, changing the flow path from the northern outlet into the eastern outlet and created a fluvial hanging valley in the northern wall of the distal part of Lethe Vallis. This also promoted erosion in the anastomosing channels between the streamlined islands. Because the contact elevation in sub-basin 3 is at about the same level as the flow at the base of the channel above cataract 3, we speculate further that the formation of cataracts 3 and 4 was also related to this drop in base level.
- Finally, the supply of liquid to the inlet of Lethe Vallis dropped and the whole channel became relict. This was either because of a loss of the supply to the Western Elysium Basin itself due to the cessation of the Athabascan flooding event or because of a sudden drop in liquid level caused by drainage of the main basin to the SW. We suggest that zone 2a of the Lethe Basin was the last fluid-filled part of the system, although the absence of any distinct channel just upstream of cataract 2 in zone 2b suggests that a small standing body of water might have also been present here.

The data presented indicate that flood-flow routing comprised the filling of a series of basins, each of which overfilled and then spilled into the next adjacent basin by the carving of a spillover channel. The filling and spilling of flood flows through a complex set of mini-basins connected by channel segments provides clear evidence of the complexity of flow routing in the wider Athabasca–Elysium–Lethe system. The geomorphology requires flood flows to have sequentially overfilled adjacent basins one after another. Moreover because the overall fill and spill occurs at the SE margin of the Elysium Basin, it requires that the Western Elysium Basin must have been (at least transiently) filled with floodwaters forming an extensive lake.

Is the Lethe system filled with recent flood lavas?

Although, as described above, the nature of the platy-ridged-polygonized material filling is still debated, until recently both the lava and fluvial–periglacial interpretations held that the channels linking the basins in the Elysium Complex were all carved by fluvial floods. Recently, though, Jaeger *et al.* (2010) speculated that lava interpreted to have formed the uppermost surface textures could also have carved the channels. This leads to three possible models for the formation of the Lethe Vallis: (i) very fluid lava carved Lethe and formed the platy-polygonized terrain (as suggested by Jaeger *et al.* 2009); (ii) water carved the Lethe channels and basins, and created the various erosional landforms; later, fluid lavas reoccupied the channels and formed the platy-ridged-polygonized terrain (e.g. Keszthelyi *et al.* 2000; Berman & Hartmann 2002; Plescia 2003; Keszthelyi *et al.* 2004*b*; Jaeger *et al.* 2007) or (iii) water carved the channels, and debris or ice in the flows was responsible for the formation of the platy-ridged-polygonized terrain; the current surface is either a sublimation lag over an ice-rich layer (Murray *et al.* 2005) or a mostly desiccated post-periglacial environment (Rice *et al.* 2002; Page 2007; Balme & Gallagher 2009; Balme *et al.* 2009), with the source for the ground ice being the floods that carved the channels.

The erosion of Lethe Vallis by flowing lava is unlikely. The low gradient and shallow depth of the channel mean that flow of a viscous, non-Newtonian fluid such as lava will be even slower than that of a low-viscosity, Newtonian medium such as water. For any reasonable viscosity of lava (even as low as 50–500 Pa s as suggested by Vaucher *et al.* 2009*b*), flow in Lethe Vallis would have been laminar, not turbulent and unable to carve deep channels or create streamlined bedrock erosional remnant landforms. Likewise, the Lethe system's evolution as a 'fill and spill' system, and the presence of cataracts, knickpoints and terraces, also argues for it being a fluvial, as opposed to a lava, system. In fact, Lethe hosts direct analogues of many of the landforms indicative of catastrophic fluvial flooding seen on Earth (Carling *et al.* 2009*a*, *b*). This strongly argues against the purely volcanic model.

The question of whether Lethe is a fluvial system that was then filled by later lavas can also be examined using the observations of landform and topography presented here. First, the surface textures within the highstand contact are consistent throughout Lethe. Thus, if they represent lava then the lava must essentially have behaved in the same way as the previous fluvial floods – filling and spilling one basin at a time through the system, and occupying at least as great an extent as the prior fluvial flood(s). Given the presence of pre-existing spillways, channels and cataracts – especially within the Lethe Basin – this seems impossible, because the fluid would have had to have ponded behind non-existent divides, previously breached by fluvial erosion. Secondly, the platy-ridged textures extend onto the tops of streamlined islands and terraces, yet the cataracts and parts of the channel floor appear to be clear of this material. There certainly does not appear to be evidence for frozen 'lava falls'. This suggests that the rubbly texture represents deposition of material at the flow margins where flow is sluggish or where surface material (ice-rich debris in the case of a fluvial interpretation or lava crust in a volcanic one) 'grounds' as the flow shallows. This seems possible only if lava can act like water – freezing solid and persisting in some places, but remaining liquid and ultimately disappearing completely in others. Thirdly, post-depositional modification of the platy-ridged surfaces within the channel rules out a simple 'fluvial then lava' model. Multiple fluvial–lava–fluvial episodes are required which, although they cannot be ruled out, do not find obvious supporting evidence in the morphology of the system.

The third hypothesis, that fluvial flooding is responsible for both channel erosion and the platy-ridged-polygonized terrain, is consistent with many observations. The cataract inner-channel systems are very similar to those found on Earth (see, e.g. figs 5.11 & 5.13 of Baker 2009), although on a smaller scale. The headward erosion in plates to form 'arching', as seen in Figure 6, is common in terrestrial sea ice (Sodhi 1977). Sorted patterned-ground landforms, which require an ice-rich substrate, are seen on the banks of Lethe Vallis (Balme *et al.* 2009) and are replicated in terrestrial cold regions. If the chevron and crescentic forms described above are, indeed, alluvial forms then they also suggest that there was no later-stage lava inundation, as these forms exist between rubbly-ridged terrains yet are not themselves superposed by a texture that could be interpreted as lava. Instead, they are gouged by a series of flow-parallel furrows and ridges that could be 'tool marks' caused by the scraping of clasts or ice floes carried by the flood. Finally, the topographical profiles of the contacts and channel thalweg are also consistent with a fluvial 'fill and spill' scenario, implying that the platy-ridged-polygonized texture represents the extents of fluvial, not lava, flooding.

Summary

The data presented here (summarized in Fig. 21) demonstrate that catastrophic fluvial flooding

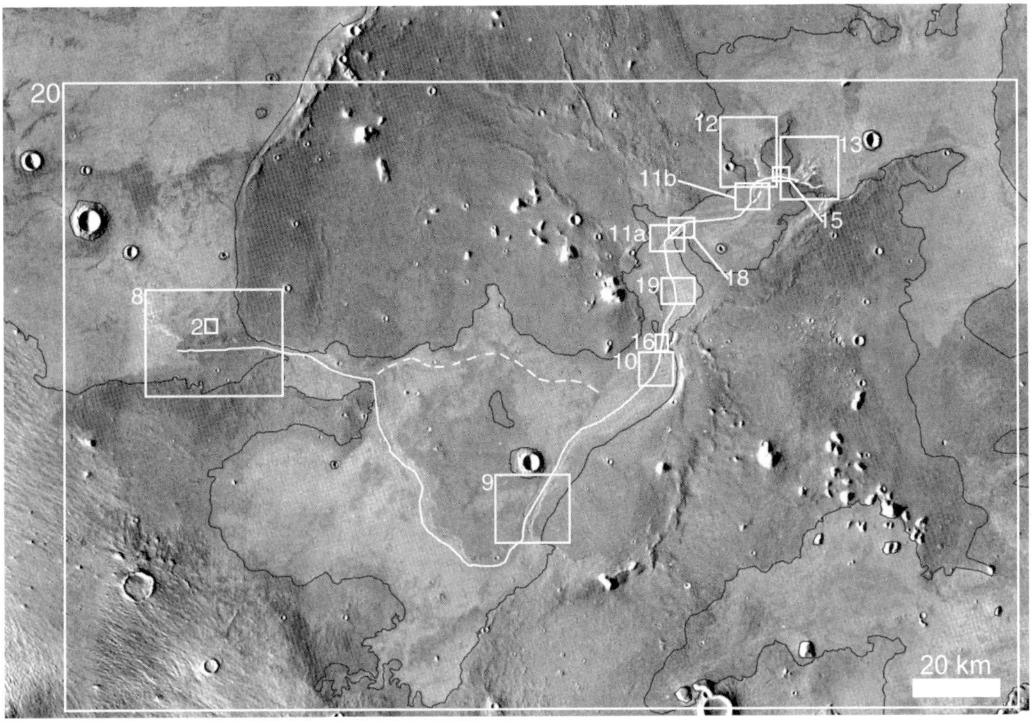

Fig. 21. Geographical context of the figures described in this chapter. The extent of this figure is about the same as that of Figures 4 & 5. The white boxes with labels indicate the extents of the figures. The background image is a mosaic of THEMIS daytime thermal infrared images. Image credit: NASA/JPL/ASU.

formed the morphology of Lethe Vallis. Despite being a smaller, lower-gradient system than many Martian outflow channels, it contains a very similar assemblage of landforms, including streamlined islands, hanging channels, cataracts and possible fluvial dunes or antidunes. In particular, the array of landforms seen within Lethe is almost identical to that within Athabasca Vallis, the order-of-magnitude larger outflow channel that was the primary spillway for floods from the Cerberus Fossae fracture system.

Lethe Vallis represents a linkage between the over-spilling Main Western Elysium Basin and a topographically lower sub-basin. The Lethe system evolved as a series of basins that filled, over-spilled and then catastrophically drained into the next, topographically lower, basin. At least three such 'fill and spill' events occurred as the flow front progressed, following the regional slope. The channel that formed within the linked basins as the system evolved had a formative discharge of 1×10^4–5×10^4 m^3 s^{-1}. It is likely that after the formative period of 10–50 days there was a period of more quiescent, lower discharge through Lethe Vallis. The topographical data and the morphology of the distributary systems suggest that Lethe debouched into a basin already filled with water. Thus, the evolution of Lethe Vallis seems to have been shaped both by catastrophic fluvial erosion in the 'fill and spill' period and by later base-level changes as the basin in which it terminated drained.

The conclusion that Lethe is a flood-carved 'fill and spill' system has an important wider impact, because it demonstrates that the Main Western Elysium Basin must have contained a substantial lake that persisted for at least as long as the time required for Lethe to form. We speculate that this lake, which must have been at least 500 km across, formed a reservoir that supplied Lethe Vallis with water for at least several weeks and, perhaps, even longer.

The morphology and topography of Lethe Vallis are inconsistent with the interpretation that lava flows created the entire system. Although scenarios can be proposed in which multiple episodes of fluvial and low-viscosity lava floods occurred, the simplest explanation, and the one that most closely matches the morphology of the observed terrain, is one of fluvial flooding only. It remains to be explained exactly what processes formed the platy-ridged-polygonized surfaces, but we speculate that they were created by a combination of primary processes during the formation of Lethe and secondary processes by which the ice-rich deposits were modified during the millions of years of exposure to the Martian climate after deposition.

There are few, if any, terrestrial analogues that can be used to explain the fine-scale morphologies generated by a Martian megaflood. The thin atmosphere and extreme cold are particularly difficult to account for and might play a key role in the final morphology of the deposits. In addition to simple fluvial erosion and deposition that occurs in terrestrial megafloods, other processes might have played a role including: the formation and motion of sediment-rich or clast-rich ice-rafts; the loss of water by freezing and boiling to form hyper-concentrated flows or non-Newtonian debris flows; and the formation of massive and crystalline ice within the turbulent floods. Post-depositional modification processes (including thaw, sublimation and aeolian deflation) are perhaps more easy to find terrestrial analogues for, especially in cold-climate patterned grounds on Earth. Understanding the relationships between primary and secondary morphogenesis in Martian megaflood systems remains a challenge.

We thank NASA's Mars Reconnaissance Orbiter (MRO) HiRISE and context camera (CTX) teams for making the imagery data available. This work was funded by the UK Science and Technology Facilities Council (STFC) through an Aurora Fellowship (M. R. Balme), and STFC Astronomy grants ST/F003099 (S. Gupta) and PP/C502622/1 (J. Murray). We thank D. Page and S. Conway for helpful discussions. The comments of two anonymous reviewers were very helpful and improved the manuscript.

References

BAKER, V. R. 2001. Water and the Martian landscape. *Nature*, **412**, 228–236.

BAKER, V. R. 2009. Channeled Scabland morphology. *In*: BURR, D. M., CARLING, P. A. & BAKER, V. R. (eds) *Megaflooding on Earth and Mars*. Cambridge University Press, Cambridge, 65–77.

BALME, M. R. & GALLAGHER, C. 2009. An equatorial periglacial landscape on Mars. *Earth and Planetary Science Letters*, **285**, 1–15.

BALME, M. R., GALLAGHER, C., PAGE, D. P., MURRAY, J. B. & MULLER, J.-P. 2009. Sorted stone circles in Elysium Planitia, Mars: implications for recent Martian climate. *Icarus*, **200**, 30–38.

BALME, M. R., GALLAGHER, C., PAGE, D. P., MURRAY, J. B., MULLER, J.-P. & KIM, J.-R. 2010. The western Elysium Planitia palaeolake. *In*: CABROL, N. & GRIN, E. A. (eds) *Lakes on Mars*. Elsevier, Amsterdam, 275–306.

BENITO, G. 1997. Energy expenditure and geomorphic work of the Cataclysmic Missoula flooding in the Columbia River Gorge, USA. *Earth Surface Processes and Landforms*, **11**, 457–472.

BERMAN, D. C. & HARTMANN, W. K. 2002. Recent fluvial, volcanic and tectonic activity on the Cerberus plains of Mars. *Icarus*, **159**, 1–17.

BRACKENRIDGE, G. R. 1993. Modern shelf ice, Equatorial Aeolis Quadrangle, Mars. *In*: *Proceedings of the 24th Annual Lunar and Planetary Science Conference, March 15–19, 1993, League City, Texas*. Lunar and Planetary Institute, Houston, TX, Abstract 175.

BURR, D. M., CARLING, P. A., BEYER, R. A. & LANCASTER, N. 2004. Flood formed Dunes in Athabasca Vallis, Mars: morphology, modeling and implications. *Icarus*, **171**, 68–81.

BURR, D. M., MCEWEN, A. S. & SAKIMOTO, S. E. H. 2002. Recent aqueous floods from the Cerberus Fossae, Mars. *Geophysical Research Letters*, **29**, 1013, doi: 10.1029/2001GL013345.

CARLING, P. A., BURR, D. M., JOHNSON, T. F. & BRENNAND, T. A. 2009*a*. A review of open-channel Megaflood depostional landforms on Earth and Mars. *In*: BURR, D. M., CARLING, P. A. & BAKER, V. R. (eds) *Megaflooding on Earth and Mars*. Cambridge University Press, Cambridge, 33–49.

CARLING, P. A., HERGET, J., LANZ, J. K., RICHARDSON, K. & PACIFICI, A. 2009*b*. Channel-scale erosional bedforms in bedrock and in loose granular material: character, processes and implications. *In*: BURR, D. M., CARLING, P. A. & BAKER, V. R. (eds) *Megaflooding on Earth and Mars*. Cambridge University Press, Cambridge, 13–31.

HARTMANN, W. K. & BERMAN, D. C. 2000. Elysium Planitia Lava flows: crater count chronology and geological implications. *Journal of Geophysical Research*, **105**, 15,011–15,025.

JAEGER, W. L., KESZTHELYI, L. P., MCEWEN, A. S., DUNDAS, C. M. & RUSSELL, P. S. 2007. Athabasca Valles, Mars: a lava-draped channel system. *Science*, **317**, 1709–1711.

JAEGER, W. L., KESZTHELYI, L. P. *ET AL.* 2010. Emplacement of the youngest flood Lava on Mars: a short, turbulent story. *Icarus*, **205**, 230–243.

KEHEW, A. E., LORD, M. L., KOSLOWSKI, A. L. & FISHER, T. D. 2009. Proglacial megaflooding along the margins of the Laurentide Ice Sheet. *In*: BURR, D. M., CARLING, P. A. & BAKER, V. R. (eds) *Megaflooding on Earth and Mars*. Cambridge University Press, Cambridge, 104–127.

KESZTHELYI, L., MCEWEN, A. S. & THORDARSON, T. 2000. Terrestrial analogs and thermal models for Martian flood lavas. *Journal of Geophysical Research*, **105**, 15,027–15,050.

KESZTHELYI, L., BURR, D. M. & MCEWEN, A. S. 2004*a*. Geomorphologic/thermophysical mapping of the Athabasca Region, Mars, using Themis Infrared Imaging. *In*: *Proceedings of the 35th Annual Lunar and Planetary Science Conference, March 15–19, 2004, League City, Texas*. Lunar and Planetary Institute, Houston, TX, Abstract 1657.

KESZTHELYI, L., THORDARSON, T., MCEWEN, A. S., HAACK, H., GUILBARD, M.-N., SELF, S. & ROSSI, M. J. 2004*b*. Icelandic analogs to Martian flood lavas. *Geochemistry, Geophysics, Geosystems*, **5**, Q11014, doi: 10.1029/2004GC000758.

KLEINHANS, M. G. 2005. Flow discharge and sediment transport models for estimating a minimum timescale of hydrological activity and channel and delta formation on Mars. *Journal of Geophysical Research*, **110**, E12003, doi: 10.1029/2005JE002521.

McEwen, A. S., Edgett, K. S., Malin, M. C., Keszthelyi, L. & Lanagan, P. D. 1998. Mars global surveyor camera tests the Elysium Basin controversy: it's lava, not lake sediments. *Geological Society of America Annual Meeting Abstracts with Programs*, **30**, A402.

Morton, R. A. 1978. Large-scale rhomboid bedforms and sedimentary structures associated with hurricane washover. *Sedimentology*, **25**, 183–204.

Murray, J. B., Muller, J.-P. & The HRSC CoInvestigator Team 2005. Evidence from the Mars express high resolution stereo camera for a frozen sea close to Mars' Equator. *Nature*, **434**, 352–355.

Neumann, G. A., Lemoine, F. G., Smith, D. E. & Zuber, M. T. 2003. The Mars Orbiter Laser Altimeter Archive: final precision experiment data record release and status of radiometry. *In*: *Proceedings of the 34th Annual Lunar and Planetary Science Conference, March 17–21, 2004, League City, Texas*. Lunar and Planetary Institute, Houston, TX, Abstract 1978.

Page, D. P. 2007. Recent low-latitude freeze–thaw on Mars. *Icarus*, **189**, 83–117.

Plescia, J. B. 1990. Recent flood lavas in the Elysium region of Mars. *Icarus*, **88**, 465–490.

Plescia, J. B. 2003. Cerberus Fossae, Elysium, Mars: a source for lava and water. *Icarus*, **164**, 79–95.

Rhoads, B. L. 1987. Stream power terminology. *The Professional Geographer*, **39**, 189–195.

Rice, J. W., Parker, T. J., Russell, A. J. & Knudsen, O. 2002. Morphology of fresh outflow channel deposits on Mars. *In*: *Proceedings of the 33rd Annual Lunar and Planetary Science Conference, March 11–15, 2002, Houston, Texas*. Lunar and Planetary Institute, Houston, TX, Abstract 2026.

Scott, D. H., Rice, J. W. & Dohm, J. M. 1991. Palaeolakes and Lacustrine Basins on Mars. *In*: *Proceedings of the 22nd Annual Lunar and Planetary Science Conference, March 18–22, 1991, League City, Texas*. Lunar and Planetary Institute, Houston, TX, 1203–1204.

Smith, D. E., Zuber, M. T. *et al.* 2001. Mars Orbiter Laser Altimeter: experiment summary after the first year of global mapping of Mars. *Journal of Geophysical Research*, **106**, 23,689–23,722.

Sodhi, D. S. 1977. *Ice Arching and the Drift of Pack Ice Through Restricted Channels*. United States Army Cold Regions Research and Engineering Laboratory (CRREL) Report, **77-18**.

Vaucher, J., Baratoux, D., Mangold, N., Pinet, P., Kurita, K. & Grégoire, M. 2009*a*. The volcanic history of central Elysium Planitia: implications for Martian magmatism. *Icarus*, **204**, 418–442.

Vaucher, J., Baratoux, D., Toplis, M. J., Pinet, P., Mangold, N. & Kurita, K. 2009*b*. The Morphologies of volcanic landforms at central Elysium Planitia: evidence for recent and fluid lavas on mars. *Icarus*, **200**, 39–51.

Warner, N., Gupta, S., Muller, J.-P., Kim, J.-R. & Lin, S.-Y. 2009. A refined chronology of catastrophic outflow events in ares vallis, Mars. *Earth and Planetary Science Letters*, **288**, 58–69.

Wilson, L., Ghatan, G. L., Head, J. W. & Mitchell, K. L. 2004. Mars outflow channels: a reappraisal of the estimation of water flow velocities from water depths, regional slopes, and channel floor properties. *Journal of Geophysical Research*, **109**, E09003, doi: 10.1029/2004JE002281.

Wilson, L., Bargery, A. S. & Burr, D. M. 2009. Dynamics of fluid flow in martian outflow channels. *In*: Burr, D. M., Carling, P. A. & Baker, V. R. (eds) *Megaflooding on Earth and Mars*. Cambridge University Press, Cambridge, 290–311.

Woodford, A. O. 1935. Rhomboid ripple marks. *American Journal of Science*, **29**, 518–525.

Geologically recent water flow inferred in channel systems in the NE Sulci Gordii region, Mars

M. C. TOWNER[1]*, C. EAKIN[1], S. J. CONWAY[2] & S. HARRISON[2]

[1]*Department of Earth Sciences and Engineering, Impacts and Astromaterials Research Centre, South Kensington Campus, Imperial College, London SW7 2AZ, UK*

[2]*Planetary Surfaces Research Team, CEPSAR, Open University, Walton Hall, Milton Keynes MK7 6AA, UK*

**Corresponding author (e-mail: m.towner@ic.ac.uk)*

Abstract: A series of fluid-carved channels in the Sulci Gordii region of Mars were investigated. Numerous channel networks exist in Sulci Gordii, part of the Olympus Mons aureole, and this area comprises some of the youngest volcanic terrain on Mars. The channels ranged in length from 43 to 155 km, with widths of 128–288 m. The morphology of the channels was analysed assuming both lava and water as possible agents. For three of the four channels studied, water appears to be the likely agent, while one channel is probably lava-formed. For the water-formed channels, discharge rates were estimated at 8000–36 000 $m^3 s^{-1}$. The lava channel was probably formed from short-lived episodic activity by a low-viscosity lava. The age of the channels and surrounding area was estimated using crater counting to be 100 Ma. Water has appeared to have flowed for almost 150 km under the climatic conditions at this time. There is some evidence for later tectonic activity, possibly as recent as 10 Ma, but crater-dating accuracy was limited by the lack of high-resolution images of some areas. Sulci Gordii is therefore a dynamic site with evidence of hydrological and volcanic activity extending into the recent geological past.

The Tharsis region dominates the western hemisphere of Mars, containing volcanic constructs interpreted as shield volcanoes (Masursky *et al.* 1972; McCauley *et al.* 1972; Carr 1973), and surrounded by grabens and wrinkle ridges (see Mouginis-Mark *et al.* 1992 for a review; see also Solomon & Head 1982; Smith *et al.* 1999; Zuber *et al.* 2000). The Tharsis region of Mars has been geologically mapped based on *Viking Orbiter* data (Scott 1981).

The Tharsis volcanoes were formed by extensive volcanic activity that started in the Noachian Period, and the last major clearly recorded stage of tectonic activity was during the Middle–Late Amazonian, based on crater size–frequency distributions (Neukum & Hiller 1981; Anderson *et al.* 2001). More recent activity appears to have been very localized and episodic (Neukum *et al.* 2004). These smaller, more local, features may provide valuable information on the more recent distribution of magma. In addition, evidence of past fluvial activity is also seen in this region, which has been inferred to be a result of the volcanic activity (Mouginis-Mark & Christensen 2005).

The Sulci Gordii feature is a relatively small, curved aureole deposit approximately 400 km due east of the summit of Olympus Mons (see inset Fig. 1) in the Tharsis region; it consists of ridges and incised valleys, with evidence of fluvial and tectonic activity (Mouginis-Mark & Christensen 2005).

We investigate in detail a series of channels in Sulci Gordii, where the interaction of apparently liquid-cut channel systems with tectonic activity provides an interesting opportunity for morphostratigraphic analysis. An area to be studied was defined that contained several channel networks and strong evidence of tectonic activity in the form of a large graben. Figure 1 is a context image giving an overview of the northern part of Sulci Gordii, highlighting the area of study with a rectangle. The area is approximately 120 km wide by 250 km long, and is centred at 233.3°E, 19.3°N.

Methodology

A wide range of remote-sensing images of the Martian surface was used in the production of a detailed map of the study area. A geological map of the area was created to allow for an in-depth examination of the surface features, such as outlining the morphology of the channels and investigating the relationships between structures (Fig. 2).

Relevant images of the area are listed in Table 1. Some coverage is offered by the *Mars Global Surveyor* Mars Orbiter Camera (MOC) (Malin

From: Balme, M. R., Bargery, A. S., Gallagher, C. J. & Gupta, S. (eds) *Martian Geomorphology*. Geological Society, London, Special Publications, **356**, 229–256.
DOI: 10.1144/SP356.12 0305-8719/11/$15.00

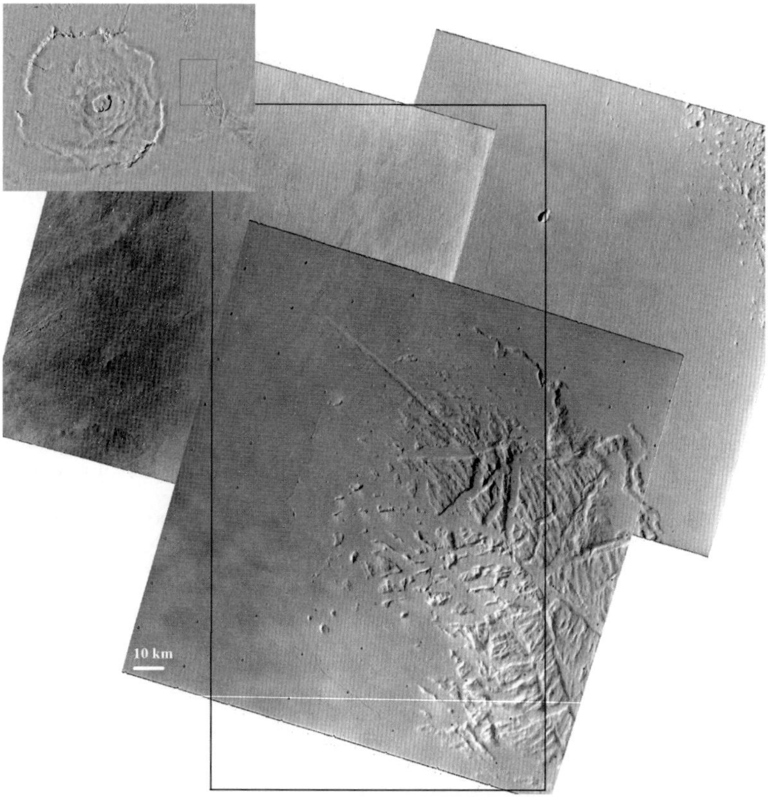

Fig. 1. Inset shows MOLA shaded relief of Olympus Mons showing the location of Sulci Gordii. The study area is outlined by the box on the main image. Background images for study area are *Viking* images F046b37, F046b46 and F046b48. Image credit: NASA; see prelim viii for acronym definitions.

et al. 1992) and the *Mars Reconnaissance Orbiter* Context Imager (CTX) (Malin *et al.* (2007)) at the highest resolution available, typically 3 and 6 m per pixel, respectively. However, the majority of mapping was completed using *Mars Odyssey* Thermal Emission Imaging System (THEMIS) visible-wavelength images (Christensen *et al.* 2004), as they covered the broadest region at a moderate resolution (about 20 m per pixel). *Viking* and wide-angle MOC images covered the largest continuous area, providing a regional overview. Infrared THEMIS data were used to infer the relative thermal properties of units, although the resolution of 100 m per pixel precludes detailed study. Images were processed initially using the USGS Integrated Software for Imagers and Spectrometers (ISIS) (http://isis.astrogeology.usgs.gov), and projected using ArcGIS 9.2 software from Environmental Systems Research Institute, Inc. (ESRI). A simple sinusoidal projection based on the Mars geoid was used. Superposition relationships for channels, faults and impact structures were used to provide insight into the sequence of emplacement events. Comparison with terrestrial morphology (primarily basaltic flow features) gives information concerning the likely modes of putative lava eruption and flow.

In addition to imagery, the *Mars Global Surveyor* Mars Orbiter Laser Altimeter (MOLA) (Zuber *et al.* (1992)) provides elevation data that can be utilized to model the topography of the area, at about 450 m resolution. MOLA measurements were also used to generate profiles of the gradient along the channels. As well as the long profile, the bankfull widths and depths of each of the channels were measured at intervals along its length. Around 45 cross-sections were gathered for each channel, the spacing between which was determined by the total length of the channel and the quality of the image at measurement locations. The width was directly measured, where the definition of the channel edge was chosen as the initial appearance of deviation from the surrounding planar surface. The depth was determined using shadow length

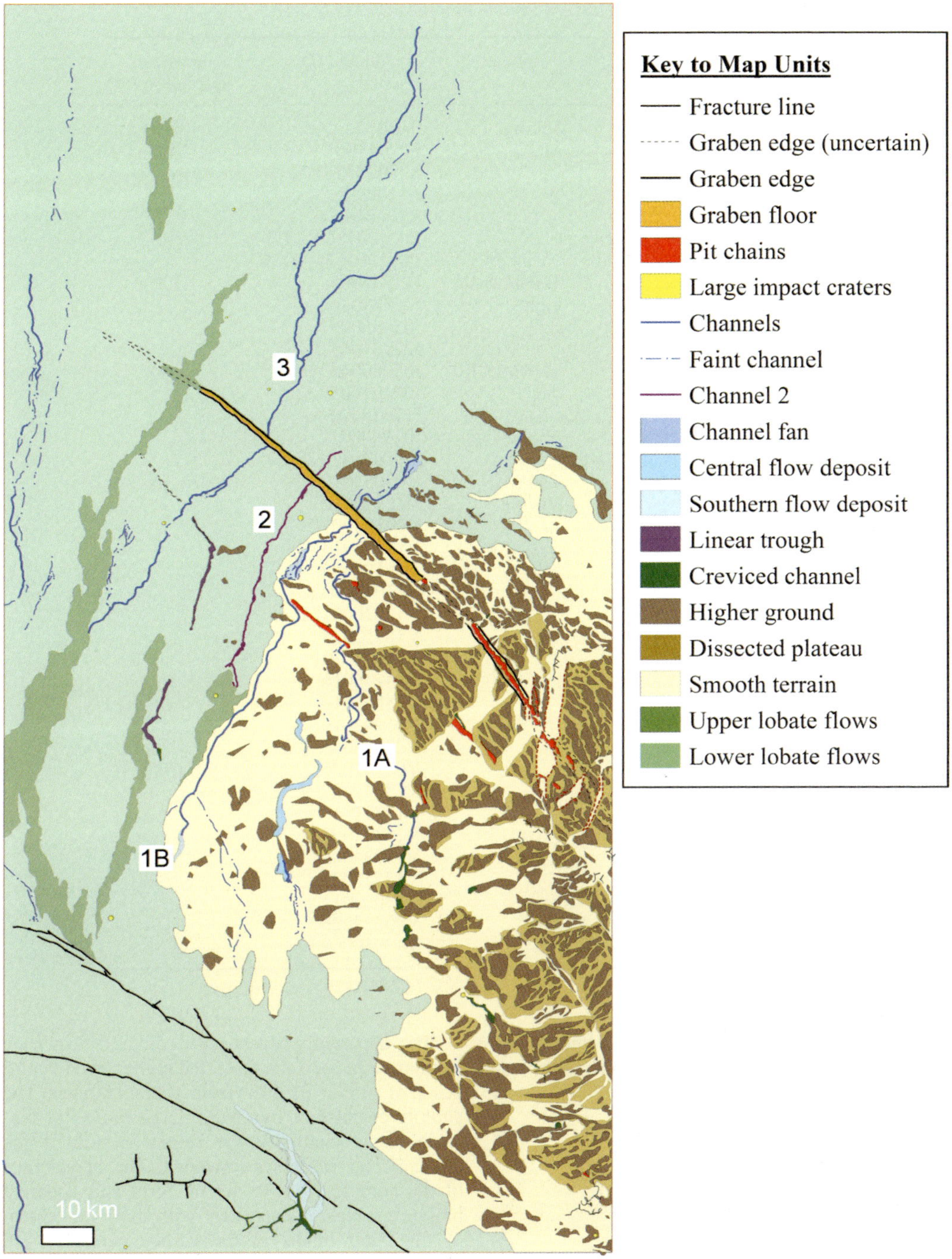

Fig. 2. Map of Sulci Gordii (centred at 126.7°W, 19.3°N) showing the general spatial relationship of units and the location of each of the three channel systems. Numbers 1–3 on the map represent the number assigned to each channel.

Table 1. *A list of all orbital images used in mapping the study area*

Orbiter	Imager	Image ID	Average resolution (m)
Viking Orbiter 2	Viking	046B37	135
Mars Global Surveyor	MOC	S0701786	260
		S0502035	
		E1100996	6
		R1003495	5
Mars Reconnaissance Orbiter	CTX	P02_001867_1979	5
		P07_003634_1955	
Mars Odyssey	THEMIS IR night	I11245005	100
		I06939016	
		I05466009	
		I22901005	
	THEMIS IR day	I05060044	100
		I02401005	
		I17628020	
		I02064002	
		I09042011	
	THEMIS Visible	V09953012	18
		V12424012	
		V20124009	
		V19188014	
		V19500029	
		V13048009	
		V13310007	
		V18252023	
		V18876012	
		V14246016	
		V18564014	
		V25902019	
		V22258045	
		V02401006	
		V13647007	
		V12711010	
		V13335007	
		V13934009	
		V14558021	
		V13023011	
		V05422026	
		V13959009	

measurements in combination with the known sun incidence angle. Other properties such as the cross-sectional area could then be calculated from these measurements.

Observations

Study area

The area of study is shown in Figure 1. From this, a unit map, illustrating the different surface units and channel locations, is shown in Figure 2. This area has also been mapped previously at low resolution, based on *Viking Orbiter* images (Scott 1981; Tanaka *et al.* 1992).

The apparently oldest unit, coloured brown on the map (Fig. 2), is the dissected higher terrain composed of aureole deposits from Tharsis Montes. The aureole consists of two parts: large blocks or plateaus (with smaller linear ridges on top of the plateau) and smoother lowlands. The appearance of the aureole, with its lobate form and apparent pressure ridges, is consistent with mass movement of the outer flanks of the edifice (Lopes *et al.* 1982), which may include ice-lubricated gravitational spreading (Tanaka 1985) and multiple pyroclastic eruptions (Morris 1982). Cratering is light across the whole study area, indicating a relatively young age for all units.

Faulting has occurred within the aureole and a large graben, 1–2 km wide, can be seen trending

NW–SE across the study area. Surrounding the higher ground is smooth lowland terrain. Outside this terrain are plains composed of many lobate flows. The uppermost of these flows have been mapped as mid-green in Figure 2.

The general area surrounding Sulci Gordii is mapped here as lower lobate flows and is understood to be extensive lava plains (Scott 1981), being extensively covered in sinuous lava deposits. The plains are Amazonian in age (Dohm *et al.* 2008) and represent some of the youngest lava flows in the Tharsis region (Mouginis-Mark 1989).

The presence of slope streaks is indicative of dust mantling, primarily on the smooth terrain unit adjacent to the dissected plateaus. This is supported by the relatively low thermal inertias seen for this area (Mellon *et al.* 2000). In addition, some small aeolian ripples are seen in the graben trench, but the lobate flow units and much of the smooth terrain appears to have no significant dust coating.

These lava plains are cut by numerous channel systems. Mouginis-Mark (1989) interpreted the region between Olympus Mons and Ceraunius Fossae, including the study area, as a site of recent water release. Mouginis-Mark & Christensen (2005) demonstrated further evidence of water release and flooding within the same aureole segment covered by the study area. Furthermore, Dohm *et al.* (2008) presented the fractures, faults, graben and structurally controlled pit crater chains that intersect the young aureole deposits of Olympus Mons and other Tharsis shield volcanoes as evidence for recent tectonic activity on Mars. This makes Sulci Gordii an intriguing study area, displaying such recent morphogenetic activity. By looking at surface features such as channels, we hope to shed more light on the relative chronology of this activity, and discover more about the conditions and processes occurring on Mars at this time.

Channels

Within the designated study area, three main channel systems were studied in detail, as shown in Figure 2. The general down-slope trend of all the channels is towards the north, matching the current regional/local topography. This trend can be seen in Figure 3, which clearly demonstrates how the channels cross perpendicular to the current-day contour lines following the steepest gradient. The highest topography is located towards the south of the figure, and this is taken to be the source area of the channels. This source is supported by the study of streamlined islands in the channels, discussed in more detail later. We describe the channel referring to ‘proximal’ as near the source regions, and ‘distal’ at the terminal points.

We describe each channel in turn, highlighting relevant features.

Channel 1A. The course of Channel 1 A is through over 70 km of the Olympus Mons aureole materials. The channel has a mean width of 197 m and a mean depth of 24 m along its profile. The source area (233.7°E, 18.4°N) appears to be a linear trough structure that splits in two as shown in Figure 4. The channel continues on north of the area represented in Figure 4. For most of its length the channel is confined to its banks and rarely avulses. The channel is not anabranching, but there are locations where other fainter channels can be seen adjacent to the main path (for example, see Fig. 5). This observation suggests several scenarios: some change in the channel evolution with time, with avulsion hinting at either an extended period of activity or representing multiple events, or alternatively substantial unconfined flow eroding multiple channels, such that the largest channel is incised deeply as the flow declined.

There are also many breaks in the channel path such as at the top of Figure 5, where it appears as if the channel loses definition. These frequently occur next to the aureole blocks, so it appears likely that mass-wasting deposits sourced from weathered aureole material have been deposited over the channels, obscuring parts.

Also of interest is the location shown in Figure 6, where the channel is cross-cut by a linear depression extending from a circular pit. There the channel is also obscured by mass-wasting detritus for a length of approximately 2 km.

The terminus of the channel coincides with the major graben (Fig. 7). The channel becomes fainter but can be traced approaching the area where the major graben has later formed as increasingly disordered flow, with multiple braided, shallow channels. There are hints of intermittent or sheet flow on the plains on the other side of the graben in the form of very faint short traces that could be interpreted as several small, shallow channel segments. However, these are insufficiently well defined to confirm that this is a continuation of the same channel. Prior to intersecting the graben, channels 1A and 1B (discussed in the next subsection) approach each other. At their closest they are only 4 km apart.

At the bottom of Figure 7, a short section of levees can be seen along the edge of the channel. This is the only location on Channel 1A where such a feature is present.

Channel 1B. Channel 1B is within the same geological context as Channel 1A but 20 km further west. It has a total length in excess of 100 km. The morphology is similar to Channel 1A; there is one main well-defined channel, with smooth edges and

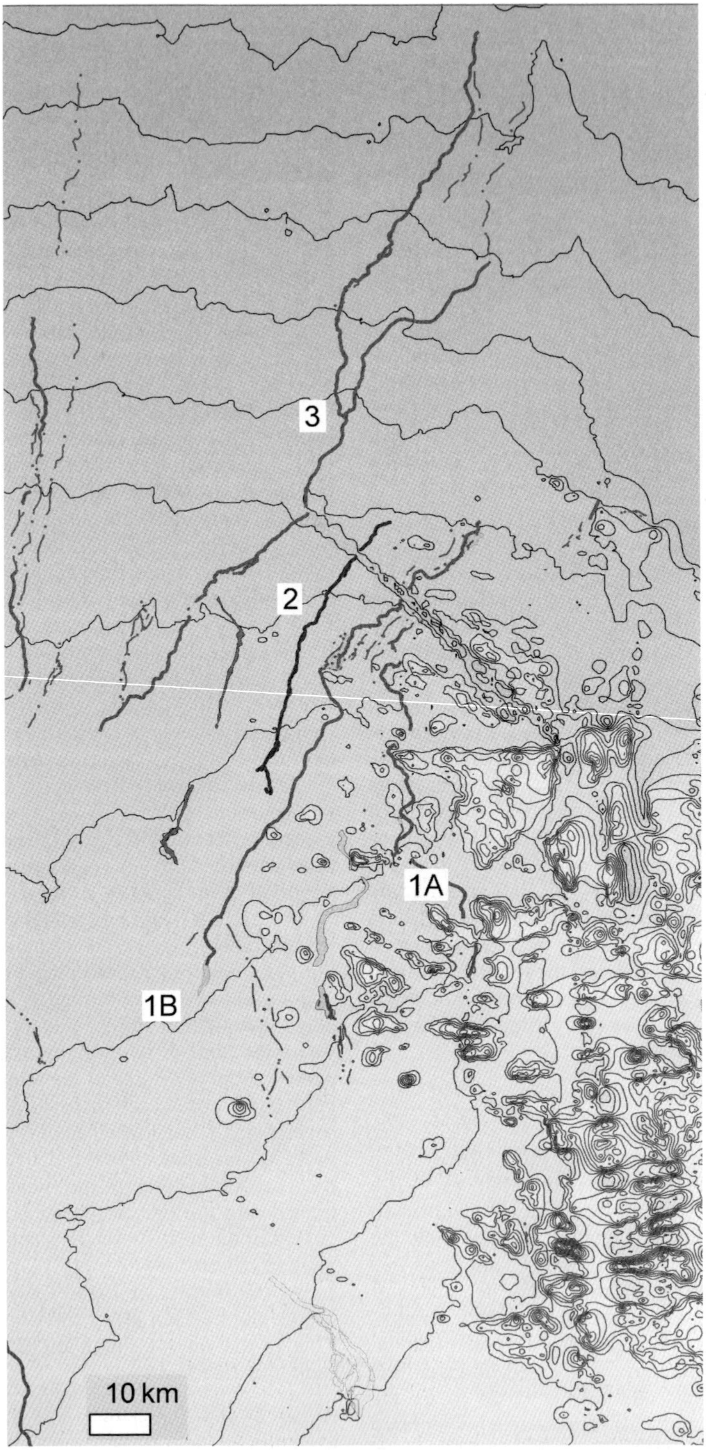

Fig. 3. The relationship between the channel systems and topography. The black contour lines have a 100 m separation. The background is MOLA gridded data. Greyscale represents elevation, where light is highest and dark is lowest.

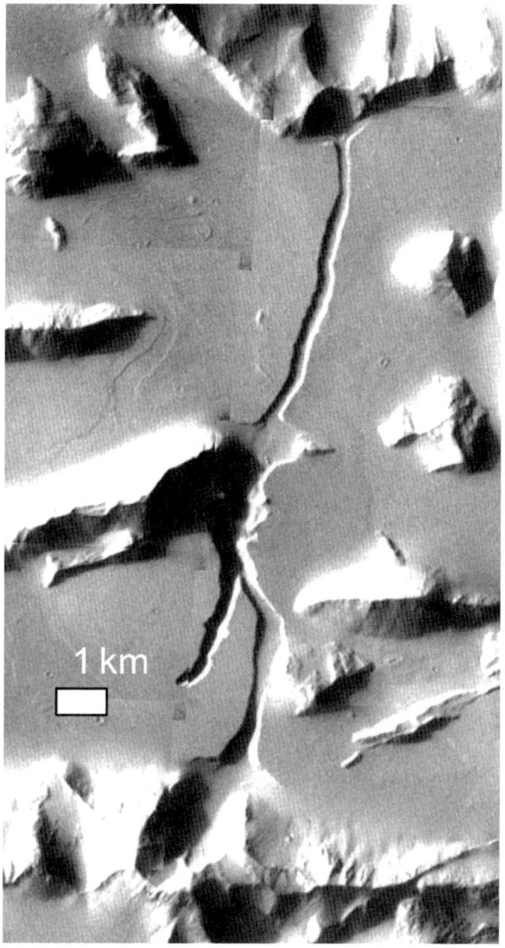

Fig. 4. Source region of Channel 1A. THEMIS image V12711010 located at 233.65°E, 18.41°N. Image credit: NASA/JPL/ASU.

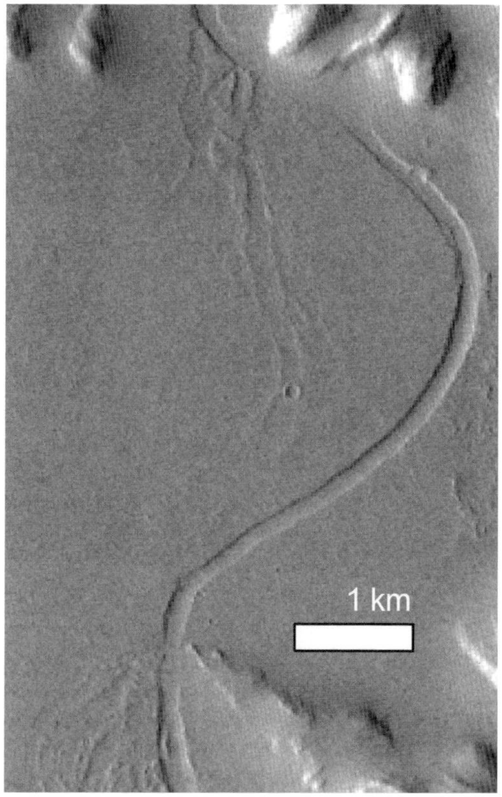

Fig. 5. Multiple faint, poorly incised channels are cross-cut by the main Channel 1A. The faint channel patterns are assumed to be traces of flow prior to the formation of the main channel. THEMIS image V09953012 located at 233.48°E, 18.93°N. Image credit: NASA/JPL/ASU.

few visible depositional features. However, there are more examples of branching than in 1A. The mean width is 192 m and the mean depth is 18 m (considering the main channel only and excluding the few overbank deposits).

The source area for Channel 1B is similar in character to that for Channel 1A. As seen in Figure 8, the area is an extended vent exhibiting diffuse effusion, with this flow forming a well-defined channel 20 km from the vent. The channel head traces to no other source regions, but flow between the channel head and this vent appears diffuse and sheet-like in nature.

In the medial reaches, where Channel 1B approaches Channel 1A, the channel branches out in several, smaller, distributory networks, as labelled on Figure 9. Further to the north, a few larger but fainter channels can be seen that eventually converge with the main channel. This region shows evidence of reduced flow velocity, demonstrating a shallower gradient and higher sinuosity.

Towards the distal part of the channel, it is disrupted by the large graben. The graben has severed the channel, creating a gap, so it is clear that the graben post-dates channel formation. The channel intersections at the graben edges can be aligned using only extensional motion, indicating little or no strike–slip activity by the graben faults.

At the distal end of Channel 1B, the morphology becomes more chaotic and fans out in several smaller streams before the channel proper ends abruptly between two blocks of rough material. Apparent chaotic features and sheet flow appears to continue some distance further, as indicated by braiding, multiple channel islands and stubby, incomplete dendritic structures. This relationship can be seen in Figure 10.

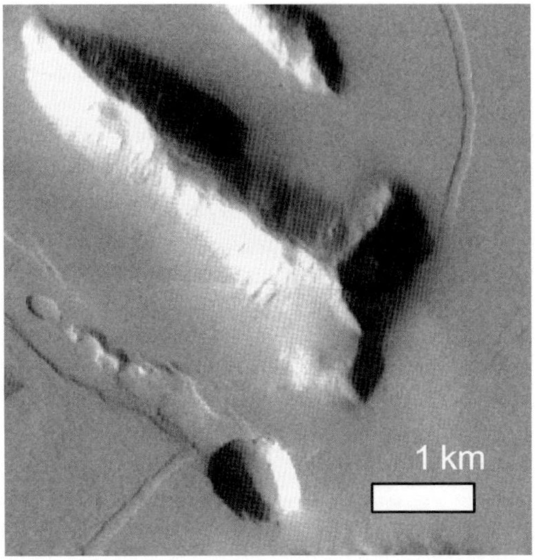

Fig. 6. Image showing location where the Channel 1A is cross-cut by a pit. THEMIS image V09953012 located at 233.47°E, 19.17°N. Image credit: NASA/JPL/ASU.

Channel 2. Channel 2 appears to have a different morphology compared with channels 1A and 1B. It is much shorter at around 50 km and also straighter, with a sinuosity value of 1.07.

The channel is located in the plains to the west of channels 1A and 1B, where there is no aureole material to impede its flow. The channel itself has a less clearly structured appearance compared to channels 1A and 1B, with rough banks and a frequently varying cross-section, as illustrated by Figure 11. This is different from the previously noted long, smooth, narrow channels. In many places on either side of the channel there are raised banks that vary in thickness and width. The channel also partly flows along the centre of one of the lobate lava sheets mapped in Figure 2.

Channel 2 is significantly wider and shallower than channels 1A and 1B. The mean width (within the channel banks) is 288 m, an increase of 50% over the other channels. The mean depth is 14 m, compared with 24 m and 18 m for channels 1A and 1B, respectively. Islands with rounded edges occur in the channel (an example has been

Fig. 7. The end of Channel 1A where its last trace can be seen. The channel flows from bottom left, and appears to fade into more open flow. The graben edge is seen in the top right. There is no clear evidence of channel flow on the other side of the graben, but traces of sheet flow can be identified. The graben edge shows no evidence of downcutting, as would be present if there was significant fluid flowing from the channel into the graben. THEMIS image V09953012 located at 233.46°E, 19.46°N. Image credit: NASA/JPL/ASU.

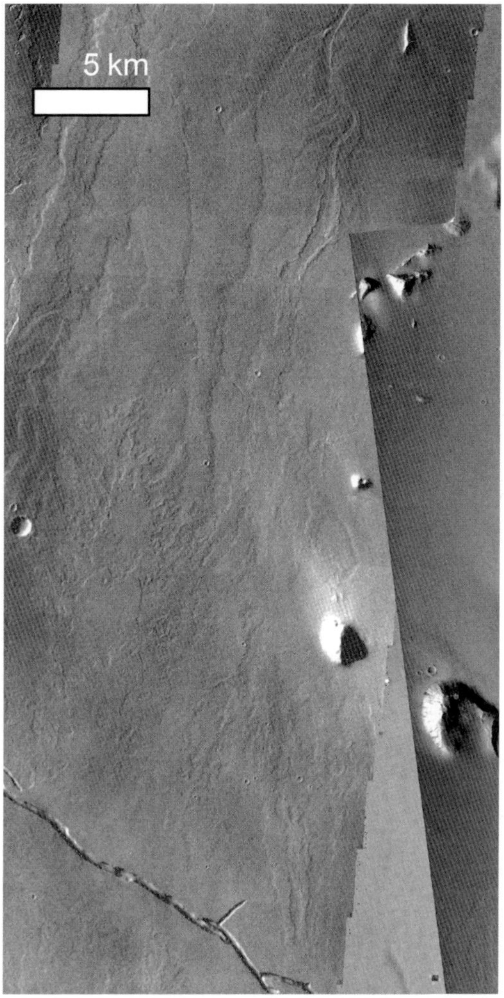

Fig. 8. Source area of Channel 1B. THEMIS image V12424012 located at 232.93°E, 18.58°N. Image credit: NASA/JPL/ASU.

highlighted in Fig. 11). The channel appears to be sourced from (possibly under) a lobate lava sheet, which is in turn part of a large series of sheet lavas, covering the western parts of the study area. On top of this particular lobe is a relatively large impact crater, 800 m in diameter (Fig. 11).

The distal part of the channel is cross-cut by the graben, in a manner similar to Channel 1B. On the south side of the graben the channel is clearly seen, but it terminates gradually (Fig. 12). This channel is significantly shorter than 1A or 1B and is different in character, being less sinuous and having a more variable cross-section.

Channel 3. Channel 3 is the longest channel in the study area, at 140 km. It has a similar morphology to channels 1A and 1B in that it is long, narrow and has an unbroken profile with no apparent steps in altitude. As can be seen in Figure 2, there are several places where the channel bifurcates. In comparison with the sparser, more asymmetric bifurcation of 1A and 1B, both branches are well defined and deeply incised. At the junctions, there is no obvious cross-cutting, which suggests simultaneous flow in both branches. Channel 3 is narrow, as with 1A and 1B, with a mean width of 128 m, but it is somewhat shallower than the other channels, with a mean depth of 11 m.

The source of the channel appears concealed by a lobate sheet that has been post-deposited (Fig. 2). This particular sheet complex sits above many of the other structures in the study area including the major graben.

There is one high-resolution MOC image of a portion of Channel 3. From this, it is possible to identify streamlined islands that sit in the centre of the channel, as shown in Figure 13. These are rounded at the southern end and tapered on the northern, indicating that the flow direction is from the south to the north.

An interesting cross-cutting relationship is shown in Figure 14. On the left, part of the channel is obscured by ejecta from a crater, which also appears to have impacted into a lobate flow (hence, the channel must predate the lobate lava). Both the lobate flow and the channel overlie a NW–SE linear feature (which is subparallel to the major graben). On the top right, the channel can be seen to split (although it rejoins just off image). In the bottom left, before vanishing, the channel appears to deflect to the left and multiple banks are seen, indicating the possibility of episodic flow in the same channels. Apart from these locations mentioned, the flow was mostly uninterrupted for the entire length, as shown by a lack of breaks in the channel. However, the channel is also interrupted by the large graben.

The channel terminates in the northern plains, as shown in Figure 15. It becomes quite undistinguished between many other similar but shorter channels that interweave complexly between the lobate sheets that comprise the lava plains.

Quantitative analysis

Width and depth

Table 2 presents the channel geometric properties, as measured from the MOLA-derived topography and image data. Channel depths were calculated via the measurement of shadowing within the channels using image data, and calculation of the sun angles. The long profiles were plotted for each channel, and the results are shown in Figure 16.

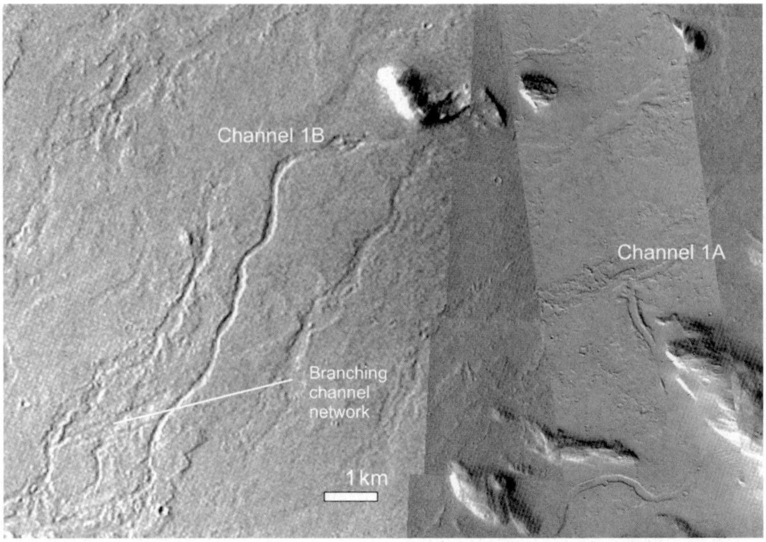

Fig. 9. Channel 1B shows evidence of braiding in this region close to Channel 1A. THEMIS image V13335007 located at 233.30°E, 19.41°N. Image credit: NASA/JPL/ASU.

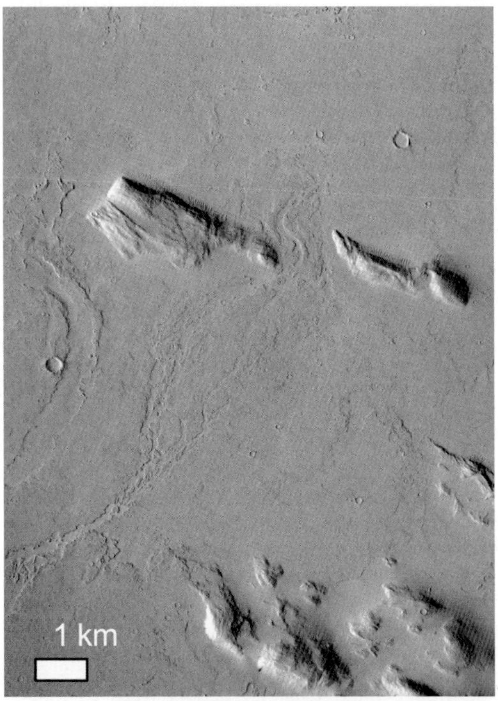

Fig. 10. The terminus of Channel 1B, illustrating the branching out of the main channel into smaller networks. CTX image P15_006970_1979_XN_17N126 located at 233.66°E, 19.72°N. Image credit: NASA/JPL/MSSS.

Channel 1A has a rectilinear cross-section, with an average down-slope gradient of approximately 0.25. This profile was plotted using gridded MOLA data, which has a coarse pixel size of 463 m. This used of gridded data has resulted in several anomalies along the profile owing to the proximity of large features at different elevations; for example, the aureole blocks. Similar artefacts are seen in the other elevation plots. The bankfull channel width as a function of distance is plotted in the middle column of Figure 16, while the channel depth is shown in the rightmost column. The width of Channel 1A is generally narrower downstream, ranging from around 400 to 100 m. The depth is more constant, mostly ranging between 10 and 50 m, although there is also a very slight decreasing trend downstream. Hence, the cross-sectional area of the channel also reduces downstream, mirroring the width.

Channel 1B presents a convex longitudinal profile. Plots of channel width and depth are both more randomly distributed than in Channel 1A. On average, the width ranges from 100 to 250 m, and the depth varies between 5 and 35 m. The width has no apparent trend, except that it becomes more variable after 60 km from the source, which corresponds to the region shown in Figure 9.

Channel 2 has a convex long profile, similar to Channel 1B. The width is highly variable, ranging from 100 to 750 m across. The width of the channel appears to converge towards a narrower value of around 150 m. The depth is fairly constant at 10–20 m along the entire length. The cross-sectional

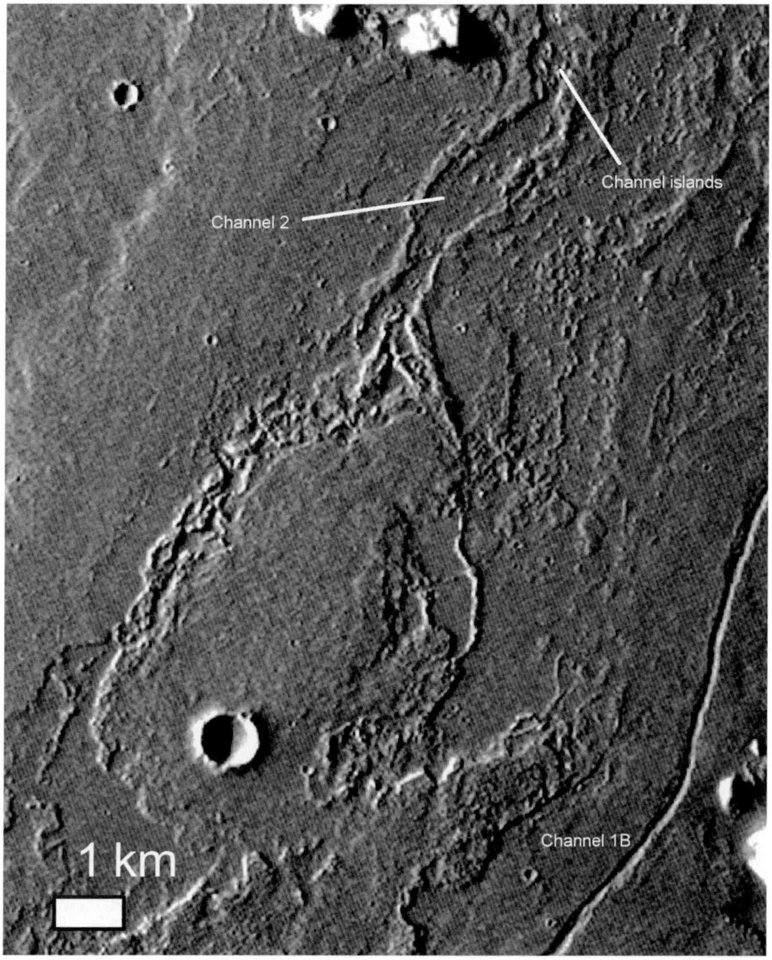

Fig. 11. The source area of Channel 2. The source of the channel is obscured from view by overlying material. The difference in morphology between wider, rough Channel 2 and the long, smooth, narrow Channel 1B nearby is clearly visible. Examples of the small rounded channel islands are shown towards the top of the image. THEMIS image V13647007 located at 233.12°E, 19.20°N. Image credit: NASA/JPL/ASU.

area of the channel is, therefore, reducing downstream, driven by the change in width.

The gradient of Channel 3 is also convex and its gradient, therefore, increases downstream, similar to channels 1B and 2. The width measurements are scattered between 50 and 250 m, with no noticeable downstream trend. The depth is more consistent than the channel width, with the majority of values lying between 5 and 15 m. As a consequence of these data, the cross-sectional area profile is also consistently constant.

Channel 1A has a sinuosity of 1.34 (channel path length/straight line distance) due to the channel deviating around the aureole blocks. Channel 1B also has a similar sinuosity value of 1.23. As already noted, Channel 2 has a very different morphology from the previous two channels; it is much shorter at around 50 km and also much straighter, with a sinuosity of 1.07. This is similar to Channel 3, which has a sinuosity of 1.09. For Channel 2 and Channel 3, this low value represents the low-lying plains over which the channels flowed, compared with the blocky aureole material around which channels 1A and 1B have to deviate.

Flow velocity, discharge rates and event duration

The fluid that formed the channels is unknown; hence, we apply a multiple working hypotheses approach, where the two primary candidates – water and lava – are both considered (Chamberlin

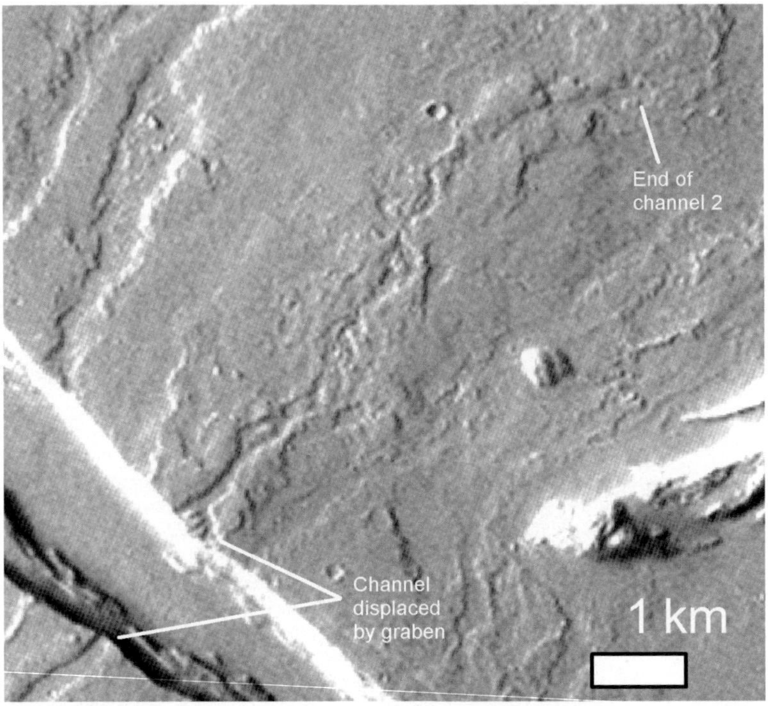

Fig. 12. The distal part of Channel 2, showing its intersection with the graben and its gradual disappearance. THEMIS image V13335007 located at 233.39°E, 19.74°N. Image credit: NASA/JPL/ASU.

1931). For each fluid it is possible to estimate flow discharge rates in the channels.

However, when estimating flow rates for either fluid, the local gradient of the channel is a required measurement. From the convexity of the underlying surface noted earlier, it is possible that there has been some post-emplacement distortion that might alter the derived velocities and flow rates. However, no post-emplacement distortion is apparent in the current observations, in the form of faults or flexure.

Water flow. Flow discharge rates were calculated using the method described by Wilson *et al.* (2004), which is summarized here. The potential flux, or discharge ($m^3 s^{-1}$), is calculated for each channel system along its profile. Discharge, the volume of water flowing through the channel per second, is calculated using the following equation:

$$Q = AV \quad (1)$$

where Q is the discharge ($m^3 s^{-1}$), A is the channel cross-sectional area (m^2) and V is the velocity of the water ($m s^{-1}$). As recommended by Wilson *et al.* (2004), the Darcy–Weisbach equation was used, rather than the Manning equation, to determine the velocity. The velocity depends on the gradient (S, dimensionless) and the hydraulic radius (R, in metres) of the channel. The hydraulic radius is equivalent to the cross-sectional area divided by the wetted perimeter of the channel:

$$R = wd/(w + 2d) \quad (2)$$

where w is the channel width (in metres) and d is the channel depth (in metres). Both of these parameters were measured at intervals along the channel profile. The Darcy–Weisbach equation scaled to Mars (Wilson *et al.* 2004) has the form:

$$V = \sqrt{(8g_M RS/f_c)} \quad (3)$$

where g_M is the acceleration due to gravity on Mars ($3.71 m s^{-2}$) and f_c is the dimensionless friction factor given by:

$$\sqrt{(8/f_c)} = 8.46(R/D_{50})^{0.1005}. \quad (4)$$

Equation (4) is specific to a sandy bed, as the friction factor is dependent on the type of surface. A sandy

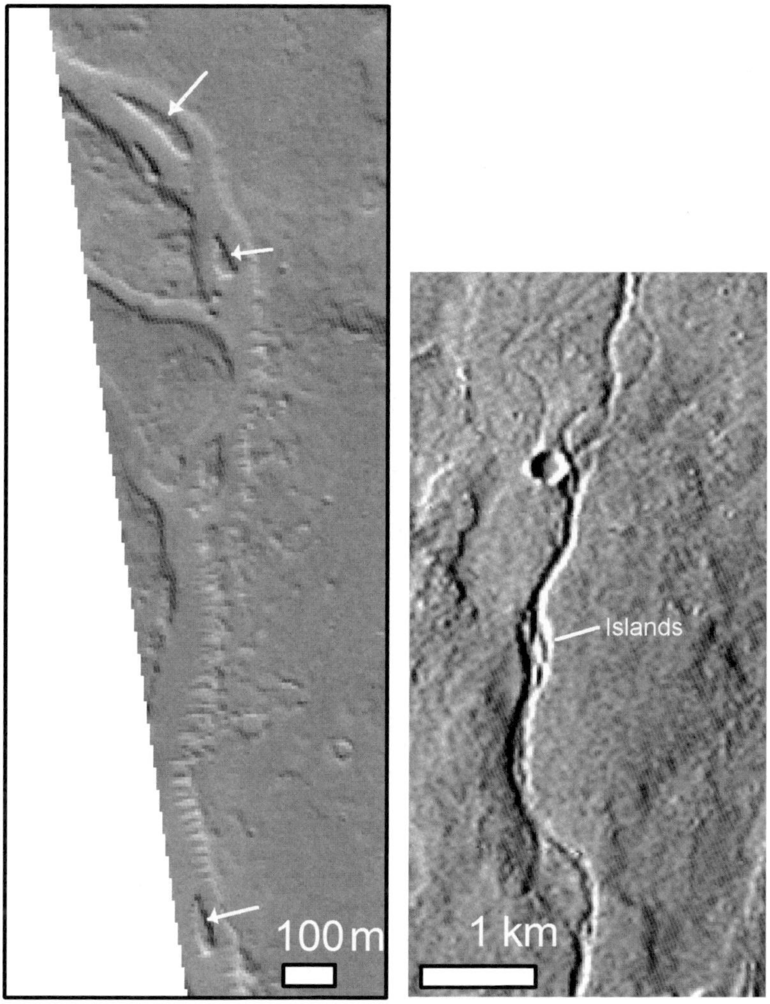

Fig. 13. Sections of Channel 3 illustrating the presence of streamlined islands indicated by the arrows. The inferred flow direction is towards the north of the image. (**a**) MOC image E1100996 located at 233.30°E, 20.32°N. (**b**) THEMIS image V13647007 located at 233.30°E, 20.38°N. Image credit: NASA/JPL/ASU.

bed is assumed to represent the most probable resemblance to the Martian surface. D_{50} is the median grain size, for which a value of 0.064 m was used. This value was determined by Wilson *et al.* (2004) from data on clast-size distributions in Martian channels taken from the *Viking* (Golombek & Rapp 1997) and *Pathfinder* (Golombek *et al.* 2003) landing sites. The gradient was estimated at the same points along the channel where width and depth measurements had been taken.

The results of these calculations for each of the channels are shown in Table 3. The velocities calculated are in the range 5–8.5 m s^{-1}, and the discharge rates calculated are of the order of tens of thousands of cubic metres per second ($m^3 s^{-1}$).

Channel 3 has, by far, the lowest bankfull discharge, at less than 8500 $m^3 s^{-1}$. Discharge rates are heavily dependent on the cross-sectional area used. For all channels, the depth downstream is roughly constant; hence, a plot of cross-sectional area against distance closely resembles that of the channel width.

Lava flow. Lava flow channels are clearly seen in other regions of the mapping area, away from the channel systems; for example, in Figure 17. However, these are morphologically distinct compared with channels 1A, 1B and 3. To investigate putative lava-flow properties, we used the classic equation first proposed by Jeffreys (1925), modified

Fig. 14. The relationship between Channel 3, running from SW to NE, which cross-cuts a linear feature that is also overlain by lobate flows. THEMIS image V13959009 located at 232.90°E, 19.56°N. Image credit: NASA/JPL/ASU.

for narrow flows. This equation relates the flow velocity, u, to lava and channel physical properties:

$$u = \frac{g\rho h^2 \sin\theta}{4\eta} \quad (5)$$

where g is gravitational acceleration for Mars, θ is the slope angle, ρ is the lava density, h is flow thickness and η is flow viscosity. Flow thickness is unknown. However, bankfull discharge is unlikely. In this case we assume a half-full channel as a reasonable estimate.

In addition, we assumed that the lava is a Bingham fluid (e.g. Zimbelman & Gregg 2000), in which the yield strength (τ_y) of the lava is given by:

$$\tau_y = \rho g h \sin\theta. \quad (6)$$

Basaltic lavas have a well-described viscosity. Values can be generated for Martian lavas based on the chemical compositions measured by lander spacecraft. For example, Williams *et al.* (2005) used typical viscosity (μ) values of the order of approximately 1000 Pa s.

Lava densities are unknown but, again, we assume a basaltic lava, which has a relatively narrow range of densities. We adopt the density of typical basaltic lava, about 1500 kg m^{-3} (which includes a gaseous component compared with solid rock values of about 2700 kg m^{-3}).

The above solutions are usually for a broader channel, or a sheet flow, with a modification to the denominator in Equation (5) to account for a narrow channel. A comparison of solutions for channelled lava flow is discussed by Sakimoto & Gregg (2001), who found that a Newtonian-fluid, rectangular-channel-flow model gave a good fit in a variety of conditions, whilst being computationally relatively easy. Hence, we also apply this channelized Newtonian flow model, as described also by Tallarico & Dragoni (1999) and applied to various other terrestrial situations by Sakimoto & Gregg (2001). Q, the volume flow rate, is given by:

$$Q = \frac{\rho^2 g \sin\theta}{3\eta} a h^3 \left(1 - \frac{384h}{\pi^5 a} \sum_{i=1,3,5...} \frac{\tanh(i\pi a/4h)}{i^5}\right) \quad (7)$$

where a is the channel width and the other terms are as defined earlier. (The symbols a and h are transposed between Sakimoto & Gregg's (2001) equation 3, and Tallarico & Dragoni's (1999), equation 15; we follow the convention of Tallarico & Dragoni 1999.)

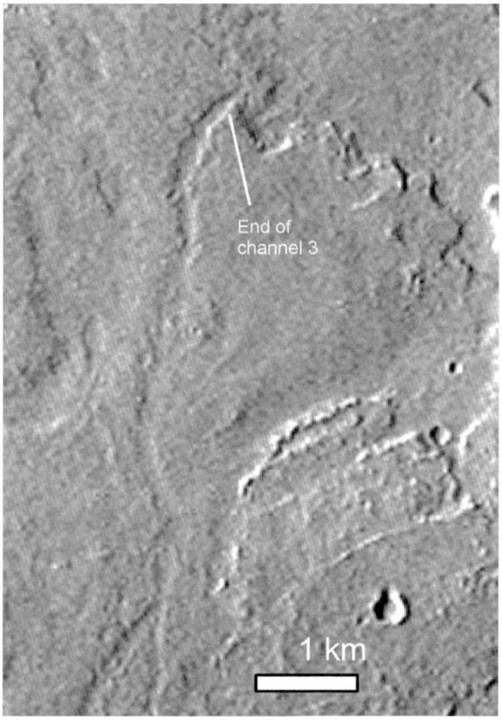

Fig. 15. The area where Channel 3 comes to an end. The channel is rather faint and hard to distinguish. THEMIS image V13335007 located at 233.69°E, 21.17°N. Image credit: NASA/JPL/ASU.

Interestingly, the lack of levees seen in channels 1–3 would imply that basal erosion is the dominant formation mechanism, somewhat different from the 'usual' lava behaviour of levee formation and channelization between them (see, e.g. cross-sections in Baloga *et al.* 2003 for a typical leveed Martian lava channel). Erosional lava channel models have been presented in the past by, for example, Williams & Lesher (1998), and have been applied to multiple situations on various planetary bodies; for example, channel flows on Io (Schenk & Williams 2004). Williams *et al.* (2005), when studying Hecates Tholus, gave erosion rates of the order of tens–hundreds of centimetres per day for flows of similar slope and comparable length to channels 1–3. This calculation generated eruption durations of the order of 50 days for their situation, which is not unreasonable when compared with terrestrial events. The model by Williams *et al.* (2005) is complex to apply without prior work. However, Kerr (2001) also derived thermal erosion models of down-cutting velocity by laminar lava flows. The equations from Kerr (2001) are relatively easily to apply to this situation, to give down-cutting velocity estimates at points

Table 2. *A summary of the morphological data collected for each of the channel systems*

Channel number	Approximate length (m)	Average width (m)	Width range (m)	Width inter-quartile range (m)	Average depth (m)	Depth range (m)	Depth inter-quartile range (m)	Elevation maximum (m)	Elevation minimum (m)	Elevation range (m)
1A	93 000	197.1	275.2	101.8	23.6	112.6	8.6	1112.8	708.1	404.7
1B	109 000	191.7	167.2	28.0	18.2	34.6	8.8	889.9	530.5	359.4
2	54 000	288.0	407.3	105.5	13.7	11.5	3.5	839.1	595.3	243.8
3	147 000	128.1	163.5	23.8	11.3	16.8	3.4	388.8	127.4	261.4

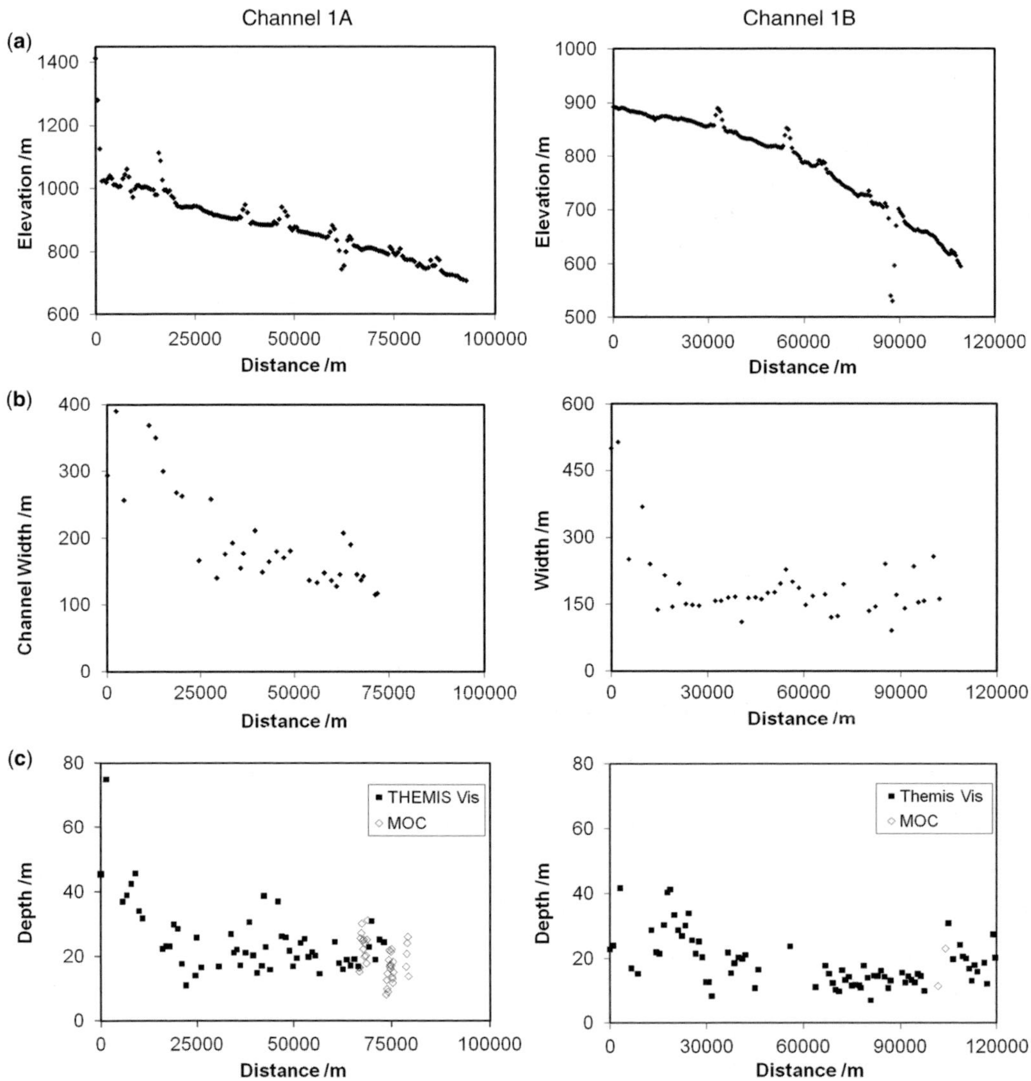

Fig. 16. (**a**) The elevation of each channel against the distance from the source. (**b**) Channel width with distance. (**c**) Channel depth with distance.

along the channel length. The down-cutting velocity, V, is given by equation (13) of Kerr (2001):

$$V = \frac{1}{\Gamma\left(\frac{4}{3}\right) \cdot S}\left(\frac{2U\kappa^2}{9dx}\right)^{\frac{1}{2}} \quad (8)$$

where S is a Stefan constant for ground–lava heat flow, U is the flow velocity (given by Equation 5), κ is the lava thermal diffusivity, $\Gamma(4/3)$ is the gamma function with an argument of $4/3$ (which evaluates to 0.8929795), d is the flow depth (h in previous equations), and x is the distance along flow. See Kerr (2001) for full derivations and explanation of terms. The values for basalt are provided in Kerr's paper and can be applied directly to the situation here. In terrestrial comparisons, such as on Hawaii, at 100 m from the vent, this equation gives 8.1 cm day^{-1} of down-cutting, which compares well to the actual Hawaiian *in situ* measured values (Kerr 2001).

We apply all of these models to this case and present the results in Table 4. We assume the lava parameters to be those used in the Kerr paper for terrestrial Hawaiian basalt. However, when

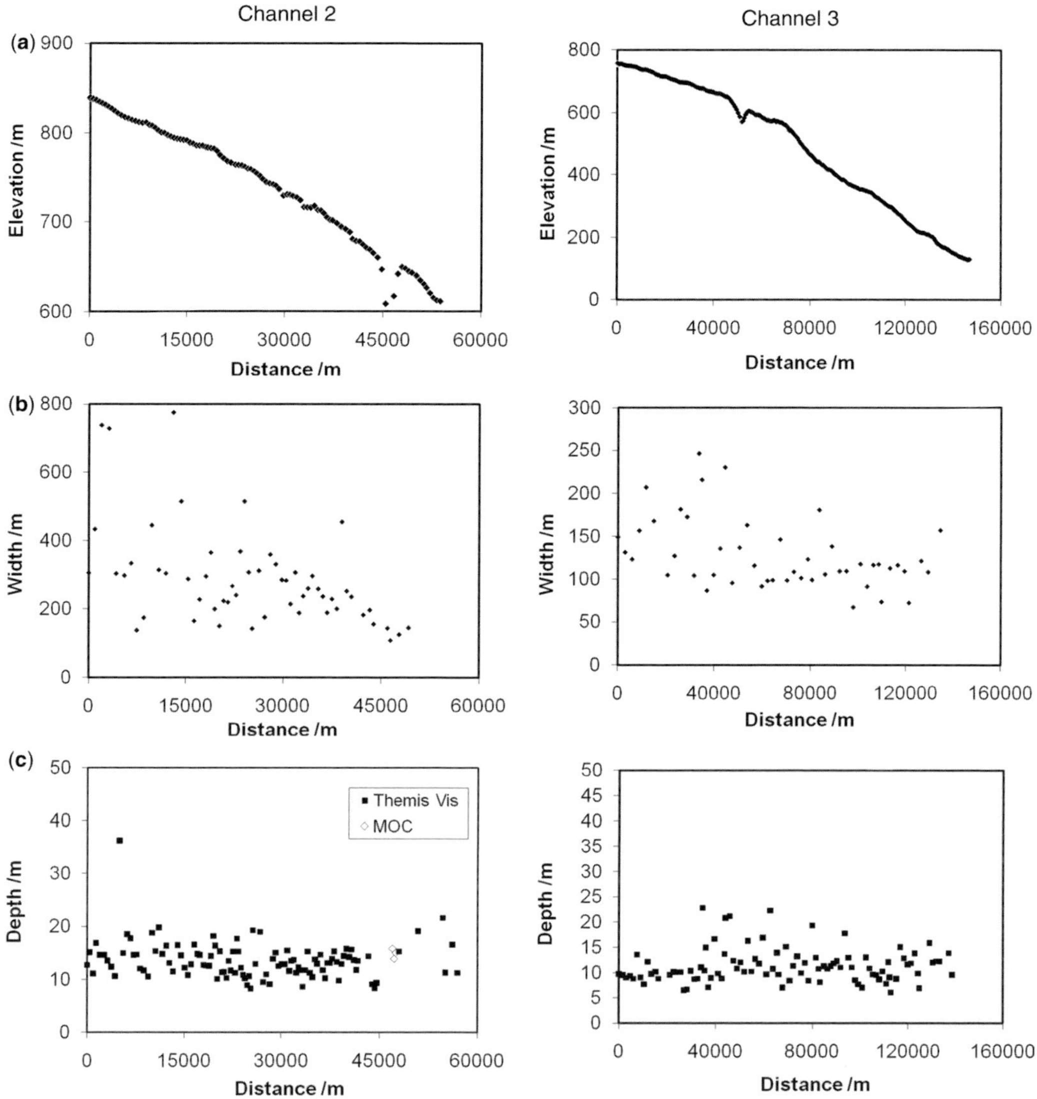

Fig. 16. *Continued.*

Table 3. *Average, minimum and maximum values of velocity, cross-sectional area and discharge of each of the channel systems*

Channel number	Average velocity (m s^{-1})	Minimum velocity (m s^{-1})	Maximum velocity (m s^{-1})	Average cross-sectional area (m^2)	Average discharge (m^3 s^{-1})	Minimum discharge (m^3 s^{-1})	Maximum discharge (m^3 s^{-1})
1A	6.45	3.68	16.70	4705.7	32 890.3	4301.8	141 476.0
1B	5.32	3.36	8.85	2727.9	15 229.4	4402.2	57 087.0
2	6.75	5.41	8.08	3393.8	23 429.8	6895.5	56 039.2
3	6.19	3.60	9.40	1346.5	8432.0	2104.8	27 899.6

Note: Velocity and discharge measurements are based on water flow.

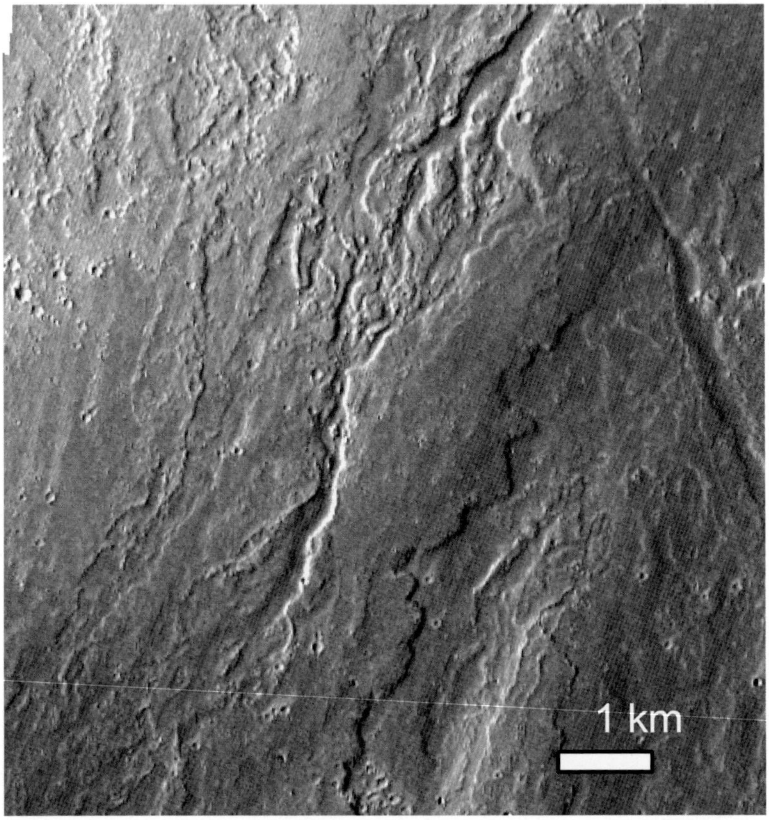

Fig. 17. Areas of lava channel in the NW, showing typical morphologies for this flow unit. From THEMIS image V19500029, located at 232.85°E, 20.34°N. Image credit: NASA/JPL/ASU.

considering the down-cutting velocity (and, hence, the event duration), we only examine two possible viscosities, as a range of values are used in the literature, those of 100 and 1000 Pa s. Compared with terrestrial eruptions, we obtain relatively low down-cutting velocities and, hence, long eruption durations. The volume flow rates derived here are high compared with 'average' Hawaiian style flows (of the order of 100 $m^3 s^{-1}$), but they are still within physically reasonable ranges (the highest recorded terrestrial flow rate, of *c.* 4000 $m^3 s^{-1}$, is from the 1783–1784 Laki eruption in Iceland: Self *et al.* 1997; Hiesinger *et al.* 2007). Flow velocities are somewhat slower than terrestrial basalt values, which are of the order of 10 $m s^{-1}$ (Sakimoto & Gregg 2001 and references

Table 4. *Flow properties of the channel features, assuming lava is the active fluid*

Channel number	Flow velocity ($m s^{-1}$)	τ (Pa)	Q ($m^3 s^{-1}$)	V* ($cm day^{-1}$)	Duration (days)	V* ($cm day^{-1}$)	Duration (days)
				μ = 1000 Pa s		μ = 100 Pa s	
1A	3.2	570	3997	2.2	1070	4.5	523
1B	1.5	334	1373	1.5	1215	3.3	552
2	1.2	343	1214	1.5	911	3.3	414
3	0.3	112	119	0.9	1258	1.8	629

*Indicates that down-cutting velocity is estimated at a distance of 10 km from source. Q, volume flow rate.

therein) for channels around 10 m wide. So our values are noticeably low given that the channel widths here are around 100 m. We have assumed a channel filling of 50%, as mentioned earlier; if filling is increased to 75%, the yield strength and the volume flow rates increase by about 50%. This results directly from the model equations used. The flow rate would be approximately doubled, which would bring the values closer to terrestrial values.

Age determination

Relative stratigraphical order

The cross-cutting relationships between geomorphological features and their surroundings give relative dating information. From this it is possible to build up a picture of how the features formed and how the area in general evolved over time. The main concern in this report is the four channel systems. When looking at the map in Figure 2, it is possible to see that the channels were incised over the smooth and rough lowland terrain. This indicates either that the terrain was in place before the channel appeared, and is therefore older, or synchronous if the channel flow is the source of the material forming the substrate.

Channels 1A and 1B also flow around the blocks and plateaus of aureole deposits as discussed, implying that they too are relatively old. In some places, however, the channel has been covered by smooth lowland material; for example, as shown at the top of Figure 5 and in Figure 6. These gaps in the channel path generally occur next to the aureole blocks. This relationship suggests dry mass wasting and that eroded material from these blocks is deposited on top of the channel. This process must have continued subsequent to channel formation.

A dominant structure across the study area is the large 1–2 km-wide graben that trends NW–SE. The channels that pass near the graben are cross-cut by it; there are, however, no traces of the channels or sedimentation on the graben floor. Although the graben cuts through most of the background material, in one location at the NW, a lobate flow has passed over the top. This is interpreted to be a lava-sheet flow (Fig. 18). When the flow travelled over the graben it spread out laterally for several kilometres, occupying the trench created by the graben. This entire lava flow has been mapped in Figure 2 as an uppermost lobate flow. This flow also cross-cuts Channel 3 at its origin, and a different flow in the same complex has concealed a portion of Channel 2. This indicates, therefore, that volcanism and the production of lava flows

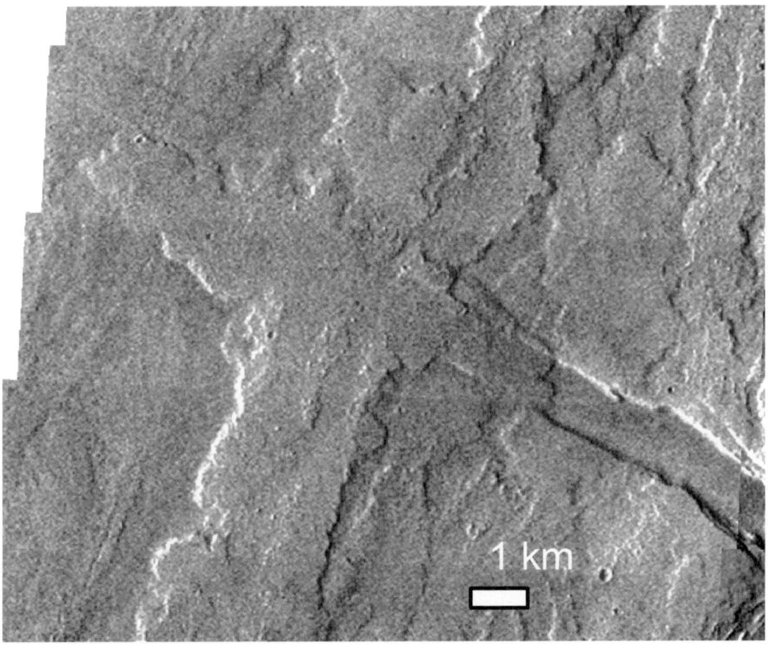

Fig. 18. Details of the filling of the graben by the uppermost lobate flow. THEMIS image V13647007, located at 232.92°E, 20.04°N. Image credit: NASA/JPL/ASU.

was taking place prior to, during and following the formation of the channels and graben.

At the SE end of the graben, there is a transformation into a chain of pits. These are possibly volcanic in origin, maybe as a result of upwelling beneath the extensional fault, that is, dyke intrusion. They cut through the blocks and plateaus of aureole material. In between the blocks, however, there has been some displacement of this chain, implying movement following emplacement. A close-up portion of the map containing the pit chains is shown in Figure 19. A second, more southerly, sub-parallel pit chain is also evident, but this shows less apparent displacement. In Figure 6, as previously shown, a similar structure cross-cuts Channel 1A. This cross-cutting chain is a continuation of the southerly segmented pit chain branch shown in Figure 19. This southerly branch has not cut through as much of the aureole material as the one to the north. Overall, this southern pit chain structure must be relatively young, as it cross-cuts a channel, post-dating that, and also post-dating the

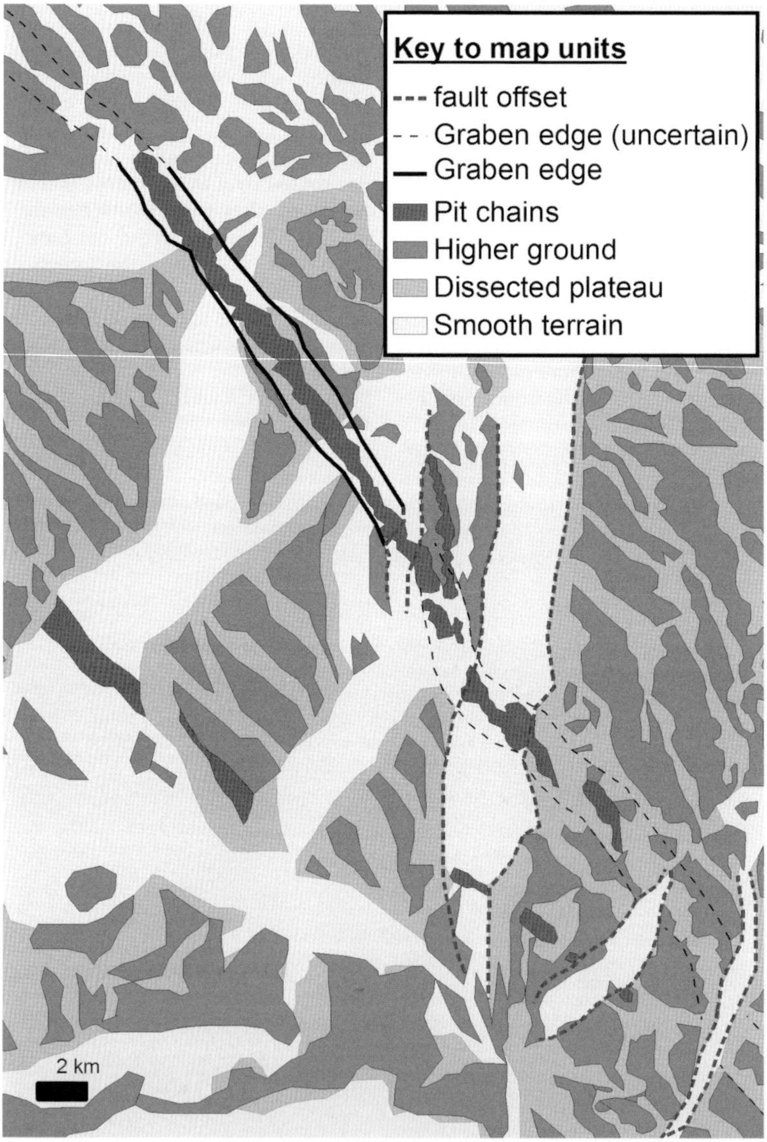

Fig. 19. Close-up map of the mid-east region of Figure 2. The map highlights the spatial coverage of pit chains and the interaction with their surroundings. The location of the centre of the map: 234.09°E, 18.87°N.

graben-forming pit chain and the subsequent deformation. It appears likely then that these two subparallel pit chains are separated in age and may represent two episodes of activity.

Crater counting

Crater counting can be used to estimate the absolute age of Martian geological features, using suitable calibrated models of crater formation rates during solar system history. The most commonly used models at present are those published by Hartmann & Neukum (2001) and Hartmann (2005). See also Hartmann *et al.* (1999) for the application to recent volcanism.

In this study area, only THEMIS visible, MOC and CTX images were of suitably high-enough resolution to permit crater counting, and the coverage is incomplete. Measurements for a surface unit were taken from the same area or the closest possible coverage of the same unit for the different image types. During the counting process and plotting of the data, quality control was enforced by discarding any craters that are less than six pixels across, as their diameters cannot be confidently resolved.

Crater-counting plots with isochrons are shown in Figure 20. The general results show most features to be around 100 Ma old. None of the units appears older than several hundred million years. This corresponds to late Amazonian age. (Typical uncertainty estimates for age dating in this epoch are believed to be of the order of 100% absolute, but relative dating is considered more reliable.) The relatively young ages derived are in agreement with previous work by Mouginis-Mark & Christensen (2005) and Dohm *et al.* (2008), stating that the aureole deposits have been the site of recent hydrological and tectonic activity. Figure 20(d) shows some reduction in crater counts at small sizes, which is indicative of resolution-limited data, although one should not discount the possibility of erosional and resurfacing processes removing craters. The data show the pit chain to be the youngest feature in the area (Fig. 20a) at around 10 Ma old. This result implies very recent tectonic activity on Mars; however, the value should be interpreted with caution, as the pit chain is a relatively small area and difficult to date accurately. In addition, it is a topographical low and may have accumulated sediments that could obscure craters, skewing the age estimate. However, the pit chain is also expected to be the youngest feature present, based on consideration of the stratigraphical order, as it cross-cuts many other features.

The smooth lowland terrain is dated at approximately 100 Ma (see Fig. 20b), but this terrain may, in fact, be older owing to resurfacing by wind-blown sediment. It has been previously mentioned how interruptions in the channels occur next to some of the aureole blocks (see Fig. 5), most likely as a result of erosion and subsequent deposition on top of the nearby channels. The smooth material surrounds the aureole deposits. Aeolian features, such as ripples and dunes, are visible in CTX images of the area and cover significant portions of the smooth surface, implying continual reworking of the sediment. This would result in a younger age because the original forms are dominated by aeolian erosion and burial under aeolian deposits.

The age of the graben is constrained at 100 Ma (Fig. 20c). This is older than the pit chain despite the two features being structurally subparallel. According to Figure 20(e), the plateaus or large blocks of aureole material also appear to have an age of 100 Ma . The higher ground on top of the plateaus has a similar age, as shown in Figure 20(f). In both cases, however, only small areas have been imaged in sufficient detail to be datable, so there must be some uncertainty. Figure 20(g) represents the rough lowland material. This is one of the oldest surfaces in the study area at several hundred million years old, although the data have been somewhat affected by curvature, as previously discussed. As can be seen from Figure 20(h), one could assert that the uppermost lobate flow is relatively young, but in reality it lacks sufficient craters to realistically date it. According to the stratigraphical order this lobate flow is expected to be one of the younger units present, as it cross-cuts the graben and Channel 3. The statistics are poor in part because it is only observed in low-resolution THEMIS images, meaning that smaller crater populations could not be mapped.

Combining the crater dating with the relative stratigraphical order, the basic dating order for this region is summarized in Table 5, with the oldest at the base of the list.

What is notable is the compressed timeline for the formation of the channels and the underlying rock on which they sit; they are essentially the same age, within the uncertainty of dating methodology.

Discussion

Channel characteristics and fluid

Liquid water (or brine) and low-viscosity lava are the two most likely fluid candidates that could have carved these channels on Mars. The different channel morphologies, as described in the previous sections and seen in the surrounding context, would advocate that there are both lava and water channels in the study area. This is a reasonable suggestion

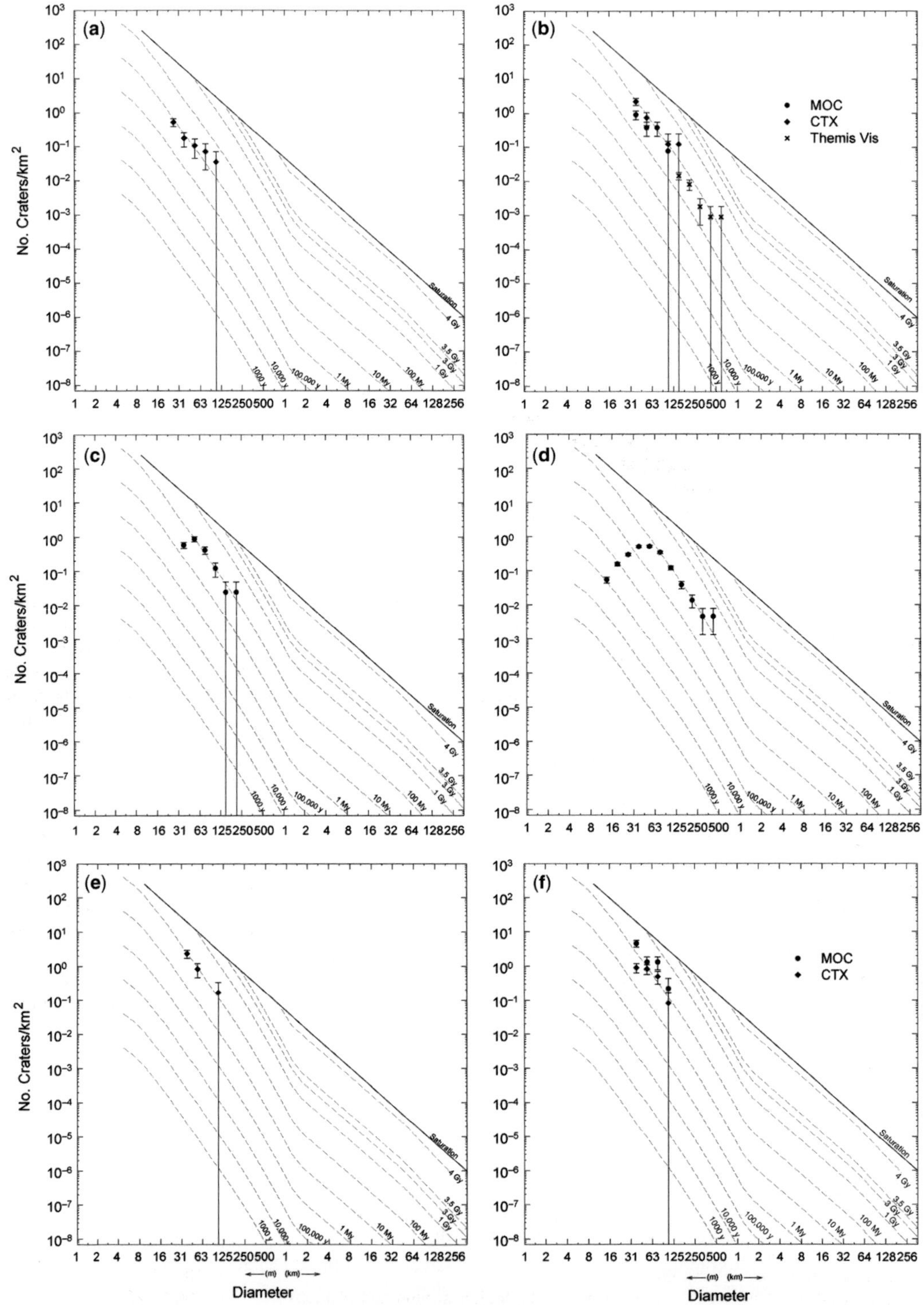

Fig. 20. Crater-count plots showing the approximate ages corresponding to mapped features in Figure 2. (**a**) The pit chain; (**b**) smooth lowland terrain cross-cut by Channel 1A; (**c**) graben; (**d**) lower lobate flows, cross-cut by Channel 3; (**e**) plateau; (**f**) higher ground; (**g**) rough lowland terrain; and (**h**) uppermost lobate flow.

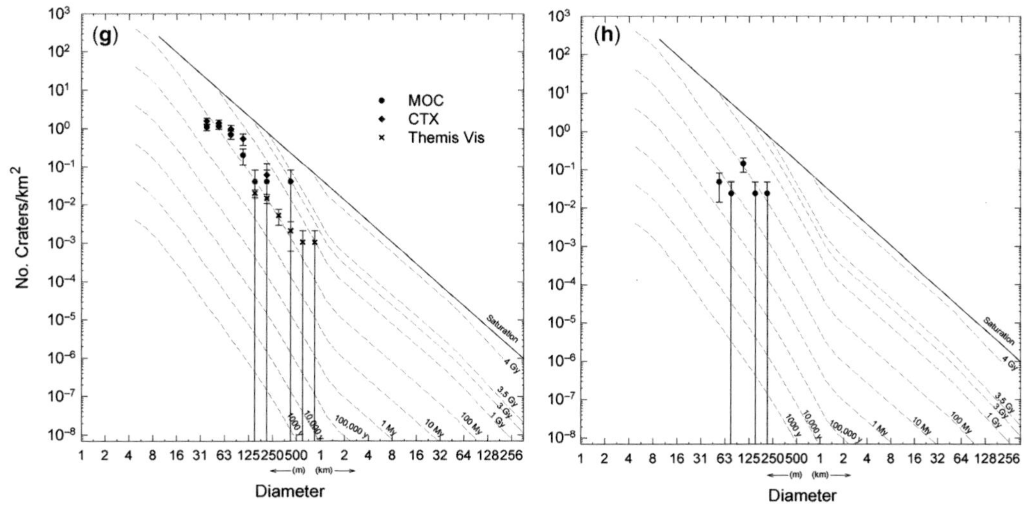

Fig. 20. *Continued.*

considering the general locale has already been noted as a site of recent volcanic and hydrological activity (Mouginis-Mark 1989; Dohm *et al.* 2008).

The channel lengths are comparable to channel studies within the Tharsis region by other authors. For example, Hiesinger *et al.* (2007) recorded lengths of up to 38 km, and longer channels have been noted by others; Zimbelman (1998) investigated one with a 250 km length; Baloga *et al.* (2003) studied a channel on Pavonis Mons 175 km in length, and Garry *et al.* (2007) described a 690 km-long major flow near Ascreus Mons.

The channels show little or no braiding, consisting essentially of one long, single, low-sinuosity channel. There are some locations, however, where older branches are cross-cut by the newer, more deeply incised channel, as seen in Figure 5 for Channel 1A. Although these are not common, they suggest that flow occurred on multiple occasions, perhaps in short periodic bursts.

The channels here, especially 1A and 1B, have few or no tributary branches. Instead, they form one long, smooth channel. This feature is indicative of a point source for the channels instead of precipitation across a drainage network, as for terrestrial rivers. The cross-sectional area of the channels decreases downstream, indicating a reduction in erosion. We interpret this as a reduction in discharge. This loss could be by evaporation, freezing or soaking into the substrate.

In Channel 1A and to a lesser extent Channel 1B, the cross-sectional area weakly decreases in size away from the source. This is the opposite of terrestrial perennial flowing rivers, where the channels become wider and deeper downstream as more and more tributaries feed into the network. However, such a trend is seen in ephemeral dry land rivers or episodic lava eruptions (Bull & Kirkby 2002).

Assuming that the formative fluid is water, the average velocity of water flow in the channels

Table 5. *Crater counting of all areas produces essentially the same age of approximately 100 Ma*

Feature	Age
Formation of southern pit chain*	Youngest *c.* 10 Ma?
Uppermost lobate flow*	
Offset distortion of eastern end of the graben?*	
Extensional faulting – graben and associated pit chain	*c.* 100 Ma
Formation of channel systems	
Lowland terrain – smooth	*c.* 100 Ma
Olympus Mons aureole deposit (map units: higher ground and plateau)	*c.* 100 Ma
Lowland terrain – rough – lava plains	Older, *c.* 100 Ma

*Indicates that relative dating of these features is by crater counting only, not by stratigraphic means, as there is no overlap.

was calculated to be 5–9 m s^{-1} using the Darcy–Weisbach equation. However, the average discharge rates are high at 8000–36 000 m^3 s^{-1}. This is because the channels all have large cross-sectional areas of several thousand square metres. For water flow, we have assumed a bankfull discharge, so the values should be considered as close to the maximum possible. Discharge rates of this order are observed on Earth and are comparable in scale to some of the larger events studied; for example, the Brahmaputra River in Bangladesh (Vörösmarty *et al.* 1996).

Considering lava as the formative fluid, observations of terrestrial basaltic flows tend to indicate that channels (as opposed to lava tube morphologies) are more indicative of higher viscosity, more episodic, eruptions at generally higher effusion rates (Sakimoto *et al.* 1997; Calvari *et al.* 2005; Bleacher *et al.* 2007). If this is the case, it would tend to imply that these flows formed relatively quickly. Estimated lava-flow velocities are somewhat low compared with terrestrial eruptions, and the effusion rates are very high. This is primarily a result of the low slope angles. However, this result appears somewhat at odds with the appearance of the channels and modelling, which implies a low (or very low) lava viscosity. Down-cutting rates are also lower than might be expected, again as a result of the low velocity, which results in long eruption times. Very-low-viscosity lava is likely to be basic or ultrabasic and to have a high temperature (e.g. terrestrial komatiite, or picrobasalt as noted at the Gusev Crater by MER Spirit: Williams *et al.* 2005), so it down-cuts relatively effectively. However, all values for lava modelling, while somewhat extreme, are within those observed during (usually exceptional) terrestrial events.

For Channel 3, the cross-sectional area is roughly constant, with less than 10% variation observed along the entire channel length, and it appears possible that this channel was formed by one event. For all channels, there appears little change in the channel morphology with length, apart from the slight decrease in dimensions already noted.

A striking feature of the channels is their convex long profiles. This convex shape is representative of the underlying topography of the region. The channels originate from within or nearby the aureole deposits that form the highest topography in the region. Overall, the aureole deposits form a dome-like shape. This is likely to be a result of tectonic uplift, related to underlying volcanic processes. The channels flow over the edge of this bulge out onto the lowland lava plains to the north, creating a convex profile. Channel 1A is the only channel with a more linear profile. This shape may be because the channel is confined to the area surrounding the aureole deposits, where it is flatter, never reaching the side of the bulge. This convex gradient profile may also partly explain why the cross-sectional area of the channels decreases downstream. If the gradient is increasing, theoretically so will the flow velocity; therefore, a smaller area is required to maintain the discharge rate. This could imply little or no loss of fluid along the channel lengths.

In all channels there is no clear sign of deposits at the channel terminus. This absence could be the result of discharge into an open area/basin or due to loss of fluid along the length of flow. In Figure 10, Channel 1B does not gradually fade, as would be expected for example if more and more water evaporated, but it finishes relatively abruptly. It is possible that the topographically distinct channel may have been buried by post-formation processes or fluid flow transitioned to sheet flow, as an alternative to fluid loss (evaporated, sublimed or sunk into the substrate).

Based on these flow estimates and the morphology seen, channels 1A, 1B and 3 are interpreted to be fluvial. Channel 2 is interpreted to have been created by lava.

Channels 1A, 1B and 3 all have similar characteristics that are indicative of flowing water. They are all longer than 90 km, requiring a very fluid material. The channels appear narrow and maintain the same shape for most of their course. The banks of the channels are smooth and continuous, giving the impression of slow erosion by a low-viscosity liquid, with almost no sign of levees. The smooth, streamlined nature of the islands in Channel 3, as shown in Figure 13, is indicative of an extremely low-viscosity fluid, such as running water. Channel 2 has a somewhat different morphology to those previously discussed and is much more characteristic of a lava channel, as might be seen during a large basaltic eruption. The channel is much shorter, at 54 km. It is also significantly wider and more chaotic than the others, with rough, bulbous banks. At nearly 300 m wide, there are flat-topped levees or raised banks up to 500 m wide; however, the width varies considerably. These can be seen next to the channel in Figure 12, just at the channel–graben intersection. The levees are irregular and not smooth; instead, they appear to be the solidified margins of a lava channel. The channel shape is also highly variable and, overall, this short channel has the appearance of having been formed by a higher-viscosity fluid such as lava. The lava modelling estimates, however, show little difference in character between the channels, but in these cases the viscosity of the lava is an input to the model rather than a solution. Channel 2 also appears to run down the centre of one of the lava sheets/lobes that make up the lava plains. It is located within the rough lowland terrain amongst the thick lava sheets.

In a wider context, in the eastern region of the study area there is more evidence of fluid activity: Figure 21 shows benches around the edges of aureole blocks that record the highstand level. Similar features were identified close to this eastern region by Mouginis-Mark & Christensen (2005). They linked these features to flooding of the area between the aureole blocks by water effluxes from a channel. If the area has been flooded, this would begin to explain why the landscape is so smooth, but it may cast some doubt on the absolute ages derived from the crater counting for these regions. Sixteen similar high water marks were identified in the surrounding aureole deposits using the same CTX image. It appears likely that more of these features exist further to the west near the channels, but there are no high-resolution images covering the area from which one could identify such fine markings.

Figure 13 of Mouginis-Mark & Christensen (2005) shows channels that they interpreted as water flow within the same segment of aureole deposits as our study area, but 60 km to the south. This supports the interpretation of channels 1A and 1B as young Martian water channels. Channel 3 has similar characteristics to these channels, so we believe that it likely that it is also of fluvial origin despite being located further to the west in the lava plains.

Formation history

None of the channel systems displays dendritic patterns that would reflect surface runoff of precipitation. The channel systems are interpreted as representing a short-lived event of volcanic origin with associated hydrothermal activity (with, possibly, the melting of surface ice). This event appears comparable in size to the activity noted by Mouginis-Mark & Christensen (2005). The water channels are most probably a result of hydrothermal activity owing to the volcanic setting in which they are located. Ice melted in a confined subsurface aquifer can be suddenly released as an outburst of groundwater. The source area of Channel 1A (shown in Fig. 4) is a long crevice in the ground, extending from an irregular pit feature in the aureole material. The source areas for the other aqueous channels are not as clear, in part, owing to the lack of high-resolution image coverage. Given that Channel 2 has an effusive lava signature, the lava responsible must have had a very low viscosity and high temperature to account for the depth of the channel, assuming plausible eruption timescales. This would imply a komatiite-like character of lava, which has also been suggested for late episodic Olympus Mons aureole eruptions.

Bleacher *et al.* (2007) noted a general trend across Tharsis Montes similar to the behaviour of the Hawaiian shield volcanoes, whereby long-lived high-volume lavas buffered by a large magma chamber produce predominantly tube-fed flows (tholeiitic in Hawaii), which over time gives way to shorter-lived, episodic, lower-volume, unbuffered flows that display more channels than tubes (alkalic lavas in Hawaii). The channel systems that we observe probably represent this later episodic activity, indicating that activity in the Tharsis

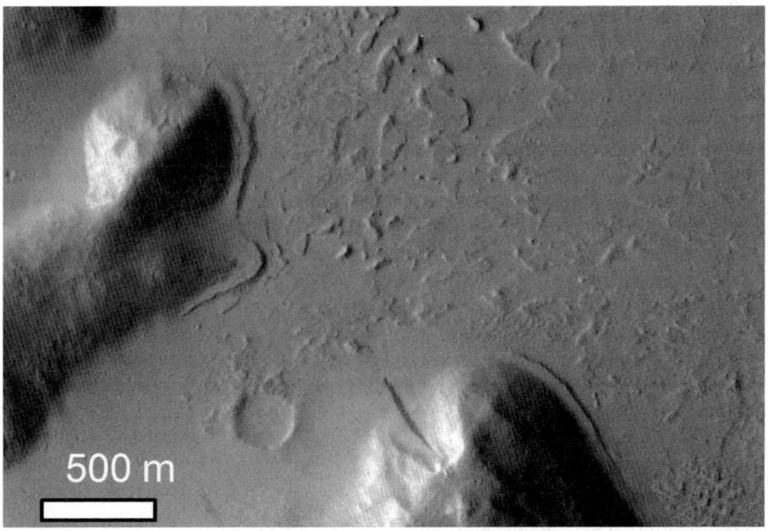

Fig. 21. Curved bench features around the edges of aureole blocks indicative of highstand water marks. CTX image P02_001867_1979, located at 234.27°E, 18.38°N. Image credit: NASA/JPL/MSSS.

region in the late Amazonian was declining from earlier peaks.

The presence of these channels confirms that volcanic activity continued into the recent geological past on Mars. It specifically indicates the aureole deposits of Olympus Mons as young dynamic areas, as previously suggested by Mouginis-Mark & Christensen (2005) and Dohm *et al.* (2008). Basilevsky *et al.* (2006) also observed volcanic and fluvial activity, possibly as recent as 25–40 Ma, and possibly younger, in the general locale on the SE flank of Olympus Mons. That activity probably post-dates the channels studied for this research but, given the uncertainties in our crater dating, temporal overlap is certainly possible. Neukum *et al.* (2004) also noted very recent activity on the Olympus Mons flanks at about 25 Ma. The Olympus Mons caldera (which is a composite structure of five smaller calderas) has an estimated age of 100 ± 50 Ma, so the whole caldera surface was formed in a relatively short period of time. This date matches the ages for the surface and channels seen here (Neukum *et al.* (2004)).

The formation of fluvial channels suggests that climatic conditions on Mars at this time, around 100 Ma, were such that the surface was able to support liquid water that could flow for almost 150 km before evaporating or absorbing into the subsurface (Wallace & Sagan 1979; Carr 1983; Goldspiel & Squyres 2000).

The southern pit chain appears to be the stratigraphically youngest feature and the youngest by crater counting, possibly around 10 Ma. This may represent much younger activity than the channel-forming events, but caution is called for as post-formation morphological or structural alteration cannot be ruled out and may skew the derived date.

Conclusions

The Sulci Gordii is a relatively young region of the Martian surface containing many channel networks. We have studied in detail four channel systems, with lengths ranging from 54 to 147 km and average widths between 128 and 288 m. These channels are cross-cut by a more recent graben.

Analysis of the channels formed by either lava or water, supported by the morphological features, indicates that one channel was most likely to have been formed by low-viscosity lava, whereas the rest were probably formed by water (or brine). These water-formed channels were identified based on their long, fairly constant morphologies, streamlined islands and location within an area believed to be a site of flooding. At least one of the channels originated from a point source that appears to have been the result of hydrothermal activity. The lava-formed channel was thus identified because of its rough banks, irregular shape, solidified margins and short length.

Using crater counting, an age of the order of 100 Ma was placed on the channels, indicating relatively recent formation. Crater dating of overlying and underlying features gives the same dates, indicating that that this was a relatively short-lived episode.

There is evidence of more recent activity in the region; for example, burial of some stretches of the channels owing to dry mass wasting from nearby topographical highs. In addition, a more recent tectonically formed pit chain was tentatively dated as being much younger, approximately 10 Ma.

Overall, these observations show that the Sulci Gordii region experienced hydrological, tectonic and volcanic processes that operated in the not-too-distant geological past, along with the climate capability to support liquid water flow over great distances on its surface.

This research has made use of the USGS Integrated Software for Imagers and Spectometers (ISIS). M. C. Towner and S. Harrison were funded by the UK Science and Technology Facilities Research Council. S. J. Conway was funded by the UK Natural Environment Research Council. C. Eakin was funded by a Nuffield Foundation Undergraduate research Bursary. Rossman P. Irwin III and one anonymous reviewer are thanked for many helpful revisions and suggestions.

References

ANDERSON, R. C., DOHM, J. M. *ET AL.* 2001. Primary centers and secondary concentrations of tectonic activity through time in the western hemisphere of Mars. *Journal of Geophysical Research*, **106**, 20 563–20 585.

BALOGA, S. M., MOUGINIS-MARK, P. J. & GLAZE, L. S. 2003. Rheology of a long lava flow at Pavonis Mons, Mars. *Journal of Geophysical Research*, **108**, 5066.

BASILEVSKY, A. T., WERNER, S. C. *ET AL.* 2006. Geologically recent tectonic, volcanic and fluvial activity on the eastern flank of the Olympus Mons volcano, Mars. *Geophysical Research Letters*, **33**, L13201.

BLEACHER, J. E., GREELEY, R., WILLIAMS, D. A., CAVE, S. R. & NEUKUM, G. 2007. Trends in effusive style at the Tharsis Montes, Mars, and implications for the development of the Tharsis province. *Journal of Geophysical Research*, **112**, E09005, doi: 10.1029/2006JE002873.

BULL, L. J. & KIRKBY, M. J. 2002. *Dryland Rivers: Hydrology and Geomorphology of Semi-arid Channels*. Wiley, New York, 398 pp.

CALVARI, S., SPAMPINATO, L. *ET AL.* 2005. Chronology and complex volcanic processes during the 2002–2003 flank eruption at Stromboli volcano (Italy) reconstructed from direct observations and surveys with a handheld thermal camera. *Journal of Geophysical*

Research – Solid Earth, **110**, B02201, doi: 10.1029/2004JB003129.

CARR, M. H. 1973. Volcanism on Mars. *Journal of Geophysical Research*, **78**, 4049–4062.

CARR, M. H. 1983. Stability of streams and lakes on Mars. *Icarus*, **56**, 476–495.

CHAMBERLIN, T. C. 1931. The method of multiple working hypotheses. *Journal of Geology*, **39**, 155–165.

CHRISTENSEN, P. R., JAKOSKY, B. M. *ET AL.* 2004. The Thermal Emission Imaging system (THEMIS) for the Mars 2001 Odyssey mission. *Space Science Reviews*, **110**, 85–130.

DOHM, J., ANDERSON, R. *ET AL.* 2008. Recent geological and hydrological activity on Mars: The Tharsis/Elysium corridor. *Planetary and Space Science*, **56**, 985–1013.

GARRY, W. B., ZIMBELMAN, J. R. & GREGG, T. K. P. 2007. Morphology and emplacement of a long channeled lava flow near Ascraeus Mons Volcano, Mars. *Journal of Geophysical Research*, **112**, E08007, doi: 10.1029/2006JE002803.

GOLDSPIEL, J. M. & SQUYRES, S. W. 2000. Groundwater sapping and valley formation on Mars. *Icarus*, **148**, 176–192.

GOLOMBEK, M. P. & RAPP, D. 1997. Size-frequency distributions of rocks on Mars and earth analogue sites: implications for future landed missions. *Journal of Geophysical Research*, **102**, 4117–4129.

GOLOMBEK, M. P., HALDEMANN, A. F. C. *ET AL.* 2003. Rock size-frequency distributions on Mars and implications for Mars exploration Rover landing safety and operations. *Journal of Geophysical Research – Planets*, **108**, 8086–8086.

HARTMANN, W. K. 2005. Martian cratering 8: isochron refinement and the chronology of Mars. *Icarus*, **174**, 294–320.

HARTMANN, W. K. & NEUKUM, G. 2001. Cratering chronology and the evolution of Mars. *Space Science Reviews*, **96**, 165–194.

HARTMANN, W. K., MALIN, M. *ET AL.* 1999. Evidence for recent volcanism on Mars from crater counts. *Nature*, **397**, 586–589.

HIESINGER, H., HEAD, J. W. & NEUKUM, G. 2007. Young lava flows on the eastern flank of Ascreus Mons: rheological properties derived from the High Resolition Stereo Camera (HRSC) images and Mars Orbiter Laser Altimeter (MOLA) data. *Journal of Geophysical Research*, **112**, E05011, doi: 10.1029/2006JE002717.

JEFFREYS, H. 1925. The flow of water in an inclined channel of rectangular section. *Philosophical Magazine Series, Series 6*, **49**, 793–793.

KERR, R. C. 2001. Thermal erosion by laminar lava flows. *Journal of Geophysical Research*, **106**, 26 453–26 466.

LOPES, R., HILLER, K., NEUKUM, G. & GUEST, J. E. 1982. Further evidence for a mass movement origin of the Olympus Mons aureole. *Journal of Geophysical Research*, **87**, 9917–9928, doi: 10.1029/JB087iB12p09917.

MALIN, M. C., DANIELSON, G. E., INGERSOLL, A. P., MASURSKY, H., VEVERKA, J., RAVINE, M. A. & SOULANILLE, T. A. 1992. The Mars observer camera. *Journal of Geophysical Research*, **97**, 7699–7718.

MALIN, M. C., BELL, J. F., III. *ET AL.* 2007. Context camera investigation on board the Mars reconnaissance Orbiter. *Journal of Geophysical Research – Planets*, **112**, E05S04.

MASURSKY, H., BATSON, R. M. *ET AL.* 1972. Mariner 9 television reconnaissance of Mars and its satellites: preliminary results. *Science*, **175**, 294–305.

MCCAULEY, J. F., CARR, M. H. *ET AL.* 1972. Preliminary mariner 9 report on the geology of Mars. *Icarus*, **17**, 289–327.

MELLON, M. T., JAKOSKY, B. M., KIEFFER, H. H. & CHRISTENSEN, P. R. 2000. High-resolution thermal inertia mapping from the Mars global surveyor thermal emission spectrometer. *Icarus*, **148**, 437–455.

MORRIS, E. C. 1982. Aureole deposits of the martian volcano Olympus Mons. *Journal of Geophysical Research*, **87**, 1164–1178.

MOUGINIS-MARK, P. J. 1989. Recent water release in the Tharsis region of Mars. *Icarus*, **84**, 362–373.

MOUGINIS-MARK, P. J. & CHRISTENSEN, P. R. 2005. New observations of volcanic features on Mars from the THEMIS instrument. *Journal of Geophysical Research – Planets*, **110**, E08007, doi: 10.1029/2005JE002421.

MOUGINIS-MARK, P. J., WILSON, L. & ZUBER, M. T. 1992. The physical volcanology of Mars. *In*: KIEFFER, H. H., JAKOSKY, B. M., SNYDER, C. W. & MATTHEWS, M. (eds) *The Physical Volcanology of Mars*. University of Arizona Press, Tucson, AZ, 424–452.

NEUKUM, G. & HILLER, K. 1981. Martian ages. *Journal of Geophysical Research*, **86**, 3097–3121.

NEUKUM, G., JAUMANN, R. *ET AL.* & THE HRSC CO-INVESTIGATOR TEAM. 2004. Recent and episodic volcanic and glacial activity on Mars revealed by the High Resolution Stereo Camera. *Nature*, **432**, 971–979.

SAKIMOTO, S. E. H. & GREGG, T. K. P. 2001. Channeled flow: Analytic solutions, laboratory experiments, and applications to lava flows. *Journal of Geophysical Research*, **106**, 8629–8644.

SAKIMOTO, S. E. H., CRISP, J. & BALOGA, S. M. 1997. Eruption constraints on tube-fed planetary lava flows. *Journal of Geophysical Research*, **102**, 6597–6613.

SCHENK, P. M. & WILLIAMS, D. A. 2004. A potential thermal erosion lava channel on Io. *Geophysical Research Letters*, **31**, L23702.

SCOTT, D. H. 1981. *Map Showing Lava Flows in the Northwest Part of the Tharsis Quadrangle of Mars*. United States Geological Survey Miscellaneous Investigations Series Map, **I-1266**.

SELF, S., THORDARSON, T. & KESZTHELYI, L. 1997. Emplacement of continental flood basalt lava flows. *In*: MAHONEY, J. J. & COFFIN, M. F. (eds) *Large Igneous Provinces: Continental, Oceanic, and Planetary Flood Volcanism*. American Geophysical Union, Geophysical Monograph, **100**, 381–410.

SMITH, D. E., ZUBER, M. T. *ET AL.* 1999. The global topography of Mars and implications for surface evolution. *Science*, **284**, 1495–1503.

SOLOMON, S. C. & HEAD, J. W. 1982. Evolution of the Tharsis Province of Mars: The importance of heterogeneous lithospheric Thickness and Volcanic construction. *Journal of Geophysical Research*, **87**, 9755–9774.

TALLARICO, A. & DRAGONI, M. 1999. Viscous Newtonian laminar flow in a rectangular channel:

application to Etna lava flows. *Bulletin of Volcanology*, **61** 40–47.

TANAKA, K. L. 1985. Ice-lubricated gravity spreading of the Olympus Mons aureole deposits. *Icarus*, **62**, 191–206.

TANAKA, K. L., SCOTT, D. H. & GREELEY, R. 1992. *Global Stratigraphy. Mars*. University of Arizona Press, Tucson, AZ, 345–382.

VÖRÖSMARTY, C. J., FEKETE, B. & TUCKER, B. A. 1996. *River Discharge Database, Version 1.0 (RivDIS v1.0), Volumes 0 through 6. A Contribution to IHP-V Theme 1*. Technical Documents in Hydrology Series. UNESCO, Paris.

WALLACE, D. & SAGAN, C. 1979. Evaporation of ice in planetary atmospheres – Ice-covered rivers on Mars. *Icarus*, **39**, 385–400.

WILLIAMS, D. A. & LESHER, C. M. 1998. Analytical/numerical modeling of the emplacement and erosional potential of Archean and Proterozoic komatiitic lavas. *In*: *Proceedings of the 29th Annual Lunar and Planetary Science Conference, March 16–20, 1998, Houston, Texas*. Lunar and Planetary Institute, Houston, TX, Abstract 1431.

WILLIAMS, D. A., GREELEY, R., HAUBER, E., GWINNER, K. & NEUKUM, G. 2005. Erosion by flowing martian lava: new insights for hecates tholus from Mars express and MER data. *Journal of Geophysical Research – Planets*, **110**, E05006, doi: 10.1029/2004JE002377.

WILSON, L., GHATAN, G. J., HEAD, J. W. & MITCHELL, K. L. 2004. Mars outflow channels: A reappraisal of the estimation of water flow velocities from water depths, regional slopes, and channel floor properties. *Journal of Geophysical Research*, **109**, E09003, doi: 10.1029/2004JE002281.

ZIMBELMAN, J. R. 1998. Emplacement of long lava flows on planetary surfaces. *Journal of Geophysical Research – Planets*, **103**, 27 503–27 516.

ZIMBELMAN, J. R. & GREGG, T. K. P. 2000. *Environmental Effects on Volcanic Eruptions: From Deep Oceans to Deep Space*, Kluwer Academic/Plenum, New York.

ZUBER, M. T., SMITH, D. E. *ET AL*. 1992. The Mars observer laser altimeter investigation. *Journal of Geophysical Research – Planets*, **97**, 7781–7797.

ZUBER, M. T., SOLOMON, S. C. *ET AL*. 2000. Internal structure and early thermal evolution of Mars from Mars global surveyor topography and gravity. *Science*, **287**, 1788–1793.

Layering and degradation of the Rupes Tenuis unit, Mars – a structural analysis south of Chasma Boreale

T. KNEISSL*, S. VAN GASSELT, L. WENDT, C. GROSS & G. NEUKUM

Institute of Geological Sciences, Planetary Sciences and Remote Sensing, Freie Universitaet Berlin, Malteserstrasse 74-100, D-12249 Berlin, Germany

**Corresponding author (e-mail: Thomas.Kneissl@fu-berlin.de)*

Abstract: The circum north-polar Rupes Tenuis unit forms the polar-proximal basal stratigraphical and morphological units that delineate the north polar cap between 180° and 300°E. In the region of the mouth of the Chasma Boreale re-entrant, the Rupes Tenuis unit is likely to extend further southwards into the northern plains. This is suggested by the occurrence of isolated remnants that have been interpreted as basaltic shield volcanoes, maar craters or mud volcanoes in the past.

As key elements of this study, we assessed the quantitative characteristics of this unit using layer attitudes derived from high-resolution images and terrain-model data, and by performing cross-correlations of prominent layers whose outcrops are observed at eight cone-like remnants. The identification and unambiguous correlation of characteristic layers across the study area provided a reasonable basis for introducing at least three additional stratigraphical subunits of the Rupes Tenuis unit. Extrapolation of altitude data indicates a gentle southward dip of remnant layers, suggesting that the unit had a much larger areal extent in Martian history. The palaeo-layer contact between two subunits of the Rupes Tenuis unit correlates well with elevation values of the Hyperborea Lingula surface. Both results disagree with an interpretation of a volcanic origin for isolated mesas but underpin that they are erosional relicts of the Rupes Tenuis unit. Average erosion rates of $2.5 \times 10^{-4} \pm 4 \times 10^{-5}$ mm year^{-1} are relatively high when compared to Amazonian rates but are not exceptional for areas undergoing deflation. They also corroborate the idea of aeolian denudation of the Rupes Tenuis unit.

The north polar plateau, Planum Boreale, is surrounded and probably underlain by the Vastitas Borealis plains (Tanaka *et al.* 2005; Picardi *et al.* 2005; Phillips *et al.* 2008), and reaches a thickness of up to 2.5 km (Herkenhoff & Plaut 2000). Between 180° and 300°E Planum Boreale is circumscribed by the abrupt scarp of Rupes Tenuis, which reaches a height of approximately 1000 m (Tanaka *et al.* 2005). East of Rupes Tenuis, the Chasma Boreale re-entrant transitions into a low plateau called Hyperborea Lingula. This lobate structure extends approximately 100 km southwards into the Vastitas Borealis plains, with a height of between 200 and 350 m (Tanaka *et al.* 2008) and a slope of about 5° at the scarp margin (Fishbaugh & Head 2002). South of this lobate structure, at approximately 77°N, the isolated Escorial Crater mesa rises more than 700 m above Vastitas Borealis. The isolated mounds of Abalos Colles are located south of the Rupes Tenuis scarp and west of the Escorial Crater. These conical and flat-topped knobs occur in the height range of several tens of metres up to several hundreds of metres, with diameters of more than 20 km.

The Rupes Tenuis unit (ABrt) (Tanaka *et al.* 2008) is exposed along the whole Rupes Tenuis scarp and at locations along the floor of the Olympia Cavi. According to recent polar mapping efforts by Tanaka *et al.* (2008), the Rupes Tenuis unit probably also forms the Escorial Crater plateau (see Fig. 1). Following Tanaka *et al.* (2008), the Rupes Tenuis unit (ABrt) stratigraphically constitutes the lower part of the Mars north polar basal unit (Fishbaugh & Head 2005), also termed the Platy unit (Byrne & Murray 2002), the Scandia Region unit (Tanaka *et al.* 2005) or Scandia materials (Tanaka 2005). A summary of past and current units can be found in Tanaka *et al.* (2008). The upper part of the basal unit is composed of the Planum Boreum Cavi unit (ABb_c) (Tanaka *et al.* 2008), which is probably the source for the eroded material that builds up the circumpolar erg (Breed *et al.* 1979; Greeley 1979; Thomas & Weitz 1989; Byrne & Murray 2002; Tanaka *et al.* 2008). However, at the terminus of Chasma Boreale and along nearly the whole Rupes Tenuis scarp, the Rupes Tenuis unit (ABrt) is directly superimposed by the Planum Boreum 1 unit (ABb_1) that belongs to the high-albedo polar layered deposits (PLD) (Tanaka *et al.* 2008). Thus, the Planum Boreum Cavi unit was not emplaced over the whole of the Rupes Tenuis unit or it was partly removed by erosion. ABrt stratigraphically overlies the Vastitas Borealis interior unit (ABv_i), which represents the oldest unit in the north polar region and defines the base of the Lower Amazonian series (Tanaka *et al.* 2005).

From: BALME, M. R., BARGERY, A. S., GALLAGHER, C. J. & GUPTA, S. (eds) *Martian Geomorphology*. Geological Society, London, Special Publications, **356**, 257–279.
DOI: 10.1144/SP356.13 0305-8719/11/$15.00

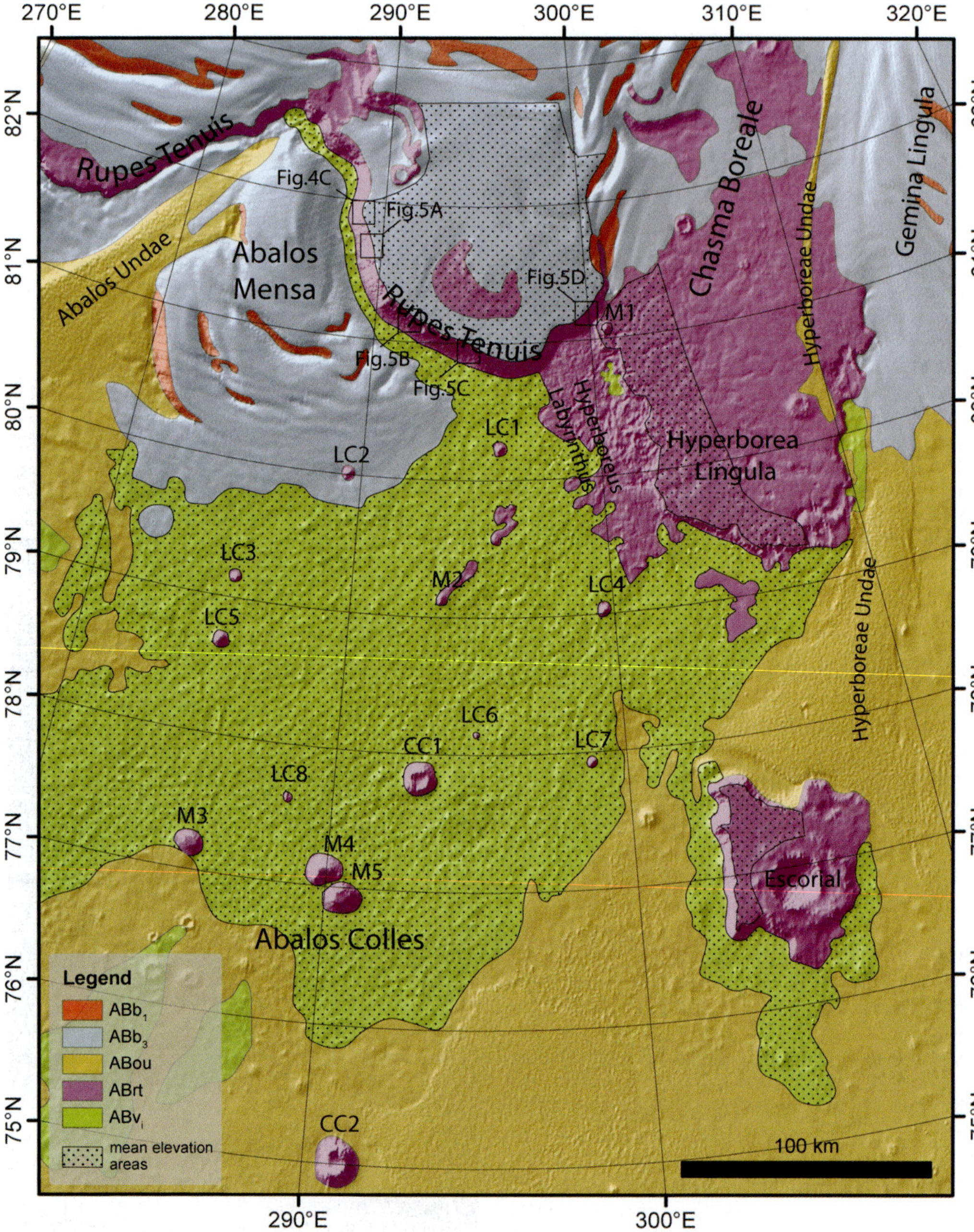

Fig. 1. Geological map of the mouth of the Chasma Boreale region modified after Tanaka *et al.* (2008). Map draped over a MOLA hillshade (*c.* 150 m per pixel). Locations of Figures 5a, d and 9c are marked as denoted boxes. Dotted areas were used for the determination of mean elevations of large-scale surface features. Abbreviations: LC, layered cone; M, mesa; CC, cratered cone.

In contrast to the finely layered, bright PLDs that superpose the Rupes Tenuis unit at the terminus of Chasma Boreale, the low-albedo Rupes Tenuis unit shows a characteristic irregular, thick, plate-like structure in cross-section (Byrne & Murray 2002). Tanaka *et al.* (2008) identified approximately 20 individual layers in the Rupes Tenuis unit, each one ranging from tens of metres to approximately 100 m thick, showing predominantly horizontal–subhorizontal layering. Individual layer surfaces are locally fractured at the metre-scale, and are commonly eroded to knobs at the margins of thick layers (Tanaka *et al.* 2008). In addition, the Rupes Tenuis unit shows evidence of extensive erosion along the scarps of Rupes Tenuis (Tanaka *et al.* 2008; Warner & Farmer 2008*a*). Erosional contacts such as 'cross-beddings' (as referred to by Tanaka *et al.* 2008) or, more specifically, discontinuities – such as unconformities – have not been observed, which is suggestive of concordant layering. Consequently, Kolb & Tanaka (2001) concluded that a vertical accumulation of material dominated by precipitation and cold-trapping of dust-laden volatiles is likely. In addition, Tanaka (2005) supposed that the Rupes Tenuis unit might contain wind-transported, silt-size sediments sourced from the nearby Scandia region unit (ABs) (as mapped by Tanaka *et al.* 2005). An additional source for the particles sedimented within the ABrt unit might be the Vastitas Borealis interior unit (ABv_i) (Tanaka *et al.* 2008). However, Tanaka *et al.* (2008) noted that the grain size of particles forming the Rupes Tenuis unit cannot be determined by the analysis of geomorphological or morphometric properties of this unit because the role of volatiles and chemical precipitates within the cementing matrix is not clear. A more detailed description of the geology and extent of the Rupes Tenuis unit can be found in Tanaka *et al.* (2008).

Hyperborea Lingula, a lobe of layered material extending from the mouth of Chasma Boreale (see Fig. 1), might consist of ABrt material (Kolb & Tanaka 2001; Fishbaugh & Head 2005; Edgett *et al.* 2003; Tanaka 2005). Warner & Farmer (2008*a*) observed at least seven distinct continuous layered units at the margin of the lobate structure, some metres to tens of metres thick. In Hyperboreus Labyrinthus, located south of Hyperborea Lingula, some of the lowermost layers of Hyperborea Lingula are visible (Tanaka *et al.* 2008). They show rounded, dish-shaped forms similar to the dish-shaped layer outcrops at the Rupes Tenuis scarp (Tanaka *et al.* 2008). However, the origin of Hyperborea Lingula is still under debate. It could represent the lower part of the ABrt unit, forming an erosional remnant (Kolb & Tanaka 2001; Edgett *et al.* 2003; Tanaka *et al.* 2008; Warner & Farmer 2008*a*), but it could also be comprised of eroded material transported and deposited during a catastrophic outflow event (Fishbaugh & Head 2002). A combination of exhumed ABrt layers and deposition of eroded ABrt material is also conceivable (Fishbaugh & Head 2005). In this scenario, the lower layers of the ABrt unit within Chasma Boreale were exhumed by outflow activity and katabatic winds during the chasma formation. Subsequently, or in the course of this outflow event, these layers were covered by eroded material of the ABrt–ABb_c units. The large dune field, Hyperboreae Undae, which superposes the eastern part of Hyperborea Lingula and the area north of the Escorial Crater, is partly sourced from the ABb_c unit eroded from outcrops in Boreum Cavus, Tenuis Cavus and parts of Chasma Boreale (Tanaka *et al.* 2008).

Southwest of Hyperborea Lingula and west of Escorial Crater, the Abalos Colles form a loose collection of flat-topped knobs with an irregular/angular shape (Fig. 2), and several cratered and

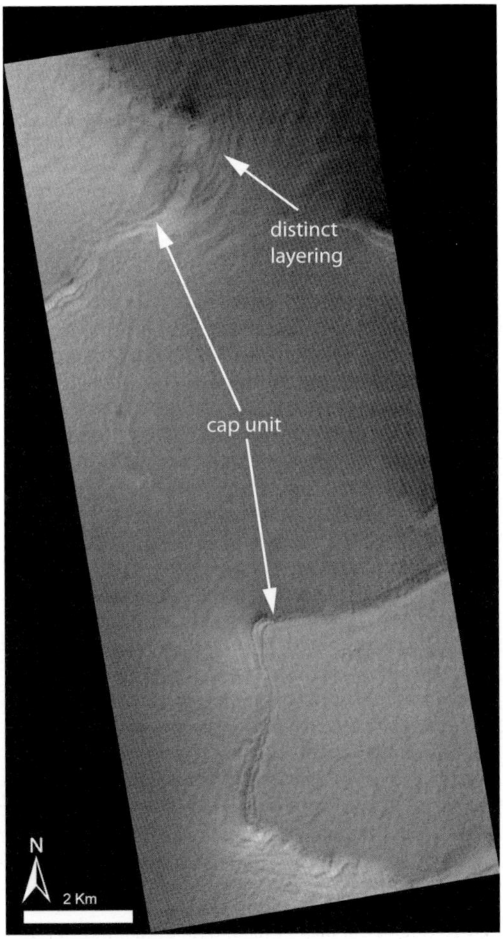

Fig. 2. HiRISE image PSP_006941_2570 of flat-topped mounds M4 and M5, being part of the Abalos Colles mound cluster. Arrows show distinct layering on the flanks and the cap layer described by Warner & Farmer (2008*b*). The image is centred at 77°N, 290.3°E; illumination is from the bottom; resolution is 0.64 m per pixel. Image credit: NASA/JPL/ASU; see prelim viii for acronym definitions.

non-cratered conical mounds (Figs 1 & 3). The isolated features superpose the Vastitas Borealis interior unit (ABv_i), with elevation ranges comparable to Hyperborea Lingula or the Escorial Crater (Fig. 4) and have been investigated in detail (e.g. Hodges & Moore 1994; Garvin *et al.* 2000; Wright *et al.* 2000; Tanaka *et al.* 2003; Tanaka *et al.* 2008; Warner & Farmer 2008*b*).

Using *Viking Orbiter* data, Hodges & Moore (1994) found cratered mounds that share slope-morphology characteristics with terrestrial maar craters, which are associated with explosive volcanism. Using topographical data provided by the *Mars Global Surveyor* (MGS) Mars Orbiter Laser Altimeter (MOLA), Garvin *et al.* (2000), Smith *et al.* (2001) and Sakimoto & Weren (2003) investigated and described those features exhibiting a central depression at the top as 'Martian Cratered Cones' (MCCs). Mean flank slope to volume and diameter relationships led to the conclusion that these features, as well as the flat-topped mounds, show a morphology similar to small terrestrial basaltic shield volcanoes scaled to Martian conditions (Garvin *et al.* 2000; Wright *et al.* 2000; Sakimoto & Weren 2003; Fagan & Sakimoto 2009). Topographical data from the High Resolution Stereo Camera (HRSC) (Neukum & Jaumann 2004; Jaumann *et al.* 2007) confirmed the morphological observations of these mound features and reinforced a possible volcanic origin (Neukum *et al.* 2005). Owing to the lack of spatially correlated volcanic landforms (i.e. characteristic linear graben structures or lava flows) and the estimated crustal thickness of the Martian lowland in the Amazonian (Nimmo & Tanaka 2005; Tanaka *et al.* 2008), the occurrence of localized, small-scale volcanic eruptions was excluded by Skinner *et al.* (2006). They concluded that sedimentary diapirism and mud volcanism from shallow depths are the only conceivable processes if volcano-like formation is hypothesized.

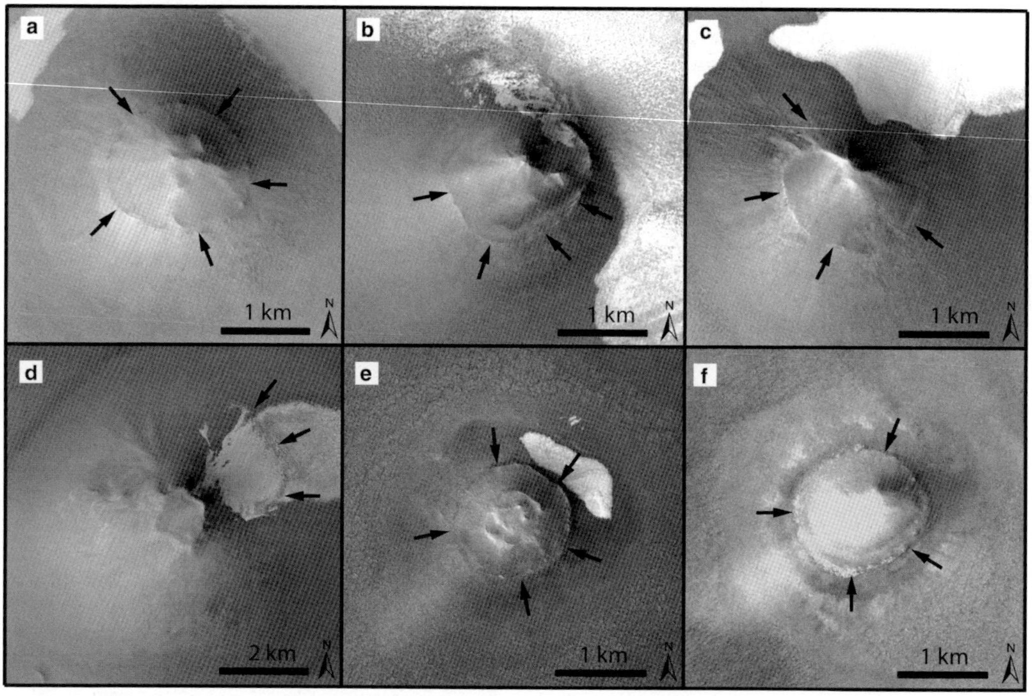

Fig. 3. Examples of layered conical features at the mouth of Chasma Boreale. Black arrows mark mapped layer contacts. (**a**) Layered cone 1 (LC1), image is part of CTX P02_001653_2623; the image is centred at 80.23°N, 295.8°E; illumination is from the lower left (6.26 m per pixel). (**b**) Layered cone 3 (LC3), image is part of CTX P22_009710_2589; image is centred at 79.16°N, 285.47°E; illumination is from the lower left (6.51 m per pixel). (**c**) Layered cone 4 (LC4), image is part of CTX P22_009551_2597; the image is centred at 79.02°N, 299.7°E; illumination is from the lower left (6.24 m per pixel). (**d**) Layered cone 5 (LC5), image is part of CTX P22_009710_2589; the image is centred at 78.68°N, 285.25°E; illumination is from the lower left (6.51 m per pixel). (**e**) Layered cone 6 (LC6), image is part of CTX P22_009525_2598; the image is centred at 78.14°N, 294.78°E; illumination is from the lower left (6.32 m per pixel). (**f**) Layered cone 8 (LC8), image is part of CTX P22_009446_2580; the image is centred at 77.62°N, 288.4°W; illumination is from the lower left (6.27 m per pixel). Image credit: NASA/JPL/MSSS.

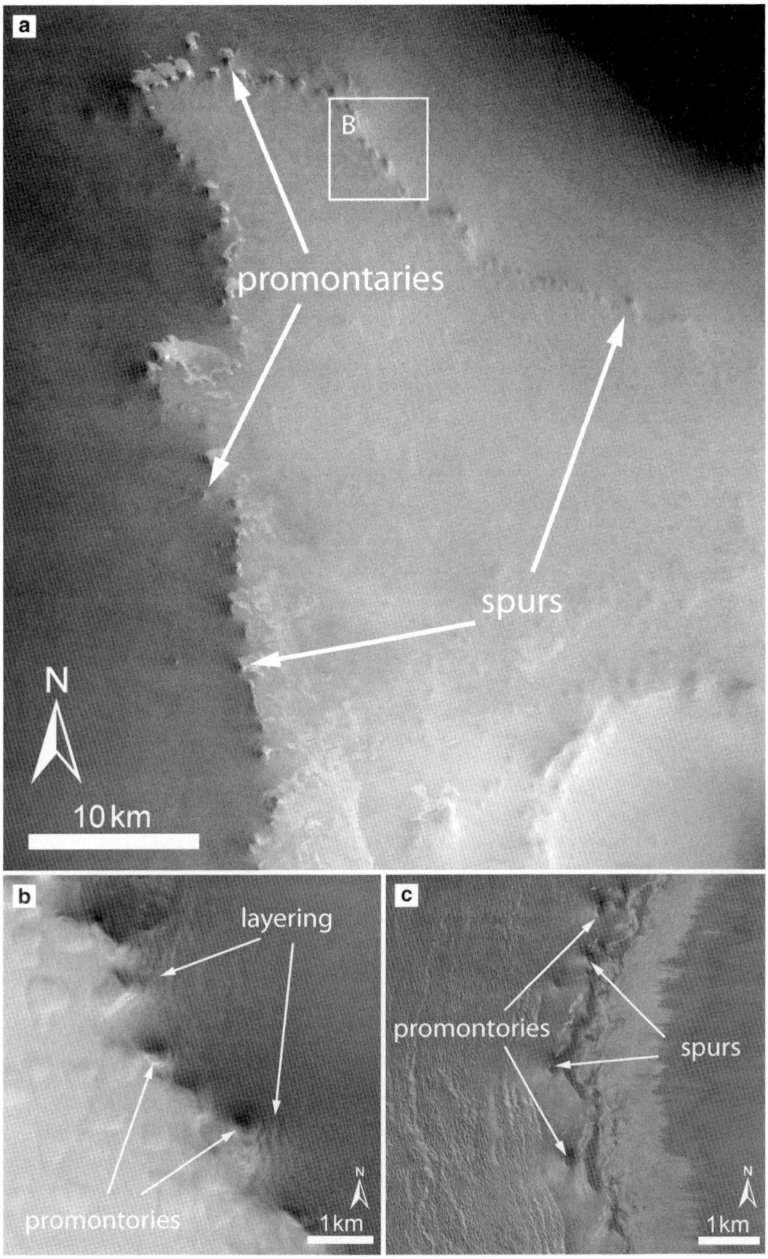

Fig. 4. Escorial Crater plateau. (**a**) Western part of the Escorial Crater plateau in HRSC image h3711_0000 (25 m per pixel). The scarp shows comparable promontories and spurs to the Rupes Tenuis scarp. The image is centred at 77.1°N, 303.7°E; illumination is from the lower right. (**b**) Close-up of the scene in (a). Promontories and distinct layering on the flank of the Escorial Crater plateau are clearly visible. The image is part of CTX P16_007191_2561 (6.23 m per pixel); illumination is from the lower left. (**c**) Comparable promontories and spurs at the Rupes Tenuis scarp. The image is part of CTX P22_009525_2598 (6.32 m per pixel). Illumination is from the lower left. The location of this frame is shown in Figure 1. Image credit: ESA/DLR/FUB, NASA/JPL/MSSS.

Tanaka *et al.* (2003, 2008), Tanaka (2005) and Warner & Farmer (2008*b*) interpreted the Abalos Colles conical mounds to be pedestal crater remnants of a once more-extensive sequence of northern plain layered material. Warner & Farmer (2008*b*) analysed mound-flank slopes and found that mounds located proximal to the present-day Rupes Tenuis scarp show similar flank slopes and morphologies as the scarp. With increasing distance from the scarp, the flank slopes and sizes of the cones decrease because of aeolian erosion, mass wasting and ice sublimation (Warner & Farmer 2008*b*). These observations are consistent with the hypothesis of Rupes Tenuis scarp retreat by katabatic winds (Warner & Farmer 2008*a*). In addition, Warner & Farmer (2008*b*) described horizontal layering on the flanks of several of these cones with a similar scale as the layering at the Rupes Tenuis scarp, supporting the hypothesis of an erosional origin of the Abalos Colles.

This work focuses on the Rupes Tenuis unit at the mouth of Chasma Boreale – that is the Rupes Tenuis scarp between Chasma Boreale and Abalos Mensa, Hyperborea Lingula – as well as the prominent mounds south of this region, and addresses the following key tasks: (i) to test the formation hypothesis of Tanaka (2005) and Warner & Farmer (2008*a*) that the conical mounds have an non-volcanic origin through analysis of topographical correlations; (ii) to test the hypotheses of Tanaka *et al.* (2008) that Hyperborea Lingula was formed by partial erosion of the Rupes Tenuis using structural analyses of the Rupes Tenuis unit; and (iii) to reconstruct a portion of the palaeo-extent of the Rupes Tenuis unit by correlating observations across the study area, and on the basis of the test results from (i) and (ii). Such a reconstruction should provide valuable information on the environmental conditions/settings during the formation of the Rupes Tenuis unit as part of the former basal unit, and for the assessment of the means of emplacement of this unit (i.e. either atmospheric or degradational) because the mechanism and contribution of processes are still being discussed.

Datasets and methodology

The study area is located between 70°–83°N and 280°–320°E. For our analysis we made use of image data obtained by the High Resolution Stereo Camera (HRSC) (0.585–0.765 μm) (Neukum & Jaumann 2004; Jaumann *et al.* 2007), the Thermal Emission Imaging Spectrometer (THEMIS VIS) (0.425–0.860 μm) (Christensen *et al.* 2004), the High Resolution Imaging Science Experiment (HiRISE) (0.550–0.850 μm) (McEwen *et al.* 2007), the Context Camera (CTX) (0.500–0.800 μm) (Malin *et al.* 2007) and the Mars Orbiter Camera (MOC) (0.5–0.9 μm) (Malin *et al.* 1992). The Mars Orbiter Laser Altimeter (MOLA) instrument (Smith *et al.* 2001) provided topographical data.

All image data were obtained from the Planetary Data System (PDS) archives as raw data. With the exception of HRSC, the image data were processed using the United States Geological Survey's (USGS's) Integrated Software for Imagers and Spectrometers (ISIS-3) system (Gaddis *et al.* 1997). HRSC data were processed using the Jet Propulsion Laboratory's (JPL's) Video Image Communication and Retrieval (VICAR) software suite (Anderson & Mann 1989; Hockey & Barnet 1994). Subsequent data ingestion and analyses were carried out in the Environmental Systems Research Institute's (ESRI's) geographical information system (GIS) ArcGIS.

Owing to the almost complete image coverage of the study area, a first-order identification and characterization of surface features was carried out on HRSC imagery with a resolution of 25 m per pixel. We used four images in our study: h1187_0000, h1264_0000, h3711_0000 and h5793_0000. In order to fill the remaining coverage gaps, five THEMIS-VIS images with a resolution of 40 m per pixel were utilized. CTX data (6.23–6.51 m per pixel), MOC images (1.53–6.19 m per pixel) and HiRISE data (33–64 cm per pixel) were incorporated in order to investigate small surface structures in detail or in order to map crater rims and layer contacts (Table 1). For large-scale topographical analyses, we used a PDS-distributed MOLA gridded digital terrain model with a resolution of 128 pixels per degree (*c.* 463 m per pixel). For closer inspection of small-scale features and for layer attitude measurements, we made use of a MOLA DTM with a resolution of 512 pixels per degree (*c.* 150 m per pixel), which we gridded from polar Precision Experiment Data Record (PEDR) MOLA tracks using the stand-alone Generic Mapping Tools (GMT) (Wessel & Smith 1995) programs and the nearest-neighbour approach. This gridding technique and the resulting errors are described in detail by (Okubo *et al.* 2004). Shallow Subsurface Radar (SHARAD) data (Seu *et al.* 2004) obtained from PDS was utilized for the analysis of the subsurface structure of the mesa surrounding the Escorial Crater. The frequency of SHARAD is centred at 20 MHz, with a 10 MHz bandwidth. The two-way range resolution amounts to 15 m divided by the square root of the real part of the permittivity of the propagation medium. The spatial resolution can vary between 300 m and 1 km in the along-track direction. Range compression and synthetic-aperture processing are performed on the ground. Radargrams were produced showing the along-track processed data frames of

Table 1. *Image data used for measurements*

Instrument	Image number	Mean date and time	Map scale (m per pixel)	Solar	
				Elevation (°)	Azimuth (°)
HRSC	h1187_0000	2004-12-21 02:29:17	25	33.3	195.0
HRSC	h1264_0000	2005-01-11 15:58:53	25	27.5	177.8
HRSC	h3711_0000	2006-11-27 20:13:10	25	22.7	124.1
HRSC	h5793_0000	2008-07-06 14:49:06	25	32.6	128.0
THEMIS VIS	V11262001	2004-06-28 16:50:40	40	20.1	272.2
THEMIS VIS	V12635004	2004-10-19 17:55:04	40	23.4	278.7
THEMIS VIS	V12847004	2004-11-06 04:50:27	40	22.4	278.2
THEMIS VIS	V10139013	2004-03-28 05:48:22	40	6.1	263.3
THEMIS VIS	V10451005	2004-04-22 22:19:59	40	11.8	263.1
MOC NA	R1900856	2004-07-11 20:16:16	1.88	30.9	199.9
MOC NA	R1701595	2004-05-18 09:10:31	1.53	24.8	196.9
MOC NA	R0101406	2003-01-28 06:21:59	6.19	15.1	57.8
MOC NA	R1800685	2004-06-08 23:18:48	3.76	26.0	197.1
MOC NA	R1900255	2004-07-04 05:50:56	4.62	15.3	51.3
MOC NA	E2301031	2002-12-22 15:26:15	3.76	31.9	200.1
CTX	B02_010237_2590	2008-10-02 04:33:44	6.30	25.9	221.6
CTX	P02_001653_2623	2006-12-03 06:41:16	6.26	20.8	214.9
CTX	P16_007191_2561	2008-02-07 19:23:07	6.23	23.8	212.5
CTX	P16_007468_2578	2008-02-29 09:31:32	6.30	25.9	212.9
CTX	P22_009446_2580	2008-08-01 13:17:29	6.27	33.3	223.6
CTX	P22_009525_2598	2008-08-07 17:02:04	6.32	31.5	223.1
CTX	P22_009551_2597	2008-08-09 17:39:32	6.24	31.7	220.7
CTX	P22_009710_2589	2008-08-22 03:00:54	6.51	30.4	228.1
HiRISE	PSP_009670_2590	2008-08-19 00:12:55	0.33	25.9	221.6
HiRISE	PSP_006941_2570	2008-01-19 07:49:46	0.64	19.8	209.0

radar power. The x-axis of a radargram is the spacecraft distance along track, while the y-axis is the range time delay. We generated the radargrams from the Reduced Data Record (RDR) dataset. Surface features located off-nadir can produce echoes that reach the radar after the nadir echo. This so-called clutter could be wrongly interpreted as subsurface reflections. The use of a surface echo simulator can help to solve this ambiguity (Russo *et al.* 2008). This program makes use of MOLA Mission Experiment Gridded Data Records (MEGDRs). Comparing the real echo and the simulated surface echo can reveal whether a given feature in a radargram is clutter or a real subsurface detection.

In order to constrain the origin of conical mounds south of Chasma Boreale, we investigated the topographical elevation relationships of prominent mound features in this region. In particular, we focused on the flat-topped Abalos Colles, on cratered cones and on non-cratered layered cones (M1–M5, CC1–CC2 and LC1–LC8 in Fig. 1). For a basic morphological characterization, we determined basal and summit elevations for all of these mounds using the elevation information of the 512 pixel per degree MOLA DTM. The summit elevations of cratered cones were determined using averaged values of topographical profiles measured along crater rims at two features. The mean elevations of the plateaus of the five flat-topped Abalos Colles mesas were determined by averaging all DTM raster-pixel values, that is, elevations below the Martian datum. Layer contacts, where present, were mapped for each conical feature in our investigation area. Mean topographical elevations of the layer contacts were determined by averaging topographical profiles along the outcrop. Elevation values of all mapped layer contacts were correlated to each other and to surrounding landforms, such as flat-topped mesas.

In order to include the topographical results of small-scale mounds into the overall analysis of our study area and to constrain the formation of Hyperborea Lingula, we extracted representative elevation data of the surfaces of large-scale landforms as a function of latitude. For this purpose, the respective areas were divided into latitudinal bins, each spanning 0.25°. This was carried out for Hyperborea Lingula, the Escorial Crater mesa, and the plateau above the Rupes Tenuis scarp between Chasma Boreale and Abalos Mensa. In order to attenuate the influence of small-scale surface features (e.g. dunes and ice), all contained DTM raster pixel values were averaged, respectively. The same procedure was applied to extract mean elevation data of unit ABv_i that is stratigraphically located beneath the Rupes Tenuis unit (ABrt), as described by Tanaka *et al.* (2008). The areas that were used for the determination of averaged elevation data are marked in Figure 1.

Layer attitude measurements of the Rupes Tenuis scarp were performed using the ArcGIS extension LayerTools described in Kneissl *et al.* (2010). This software tool uses latitude/longitude coordinates of measurement points defined on image data in combination with their elevation values extracted from an underlying digital terrain model (DTM). In particular, the locations and elevations of different measurement points along an observed outcrop of a geological layer allow the construction of planes describing the extent of the geological layer of interest in terms of planar dip and dip direction as defined by at least three measurement points. For more than three points, the best-fitting plane is computed using a one-degree polynomial fit (Kneissl *et al.* 2010). Output from this GIS extension comprises dip and dip-direction values of the interpolated planes, as well as the root-mean-square (RMS) error of the interpolation. However, we did not measure the attitudes of the individual observed layer contacts on the flanks of the conical mounds because this kind of attitude measurement, with very small distances between the interpolation points, is not very accurate owing to uncertainties of the positioning of the image data on the MOLA DTM.

Errors for layer attitudes were calculated assuming a normal distribution of measurement errors and the Gaussian error propagation for standard deviations. Individual measurement points (latitude/longitude co-ordinates and elevations) were fitted to a plane so that the sum of squared errors was minimized. Dip directions (α) and dip angles (ϕ) were derived trigonometrically from the plane's normal vector. Standard deviations for α and ϕ were obtained by partially differentiating both trigonometric equations using a conservative standard deviation of 200 m for the horizontal and 100 m for the vertical position of each measurement triplet. A figure of 200 m was chosen in order to reflect an averaged 400 m footprint representation of each MOLA shot. Errors for dip and dip direction depend primarily on the errors in pixel locations used for obtaining measurements (MOLA as the most-coarse dataset). They become relatively small when the spacing of measurement points is wide so that vertical or horizontal shifts in locations have only a minor influence on the attitude values.

For the interpolation and reconstruction of former layer surfaces (LSs) using averaged elevation values, such as mean elevations of topographical profiles (layer contacts) or flat surfaces (mesas), it is not possible to use the common workflow of the software tool. Instead of measuring on DTM surfaces (definition of interpolation points by direct extraction of individual elevation values from a DTM), we had to define interpolation points manually by co-ordinate triplets of individual surface features; that is, latitude/longitude co-ordinates and averaged elevations of layer contacts or surfaces of the flat-topped mountains. In order to reconstruct the former spatial extent of the ABrt unit, the extension additionally allows the construction of the intersection of the interpolated planes with the present-day surface; that is, the original terrain model.

For the determination of area sizes and volumes, digital elevation data of the interpolated layer surfaces and the original MOLA DTM – that is the present-day surface – were projected onto an equal-area sinusoidal map projection using the centre of our investigation region as the central meridian. By subtracting both datasets from each other a raster dataset is obtained that contains pixel values for the volume that was eroded or deposited.

Observations and morphometry results

Rupes Tenuis unit

The Rupes Tenuis unit forms the Rupes Tenuis scarp and probably underlies all of Planum Boreum (Tanaka *et al.* 2008). At the southernmost end (80.9°N, 296.4°E) of the plateau, above the Rupes Tenuis scarp between Chasma Boreale and Abalos Mensa, the Rupes Tenuis unit has a thickness of approximately 1000 m as derived from the MOLA DTM. Warner & Farmer (2008*b*) described a single approximately 100 m-thick resistant cap unit, forming the upper part of the Rupes Tenuis unit, identified in several places along the Rupes Tenuis scarp. Warner & Farmer (2008*b*) proposed a hypothesis that this cap unit is being undermined by aeolian erosion, solar ablation and mass wasting, resulting in small promontories along the whole scarp. We traced the lower layer contact of this cap unit along the scarp between Chasma Boreale and the northern end of Abalos Mensa, measuring the thickness of this unit to vary between about 100 and 200 m (Fig. 5).

We systematically performed strike and dip measurements along the Rupes Tenuis scarp

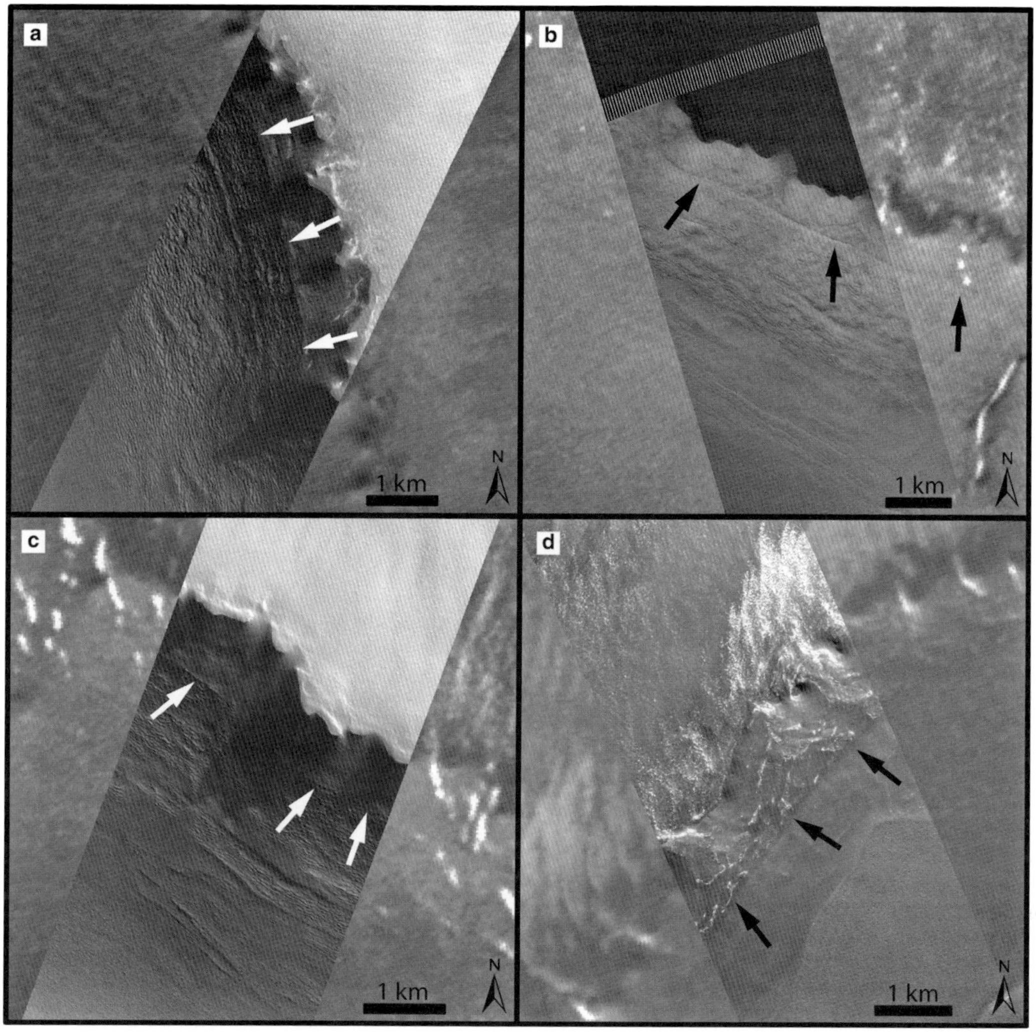

Fig. 5. Layering at the Rupes Tenuis scarp. Black and white arrows mark the layer contact between the cap unit described by Warner & Farmer (2008*b*) and the underlying ABrt unit. (**a**) Image composed of HRSC h1187_0000 superimposed with MOC image r0101406. The centre is located at 81.73°N, 289.39°E; illumination is from the upper right. (**b**) HRSC h1187_0000 superimposed with MOC image r1800685; the image is centred at 81.16°N, 291.89°E; illumination is from the bottom. (**c**) Image composed of HRSC h1187_0000 superimposed with MOC image r1900255, the image is centred at 80.93°N, 294.83°E; illumination is from the upper right. (**d**) HRSC h1187_0000 superimposed with MOC image e2301031, the image is centred at 81.21°N, 299.9°E; illumination is from the lower left. Bright pixels are over-saturated. Image credit: NASA/JPL/MSSS, ESA/DLR/FUB.

(Fig. 6 & Table 2). All measurements show comparable dip values in the range of 0.1°–0.5° ($\bar{x} = 0.3$, $\sigma = 0.1$) towards the NE (i.e. Chasma Boreale). As all measurements provide comparable dip directions in the range of 35.6°–71.6° ($\bar{x} = 46.1$, $\sigma = 12.0$) and comparable dip angles, it is conceivable that the whole ABrt unit was deposited homogeneously, that is, with a constant thickness over a large extent, and subhorizontally at least in our study area, which is best observed at the western wall of Chasma Boreale and along the walls of the narrow trough separating Abalos Mensa from the Rupes Tenuis scarp. Topographical profiles and cross-sections for these two locations are shown in Figure 7. The observed slope breaks coincide with different erodibilities. Based on these different erodibilities, the Rupes Tenuis unit can be subdivided into at least three major subunits, $ABrt_1$, $ABrt_2$

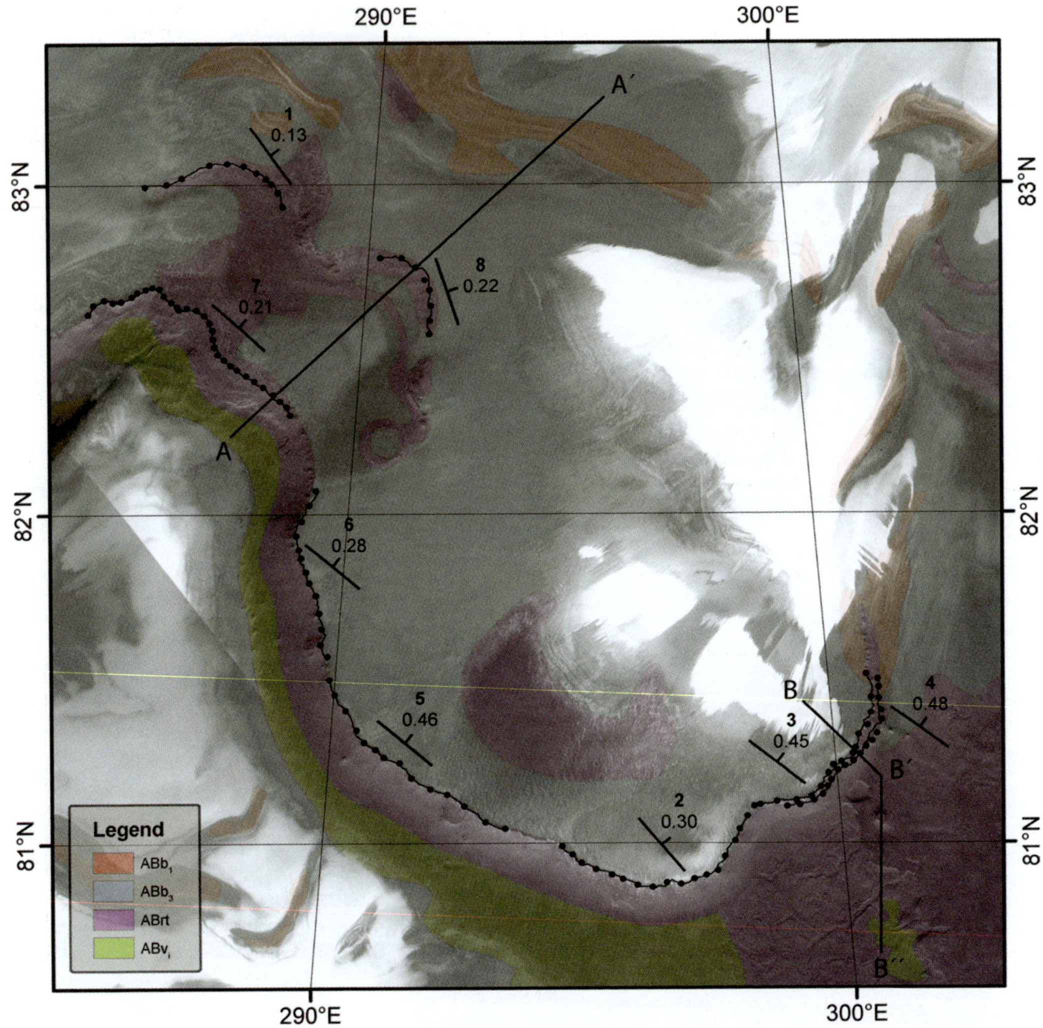

Fig. 6. Strike and dip measurements at the Rupes Tenuis unit west of the mouth of Chasma Boreale. Geological map modified after Tanaka *et al.* (2008), draped over HRSC images h1187_0000 and h1264_0000 (both 25 m per pixel). Strike and dip measurements are shown with their corresponding measurement points (black dots) and the interpolated outcrop lines. See Table 2 for corresponding errors of dip and dip direction. Cross-sections A–A′ and B–B′–B″ are shown in Figure 7.

and $ABrt_3$ (cap unit), with thicknesses of approximately 310, 300 and 150 m, respectively. Subunits, as defined by slope breaks in topographical and image data, are mostly indistinguishable on the basis of rock-unit characteristics; that is, properties such as texture, albedo and roughness. They are generally characterized by slope-parallel lineations in the decametre range. Other ABrt outcrops, stratigraphically higher than the $ABrt_3$ unit, were observed north of Abalos Mensa and interpreted to be remnants of debris-flow margins (Tanaka *et al.* 2008). Comparable strike and dip values, as for units $ABrtpt_{1-3}$, suggest that they are subhorizontal layers as well, forming two additional local units, $ABrt_4$ and $ABrt_5$, differentiated by different erodibility (Fig. 7a & Table 2).

Mounds south of Chasma Boreale

Eight conical mounds in the study area (Fig. 1) show a distinct horizontal layer contact, whose topographical elevation relationships can provide important insights into the formation mechanism of the cones. Not all mound flanks exhibit such characteristic

Table 2. *Strike/dip measurements at the Rupes Tenuis scarp and layer surface (LS) reconstructions from elevation data of layer contacts on the flanks of conical mounds (LS1) and the elevations of surfaces of flat-topped mounds (LS2).* n *refers to the number of points used for the interpolation. LS refers to layer surface interpolations*

ID	Dip (°)	Dip error (°)	Dip direction (°)	Dip direction error (°)	*n*	RMS
1	0.13	0.008	53.85	3.305	11	2.3
2	0.30	0.001	50.04	0.652	18	6.0
3	0.45	0.002	37.50	0.883	16	9.1
4	0.48	0.002	35.60	1.173	18	12.8
5	0.46	0.015	39.73	0.156	14	5.3
6	0.28	0.015	39.35	2.912	14	8.8
7	0.21	0.004	41.51	0.130	32	13.7
8	0.22	0.015	71.56	1.848	8	5.6
LS1	0.07	0.0007	182.54	0.47	7	17.20
LS2	0.12	0.0005	185.33	0.21	6	13.39

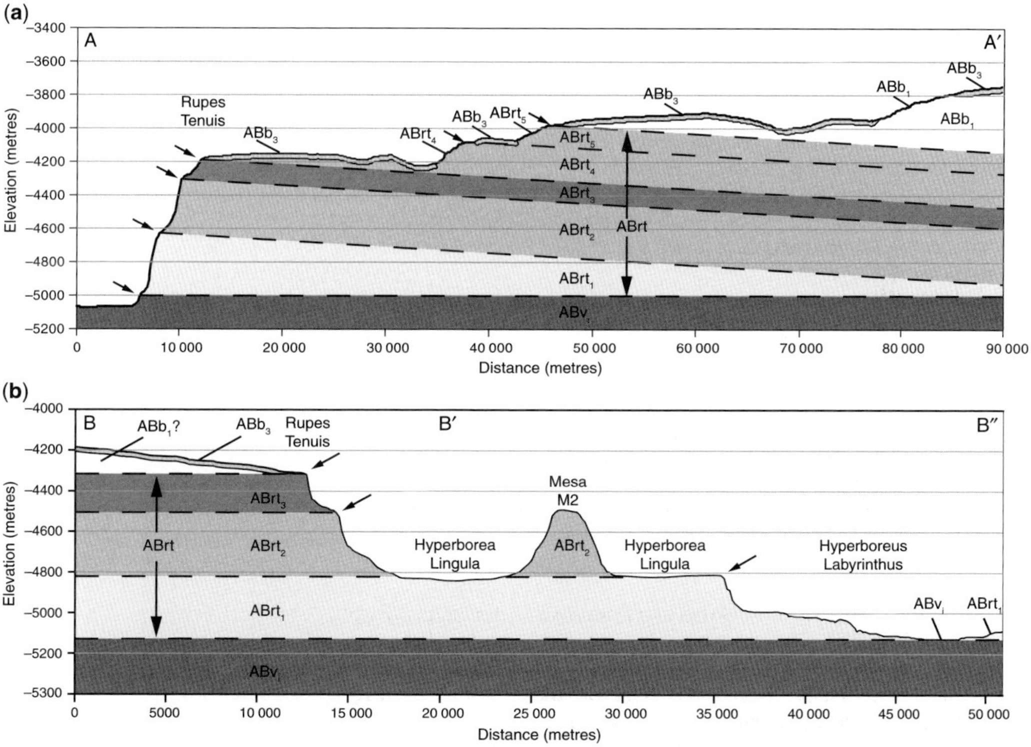

Fig. 7. Topographical profiles and inferred geological cross-sections at the Rupes Tenuis scarp using a MOLA DTM (*c.* 150 m per pixel). Arrows mark kinks interpreted as layer contacts. Locations of the cross-sections are shown in Figure 6. (**a**) Topographical profile north of Abalos Mensa. Dip angles of cross-section A–A′ correspond to the mean layer dip angles (0.2°) measured in this area (Fig. 6). Vertical exaggeration is approximately ×16.7. (**b**) Topographical profile at the mouth of Chasma Boreale. As cross-section B–B′–B″ is parallel to the mean direction of strike, the apparent dip angles are 0. Vertical exaggeration is about ×13.

contacts, which may be due to wind erosion, gravity-driven deposits and sublimation processes, as described by Warner & Farmer (2008*b*). Here, the detected sections of layer contacts were mapped and provided elevation data for further analyses.

As can be seen in Figure 8, the layer at the upper flank of LC4 appears to be very smooth in contrast to the layer at the lower flank, which shows a rough texture. The smooth texture of the upper layer, however, may be caused by mantle deposits, as commonly found in the polar regions of Mars (e.g. Mustard *et al.* 2001). Polygonal patterns occur on both the upper and lower layers. Erosion of the upper layer results in mass-wasting deposits (i.e. slumps and boulders) on the lower flanks of the mounds. Furthermore, this loose material provides the conditions for the formation of gullies (Fig. 8). The smooth texture for the upper layer and the rough texture for the lower one is generally also found at LC1, LC2, LC3, LC7 and LC8, leading to the assumption that the observed layer contacts separates the same layers for all of these conical features. Small-scale surface features, such as patterned ground or boulders, are difficult to identify at these mounds because only lower-resolution imagery (CTX) was available (in contrast to LC4 with HiRISE). At LC5 mass-wasting deposits (slumps) have heavily affected all flanks of the mound and covered most of the layer contact (Fig. 3d). This might be due to its height – that is, its actively erodible surface area – which is approximately twice that of the rest of the LCs. In addition, the summit of LC5 is not a peak, but rather a flat and smooth surface. LC6 also seems to be different to the other LCs. While there is a comparably well-defined layer contact on the flanks (marked by black arrows in Fig. 3e), the summit of the cone is more complex than the summit of other LCs. Here, two more or less circular-shaped structures that probably represent two additional layer contacts are seen at the summit.

The determined mean elevations of the layer contacts span a range between −4750 and −5140 m below datum (see Table 3). These layer-contact elevations (except the one at LC6 (−5140 m), which was excluded from this comparison because

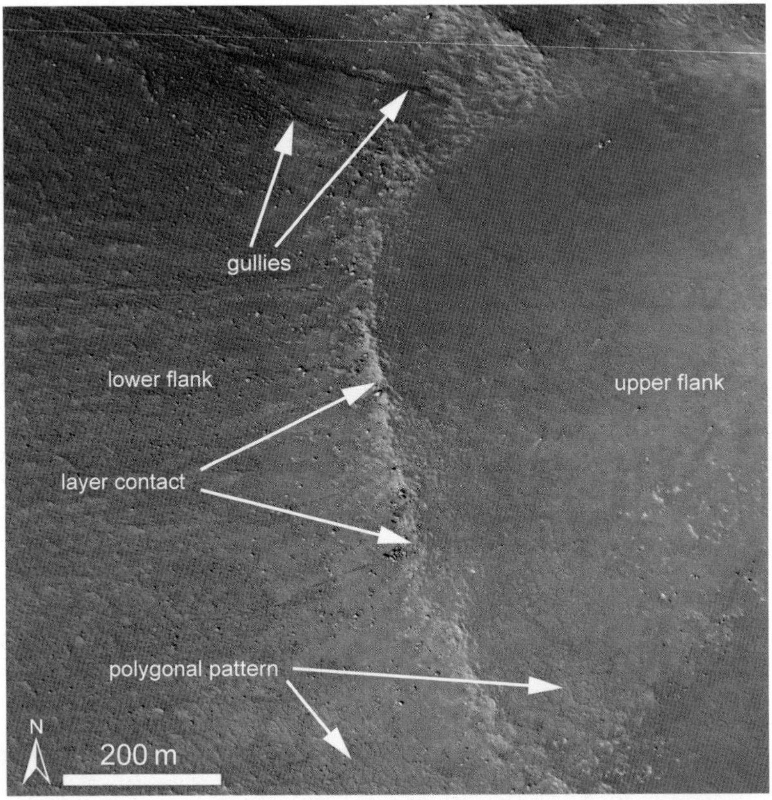

Fig. 8. Distinct layer contact on the flank of LC4 separating the smooth upper flank (right) from the rough lower flank (left). The image is part of HiRISE PSP_009670_2590 of LC4, with a resolution of 33 cm per pixel; illumination is from the left. Image credit: NASA/JPL/ASU.

Table 3. *Features analysed in this study*

ID	Layer surface	Centre co-ordinates (latitude °N/ longitude °E)	Base elevation (m)	Layer-contact elevation (m)	Calculated original surface (m)	Summit elevation (m)	Mean rim elevation (m)	Mean plateau elevation (m)	Deviation to interpolated surface (m)
LC1	1	80.23/ − 64.20	−5102	−4775		−4642			23.93
LC2	1	80.01/ − 70.60	−5072	−4850		−4805			−31.64
LC3	1	79.16/ − 74.53	−5047	−4889		−4775			−5.54
LC4	1	79.02/ − 60.30	−5109	−4894		−4795			−9.27
LC5	1–2	78.69/ − 74.75	−5050	−4887		−4421			30.82/ − 22.26
LC6	–	78.15/ − 65.22	−5168	−5130		−5098			
LC7	1	77.93/ − 61.20	−5205	−4959		−4943			5.46
LC8	1	77.62/ − 71.60	−5069	−5008		−4919			−13.77
CC1	–	77.83/ − 67.15	−5158		−4927		−4749		
CC2	–	75.00/ − 69.03	−5016		−4698		−4497		
M1	–	81.00/ − 59.24	−4821					−4497	
M2	–	79.13/ − 66.58	−5063					−4886	
M3	–	77.20/ − 74.60	−5032					−4803	
M4	2	77.13/ − 70.10	−5109					−4564	24.12
M5	2	76.93/ − 69.44	−5088					−4614	0.24
ECPN	2	77.63/ − 56.40						−4480	10.05
ECPS	2	76.63/ − 56.67						−4625	−17.91
RTS	2	80.86/ − 63.64						−4079	5.75

LC, layered cones; CC, cratered cones; M, mesas; ECPN, Escorial Crater plateau north; ECPS, Escorial Crater plateau south; RTS, Rupes Tenuis scarp.

of a different morphology of the layer above the mapped layer contact) seem to be independent of edifice size and height, as clearly observed by, for example, comparing edifices LC3 and LC5 (Fig. 9). Plotting layer-contact elevations v. latitude (Fig. 9) confirms a height correlation of observed layer contacts (except the one at LC6) and shows a slight southward dip of this layer surface (LS1) (0.07°) (see also LS1 in Table 3). In addition, layer-contact elevations of individual cones match with the averaged elevation values of the Hyperborea Lingula plateau. As the upper surface of Hyperborea Lingula shows a slight dip in the eastward direction and all investigated conical features are located west of the Lingula, we only used the western part of the plateau for our mean elevation determination (see the used areas for averaged elevation data in Fig. 1). Assuming that the surface of Hyperborea Lingula represents the layer surface of $ABrt_1$, as shown in Figure 7b, the vertical correlation of the layer contact of the conical features (except LC6) with the Hyperborea Lingula surface indicates that the observed layer contact is separating $ABrt_1$ from $ABrt_2$ within the conical features (except LC6). By intersecting the corresponding interpolated layer surface (LS1) with the original MOLA DTM, the former spatial extent of the lower layer ($ABrt_1$) could be estimated. It was found that the layer intersected the terrain model at approximately 76°N (Fig. 10).

Apart from layer-contact elevations, averaged elevation values of surfaces of the flat-topped mountains M4, M5, Escorial Crater mesa and the Rupes Tenuis scarp also seem to mark a common layer surface that was once contiguous, as previously proposed by Tanaka *et al.* (2008) and Warner & Farmer (2008*b*). This layer surface (LS2) might represent the surface of $ABrt_3$ previously described as 'cap rock' by Warner & Farmer (2008*b*). For the reconstruction of this layer surface (LS2) we have used the elevations of the Escorial Crater plateau (northern- and southernmost bin of the 1/4-degree bins on the Escorial Crater plateau; abbreviations ECPN and ECPS), M4, M5, LC5 and the southernmost elevation of the plateau above the Rupes Tenuis scarp (RTS at 80.86°N, 296.36°E) (ECPN -4480; ECPS -4625; M4-4564; M5-4614; LC5 -4421; and RTS -4079 m above datum). We used the elevation of the Rupes Tenuis scarp and the surface of the Escorial Crater mesa as we observed similar spurs and small promontories along the western scarp of the Escorial Crater mesa, which are characteristic of the Rupes Tenuis scarp (Fig. 4). It is assumed that these are probably created by aeolian erosion, solar ablation and mass wasting (Warner & Farmer 2008*b*). Similar to the

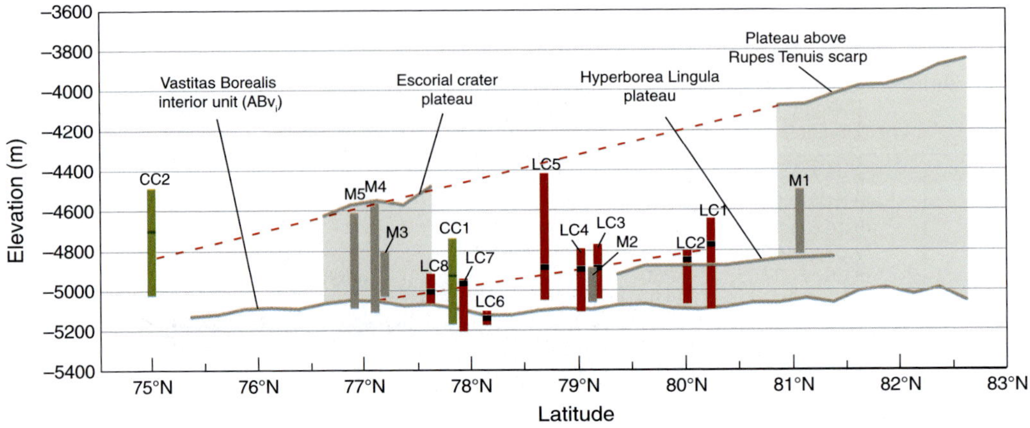

Fig. 9. Plot of feature latitude v. elevation. The red bars show summit and mean basal elevations; black rectangles represent mean elevations of the mapped layer contacts. The grey bars show mean basal and plateau elevations of flat-topped mesas. Mean crater-rim elevations and basal elevations of cratered cones are represented by light-green bars. Dark-green rectangles show the pre-impact surface elevations calculated using formulae provided by Garvin *et al.* (2002). Light grey lines represent mean elevations of large-scale surface features in the investigation area. Trend lines, that is, interpolated layer surfaces, are represented as dashed red lines. The lower layer surface (LC1) was interpolated using all mean layer-contact elevations, except for LC6. For interpolation of the upper layer surface (LS2), elevations of the Escorial cater plateau, M5, M4, LC5, and the southernmost elevation of the plateau above Rupes Tenuis scarp were used.

surface of Hyperborea Lingula, the Escorial Crater mesa dips towards the east and the conical features are located in the west. Therefore, we only used the western plateau of the mesa (ejecta blanket excluded) for our elevation determination and the interpolation of the upper layer surface (LS2) (see dotted areas in Fig. 1). The mean elevations of M4 and M5 were used for plane reconstruction, as they show a similar cap unit to the Rupes Tenuis unit (Warner & Farmer 2008*b*) (Fig. 2) and are at a comparable elevation level as the Escorial Crater mesa. The elevation value for LC5 was used because it has a more or less flat surface with a polygonal shape, different to the other conical mound forms. The reconstructed layer surface (LS2) shows a comparable dip angle of 0.12° (see LS2 in Table 3), with the plane reconstructed from the layer-contact elevations of the investigated layered cones (LS1). Similar to the lower layer surface (LS1), we have estimated the former southward continuation of the upper layer surface (LS2) by intersecting it with the original MOLA topography, showing a southward extent reaching approximately 74°N latitude (Fig. 10). For layer reconstruction we used only two of the five contacts because south of the Rupes Tenuis scarp only layer contacts at the conical mounds and the flat-topped features provide the possibility of a reconstruction of the corresponding layer surfaces towards the south.

Assuming that cratered cones CC1 and CC2 represent remnants of highly eroded impact craters, as proposed by Tanaka *et al.* (2008) and Warner & Farmer (2008*b*), the morphometric treatment by Garvin *et al.* (2002) enabled us to calculate expected rim heights of crater CC1, with a diameter of approximately 6.9 km, and CC2, with a diameter of approximately 10.2 km. For simple craters – that is, those craters with diameters, D, of less than 7 km – rim heights, h, follow the relationship:

$$h = 0.07D^{0.52} \quad (1)$$

For complex craters with D between 7 and 110 km, the following equation gives the expected rim height:

$$h = 0.05D^{0.60} \quad (2)$$

Using these calculated rim heights, 191 ± 7 m for CC1 and 201 ± 5 m for CC2 (errors correspond to an over-/underestimation of the crater diameter by *c.* 500 m) (Table 3), we are able to determine approximated elevations of the pre-impact plateaus. However, since we do not know whether the observed diameters are the actual crater diameters or the diameters of remnants of the interior crater walls, care must be taken to not over interpret our elevation data of the pre-impact plateaus. Owing to the obvious infill of the crater bowls, it is not possible to reconstruct the original crater shapes and diameters using the cavity formulae by Garvin *et al.* (2002). If we have not measured the actual crater diameter, but rather the diameter of interior

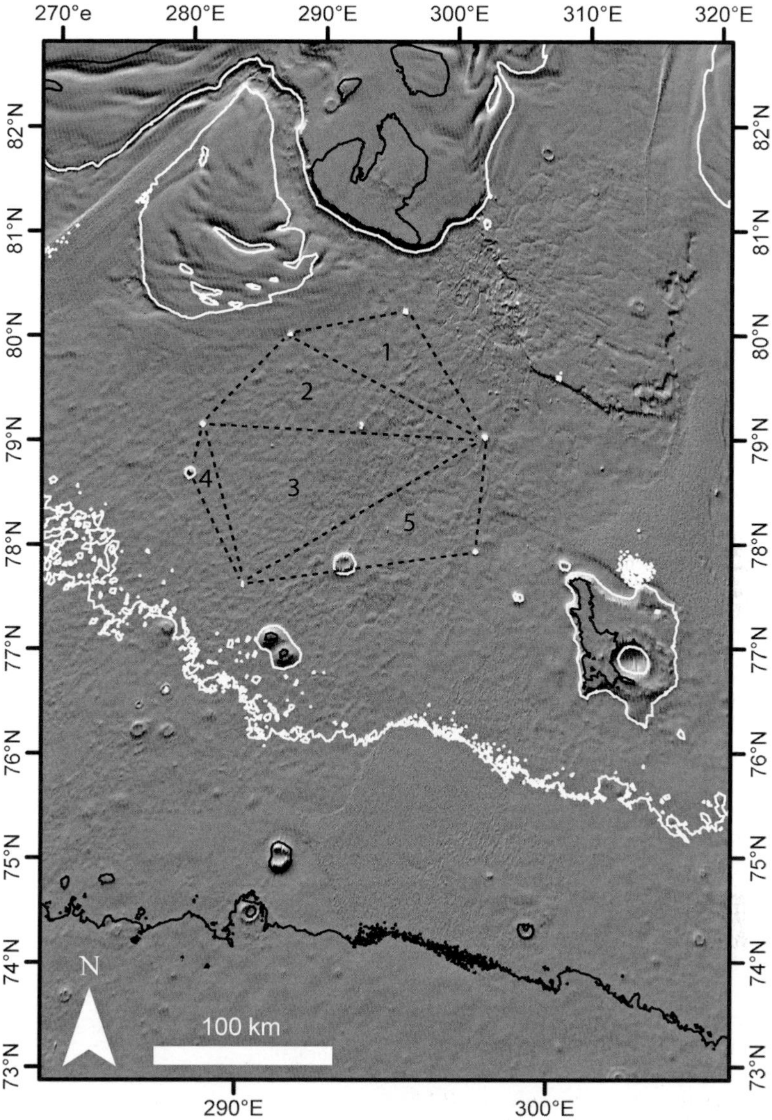

Fig. 10. Calculated southward extent of the interpolated layer surfaces, defined by intersection with the MOLA DTM. White lines show the extent of the lower interpolated layer surface (LS1) based on layer-contact elevations measured at the margins of the layered cones (LC1–LC5, LC7 and LC8). Black lines show the former extent of the upper layer surface (LS2), defined by the southernmost mean elevation of the plateau above the Rupes Tenuis scarp, the summit elevation of LC5, the mean elevations of the Abalos Colles mesas M4 and M5, and the elevation of the Escorial Crater plateau (Table 3 & Fig. 1). Dashed black lines show the triangular areas used for volume determinations. The background is a hillshade representation of the MOLA DTM.

crater walls – that is, the measured diameter is smaller than the original diameter – the determined elevations of pre-impact plateaus have to be corrected upwards. The calculated elevations for the measured diameters are marked in Figure 9. The calculated pre-impact elevations have not been used for the reconstruction of the former layer surfaces because there are major uncertainties in their determinations. Furthermore, the elevation of the pre-impact surface of CC1 does not necessarily correspond to LS1, as the impact could have happened during the erosion of $ABrt_{2-3}$.

The intersection of the extrapolated, slightly southwards-dipping, lower layer surface (LS1) with the original MOLA DTM shows its former spatial extent (see Fig. 10). As the Escorial Crater mesa is cut by this extrapolated layer surface, it should show a similar distinct layer contact to that observed at the margins of the small conical features. This layer contact is not visible at the margins of the Escorial Crater mesa, which might be due to erosional/mass-wasting processes. In order to check the existence of this layer contact, we analysed the SHARAD profile R_0889901_001_ss05_700_a_b, which crosses this mesa. It shows a reflection of the anticipated layer contact in the southern part of the mesa (Fig. 11a).

However, for unknown reasons, this reflector is not clearly visible in other SHARAD tracks covering the Escorial Crater mesa. Therefore, a clutter simulation has been performed for this SHARAD track (Fig. 11b). This simulation confirms our

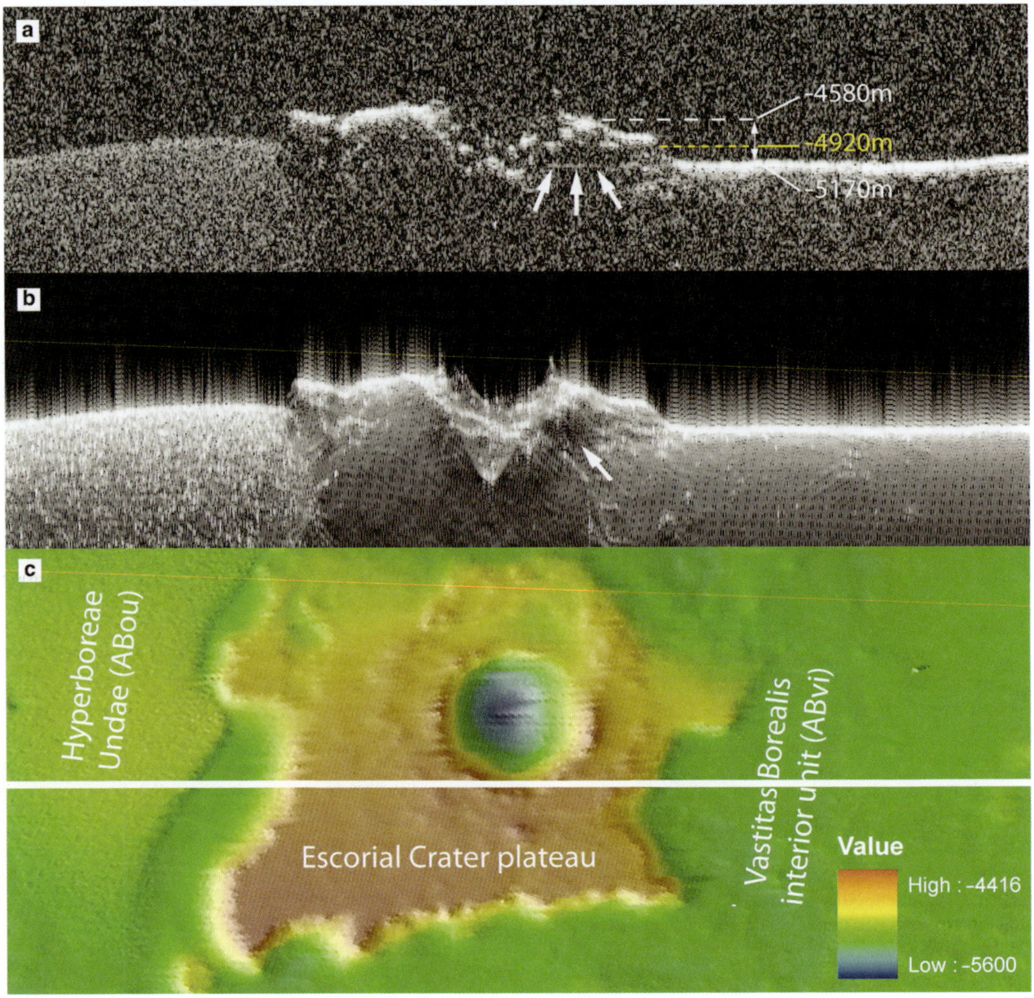

Fig. 11. SHARAD track R_0889901_001_ss05_700_a_b crossing the Escorial Crater plateau. (**a**) SHARAD radargram showing a reflector in the subsurface (marked by white arrows). Assuming a dielectric constant of 3 (pure-water ice), the true elevation of this reflector (labelled in yellow) can be calculated using MOLA DTM elevations of the Escorial Crater mesa and the surrounding plains. (**b**) Surface clutter simulation for the SHARAD track using the MOLA DTM. At the position of the reflector (white arrow), no reflection due to surface clutter appears, suggesting that the reflector is real. (**c**) Colour-coded MOLA hillshade with SHARAD ground track superimposed (white line). Image credit: NASA/ESA/JPL-Caltech/ASI/University of Rome/Washington University in St. Louis, NASA/JPL/MOLA Science Team.

Table 4. *Morphometric values for triangulated cells*

Cell number	Area (km^2)	Removed volume (km^3)	Erosion rate ($mm\ a^{-1}$)
1	2606.80	2309.49	0.00030
2	4212.57	3385.66	0.00027
3	7112.76	5073.52	0.00024
4	646.70	404.91	0.00021
5	4160.31	3003.41	0.00024

observation, as the reflection is not present in the simulation.

Owing to different dielectric constants of the mesa material and the Martian atmosphere, this reflection appears with the same time delay as the surrounding plains in the shown radargram. For the mesa material, we assumed a dielectric constant of 3 as a lower estimate comparable to that of the PLD (Picardi *et al.* 2005), and a dielectric constant of 1 for the Martian atmosphere, which is close to a vacuum. With these values, the vertical position of the reflector within the radargram can be shifted to its actual position using the following formula (Carter *et al.* 2009):

$$t = \frac{D \cdot 2\sqrt{E_r}}{c} \quad (3)$$

where t is the pulse travel time (in s), D is the depth (m), E_r is the relative dielectric constant of the material and c is the speed of light ($m\ s^{-1}$). The calculated relative vertical position of the reflector in the radargram is shown in Figure 11a. In the case of a mesa made of material with a lower content of water ice mixed with dust or sediments, this reflector elevation would have to be corrected upwards. Whatever the exact constant, these calculations show that the observed reflector is not the basal layer of the surrounding plains.

In order to assess the volume of removed material, reconstructed palaeo-surfaces were generated using topographical nodes observed at layer outcrops. Today's topography, as observed by the MOLA instrument, was subsequently substracted from the upper palaeo-surface (LS2) ($ABrt_1$) so that proper volume estimates could be made (Table 4). Volumetric values were obtained for each raster cell and, rather than working on an extrapolated surface covering an arbitrarily defined area, only the surface defined through the co-ordinate triplets for our observations was used for volume assessments. These topographical points are delineated by a convex hull which describes the actual measurement area. Such an approach is needed for a proper assessment as an extrapolation of a layer beyond the definition of points would have introduced new uncertainties and errors. Furthermore, in order to determine whether a southward thinning of the reconstructed palaeo-surface is detectable, measurement nodes were triangulated so that volume estimates could be made for each triangulated area. The sum of individual volumes corresponds to the total sum of removed material, and the deviation from the average volume normalized to the measured area indicates whether there are directional, that is latitudinal or longitudinal, dependencies.

The surface as seen today is part of the ABv_i unit, which was determined stratigraphically to be Late Hesperian–Early Amazonian (Tanaka *et al.* 2005) and defines a lower boundary of 3 Ga for our age estimate (Hartmann & Neukum 2001). As a first-order and conservative approximation, we assume that removal of the palaeo-plateau started directly after its formation and continued until recently, as the lack of a substantial number of impact craters suggests. However, since the current MOLA surface does not necessarily represent the actual surface on which the Rupes Tenuis unit was deposited, the determined volume corresponds to the sum of the removed material of the Rupes Tenuis unit plus the material of the Vastitas Borealis interior unit that was deflated after the Rupes Tenuis unit was completely removed. For the part of our study area that is delineated by observation points, erosion rates of the single triangulated areas (Table 4) are in the range of $2.5 \times 10^{-4} \pm 3 \times 10^{-5}\ mm\ a^{-1}$ with a slightly higher rate at the location of the northern cell number 1 (Table 4).

Summary and discussion

All layer-attitude measurements at the Rupes Tenuis scarp show comparable dip values in the range of 0.1°–0.5° ($\bar{x} = 0.3$, $\sigma = 0.1$) and dip directions in the range of 35.6°–71.6° ($\bar{x} = 46.1$, $\sigma = 12.0$) towards the NE (Table 2), indicating that the whole Rupes Tenuis unit was sedimented homogeneously and as layered strata of constant thickness.

Distinct slope breaks in topographical profiles drawn across the Rupes Tenuis scarp at the western wall of Chasma Boreale and at the narrow trough north of Abalos Mensa indicate different levels of erodibility of layers cropping out. Therefore, the Rupes Tenuis unit at the scarp can be subdivided into at least three distinct subunits, $ABrt_1$–$ABrt_3$ (see Fig. 7). At the margins of LC 1–LC8, south of the Rupes Tenuis scarp, several distinct layer contacts were observed. The elevations of the layer contacts of LC1–LC5, LC7 and LC8 are correlated with each other (Fig. 9), and their elevations are in agreement with the mean elevation of the western part of the Hyperborea Lingula surface. The analysis of SHARAD track R_0889901_001_ss05_700_a_b taken over the Escorial crater plateau provides evidence for a similar and horizontal subsurface layer contact (Fig. 11) situated within the mesa. Another correlation of elevations was observed between the mesa surfaces of M4 and M5, the summit elevation of LC5, and the surface of the Escorial Crater mesa.

Correlation of observed layer contacts of the layered cones indicates that these contacts, except the one at LC6 that was excluded from this comparison due to its different morphology, describe a continuous layer surface probably corresponding to the surface of $ABrt_1$. In addition, surface elevations of the flat-topped features in combination with the southernmost elevation value of the plateau located above the Rupes Tenuis scarp might define another continuous surface of a former plateau, supposedly the surface of $ABrt_3$. The reconstruction of these two layer surfaces provides information on their attitudes, showing a slight slope of the surfaces towards the south (0.07° and 0.12°).

The topographical correlation of the stratigraphical structure at different locations of the Rupes Tenuis scarp and of unit ABrt (Fig. 7) – as well as the observed elevations of flat-topped mounds and layer contacts at several conical mounds south of the scarp – suggest that these mounds are composed of Rupes Tenuis unit material. This has also been proposed by, for example, Tanaka *et al.* (2003) and Warner & Farmer (2008*a*). The observed horizontal layering is not considered typical for volcanic constructs and a correlation of layer contacts at margins of mounds that have a spacing of more than 100 km is also in disagreement with a common volcanic origin, as proposed by, for example, Garvin *et al.* (2000) and Sakimoto & Weren (2003).

The correlation of layer-contact elevations with the determined mean elevation of the Hyperborea Lingula surface also indicates that Hyperborea Lingula is composed of the same unit that makes up the lower part of the layered conical mounds, as also proposed by Kolb & Tanaka (2001), Edgett *et al.* (2003) and Warner & Farmer (2008*a*). This leads to the conclusion that deposition of material in the course of catastrophic outflow events related to Chasma Boreale (Fishbaugh & Head 2002) or a combination of exhumation of the lower ABrt layers together with outflow deposits (Fishbaugh & Head 2005) are unlikely.

The elevation of the surface of Hyperborea Lingula, as well as the elevations for correlated layer contacts, could suggest that remnant mounds are genetically related to a Chasma Boreale outflow event. However, the observation that mound material is found above the vertical extent of the layer contact excludes the outflow theory.

Attitude measurements of the Rupes Tenuis layers at the Rupes Tenuis scarp slightly dip towards the NE. The extrapolated layer surfaces south of the scarp, however, show slight southward dips. In addition, the surfaces of the Hyperborea Lingula, as well as the plateau of the Escorial Crater, dip towards the east. Combining these observations, the former 'palaeo-plateau' might have formed as a slightly dome-shaped surface – at least for the upper layer surface (LS2). However, dip angles are relatively small and, despite the good correlation, care must be taken not to overinterpret such measurements. As we could only use five interpolation points for the reconstruction of the upper surface, LS2 (Escorial cater plateau, M5, M4, LC5 and the southernmost elevation value of the plateau above the Rupes Tenuis scarp), the interpolation of a dome-shaped structure via higher-degree polynomial interpolation would have been too speculative. A planar surface reconstruction for the points south of the Rupes Tenuis scarp is a more conservative approach that was followed in order to determine the former southward extent of the layers (shown in Fig. 10) as well as for the two-dimensional elevation fits in the latitude-elevation plot (Fig. 9).

The layer contact at LC6 has an elevation more than 100 m below the suggested lower layer surface (LS1) (Fig. 9). However, since it shows a different morphology above the mapped layer contact, that is a complex summit probably showing additional layers, the mapped layer contact on the flanks might represent the layer contact between the lower Rupes Tenuis unit $ABrt_1$ and the underlying Vastitas Borealis interior unit ABv_i.

In summary, all observations are in agreement with the assumption that the Rupes Tenuis unit extends further southwards and was, perhaps, even deposited as a slightly up-doming unit reaching far beyond the Escorial Crater mesa. Owing to the suggested extensive aeolian denudation and solar ablation (Warner & Farmer 2008*b*), much of the material was removed and only isolated remnants of that unit, such as the conical mounds and the Escorial Crater mesa, were left behind (e.g. Tanaka *et al.* 2003; Warner & Farmer 2008*b*).

Hyperborea Lingula is composed of lower ABrt material, as layer correlations indicate, and was therefore not emplaced in the course of outflow events.

An erosion rate of approximately 250 nm a^{-1}, as determined for the removed material south of Chasma Boreale – that is, Rupes Tenuis subunits $ABrt_{1-3}$ plus the deflated Vastitas Borealis interior unit – are considerably larger than average values given for Martian mid-latitudes with approximately 0.02–100 nm a^{-1} for the timespan of 3.1–0 Ga (Golombek *et al.* 2006). As the polar environment, however, cannot be directly compared to the mid-latitudes and the landing-site conditions, the derived erosion rates are conceivable. However, our estimates are a conservative approach considering continuing layer degradation over about the last 3 Ga without the accumulation of new material. New insights are expected from analysis of the *Phoenix Lander* data, although the proper derivation of erosion rates might be difficult.

Although the layer reconstruction provides a consistent picture, the formation of remnant mounds needs further discussion in order to explain their isolated distribution. Tanaka *et al.* (2008) proposed a scenario in which the conical mounds represent remnants of highly eroded impact craters. Warner & Farmer (2008*b*) noted that the conical mounds without summit craters might also be random remnants produced by irregular scarp retreat, even without a protective cover. We favour the impact-crater scenario, at least for the layered conical mounds (except for LC5) and the cratered cones (all CCs) investigated in this study. The flat-topped mounds, LC5 (also showing a flat top), and several smaller mounds in this region, however, might be randomly isolated by irregular scarp retreat. We have adjusted the impact-crater scenario in order to explain the layering on the flanks of the mounds. Here, impact cratering occurs in ice-rich materials that contained multiple layers at a time when the layered Rupes Tenuis unit (ABrt) was laterally and vertically more extensive (Fig. 12a, b). Patchy impact ejecta and atmospheric dust within the impact crater act as a protective cover that prevented subsurface water ice from undergoing sublimation or rapid vapour diffusion (Fig. 12c). Sublimation of water ice from crater-surrounding areas that were unprotected subsequently left behind a positive relief feature showing a summit crater (Fig. 12d). As each ice-rich layer is slightly different in material, ongoing sublimation produces distinct layer contacts at the margin of the mound (Fig. 12e). The sublimation and degradation of individual features could finally have removed the summit depressions and left behind conical remnant mounds showing multiple layers (LC1–LC8: Fig. 12f).

Common pedestal-crater plateaus in the mid to highlatitudes of Mars (30°–60° north and south) typically have an elevation of about 20–80 m above the surrounding plains (Kadish *et al.* 2009). Although Kadish *et al.* (2008) described pedestal craters in Utopia Planitia and Malea Planum showing plateau elevations of approximately 100 m above the surrounding plains, these values are clearly smaller than the thickness of layered material that has been removed from our study region (*c.* 500 m in the Escorial Crater plateau to *c.* 1000 m at the Rupes Tenuis scarp) (see Fig. 9). This substantial thickness of removed material might be the reason for the unusual shape of these highly eroded pedestal craters, as the ongoing ablation of the surrounding plains removes the typical plateau-like character of common pedestal craters.

In the study area, both uncratered and cratered conical features were observed in close vicinity to each other. They are considered to represent different erosional stages (Fig. 12e, f). Such differences might be related to impact-crater size, and associated ejecta thickness and distribution, which consequently led to differences in local degradation rates.

The reason for the massive disintegration of the layered material is not known. However, the unusual high density of layered mounds in the direct vicinity of the mouth of Chasma Boreale suggests that the formation mechanism of Chasma Boreale is closely associated with the removal of material south of it. There are three major hypotheses for the formation of Chasma Boreale.

The first hypothesis is the formation by basal melting leading to a catastrophic outflow event, suggested by putative evidence for fluvial landforms, such as depositional bars, cataracts, a large-scale sinuous depression at the margin of the north polar cap or the lobate structure, Hyperborea Lingula, at the mouth of Chasma Boreale (Clifford 1980, 1987; Benito *et al.* 1997; Anguita *et al.* 1998; Fishbaugh & Head 2002). However, the analysis of high-resolution imagery and topographical data (MOC, THEMIS, MOLA) could not confirm these fluvial features (Warner & Farmer 2008*a*).

The second hypothesis is a formation related to wind erosion and wind-enhanced ablation that is supported by, for example, frost streaks, sand dunes and yardangs (Howard 1978, 1980; Warner & Farmer 2008*a*). These morphological features are obvious evidence for the competency of wind to transport frost and sediment or to erode layered materials in this region (Howard 2000). Warner & Farmer (2008*a*) proposed a model in which the formation of Chasma Boreale began as strong katabatic winds developed down the slope of a pre-existing NE-trending north polar scarp. The combination of aeolian erosion and solar ablation of ice-rich material along a topographical discontinuity in the

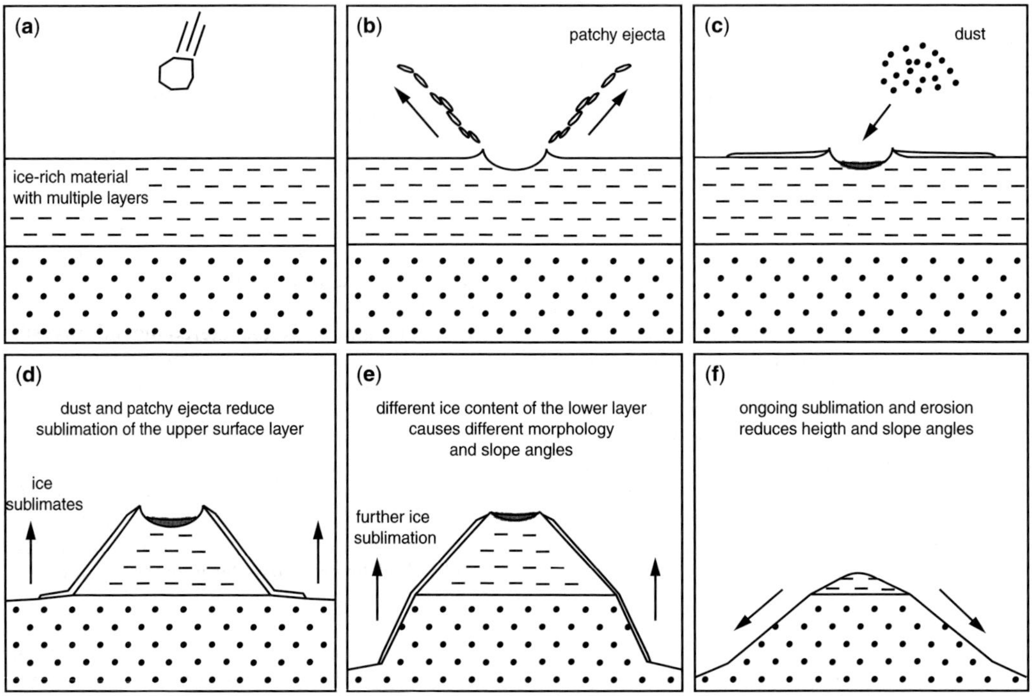

Fig. 12. Proposed formation mechanism of layered, conical edifices at the mouth of Chasma Boreale. (**a**) Impact into ice-rich ABrt material containing multiple layers with varying ice content. (**b**) Impact distributes patchy ejecta, forming a protective cover. (**c**) Accumulation of atmospheric dust inside the crater. (**d**) Ejecta and dust reduce the sublimation rates of the water ice in the upper layer. Sublimation of water ice from the surrounding area creates a conical mound. (**e**) Varying ice/dust content of different layers results in different morphologies and slope angles at the margins of the conical mound. This stage possibly corresponds to cratered cones CC1 and CC2. (**f**) Ongoing sublimation and erosion decreases the height of the conical mound, ultimately leading to disappearance. The result is a remnant cone with multiple layers. This stage possibly corresponds to LC1–LC4 and LC6–LC8 (see examples in Fig. 3a–c, e & f).

polar cap might have led to the development of a parallel scarp basal depression that was deepened, widened and lengthened by continuous down-scarp katabatic winds (Warner & Farmer 2008*a*). Modern scarp-basal depressions are observed in the topography data beneath the Rupes Tenuis scarp and along equatorial-facing scarps within Chasma Boreale (Warner & Farmer 2008*a*). These features, along with the scarp-proximal layered remnants described in our analysis, may represent evidence for the long-term retreat of the margins of the polar cap.

Recently, a third hypothesis for the formation of Chasma Boreale has been proposed by Holt *et al.* (2010). Based on SHARAD data, Holt *et al.* (2010) suggested that Chasma Boreale is a long-lived feature that formed as a result of non-uniform air-fall (ice and dust) accumulation, instead of large-scale erosion and removal of significant volumes of PLD. In their model, the Early–Middle Amazonian-age basal unit was eroded in the region of present-day Chasma Boreale before the Middle–Late Amazonian PLDs were accumulated. The lower part of the PLDs (radar unit PLD1 in Holt *et al.* 2010) may have accumulated within a linear depression that was created during erosion of the basal unit. Holt *et al.* (2010) suggested that PLD1 might have been eroded by katabatic winds, perhaps with some contribution from solar ablation, forming a proto-Chasma Boreale. From the SHARAD data, the upper part of the PLDs (their radar unit PLD2) drapes the dome-shaped erosional surface of a proto-Gemina Lingula, but is not present within the proto-Chasma Boreale. This suggests, in contrast to the erosion hypothesis, that modern-day Chasma Boreale exists partially due to a lack of air-fall accumulation in the Middle–Late Amazonian.

Although this formation theory for Chasma Boreale differs in the detail of formation from the theory proposed by Warner & Farmer (2008*a*), the model of erosion for the basal unit (including the Rupes Tenuis unit) and the lower part of the PLDs by katabatic winds and solar ablation is consistent across both hypotheses. However, modern katabatic winds and solar ablation on scarp slopes are likely to have been responsible for: the undermining and retreat of the modern scarps of Rupes Tenuis; the formation of conical mounds and promontories that are still attached to their host scarp; and the formation of the narrow troughs north of Abalos Mensa and along the equatorial-facing scarps of Chasma Boreale (Warner & Farmer 2008*a*). This indicates that scarp retreat and PLD erosion is an ongoing process that has operated at least during the Late Amazonian. Analyses of modern wind directions through mapping of aeolian features, such as wind streaks, sand dunes and active dust storms, are consistent with a katabatic wind-related formation of steep polar scarps in the Chasma Boreale region (Howard 2000; Warner & Farmer 2008*a*). A continuous retreat of the south-facing Rupes Tenuis scarp by katabatic winds and solar ablation is therefore a reasonable mechanism for the removal of layered material that once might have had a southward extent to approximately 74°N (Warner & Farmer 2008*a*).

Conclusions

The Rupes Tenuis unit might once have formed a continuous ice-rich 'paleo-plateau' extending southwards from today's Rupes Tenuis scarp (Tanaka *et al.* 2003, 2008; Warner & Farmer 2008*b*). Mounds located south of the Rupes Tenuis scarp are likely to represent erosional remnants of these layers rather than being isolated volcanic edifices, as proposed by, for example, Hodges & Moore (1994) and Garvin *et al.* (2000), for which there is little observational evidence in the surroundings.

This work investigated particular topographical relationships between several major surface features of this region and the Rupes Tenuis scarp itself. The results and observations lead to the following conclusions.

- Abalos Colles formed as remnants of extensive denudation of the Rupes Tenuis stratigraphic unit, probably as a result of local differences in erodibility caused by impact events and ejecta blankets that formed a protective cover, as previously proposed by Tanaka *et al.* (2008) and Warner & Farmer (2008*b*).
- The Abalos Colles show distinct layering. The correlation in terms of low RMS errors of observed layer-contact elevations indicates that they represent remnants of a once-continuous layer surface of the Rupes Tenuis unit. The correlation with the mean elevation of the Hyperborea Lingula suggests that this lobate structure is an erosional product, as proposed by Kolb & Tanaka (2001), Edgett *et al.* (2003), Tanaka *et al.* (2008) and Warner & Farmer (2008*a*), rather than a sedimentary product placed by an outflow event out of Chasma Boreale, as proposed by Fishbaugh & Head (2002).
- The origin of layered cones by degradation/denudation of the surrounding areas is supported by observations by SHARAD, showing a comparable interior layer contact within the Escorial Crater mesa.
- The existence, distribution and correlation of layer contacts at the margin of several edifices indicate a non-volcanic origin because horizontal layering is not typical for volcanic landforms.
- Conical cratered and uncratered mounds investigated in this study might represent different erosional stages of raised impact craters, which might be related to the impact crater sizes.
- The attitudes of Rupes Tenuis scarp layers and the interpolated layer surfaces indicate a slight up-doming of the once more southward-extended Rupes Tenuis unit. However, that observation does not affect the correlation of individual layers across large areas south of the Rupes Tenuis scarp.
- The slight southward dip of the reconstructed lower and upper layer surfaces results in an interpolated former southward extent to approximately 76°N for the lower layer (LS1) ($ABrt_1$) and about 74°N for the upper layer (LS2) ($ABrt_3$).
- The constant horizontal layering of the Rupes Tenuis unit over wide areas, as found in general in the whole study area, further supports the hypothesis that the formation of the Rupes Tenuis unit was mainly influenced by the vertical accumulation of material dominated by precipitation and the cold-trapping of dust-laden volatiles, as proposed by Kolb & Tanaka (2001).

We want to thank the SHARAD team for the processing of the radargram and the corresponding clutter analysis. We acknowledge K. Fishbaugh and the HiRISE experiment team for providing the high-resolution image data used in this study. We thank N. Warner and C. Fortezzo for the detailed reviews that significantly improved this paper. This research was partly supported by the Helmholtz Association through the research alliance 'Planetary Evolution and Life' and the German Space Agency (DLR), grant 50QM0301 (HRSC on Mars Express).

References

Anderson, D. & Mann, M. 1989. VICAR image processing using Unix, X Windows, and CD-ROMS. *In*: *Proceedings of the 40th Annual Lunar and Planetary Science Conference, March 1989, Houston, Texas*. Lunar and Planetary Institute, Houston, TX, Abstract, 17.

Anguita, F., Bábin, R., Benito, G., Collado, A., Gómez, D. & Rice, J. W. 1998. *Chasma Australe*

Mars: Structural Framework for a Catastrophic Outflow Origin. LPI Contributions, **953**, 1. Lunar and Planetary Institute, Houston, TX.

BENITO, G., MEDIAVILLA, F., FERNANDEZ, M., MARQUEZ, A., MARTINEZ, J. & ANGUITA, F. 1997. Chasma Boreale, Mars: a sapping and outflow channel with a Tectono-thermal origin. *Icarus*, **129**, 528–538.

BREED, C. S., GROLIER, M J. & MCCAULEY, J. F. 1979. Morphology and distribution of common 'sand' dunes on Mars – comparison with the earth. *Journal of Geophysical Research*, **84**, 8183–8204.

BYRNE, S. & MURRAY, B. C. 2002. North polar stratigraphy and the paleo-erg of Mars. *Journal of Geophysical Research – Planets*, **107**, 5044.

CARTER, L. M., CAMPBELL, B. A. *ET AL.* 2009. Shallow radar (SHARAD) sounding observations of the Medusae Fossae Formation, Mars. *Icarus*, **199**, 295–302.

CHRISTENSEN, P. R., JAKOSKY, B. M. *ET AL.* 2004. The Thermal Emission Imaging System (THEMIS) for the Mars 2001 Odyssey Mission. *Space Science Reviews*, **110**, 85–130.

CLIFFORD, S. M. 1980. Chasma.boreale (85°n, 0°w): remnant of a martian jokulhlaup?. *Bulletin of the American Astronomical Society*, **12**, 678.

CLIFFORD, S. M. 1987. Polar basal melting on mars. *Journal of Geophysical Research – Planets*, **92**, 9135–9152.

EDGETT, K. S., WILLIAMS, R. M. E., MALIN, M. C., CANTOR, B. A. & THOMAS, P. C. 2003. Mars landscape evolution: influence of stratigraphy on geomorphology in the north polar region. *Geomorphology*, **52**, 289–297.

FAGAN, A. L. & SAKIMOTO, S. E. H. 2009. Formation constraints on Martian North Polar volcanic edifices. *In*: *Proceedings of the 40th Annual Lunar and Planetary Science Conference, March 23–27, 2009, The Woodlands, Texas*. Lunar and Planetary Institute, Houston, TX, Abstract 1976.

FISHBAUGH, K. E. & HEAD, J. W. 2002. Chasma Boreale, Mars: topographic characterization from Mars Orbiter Laser Altimeter data and implications for mechanisms of formation. *Journal of Geophysical Research – Planets*, **107**, 5013.

FISHBAUGH, K. E. & HEAD, J. W. 2005. Origin and characteristics of the Mars north polar basal unit and implications for polar geologic history. *Icarus*, **174**, 444–474.

GADDIS, L., ANDERSON, J. *ET AL.* 1997. An overview of the Integrated Software for Imaging Spectrometers (ISIS). *In*: *Proceedings of the 28th Annual Lunar and Planetary Science Conference, March 17–21, 1997, The Woodlands, Texas*. Lunar and Planetary Institute, Houston, TX, Abstract 387.

GARVIN, J B., SAKIMOTO, S. E. H., FRAWLEY, J. J., SCHNETZLER, C. C. & WRIGHT, H. M. 2000. Note: topographic evidence for geologically recent near-polar volcanism on Mars. *Icarus*, **145**, 648–652.

GARVIN, J. B., SAKIMOTO, S. E. H., FRAWLEY, J. J. & SCHNETZLER, C. 2002. Global geometric properties of Martian impact craters. *In*: *Proceedings of the 33rd Annual Lunar and Planetary Science Conference, March 11–15, 2002, League City, Texas*. Lunar and Planetary Institute, Houston, TX, Abstract, 1255.

GOLOMBEK, M. P., GRANT, J. A. *ET AL.* 2006. Erosion rates at the Mars Exploration Rover landing sites and long-term climate change on Mars. *Journal of Geophysical Research – Planets*, **111**, E12S10.

GREELEY, R. 1979. Silt-clay aggregates on Mars. *Journal of Geophysical Research*, **84**, 6248–6254.

HARTMANN, W. K. & NEUKUM, G. 2001. Cratering chronology and the evolution of Mars. *Space Science Reviews*, **96**, 165–194.

HERKENHOFF, K. E. & PLAUT, J. J. 2000. Surface ages and resurfacing rates of the polar layered deposits on Mars. *Icarus*, **144**, 243–253.

HOCKEY, T. & BARNET, C. 1994. A DOS-based version of the VICAR image processing system for planetary image reduction, navigation, and measuring. *Bulletin of the American Astronomical Society*, **26**, 900.

HODGES, C. A. & MOORE, H. J. 1994. *Atlas of Volcanic Landforms on Mars*. United States Geological Survey Professional Paper, **1534**.

HOLT, J. W., FISHBAUGH, K. E. *ET AL.* 2010. The construction of Chasma Boreale on Mars. *Nature*, **465**, 446–449.

HOWARD, A. D. 1978. Origin of the stepped topography of the Martian poles. *Icarus*, **34**, 581–599.

HOWARD, A. D. 1980. Effect of Wind on Scarp Evolution on the Martian Poles. *In*: *Reports of Planetary Geology Program. NASA Planetary Geology Annual Meeting, 1980*. NASA Technical Memoirs, **82385**, 333–335.

HOWARD, A. D. 2000. The role of Eolian processes in forming surface features of the Martian polar layered deposits. *Icarus*, **144**, 267–288.

JAUMANN, R., NEUKUM, G. *ET AL.* & THE HRSC CO-INVESTIGATOR TEAM 2007. The high-resolution stereo camera (HRSC) experiment on Mars Express: instrument aspects and experiment conduct from interplanetary cruise through the nominal mission. *Planetary and Space Science*, **55**, 928–952.

KADISH, S. J., HEAD, J. W., BARLOW, N. G. & MARCHANT, D. R. 2008. Martian pedestal craters: marginal sublimation pits implicate a climate-related formation mechanism. *Geophysical Research Letters*, **35**, 16104.

KADISH, S. J., BARLOW, N. G. & HEAD, J. W. 2009. Latitude dependence of Martian pedestal craters: evidence for a sublimation-driven formation mechanism. *Journal of Geophysical Research – Planets*, **114**, 10001.

KNEISSL, T., VAN GASSELT, S. & NEUKUM, G. 2010. Measurement of strike and dip of geologic layers from remote sensing data – new software tool for ArcGIS. *In*: *Proceedings of the 41st Annual Lunar and Planetary Science Conference, March 1–5, 2010, The Woodlands, Texas*. Lunar and Planetary Institute, Houston, TX, Abstract 1640.

KOLB, E. J. & TANAKA, K. L. 2001. Geologic history of the polar regions of Mars based on Mars global surveyor data. II. Amazonian period. *Icarus*, **154**, 22–39.

MALIN, M. C., DANIELSON, G. E., INGERSOLL, A. P., MASURSKY, H., VEVERKA, J., RAVINE, M. A. & SOULANILLE, T. A. 1992. Mars Observer camera. *Journal of Geophysical Research*, **97**, 7699–7718.

MALIN, M. C., BELL, J. F. *ET AL.* 2007. Context camera investigation on board the Mars reconnaissance Orbiter. *Journal of Geophysical Research – Planets*, **112**, 5–30.

McEwen, A. S., Eliason, E. M. *et al.* 2007. Mars Reconnaissance Orbiter's High Resolution Imaging Science Experiment (HiRISE). *Journal of Geophysical Research – Planets*, **112**, 5–45.

Mustard, J. F., Cooper, C. D. & Rifkin, M. K. 2001. Evidence for recent climate change on Mars from the identification of youthful near-surface ground ice. *Nature*, **412**, 411–414.

Neukum, G. & Jaumann, R. 2004. HRSC: The High Resolution Stereo Camera of Mars Express. *In*: Wilson, A. (ed.) *Mars Express: The Scientific Payload*. **SP-1240**, 17–35.

Neukum, G., Jaumann, R. *et al.* & The HRSC Co-Investigator Team 2005. Mars: recent and episodic volcanic, hydrothermal, and glacial activity revealed by the Mars Express High Resolution Stereo Camera (HRSC). *In*: *Proceedings of the 36th Annual Lunar and Planetary Science Conference, March 14–18, 2005, The Woodlands, Texas*. Lunar and Planetary Institute, League City, TX, Abstract 2144.

Nimmo, F. & Tanaka, K. 2005. Early crustal evolution of Mars. *Annual Review of Earth and Planetary Sciences*, **33**, 133–161.

Okubo, C. H., Schultz, R. A. & Stefanelli, G. S. 2004. Gridding Mars orbiter laser altimeter data with gmt: effects of pixel size and interpolation methods on DEM integrity. *Computers & Geosciences*, **30**, 59–72.

Phillips, R. J., Zuber, M. T. *et al.* 2008. Mars north polar deposits: stratigraphy, age, and geodynamical response. *Science*, **320**, 1182–1185.

Picardi, G., Plaut, J. J. *et al.* 2005. Radar soundings of the subsurface of Mars. *Science*, **310**, 1925–1928.

Russo, F., Cutigni, M. *et al.* 2008. An incoherent simulator for the SHARAD experiment. *In*: *IEEE Radar Conference 2008, EuRAD 2008, 30–31 October 2009*. IEEE, Los Alamitos, CA, 1–4.

Sakimoto, S. E. H. & Weren, S. L. 2003. Martian north polar impacts and volcanoes: feature discrimination and comparisons to global trends. *In*: *Third International Conference on Mars Polar Science and Exploration, October 13–17, 2003, Alberta, Canada*. Lunar and Planetary Institute, Houston, TX, Abstract 8094.

Seu, R., Biccari, D. *et al.* 2004. SHARAD: the MRO 2005 shallow radar. *Planetary and Space Science*, **52**, 157–166.

Skinner, J. A., Jr., Hare, T. M. & Tanaka, K. L. 2006. Northern plains Craterforms: evidence for the accumulation and degradation of Paleo-Mantles. *In*: *Proceedings of the 37th Annual Lunar and Planetary Science Conference, March 13–17, 2006, League City, Texas*. Lunar and Planetary Institute, Houston, TX, Abstract 1476.

Smith, D. E., Zuber, M. T. *et al.* 2001. Mars Orbiter Laser Altimeter: experiment summary after the first year of global mapping of Mars. *Journal of Geophysical Research*, **106**, 23,689–23,722.

Tanaka, K. L. 2005. Geology and insolation-driven climatic history of Amazonian north polar materials on Mars. *Nature*, **437**, 991–994.

Tanaka, K. L., Skinner, J. A., Hare, T. M., Joyal, T. & Wenker, A. 2003. Resurfacing history of the northern plains of Mars based on geologic mapping of Mars Global Surveyor data. *Journal of Geophysical Research – Planets*, **108**, 8043.

Tanaka, K. L., Skinner, J. A. & Hare, T. M. 2005. *Geologic Map of the Northern Plains of Mars, scale 1:15,000,000*. United States Geological Survey Miscellaneous Investigations Series Map, **2888**.

Tanaka, K. L., Rodriguez, J. A. P. *et al.* 2008. North polar region of Mars: advances in stratigraphy, structure, and erosional modification. *Icarus*, **196**, 318–358.

Thomas, P. & Weitz, C. 1989. Sand dune materials and polar layered deposits on Mars. *Icarus*, **81**, 185–215.

Warner, N. H. & Farmer, J. D. 2008*a*. Importance of aeolian processes in the origin of the north polar chasmata, Mars. *Icarus*, **196**, 368–384.

Warner, N. H. & Farmer, J. D. 2008*b*. The origin of conical mounds at the mouth of Chasma Boreale. *Journal of Geophysical Research – Planets*, **113**, 11008.

Wessel, P. & Smith, W. H. F. 1995. New version of the Generic Mapping Tools released. *Eos, Transactions of the American Geophysical Union*, **76**, 329–329.

Wright, H. M., Sakimoto, S. E. H. & Garvin, J. B. 2000. Morphology of some small Mars north-polar volcanic edifices from Viking Images and MOLA topography. *In*: *Proceedings of the 31st Annual Lunar and Planetary Science Conference, March 13–17, 2000, Houston, Texas*. Lunar and Planetary Institute, Houston, TX, Abstract 1894.

A comparative study of interior layered deposits on Mars

MARIAM SOWE[1]*, RALF JAUMANN[2] & GERHARD NEUKUM[1]

[1]*Planetary Sciences & Remote Sensing, Institute of Geological Sciences, Free University of Berlin, Malteserstrasse 74-100, 12249 Berlin, Germany*

[2]*Institute of Planetary Research, German Aerospace Center (DLR), Rutherfordstrasse 2, 12489 Berlin, Germany*

**Corresponding author (e-mail: mariam.sowe@fu-berlin.de)*

Abstract: Interior layered deposits (ILDs) of the eastern Valles Marineris and adjacent chaos regions were analysed using high-resolution imagery, topography and spectral data in order to detect possible correlations. We find that ILDs are susceptible to erosion and weathering, as proven by their shapes (mesa, buttes), surface structures (pitted, fluted, yardangs), stair-stepped morphologies at different scales, and metre-sized boulders and talus. ILDs bear hydrated sulphates; consequently, we conclude that aqueous conditions dominated during their formation. Subhorizontal layering and parallel bedding of the ILDs could then indicate that deposition took place under low-energy aquatic conditions. Their superposition on chaotic terrain suggests that they are younger than chaotic terrain and, hence, younger than Late Hesperian. For the hydrated ILDs, which show polyhydrated on top of monohydrated sulphates, we think that formation within an evaporative body is not conceivable and we assume instead that a conversion of sulphates by post-formational humidity changes took place. As hydrated ILDs correlate well with rock fragmentation, we suppose that volume changes due to water content are responsible for rock fragmentation. Despite the different ILD settings, the basic conditions during sedimentation and erosion of ILDs could not have varied greatly because comparable mineralogies and morphologies are found among ILDs.

Interior layered deposits (ILDs) are exposed in several depressions on the Martian surface (McCauley 1978). They differ from the surrounding terrain because of their distinct layering, high albedo, morphology, high night-time thermal inertia and brightness temperature (Catling *et al.* 2006). Several researchers studied ILDs (Lucchitta *et al.* 1994; Fueten *et al.* 2005; Mangold *et al.* 2008; Rossi *et al.* 2008), which are concentrated in impact craters, various depressions of the Valles Marineris chasmata and are often associated with chaos regions (Sharp 1973), which lead into the Late Hesperian-aged outflow channels (Scott & Tanaka 1986). ILDs have been variously interpreted over the last decades, and their origin and timing is still debated. They have been proposed to be: of sedimentary origin (Malin & Edgett 2000); of volcanic origin (Chapman & Tanaka 2001); to be formed by salt diapirism (Milliken *et al.* 2007); to be related to aeolian or pyroclastic processes (Peterson 1981); and to be spring deposits (Rossi *et al.* 2008). A combination of different processes is conceivable arising from the volcano-tectonic setting of ILDs (Lucchitta *et al.* 1992) and the activity of aeolian processes (Greeley *et al.* 1992).

Hydrated sulphates, hydrated silica, phyllosilicates and hematite were detected on ILD surfaces from orbit (Gendrin *et al.* 2005; Glotch & Christensen 2005; Le Deit *et al.* 2008; Murchie *et al.* 2009; Roach *et al.* 2009).

On-site investigations by the *Mars Exploration Rover* (MER) confirmed the presence of sulphates and hematite-rich exposures in Meridiani Planum (e.g. Squyres *et al.* 2004; Clark *et al.* 2005). These findings clearly show that comparable aquatic conditions favouring mineral formation also occurred outside Valles Marineris and chaos regions (Andrews-Hanna *et al.* 2007) (Fig. 1). From a global point of view, sulphates and hematite are supposed to have formed successively after the Martian climate changed due to the cessation of volcanic activity in the Late Noachian–Early Hesperian (Bibring *et al.* 2006). Groundwater upwelling and evaporation may have dominantly contributed to the alteration of rocks (Andrews-Hanna *et al.* 2007).

Thus, analysing ILDs is an excellent opportunity to get insights into the Martian climatic conditions at the time when the sulphates and hematite were formed, which directly indicate water availability for longer time periods. Both mineral groups require specific temperatures and Eh–pH conditions for their formation (Matthes 2001) that are essential to constrain the physical and chemical surface and subsurface conditions at that time.

From: BALME, M. R., BARGERY, A. S., GALLAGHER, C. J. & GUPTA, S. (eds) *Martian Geomorphology*. Geological Society, London, Special Publications, **356**, 281–300.
DOI: 10.1144/SP356.14 0305-8719/11/$15.00

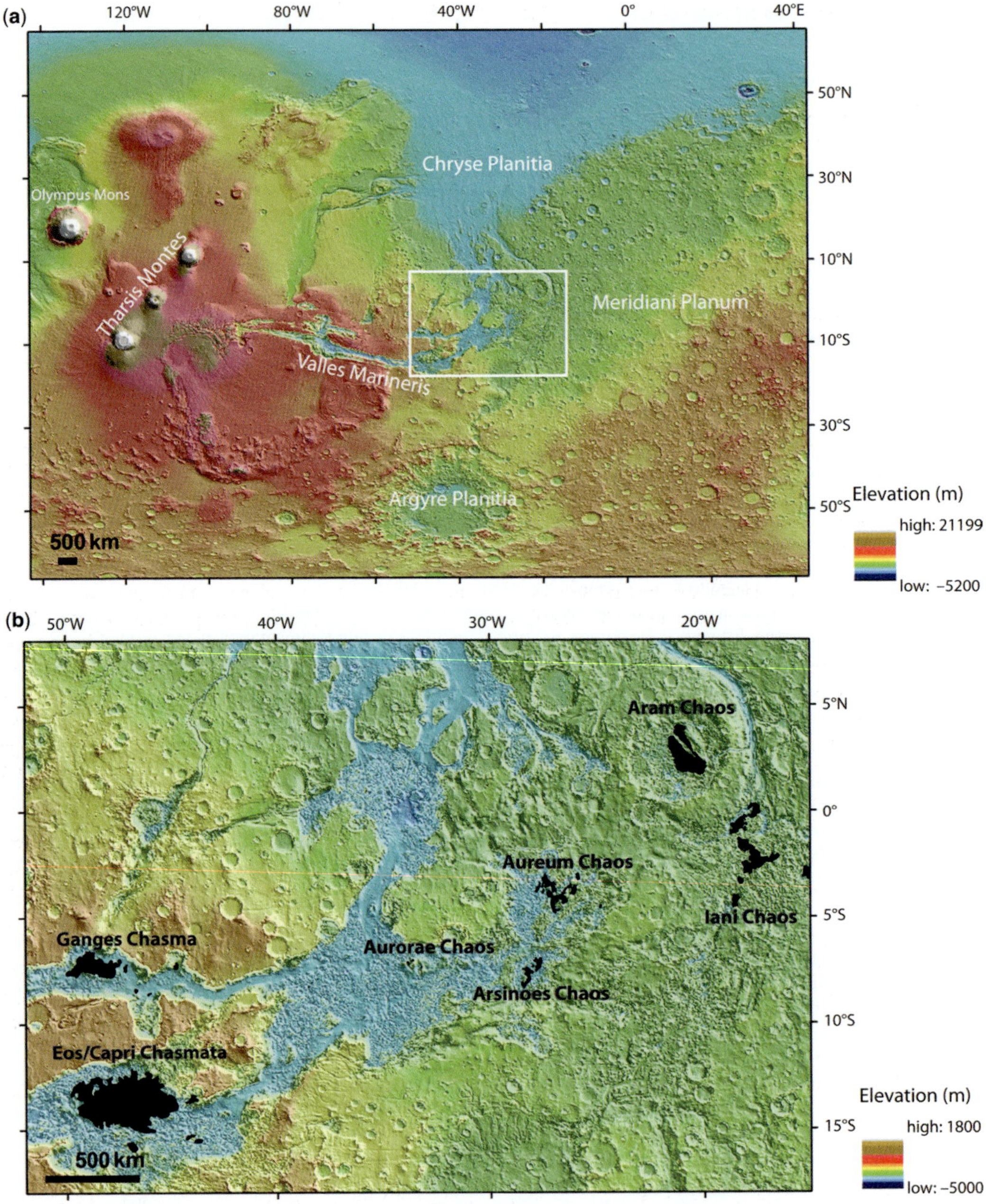

Fig. 1. (**a**) MOLA map showing the study area (box) located in the northern Martian lowlands. It comprises the eastern Valles Marineris and the adjacent chaos regions. (**b**) ILDs are marked black, and are located in the chaotic terrains of Aram, Iani, Aureum, Arsinoes and Aurorae. Further to the west, in the huge graben system Valles Marineris, there are the most prominent and largest ILDs located either in central, peripheral or enclosed troughs. Image credit: NASA/JPL/MOLA Science Team.

We characterized and compared ILDs by their morphology, albedo, elevation, thickness, material consolidation and mineralogy, and discussed potential formation processes as it is ambiguous whether the same process formed all ILDs. Therefore, we used high-resolution images and elevation data, as well as spectral information. The study area is located near the equator in the eastern Valles Marineris and the adjacent eastern chaos regions, with Ganges Chasma to the west and Iani Chaos to the east (Fig. 1a). Exposures of Ganges and Capri Chasmata, as well as in Aurorae, Arsinoes, Aureum, Aram and Iani Chaos, were analysed (Fig. 2a, b & Table 1).

Datasets and methodology

ILDs were classified based on the grey-scale brightness into low, intermediate and high albedo, using nadir images of the High Resolution Stereo Camera (HRSC: Jaumann *et al.* 2007) on board ESA's *Mars Express* (MEX), with a spatial resolution of 12.5–25 m per pixel. Albedo classes were corrected for aeolian coverage.

Imagery of the Context Imager (CTX: Malin *et al.* 2007) and the High Resolution Imaging Science Experiment (HiRISE: McEwen *et al.* 2007) on board NASA's *Mars Reconnaissance Orbiter* mission, and the Mars Orbiter Camera (MOC, Malin *et al.* 1992) on board the *Mars Global Surveyor* (MGS), were used to describe small-scale features (with a resolution of approximately 6, 0.3 and 3 m per pixel, respectively). These are essential for characterizing surface morphologies and textures.

For multi-spectral observations, HRSC and HiRISE colour channels are applied as they reveal compositional discrepancies on the surface (at a resolution of approximately 100 and 0.5 m per pixel, respectively).

We have obtained information about the consolidation of ILD material relative to its surroundings by utilizing night-time infrared images of the Thermal Emission Imaging Spectrometer (THEMIS: Christensen *et al.* 2004) on board the *Mars Odyssey* (resolution of *c.* 100 m per pixel) to obtain the brightness temperature (BT).

Bolometric night-time data of the MGS Thermal Emission Spectrometer (TES: Christensen *et al.* 2001*a*) was applied to extract the thermal inertia. Thermal inertia (TI) reflects the physical properties of the surface (density, heat capacity, thermal conductivity) and, thus, is a proxy for the material consolidation (Pelkey *et al.* 2001). A high TI value indicates a more consolidated, coarser material corresponding to bright regions that have higher surface temperatures and are able to keep the heat for a longer period of time (Jakosky *et al.* 2000). According to Putzig *et al.* (2005), rocks, bedrock, duricrusts and polar ice have TIs of more than 386 $J\ m^{-2}\ K^{-1}\ s^{-\frac{1}{2}}$, whereas sand, rock, bedrock and some duricrusts have TIs of 140–386 $J\ m^{-2}\ K^{-1}\ s^{-\frac{1}{2}}$. We have made a classification into low (304–368 $J\ m^{-2}\ K^{-1}\ s^{-\frac{1}{2}}$), intermediate (379–424 $J\ m^{-2}\ K^{-1}\ s^{-\frac{1}{2}}$) and high TI (428–498 $J\ m^{-2}\ K^{-1}\ s^{-\frac{1}{2}}$) using univariate statistics. A low TI does not necessarily mean only loose material is present, as the material covering it, such as dust and sand (Jakosky 1986; Edgett & Christensen 1991), affects the TI of the target region. However, the low spatial resolution of the instrument (*c.* 3 km per pixel) is inappropriate to describe small-scale features. The mean TI values for ILDs and their surroundings were determined and are listed in Table 1 with their respective standard deviations.

Using spectral analyses of the MRO Compact Reconnaissance Imaging Spectrometer for Mars (CRISM: Murchie *et al.* 2007), ILD mineralogies are described (Table 1). We combined HiRISE and CRISM data to test how morphology and mineralogy are related. Minerals were identified by their characteristic absorptions using the CRISM spectral library. Hydrated sulphates have absorptions close to 2.4 μm due to H_2O and OH, as well as sulphate bending overtones (Karr 1975). Monohydrated sulphate shows absorptions near 1.6 and 2.1 μm, such as kieserite ($MgSO_4 \cdot H_2O$) and szomolnokite ($FeSO_4 \cdot H_2O$). According to Bishop *et al.* (2009), kieserite has a double absorption at around 2.06 and 2.13 μm, and szomolnokite has a single absorption at approximately 2.08 μm. Polyhydrated sulphate (PHS) is identified by absorptions near 1.4 and 1.9 μm due to H_2O vibrations. PHS could be, for instance, epsomite ($MgSO_4 \cdot 7H_2O$) or copiapite $[(Fe^{2+}Fe_4^{3+}(SO_4)_6(OH)_2 \cdot 20(H_2O)]$. PHS was not clearly identified as the spectra often lack iron absorptions. Hydroxylated ferric sulphate (e.g. jarosite), as described by Bishop *et al.* (2009), has absorptions at 1.44, 1.94, 2.23 and 2.4 μm. Hematite-rich regions were identified by TES (Christensen *et al.* 2001*b*) (Table 1).

Applying digital terrain models (DTMs) of HRSC and MGS Mars Orbiter Laser Altimeter (MOLA: Smith *et al.* 2001), we looked at the elevations, thicknesses, slopes and layering geometries of ILDs. Layering geometry measurements were measured after Kneissl *et al.* (2010). Profiles were obtained from HRSC and MOLA DTMs to visualize ILD morphology and its context. Thickness measurements were derived from minimum and maximum elevations of HRSC DTMs (correlated with MOLA elevation: Gwinner *et al.* 2010), and are listed in Table 1.

Observations of ILDs

The observations of the studied ILDs will be presented in the following sections with respect to distribution, elevation, morphology and mineralogy (Figs 1–10 & Table 1). ILDs were named by their location. A more detailed work on the comparison of ILDs is found in Sowe (2009).

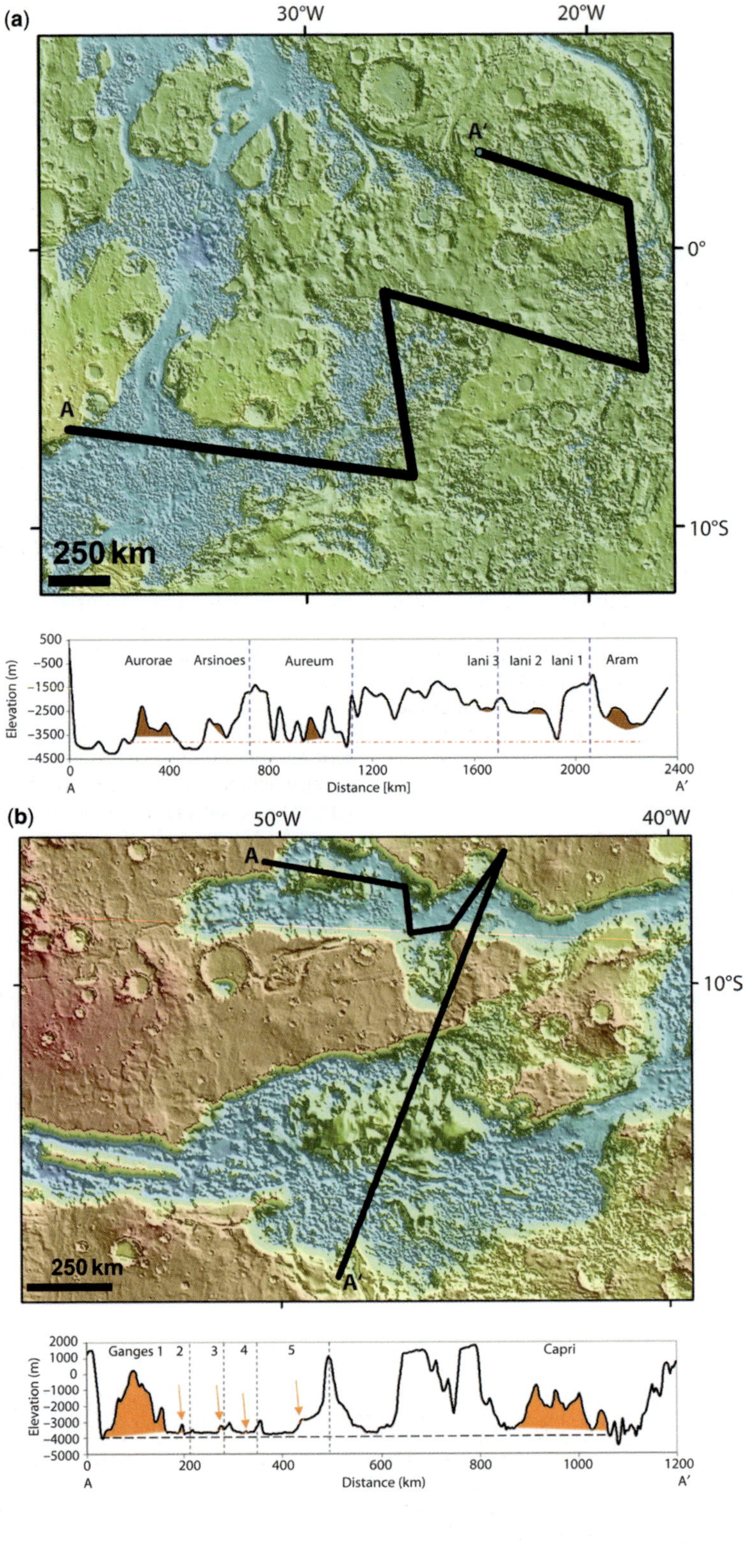
(a)
30°W
20°W
A′
0°
A
10°S
250 km
Elevation (m)
500
−500
−1500
−2500
−3500
−4500
Aurorae
Arsinoes
Aureum
Iani 3
Iani 2
Iani 1
Aram
0
400
800
1200
1600
2000
2400
A
Distance [km]
A′
(b)
50°W
40°W
A
10°S
250 km
A′
Elevation (m)
2000
1000
0
−1000
−2000
−3000
−4000
−5000
Ganges 1
2
3
4
5
Capri
0
200
400
600
800
1000
1200
A
Distance (km)
A′

Fig. 2.

ILDs in chaos regions

Aram Chaos. Aram Chaos is a 280 km-wide circular structure, which is located between the Ares Vallis outflow channel to the east and Aureum Chaos, and Iani Chaos to the SW (Fig. 1a, b).

Aram is a cliff-forming ILD that is elongated in a NW–SE direction and crosses the chaotic terrain with an extent of 120 × 140 km (Fig. 1a, b). Elevation ranges from −3700 to −2900 m (i.e., 3700 to 2900 m below Mars datum) (Table 1). The cross-section indicates the dome-like structure of the ILD (Fig. 2a). Erosional windows are present on top of the ILD that reveal its stratigraphy of disrupted chaotic terrain below a horizontal cap rock. Corresponding to the morphology of mesas (cf. Fig. 3a), the ILD features steep scarps (10°–30°) and a flatter top (2°–10°). Scarps show high albedo and appear massive (comparable to Fig. 4a, b). The overall surface is heavily pitted and grooved (Fig. 5b). The top of the ILD shows a rough-textured cap rock and a stair-stepped morphology is present (Fig. 5b). Overall weathering has affected the ILD, as is demonstrated by the presence of boulders (3–5 m in size) and talus. Spectral differences between talus, boulders, loose material and bedrock are visible on the HiRISE false colour images. PHS was found on top of monohydrated, hydroxylated sulphate and hematite within the ILD (Table 1).

Aureum Chaos. Aureum Chaos is a depression with a diameter of approximately 295 km, situated SW of Aram Chaos and east of Valles Marineris (Fig. 1a, b). It is bound by Aurorae to the SW and Arsinoes Chaos to the south (Fig. 1b). The chaotic terrain dominating its floor is superimposed by smooth, cliff-forming, light-toned material in the north to central part, which shows an approximate north–south alignment (Fig. 1b).

Aureum 1 has an irregular shape with frayed marginal parts (northernmost spot in Fig. 1b) and features a mesa-profile. It extends 7 × 15 km and is exposed from −4600 to −4100 m (Table 1). The ILD surrounds and clearly overlies mounds of chaotic material. Its overall albedo is intermediate and is higher at the scarps than on the flatter top, which shows a stair-stepped morphology, small low-albedo mesas and yardangs.

Aureum 2 features elongated mesa morphology (flat tops of 0°–5° and steep scarps of 10°–30°) and dome-like knobs (Fig. 3a). ILDs are exposed in a 34 × 60 km region. Their elevation varies between −5100 and −3300 m (Table 1). The northern part of each ILD exposure seems to have a sharp border, whereas the southern part is more frayed. Scarps and boulders are higher in albedo than top and talus (Fig. 4a, b). Concerning the stratigraphy, the lower part of the ILD is thickly bedded with a massive-appearing surface that shows angular joints (e.g. Fig. 4b). A cap rock with a rough, irregular surface that features thin bedding is on top (Figs 4a & 5b). Thinly bedded strata alternate with bouldered parts. The inner strata show between eight and 10 sequences within a total thickness of approximately 50 m (*c.* 5–6 m in thickness per sequence). Along the scarps, kieserite, PHS and hydroxylated ferric sulphate were detected (Fig. 7). Hydroxylated sulphate was found in heavily eroded high-albedo knobs, and monohydrated sulphate was mainly found below polyhydrated sulphate (Figs 4 & 7). The cap rock on top is spectrally neutral. Grey hematite detections correspond to low albedo material at the base of ILDs and are located close to the centre of Aureum Chaos (Table 1).

Iani Chaos. Iani Chaos is a large depression representing the source region of the Ares Vallis outflow channel that extends to the NW and drains into the Chryse Planitia (Fig. 1a, b). The ILDs are aligned along an axis of up to 260 km in length, in depressions measuring up to 4500 m below datum (Table 1).

Iani 1 is elongated in a NE–SW direction and features a dome-like cross-section. It measures 66 × 20 km, and is exposed at an elevation of between −4500 and −3400 m. No mesas are present there. The ILD is irregular in shape and features a sharp contact with a steep southern scarp, while other parts are flatter and frayed. ILD material has an overall high albedo (Table 1) and is surrounded by low-albedo chaotic terrain mounds (Fig. 10c), which to some extent overtop the ILD by a few hundred metres. Figure 10c shows that ILD material overlies these chaotic terrain mounds. Linear structures are oriented in a NW–SE direction, and are highlighted by settled dark aeolian material in depressions. Along this lineation,

Fig. 2. (*Continued*) MOLA profiles covering the ILDs. (**a**) Chaotic terrain ILDs: vertical lines indicate breaks in the profile. The horizontal line marks the minimum elevation (base level?), at which ILDs are exposed along the profile; here Aureum 2 is shown as a reference. Note that absolute elevations for each ILD were determined by utilizing HRSC DTMs (cf. Table 1). Accuracy: distance ±0.463 km, topography ±2 m. (**b**) Valles Marineris ILDs: arrows indicate the small exposures of Ganges 2–5. Dashed vertical lines mark breaks in the profile. The dashed horizontal line shows the minimum elevation at which ILDs are exposed, here with Ganges 1 as reference. Accuracy: distance ±0.463 km, topography ±2 m. Image credit: NASA/JPL/MOLA Science Team.

Table 1. *ILD parameters compared*

Locality Latitude (°N)/ longitude (°W)	Morphology	Relative albedo	Min./max. elevation (m)	Estimated thickness (m)	Consolidation	Mineralogy
Aram 2.9/−20.7	Mesa, dome-like; type 2	Low	−3700 ± 25 −2900 ± 25	800 ± 25	High TI: TI Ø 461 ± 50 J m^{-2} K^{-1} $s^{-\frac{1}{2}}$ (surrounding Ø 372 ± 43 J m^{-2} K^{-1} $s^{-\frac{1}{2}}$); BT 185–193K (surrounding 175–185K); boulders and talus present; group 1	Hematite as erosional lag below; monohydrated sulphate +2.23 μm feature (hydroxylated Fe-sulphate) above; PHS on top (Glotch & Rogers 2007; Lichtenberg *et al.* 2009)
Aureum 1 −3.2/−25.9	Irregular, Mesa; dark mesa unit; type 2	Intermediate	−4600 ± 12.5 −4100 ± 12.5	500 ± 12.5	Intermediate TI: TI Ø 401 ± 43 J m^{-2} K^{-1} $s^{-\frac{1}{2}}$ (surrounding Ø 354 ± 46 J m^{-2} K^{-1} $s^{-\frac{1}{2}}$); BT 198–206K (surrounding 190–206K)	No data
Aureum 2 −4.1/−26.2	Mesas and dome-like knobs, irregular; dark mesa unit; convolute bedding; type 2	Low	−5100 ± 25 −3300 ± 25	1800 ± 25	Low TI: TI Ø 368 ± 44 J m^{-2} K^{-1} $s^{-\frac{1}{2}}$ (surrounding Ø 296 ± 30 J m^{-2} K^{-1} $s^{-\frac{1}{2}}$); BT 190–218K (surrounding 185–192K); boulders and talus present; group 1	Mono- and PHS; monohydrated (sometimes with 2.23 μm feature → hydroxylated Fe-sulphate) above or interlayered with PHS; PHS sometimes with nontronite signature (Sowe 2009); hematite (Glotch & Rogers 2007)
Iani 1 −0.7/−18.5	Complex, dome-like; convolute bedding; type 1	High	−4500 ± 12.5 −3400 ± 12.5	1100 ± 12.5	High TI: TI Ø 482 ± 77 J m^{-2} K^{-1} $s^{-\frac{1}{2}}$ (surrounding Ø 344 ± 30 J m^{-2} K^{-1} $s^{-\frac{1}{2}}$); BT 203–208K (surrounding 191–205K); boulders and talus present; group 1	Monohydrated sulphate, best match szomolnokite (this study)
Iani 2 −1.6/−17.6	Terrace-like, mesa; type 2	Intermediate	−3800 ± 12.5 −3000 ± 12.5	800 ± 12.5	Low TI: TI Ø 342 ± 52 J m^{-2} K^{-1} $s^{-\frac{1}{2}}$ (surrounding Ø 297 ± 28 J m^{-2} K^{-1} $s^{-\frac{1}{2}}$); BT 195–203K (surrounding 187–195K); boulders and talus observed; group 1	Monohydrated sulphate (eastern part of ILD – this study); hematite in the same unit (Glotch & Rogers 2007)
Iani 3 −4.4/−18.5	Terrace-like, dome-like; type 2	Intermediate	−4300 ± 12.5 −3000 ± 12.5	1300 ± 12.5	High TI: TI Ø 428 ± 41 J m^{-2} K^{-1} $s^{-\frac{1}{2}}$ (surrounding Ø 308 ± 80 J m^{-2} K^{-1} $s^{-\frac{1}{2}}$); BT 191–201K (surrounding 180–201K); talus and boulders present; group 1	Monohydrated sulphate (eastern part of ILD; sometimes with 2.23 μm feature of hydroxylated Fe-sulphates – this study); hematite found within the same unit (Glotch & Rogers 2007)

Arsinoes −7.4/−27.2	Streamlined; dome-like; type 2	Low	−5200 ± 25 −3800 ± 25	1400 ± 25	Low TI TI Ø 359 ± 82 $J\,m^{-2}\,K^{-1}\,s^{-\frac{1}{2}}$ (surrounding Ø 333 ± 39 $J\,m^{-2}\,K^{-1}\,s^{-\frac{1}{2}}$); BT 180–203K (surrounding 180–195K)	Spectrally featureless to CRISM (this study)
Aurorae −7.3/−33.7	Butte, dome-like; type 2	Intermediate	−4600 ± 50 −3600 ± 50	1000 ± 50	Low TI: TI Ø 304 ± 40 $J\,m^{-2}\,K^{-1}\,s^{-\frac{1}{2}}$ (surrounding 280 ± 33 $J\,m^{-2}\,K^{-1}\,s^{-\frac{1}{2}}$); BT 185–197K (surrounding 178–182K); talus and boulders present; group 2	Monohydrated sulphate sometimes with 2.23 µm feature (hydroxylated Fe-sulphates; e.g. top – this study)
Ganges 1 −7.3/−49	Mesa, dome-like; dark mesa unit; convolute bedding; type 2	Intermediate	−4100 ± 12.5 −500 ± 12.5	3600 ± 12.5 Stair-steps of 500 m	Intermediate TI: TI Ø 424 ± 86 $J\,m^{-2}\,K^{-1}\,s^{-\frac{1}{2}}$ (surrounding 327 ± 90 $J\,m^{-2}\,K^{-1}\,s^{-\frac{1}{2}}$); BT 180–198K (surrounding 180–185K); talus and boulders present; group 1	Kieserite up to −1900 m, PHS near the top at *c.* −500 m (Sowe *et al.* 2008); hematite (Christensen *et al.* 2001*b*) as erosional lag at the base
Ganges 2 −7.4/−46.9	Streamlined, dome-like; type 1	High	−4700 ± 12.5 −4000 ± 12.5	700 ± 12.5	Intermediate TI: TI Ø 385 ± 51 $J\,m^{-2}\,K^{-1}\,s^{-\frac{1}{2}}$ (surrounding 308 ± 43 $J\,m^{-2}\,K^{-1}\,s^{-\frac{1}{2}}$); no BT data; talus and boulders observed; group 2	Spectrally featureless (this study)
Ganges 3 −8.4/−46.7	Streamlined, dome-like; type 1	High	−4700 ± 12.5 −3700 ± 12.5	1000 ± 12.5	High TI: TI Ø 498 ± 61 $J\,m^{-2}\,K^{-1}\,s^{-\frac{1}{2}}$ (surrounding 387 ± 66 $J\,m^{-2}\,K^{-1}\,s^{-\frac{1}{2}}$); BT 195–204K (surrounding 184–198K); talus observed	No data
Ganges 4 −8.6/−45.8	Streamlined, dome-like; type 1	High	−4800 ± 12.5 −4200 ± 12.5	600 ± 12.5	Intermediate TI: TI Ø 379 ± 38 $J\,m^{-2}\,K^{-1}\,s^{-\frac{1}{2}}$ (surrounding 325 ± 32 $J\,m^{-2}\,K^{-1}\,s^{-\frac{1}{2}}$); BT 185–195K (surrounding 174–185K); talus present	Spectrally featureless (this study)
Ganges 5 −7.5/−44.7	Streamlined, dome-like; dark mesa unit; type 1	High	−3800 ± 12.5 −3500 ± 12.5	300 ± 12.5	High TI: TI Ø 491 ± 68 $J\,m^{-2}\,K^{-1}\,s^{-\frac{1}{2}}$ (surrounding 436 ± 80 $J\,m^{-2}\,K^{-1}\,s^{-\frac{1}{2}}$); BT 203–215K (surrounding 196–205K); group 2	Monohydrated sulphate (this study)
Capri/Eos −13.1/−49.1	Mesa, dome-like; convolute bedding; type 2	Low	−5200 ± 12.5 −1700 ± 12.5	3500 ± 12.5	Intermediate TI: TI Ø 388 ± 65 $J\,m^{-2}\,K^{-1}\,s^{-\frac{1}{2}}$ (surrounding 364 ± 85 $J\,m^{-2}\,K^{-1}\,s^{-\frac{1}{2}}$); BT 184–204K (surrounding 183–192K); talus and boulders present; group 1	PHS above monohydrated sulphate (Roach *et al.* 2010) and hematite (Christensen *et al.* 2001*b*)

Ø, mean average; $\rightarrow$, could be.

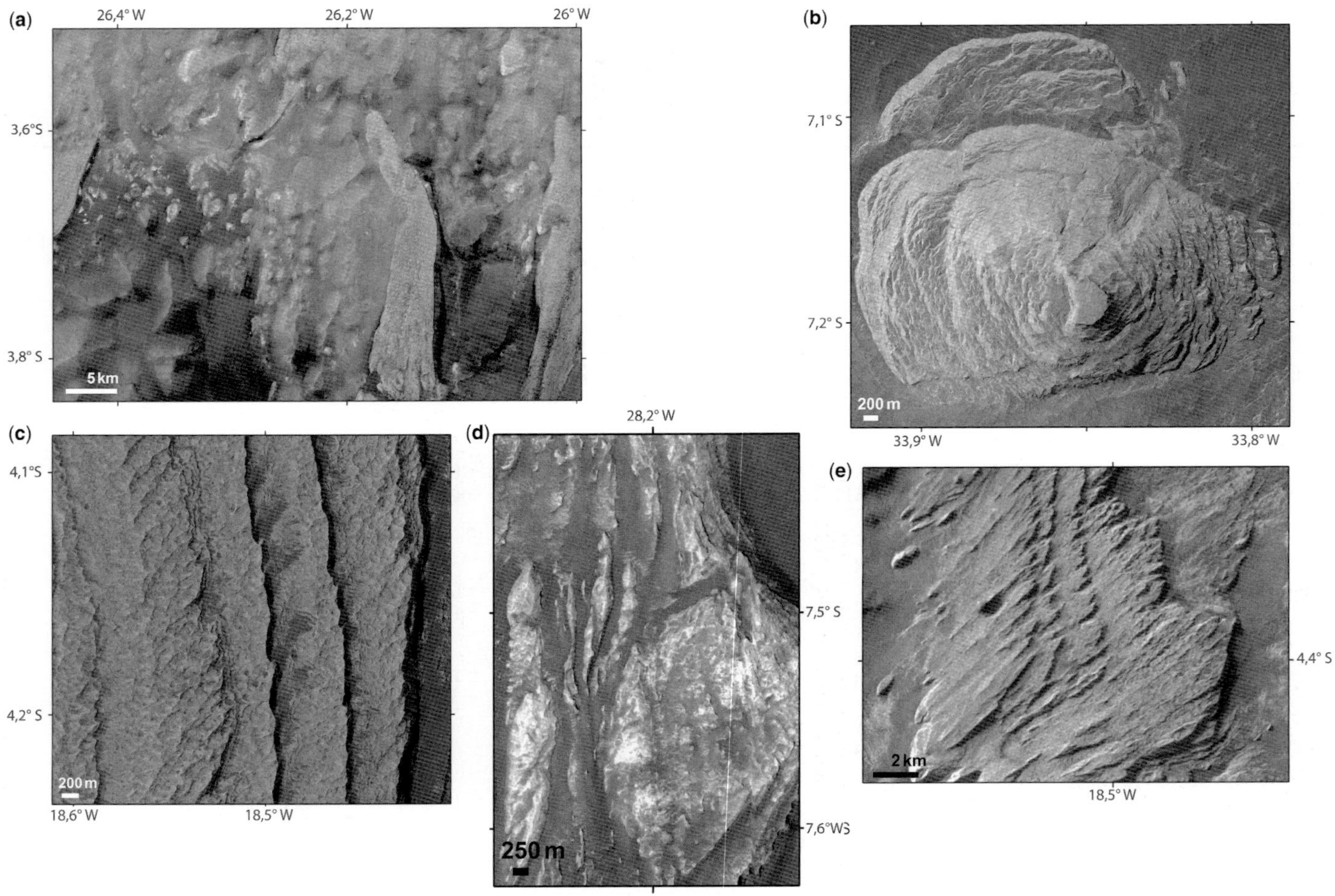

Fig. 3. Erosional morphologies present in ILDs. (**a**) Aureum 2 shows mostly extensive mesas crossing the depression in approximately north–south directions. Flat tops and steep sides (HRSC nadir h0103_0009) characterize these morphologies. (**b**) The HiRISE grey-scale image shows the westernmost ILD mound of Aurorae Chaos (PSP_007415_1730). The layering is extensive and thinly bedded up to the top. Its stair-stepped appearance shows several sequences of material. (**c**) Detail of layering in Iani 3. Note the pitted surface and the thinly bedded strata. No morphological differences were observed between layers. Dark material is present in surface depressions (HiRISE PSP_002628_1760). (**d**) The surface of Arsinoes is heavily grooved and fluted (black arrow). The ILD shows thinly bedded strata (white arrow). Steeper parts feature a high albedo; less steep parts show a lower albedo due to dark wind-blown material, mostly ripples, located in surface depressions (MOC E1300822). (**e**) Iani 3 is characterized by a SW–NE-trending surface structure. Yardangs characterize its surface (HRSC nadir h0934_0000). Image credit: ESA/DLR/FUB, NASA/JPL/MSSS.

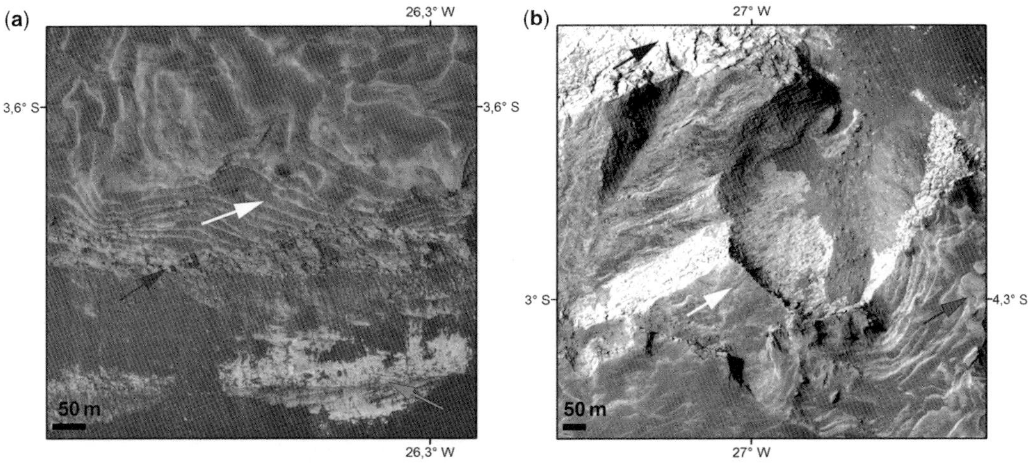

Fig. 4. Layering is observed at different scales. (**a**) On HiRISE images of Aureum 2, two morphological units are visible (PSP_004026_1765): a lower, rough and massive high-albedo unit and an upper distinctly layered low-albedo unit (grey and white arrow). Loose dark material is trapped in surface depressions of the upper unit. A stair-stepped morphology is present. Weathering is evident on scarps (black arrow); angular joints are apparent especially in the lower unit. (**b**) Discrepancies between the lower and the upper unit are present. The lower unit coincides with kieserite-rich material (black arrow) and PHS (white arrow). It features thickly bedded strata, whereas the upper unit displays distinct thinly bedded layering and stair-stepped strata (grey arrow). PHS was found on bedding planes of low-albedo regions whereas kieserite is present on high-albedo scarps of Aureum 2 (HiRISE PSP_007217_1755). Image credit: NASA/JPL/UofA.

yardangs are present (comparable to Fig. 3e). Hence, the whole surface of Iani 1 appears rough, grooved, heavily fractured and more disrupted than elsewhere (comparable to Figs 3d & 5a). Overall, layering is thinly bedded and hardly traceable on the surface. A stair-stepped morphology is present, with an average thickness of about 2–3 m per step. Likewise, other ILDs, dark talus and light-toned metre-sized (<5 m) boulders are visible at the base of steep scarps. Monohydrated sulphate was found with CRISM data (Table 1 & Fig. 7).

Iani 2 has a stair-stepped morphology and shows light-toned knobs. It is situated at elevations of between −3800 and −3000 m (Table 1), and has

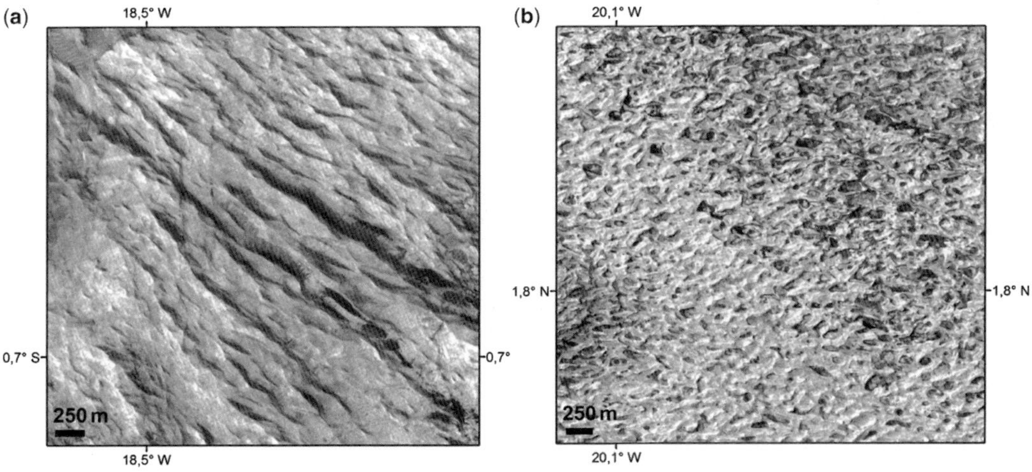

Fig. 5. Two different surface morphologies were observed on the tops of ILDs. (**a**) Surface type 1 shows an 'adjusted' surface structure, grooves and flutes are present. Layering in the upper more light-toned part is extremely fine (MOC R0900025). (**b**) Surface type 2 displays an irregular surface with sharp-edged crests (MOC E2000998). Image credit: NASA/JPL/MSSS.

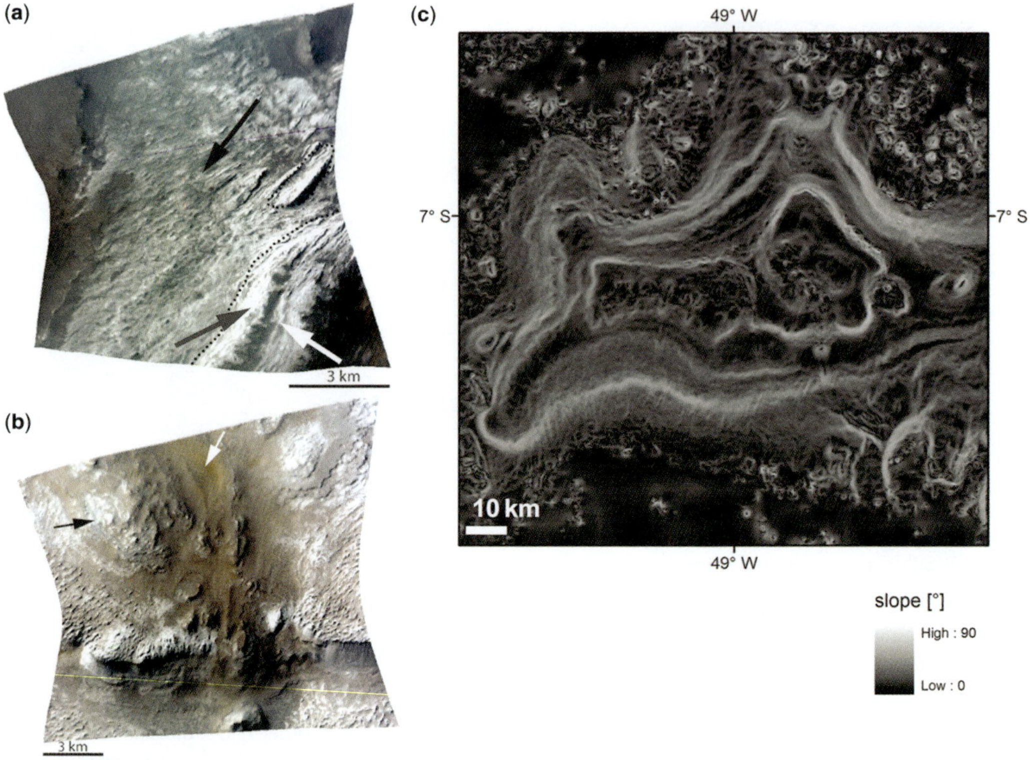

Fig. 6. Ganges 1 represents the westernmost ILD (cf. Fig. 1b). (**a**) The lowermost part of the ILD is kieserite-rich (CRISM FRT00009A1B). The spectral character of kieserite is different in the lower NW and the upper SE part, marked by the dotted line. Dark areas represent aeolian material that is mostly rippled but neither shows a pyroxene, Fe-oxide nor olivine signature. Arrows mark the regions where spectra were taken. Kieserite is present up to the elevation of −1900 m (Sowe *et al.* 2008). (**b**) The central top of the ILD with a white arrow shows HCP-rich ripples (FRT0000A3E3). The bright ILD material has a PHS signature (at −500 m) indicated by a black arrow where spectra were taken. (**c**) A stair-stepped morphology at a large scale is demonstrated. Cliff- and slope-forming strata characterize the ILD. Steep slopes (>25°) are light colours, flatter regions are coloured dark (HRSC h2211_0000). The steeper regions also have higher surface temperatures and thermal inertia values. Image credit: NASA/JPL/JHUAPL.

an extent of 22 × 33 km. The north to NW part of the ILD has a sharp contact, whereas the southern and eastern parts are frayed. The overall albedo is intermediate and, again, much higher at the scarps or on eroded knobs than at the flat top. Thinly bedded structures and a stair-stepped morphology are observed with high-albedo cliff-forming materials of high BT and TI (Table 1). An angular surface pattern with edges visibly friable is present and metre-sized boulders are deposited along scarps. Monohydrated sulphate (eastern region of ILD) and grey hematite were detected with CRISM and TES, respectively (Table 1 & Fig. 7).

Iani 3 is oriented in a north–south direction and measures 19 × 33 km at an elevation of −4300 m up to −3000 m (Fig. 1b & Table 1). It is a cliff-forming mound with distinct layering and a dome-like cross-section (Fig. 2a). Chaotic terrain is surrounded and in parts clearly overlain by ILD material (cf. Fig. 10c). To the SW, small light-toned exposures and patches in nearby knobs also show light-toned material. Its western part is higher in albedo, more eroded and elevated in contrast to the eastern part, which is more frayed and lower in elevation. The ILD has a steep SW and a flat NE slope. Overall, the ILD shows talus and boulders, in particular on the steep SW side. Yardangs oriented in a SW direction are found on the ILD top, especially in the southern part along the layers (Fig. 3e). The surface is etched and pitted, and looks rough and massive (Fig. 3c). A stair-stepped morphology and thinly bedded layers are visible (Fig. 5b); about 15 sequences were identified with an average thickness of 70 m per sequence. Layers are flat dipping downslope (<10°: Fig. 10a) and are indicative of an antiform. Monohydrated sulphate (low albedo, distinctly layered eastern part of the ILD) and grey hematite were found with CRISM and TES, respectively (Table 1).

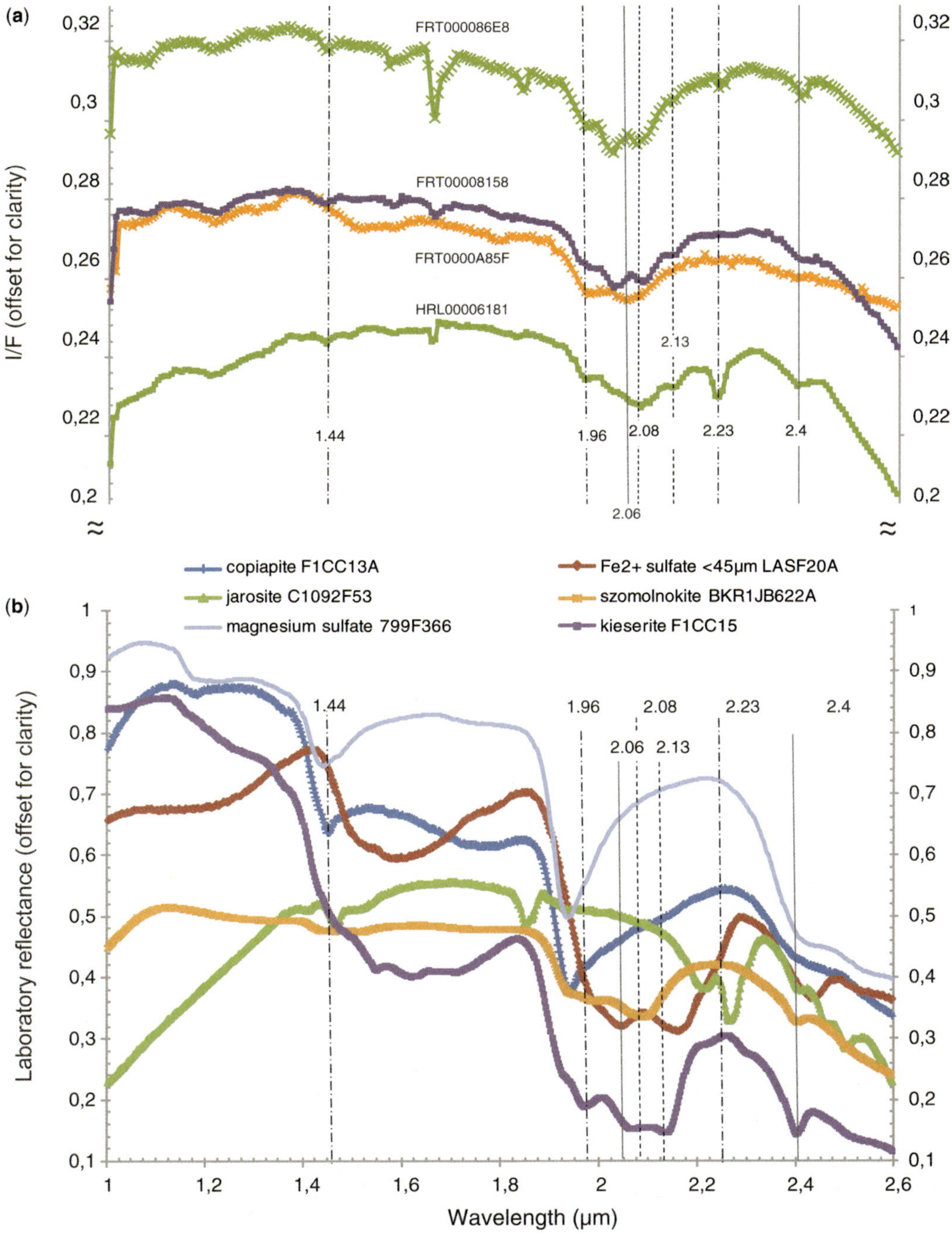

Fig. 7. (**a**) CRISM observations. Monohydrated sulphates were identified by their absorptions between 2.07 and 2.08 µm, and near 2.4 µm. Kieserite spectra show additional features at 2.13 µm. Some spectra show a 2.23 µm feature, which is due to OH. Absorptions at 1.44 and 1.96 µm could indicate the presence of hydroxylated ferric sulphates. (**b**) Laboratory spectra of different sulphate minerals for comparison.

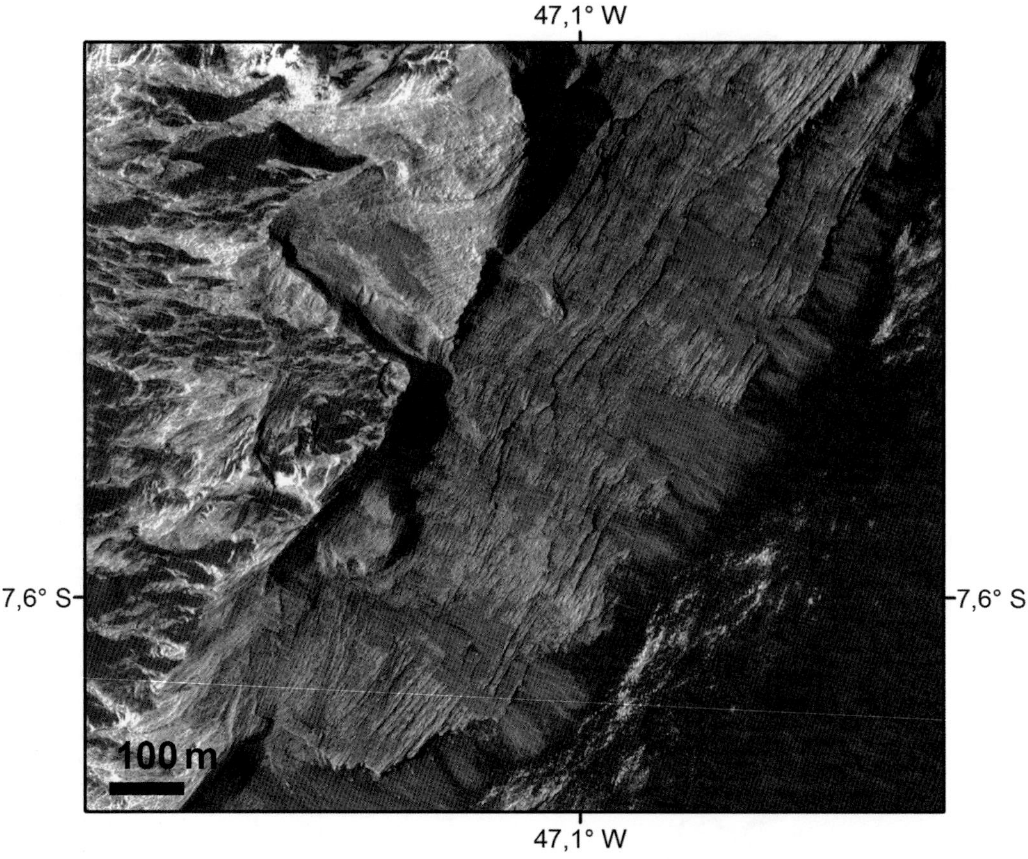

Fig. 8. HiRISE grey-scale image PSP_005952_1725 of Ganges 2. The ILD is spectrally neutral and hardly affected by rock fragmentation (group 2). Note the higher albedo of the upper part that also traps dark sandy material in surface depressions. Image credit: NASA/JPL/UofA.

Arsinoes Chaos. Arsinoes Chaos is situated at the eastern end of Valles Marineris; and is bounded by Aureum to the north, Aurorae to the west and Pyrrhae Chaos to the south (Fig. 1b). This locality is characterized by chaotic terrain and heavily eroded ILD material.

Arsinoes is elongated in a north–south direction and features a streamlined shape. Its extent is 92 × 40 km, and its elevation ranges between −5200 and −3800 m (Table 1). ILD material surrounds or overlies chaotic terrain knobs. Light-toned material patches and fragments are found on chaotic terrain knobs within Arsinoes Chaos. The northern part of the ILD seems to be more eroded than the frayed southern part. Overall, the albedo is low, especially on the top where quantities of dark loose material are trapped in surface depressions. The top appears rough and massive, shows grooves and yardangs that are oriented in a north–south direction, and thinly bedded strata have been observed (Fig. 3d; cf. Fig. 5b). The ILD does not show iron-rich or hydration features and, hence, is spectrally neutral (Table 1).

Aurorae Chaos. Aurorae Chaos is located east of the Valles Marineris, and extends toward Capri and Eos Chasmata to the west (Fig. 1a, b). To the east, it is connected to Aureum Chaos and in the north to Hydraotes Chaos. It is a large, low-lying region (−5000 m) sharply confined by the surrounding plateau (more than −600 m). Several resistant chaotic terrain knobs are exposed on its partially smooth floor. The ILD mounds occur in a region, which appears smooth owing to thick low-albedo mantling in parts.

Aurorae mainly shows butte morphologies (Fig. 3b) exposed at elevations from −4600 to −3600 m within an area measuring 45 × 22 km. The northern part seems to be more eroded, whereas the southern part is frayed. The overall

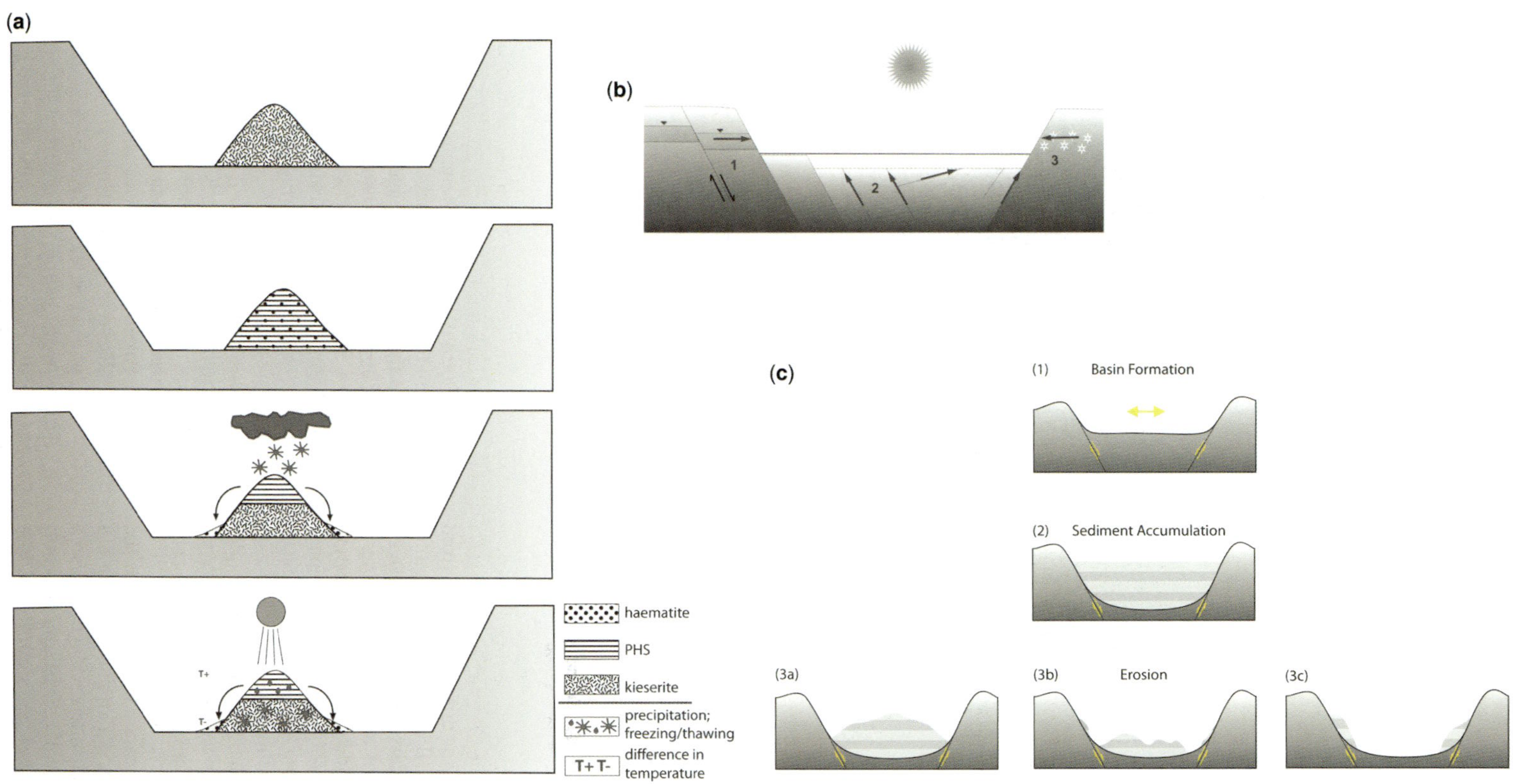

Fig. 9. (**a**) Potential conversion model of sulphates under the assumption that they did not form coevally. Water absorption then took place in the upper part of the ILD in which PHS is still exposed. The secondary hydration could have occurred via melting of a frost layer during insolation (bottom); or by precipitation, resulting in water absorption of the monohydrated sulphate-rich ILD (top). Hematite mostly occurs at the base of ILDs as a colluvial deposit. (**b**) Model of the potential input sources for deposition in a closed basin. (1) Confined aquifers that were released by erosion. (2) Subsurface flows from Tharsis, resulting in hydrothermal fluids that rise along faults. (3) Melting of a permafrost layer located in the wall rock. (**c**) ILD formation and evolution of Valles Marineris ILDs. (1) Extensional stress regime possibly caused rifting and basin (graben) evolution. (2) Accompanied sediment accumulation within the rift basin. (3) Erosion by effluent water and wind apparently affected the ILDs differently. ILDs are exposed in different locations within depressions indicating that erosion was more extensive: on rift shoulders (3a) present in Ganges 1; preferentially occurred along cracks and other contact points (3b) present in Capri/Eos, Ganges 2–4; or was more intense in the central part up to the bedrock (3c) present in Ganges 5 (Sowe 2009).

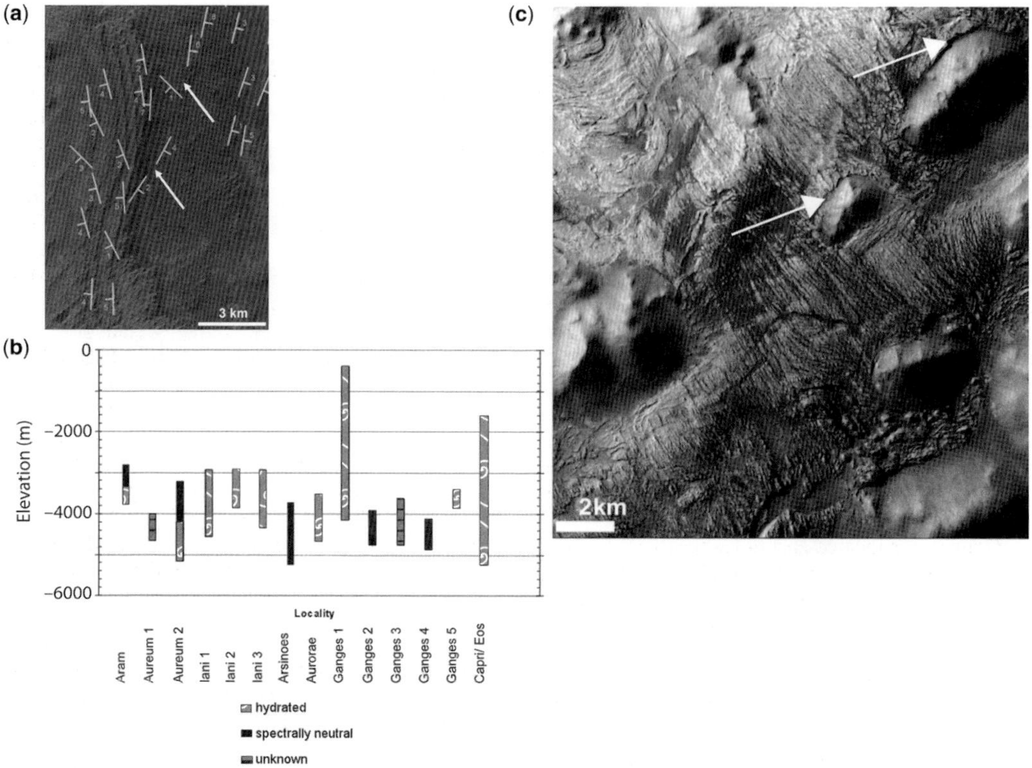

Fig. 10. (**a**) Iani3 strike and dip measurements. Note the shallow dip values and the dipping direction (arrows) in the eastern part. The layers dip by about <5° indicating subhorizontal layering. Layering seems consistent in the western part (HRSC orbit h0934_0000). (**b**) The diagram shows the extent of hydration within ILDs. Hydrated minerals were detected at the top and base of ILDs: at −5200 m and at −500 m. In the chasmata, hydrated minerals are found at higher elevations (−500 m) than in the eastern chaos regions (−3000 m), its base is comparable. Elevations are based on HRSC DTMs (cf. Table 1). (**c**) HRSC nadir image (h0934_0000) shows ILD material that clearly overlies chaotic terrain mounds in Iani 1. Arrows indicate the contact between both units. ILD material is heavily eroded and seems more susceptible to erosion than chaotic terrain. Image credit: ESA/DLR/FUB.

albedo is intermediate but it differs strongly from the low-albedo canyon floor and the chaotic material. Talus and boulders are present on steep scarps, with inclinations of up to 35°. Layering is consistent and thinly bedded with an apparent stair-stepped morphology (Figs 3b & 5b). Monohydrated sulphate was detected by CRISM and, in particular, the eroded top shows an indication of hydroxylated ferric sulphate (Figs 4b & 7).

ILDs in Valles Marineris

Ganges and Capri Chasmata are assumed to be the source regions of Simud and Tiu Valles, which are Hesperian-aged outflow channels. Ganges Chasma is located in the NE part of Valles Marineris, north of Capri Chasma and west of Aurorae Chaos (Fig. 1a, b).

Ganges Chasma. ILDs in Ganges Chasma are exposed amidst huge amounts of dark aeolian material on the chasma floor and hillside within a distance of approximately 330 km.

Ganges 1 is situated in the westernmost part of the semi-open chasma, showing an elongated shape parallel to the chasma (Fig. 1b). The ILD is exposed between −4100 and −500 m elevation, and measures 180 × 75 km. It exhibits a stair-stepped morphology (Fig. 6c). The overall relative albedo is intermediate but clearly varies within the ILD with steeper cliff-forming material of high albedo and flatter parts of lower albedo (Fig. 6c). Small, low-albedo mesas, yardangs, pits, flutes and grooves are present on the surface (cf. Figs 3c–e & 5b). Thinly bedded strata are observed throughout the whole ILD. Fine layers are best observed in the southern part. Furthermore, the flat parts show even, smooth-looking surfaces that appear disintegrated. Visible rock fragmentation due to angular joints, talus and boulders is found on the scarps. The TI is higher in the lower, more exhumed part than in the upper part (Sowe *et al.* 2008).

The lower part of the ILD has a strong kieserite signature (Fig. 6a). A transition zone characterized by a discrete layer at an elevation of about −1900 m marks the beginning of the region, in which a weak PHS signature was observed. PHS was also detected near the top of the ILD at an elevation of approximately −500 m (Fig. 6b). The dark ripples on top and in grooves show a high-calcium pyroxene signature. Hematite deposits were found off the south flank of the ILD in a low-albedo material (Christensen *et al.* 2001*b*). Five sequences with varying thicknesses were distinguished within the ILD, comprising a total thickness of about 2500 m. Layers show slopes of less than 10° downslope (Sowe 2009).

Ganges 2 is located east of Ganges 1 (Fig. 1b). It is elongated in a north–south direction and measures 22 × 8 km, and occurs at elevations of between −4700 and −4000 m. A streamlined morphology and a dome-like cross-section are present there (Fig. 2b). Its western and northern sides are partially frayed, with angular borders at the southern side. The western part seems flat, unlike the conspicuously steep SE scarp, which has thinly bedded strata, and shows talus and boulders. The overall albedo is high, with highest values in the upper part of the ILD, which looks rough and massive (Figs 5a & 8). No characteristic spectral signatures are present in CRISM, although the surface is hardly covered by aeolian material and appears freshly eroded.

Ganges 3 consists of four high-albedo blocks with streamlined morphology located south of Ganges 2. The whole extent is 6 × 6 km, and it is situated at elevations of −4700 to −3700 m. The blocks display possible slumping, especially the southern and northern blocks. Dark talus is visible on the slopes. TI is high, indicating highly consolidated material (Table 1). The top of the ILD is thinly bedded, and exhibits flutes and grooves. It is higher in albedo than the scarps, which show thinly bedded strata (cf. Fig. 5a).

Ganges 4 is characterized by two segmented blocks of light-toned, freshly eroded material. It is situated SE of Ganges 3, and is oriented in a NE direction. The ILD measures 7 × 3 km and is exposed at an elevation of between −4800 and −4200 m (Table 1). The NE part has a sharp contact indicating erosion was more intense than in the SW part. Talus is observed on the steep easternmost block. The presence of boulders is not confirmed as there are no appropriate images. The overall albedo is intermediate (Table 1) but higher in the upper ILD part. Scarps show thinly bedded strata, and the top looks massive and littered with fractures, flutes and grooves with dark aeolian material trapped within (cf. Fig 5a). To some extent, thinly bedded strata are observed. The ILD surface appears freshly eroded but is spectrally featureless to CRISM (Table 1).

Ganges 5 is exposed on the south-facing hillsides of Ganges Chasma, NE of Ganges 4. The ILD is oriented in the north–south direction, measures 15 × 8 km and is exposed at −3800 to −3500 m (Fig. 1b). Its northern and eastern parts have sharp edges, whereas the southern and western parts are frayed. Light-toned material is also observed on knobs to the west. The overall albedo is high but slightly higher on the top. Small low-albedo mesas are present on the rough-texture and grooved surface. The ILD has thinly bedded strata and shows only a few steep regions. Its TI is high, indicating highly consolidated material (Table 1). Neither talus nor boulders are observed. CRISM spectra show the presence of monohydrated sulphate.

Capri/ Eos Chasmata. Capri Chasma is connected to Coprates Chasma to the west and to Ganges Chasma in the north. Towards the east, it extends into Aurorae Chaos (Fig. 1). The ILD is enclosed by Capri Chasma in the north and by Eos Chasma to the south.

Capri/Eos shows the typical mesa morphology, with a flat top and steep slopes. It is exposed between −5200 and −1700 m elevation, and measures 250 × 150 km. The cross-section through the area (Fig. 2b) indicates a complex morphology. Scarps appear light toned and layered, while the other parts are thickly covered by dark wind-blown material and thus appear smooth. The steep lower part of the ILD is higher in albedo and TI, and shows a rough and massive surface, which displays flutes and grooves. The upper part has a stair-stepped morphology and is pitted (cf. Fig. 5b). Talus and boulders are present at the scarps. Kieserite, PHS and ferric oxides were detected in the central and western part of the mesa (Table 1), with kieserite corresponding to more massive material and PHS to smoother outcrops. Layering seems thinly bedded and consistent below the dark mantling deposit. Layers dip gently downslope in the upper regions (dip <10°) with a steeper dip in the lowermost part (<20°: Sowe 2009).

Comparison of ILDs

All studied properties have been compiled in Table 1 to highlight our main criteria for correlation. We found ILDs that are located at elevations below the surface level of the surrounding plateaus but not all of these locations correspond to the deepest regions of each respective depression (Fig. 2a, b). Thickness and geographical location (i.e. chasmata or chaos regions) do not significantly correlate (Table 1 & Fig. 9b). A remarkably large thickness of approximately 3600 m is present in both Ganges 1 and Capri/Eos, whereas the other ILDs show a mean thickness of about 1200 m.

ILDs appear as mesas or buttes and mounds (Fig. 3a, b), Aurorae Chaos with surfaces that are characterized by pits (Fig. 3c), flutes and grooves (Fig. 3d), and yardangs (Fig. 3e). We observed stair-stepped and thinly bedded strata (Fig. 3b) at different scales with sequences reaching from tens of metres to 500 m.

The tops of all ILDs are distinctly layered at a smaller scale, whereas lower parts often show large-scale layering (Fig. 4). However, the surface morphologies observed on the tops of ILDs differ (Fig. 5). We have distinguished between two morphological types that correlate quite well with hydrated mineralogy (except for Iani 1 and Arsinoes: Table 1). Type 1 is heavily fractured, fluted and grooved, with a massive and rough-appearance surface (Fig. 5a). Its albedo is high–intermediate, and it is mainly present on ILDs that show no hydrated sulphates and no hematite-rich minerals. Type 2 is characterized by a dissolved-looking surface structure that exhibits irregular surfaces and sharp-edged crests revealing underlying strata (Fig. 5b). It shows a low–intermediate albedo, a low–high TI and is mainly present on ILDs with hydrated sulphates. The morphology of surface type 2 is the spectrally neutral cap rock, which is present, for example, in Aureum and Aram (Figs 4 & 5b).

We find hydrated sulphates on comparably high-albedo scarps, which are barely covered by aeolian material (Figs 6 & 7). These outcrops have mostly local thicknesses of less than 100 m. Outcrops of monohydrated sulphate are often thickly bedded, fractured and cliff-forming, and appear more massive (Fig. 4). In contrast, PHS was detected mostly in regions of lower albedo, and appears smoother and distinctly layered (Fig. 4). In places where we detected both mono- and polyhydrated sulphate – for example, in Ganges 1 and Aureum 2 – PHS occurs above monohydrated sulphate (Fig. 6). Nevertheless, there are also regions where they occur interlayered; for example, in Aureum 2 (Fig. 4b). Hematite deposits correspond to low-albedo regions, which are smooth and either coincident with sulphate-rich materials or are located downslope (Table 1).

Metre-sized boulders and talus, indicating weathering of previously well-consolidated material (Fig. 4 & Table 1), characterize the bases of ILD scarps. We see differences within ILDs concerning the level of bedrock decomposition (Table 1). There are ILDs that are highly affected by rock fragmentation, for example, showing angular joints (group 1, Fig. 4), while others appear hardly affected (group 2, Fig. 8). No group matches with the TI classification presented in Table 1.

However, there seems to be a good correlation between rock fragmentation and hydrated ILDs as eight ILDs out of 10 match that correlation. We cannot test the correlation of the remaining four ILDs because no appropriate spectral or image data are available (cf. Table 1).

The lack of impact craters suggest ILDs have apparently young crater retention ages, which could be due to the heavy erosion they have experienced. Based on cratering model ages (Hartmann & Neukum 2001), ILD surfaces within this research area were measured to the Mid to Late Amazonian by Rossi *et al.* (2008) and Sowe (2009).

Potential formation and evolution of ILDs

ILDs are found in depressions of the chasmata of Valles Marineris (Figs 1 & 2), which is a volcano-tectonic environment (Scott & Tanaka 1986), as well as in weak crustal zones like the chaos regions. Groundwater movements due to high hydrostatic head might have been very common in the subsurface (Lucchitta *et al.* 1994; Schultz 1998). As ILDs in chaos regions mostly occur in closed basins, water sources are required to form the detected hydrated sulphates. Figure 9b shows three possibilities of how to fill the basins (no inflow from the plateau): confined aquifers (Carr 1979) could be a source for water; volcanic activity of Tharsis and its location upslope of Valles Marineris could have enabled the lengthwise movement of subsurface floods (Andrews-Hanna *et al.* 2007); or a permafrost layer within the wall rock that subsequently melted due to insolation also causing slope failure into the basin.

We detected sulphates of different hydration states: mono-, polyhydrated sulphates and hydroxylated sulphates. These findings are comparable to other ILDs; for example, in Juventae Chasma (Bishop *et al.* 2009; Wendt *et al.* 2009). Figure 9a shows the coexistence of these sulphates as confirmed by CRISM (Fig. 7a, b & Table 1). Sulphates form by evaporation or alteration during subsurface circulation, or by dehydration at acidic pH values of less than 5 (Bigham *et al.* 1996).

PHS can also form by water absorption of monohydrated sulphate at lower temperatures and in a relatively short process (Chipera & Vaniman 2007) compared to the reverse process (dehydration of PHS into monohydrated sulphate), which could take much longer and produces an amorphous PHS first, which would be stable on Mars (Vaniman *et al.* 2004). Under the current Martian surface conditions, dehydration from PHS into monohydrated sulphate is not possible because the current surface temperatures are too low to favour the respective conversion (Chipera & Vaniman 2007). According to Bishop *et al.* (2009), hydroxylated ferric sulphates, in turn, form by the dehydration of more highly hydrated materials (either mono- or PHS). It is likely that in regions where PHS is exposed above monohydrated sulphate, one sulphate formed out of the other. This is because, in an evaporating body, PHS would crystallize after monohydrated sulphate and thus would occur at the bottom of the ILD but not at the top.

Figure 9a shows how the melting of a frost layer due to insolation and by precipitation (e.g. snow) could have contributed to the formation of, for example, PHS. Water absorption would be facilitated by the observed rock fragmentation (increased surface area, Fig. 4). The preservation of monohydrated/hydroxylated sulphate implies that multiple desiccation and water supply events did not take place. More likely, warm and dry regions are assumed (Marion & Kargel 2005).

Hydrated sulphate minerals are not present in all ILDs – for example, Ganges 2, Ganges 4 and Arsinoes lack these minerals (Table 1) – that is, these ILDs are spectrally neutral to the CRISM instrument. ILDs in Ganges 2 and Ganges 4 are freshly eroded, and have high albedo and intermediate–high TI values (Table 1 & Fig. 8). Consequently, their absence cannot be explained by dust or sand coverage. Arsinoes, however, shows low TI values and a low albedo (Table 1 & Fig. 3d). Even if not dust- or sand-contaminated, there are several reasons for the non-detection of hydrated minerals in Ganges 2 and Ganges 4. For example: sulphates may, indeed, be present but anhydrous (e.g. anhydrite) or amorphous (e.g. amorphous PHS); other spectrally neutral minerals such as evaporates may be present (e.g. halite); or hydrated sulphates may be present but their amount is too small to be observed; or they do not exist there at all.

Different water sources (i.e. different compositions) may also explain differences in ILD mineralogy, that is, hydrated v. spectrally neutral ILDs (cf. Figs 6 & 8; Table 1). Layering is present throughout the ILDs at all scales (e.g. Fig. 4) and implies the occurrence of multiple events, which is ensured by the activity of large outflow channels (Andrews-Hanna & Phillips 2007) from the Late Hesperian into the Amazonian. Deposition of ILD material possibly took place under low-energy aquatic conditions, as suggested by sub-horizontal layering (Fig. 10a). With incoming water, sulphates would later be partially eroded or even dissolved and reprecipitated. Their low density lets them settle slowly, thus causing layering (Warren 2006).

Hydrated sulphates and hematite are spatially and stratigraphically closely related (Table 1), which argues for a close genetic relationship. Hence, it is possible that hematite is weathered out of Fe-rich sulphate layers and presently occurs as erosion lag (Fig. 9a).

Hematite deposits represent regions where neutral groundwater could have caused the diagenesis of Fe- and sulphate-bearing rocks and, thus, forming the hematite, which was observed for Meridiani Planum. Glotch & Christensen (2005) reported on a chronology derived from the stratigraphical relationship of hematite that gets younger from east to west: from Meridiani Planum (Noachian) via Aram Chaos (Hesperian) to Valles Marineris deposits (Hesperian–Amazonian). Therefore, the pH must have changed from acidic (sulphate formation) to neutral to form diagenetic hematite in chaos regions and Valles Marineris.

Rock decomposition into metre-sized boulders and talus is well observed on ILDs and clearly indicates weathering. We observe that rock fragmentation (i.e. the presence of angular joints and fragmentation within bedrock: Fig. 4) is correlated with hydrated ILDs. Consequently, we consider weathering, due to temperature differences, to be responsible for their fragmentation (increase/decrease of volume due to water content). Differences in rock fragmentation could be an effect of cementation, that is, group 2 (hardly affected by fragmentation) shows higher cementation and thus lower decomposition (Fig. 8). Instead, group 1 could be weaker and therefore more affected by rock decomposition (Fig. 4).

We note that laboratory experiments showed that hydration from atmospheric water is more important at the surface than at greater depths. Further, the extent of hydration penetration depends on the diffusivity and probably on the cohesiveness of materials. Hence, ILDs with advanced rock fragmentation would be more affected by hydration penetration.

The heavily eroded nature of ILDs (Fig. 3) suggests that their extent was once much greater than at present. Surface structures such as flutes, grooves and pits result from erosion by wind or water (Bourke & Viles 2007), indicating irregularities (e.g. cracks) within ILDs. Stair-stepped morphologies (Figs 4 & 6c) imply alternating strata of consolidated and less consolidated material, undergoing weathering and erosion. Changes in the depositional or erosional conditions, for example in Aureum 2, are considered responsible for these discrepancies in layering.

Figure 9c shows the present locations of ILDs in Valles Marineris but it is also applicable to other chaos regions when disregarding the rift-basin setting. It shows their present shapes, which were carved by erosion. This indicates erosion was more extensive on the rift shoulders (e.g. Ganges 1), preferentially occurred along cracks and other contact points (Capri/Eos, Ganges 2–4), or was more intense in the central part up to the base rock (Ganges 5; Fig. 3c). According to Arvidson *et al.* (1979) and Golombek & Bridges (2000), erosion was more intensive during the Hesperian (0.02 μm year^{-1}) in contrast to the Amazonian (0.00004 μm year^{-1}). Since ILDs are definitely younger than chaotic terrain (Late Hesperian: Scott & Tanaka 1986), show a young Amazonian age, are heavily eroded and needed water to form hydrated minerals (before the cessation of outflow channels in the Mid-Amazonian), this would imply that their main erosion took place before the Amazonian.

Conclusions

- We determined that ILDs show various characteristics of erosion, as proven by their overall morphology (stair-stepped, mesas, buttes, erosion mounds, knobs) and surface structure (pits, yardangs, flutes). Likewise, their morphology shows that they are more susceptible to erosion than, for example, chaotic terrain (Fig. 10c). This is shown by the presence of rock fragmentation, and of talus and boulders.
- ILD surfaces are hardly cratered and appear young but, as erosion affected ILDs considerably, these ages correspond to erosional ages and not to the age of formation (Sowe 2009).
- Hydrated sulphates within ILDs are supposed to have formed when outflow channels were active (Late Hesperian–Mid-Amazonian: Head *et al.* 2001) owing to the presence of water; consequently, ILDs are supposed to be older than the Mid-Amazonian.
- When assuming an aqueous environment, low-energy water conditions could have produced the subhorizontal parallel bedding of ILDs. In addition, superposition of ILDs on chaotic terrain (Fig. 10c) and subhorizontal layering indicate that ILDs formed after chaotic terrains (hence, are younger than Late Hesperian).
- Layering shows that periodic variations of sedimentation supply dominated. The presence of material of different consolidation, resulting in erosion into stair-steps, could be a result of density and material contrasts during sedimentation (e.g. density discrepancies between water and brine, evaporate and clastic particles).
- The stratigraphical relationship of both hydrated sulphates (PHS on top of monohydrated sulphates) could indicate that a formation within an evaporative body did not take place and that one sulphate formed secondarily, which is PHS according to Chipera & Vaniman (2007). According to Marion & Kargel (2005), regions with preserved mono- and hydroxylated sulphates show that warm and dry conditions dominated or that these deposits were protected from hydrous events. pH values from acidic to neutral could have been favoured later on during diagenetic hematite formation, which would be possible during groundwater movements as proposed by Glotch & Christensen (2005) and Andrews-Hanna *et al.* (2007).
- Groundwater movements (as mentioned in the previous point) could be the dominant processes occurring to fill the closed basins and to produce water-related minerals.
- A good correlation is observed between hydrated ILDs and rock fragmentation; hence, we consider that physical weathering due to water content is responsible for rock fragmentation as spectrally neutral ILDs lack these features.
- Similarly, we find spectrally insignificant ILDs mainly in discharge regions, often with streamlined morphology, and hence assume that erosion was more extensive and could have prevented the deposition of (low density: Warren 2006) hydrated sulphates there.
- Although the settings in which we find ILDs (chaos and chasmata) are completely different, we observe a similar mineralogy and morphology. Hence, conditions (e.g. pH value, water supply/source, temperature) during deposition of materials and erosion should have been similar.

This research was partly supported by the Helmholtz Association through the research alliance 'Planetary Evolution and Life', the German Science Foundation (DFG) through the Priority Program Mars and the Terrestrial Planets (DFG-SPP 1115, Project: Chronostratigraphy of Mars, grant: NE 212/8-3) and the German Space Agency (DLR), grants 50QM0301 and 50QM1001 (HRSC on Mars Express). We thank the HRSC Experiment Teams at DLR Berlin and Freie Universitaet Berlin, and the teams involved in MRO, MGS and MO for making their data available. T. Platz, C. Gross, G. Michael and L. Roach are much appreciated for reading and improving the manuscript. L. Wendt is acknowledged for assisting in spectral interpretation. This manuscript benefited from thorough and thoughtful reviews by F. Fueten and an anonymous reviewer.

References

Andrews-Hanna, J. C. & Phillips, R. J. 2007. Hydrological modelling of outflow channels and chaos regions on Mars. *Journal of Geophysical Research – Planets*, **112**, E08001, doi: 10.1029/2006JE002881.

Andrews-Hanna, J. C., Phillips, R. J. & Zuber, M. T. 2007. Meridiani Planum and the global hydrology of Mars. *Nature*, **446**, 163–166.

ARVIDSON, R. E., GUINNESS, E. & LEE, S. 1979. Differential aeolian redistribution rates on Mars. *Nature*, **278**, 533–535.

BIBRING, J.-P., LANGEVIN, Y. *ET AL.* & THE OMEGA TEAM. 2006. Global mineralogical and aqueous Mars history derived from OMEGA/Mars express data. *Science*, **312** 400–404.

BIGHAM, J. M., SCHWERTMANN, U., TRAINA, S. J., WINLAND, R. L. & WOLF, M. 1996. Schwertmannite and the chemical modeling of iron in acid sulfate waters. *Geochimica et Cosmochimica Acta*, **60**, 2111–2121.

BISHOP, J. L., PARENTE, M. *ET AL.* 2009. Mineralogy of Juventae Chasma: sulfates in the light-toned mounds, mafic minerals in the bedrock, and hydrated silica and hydroxylated ferric sulfate on the plateau. *Journal of Geophysical Research*, **114**, E00D09, doi: 10.1029/2009JE003352.

BOURKE, M. C. & VILES, H. A. (eds). 2007. *A Photographic Atlas of Rock Breakdown Features in Geomorphic Environments*. Planetary Science Institute, Tucson, AZ.

CARR, M. H. 1979. Formation of Martian flood features by release of water from confined aquifers. *Journal of Geophysical Research – Planets*, **84**, 2995–3007.

CATLING, D. C., WOOD, S. E., LEOVY, C., MONTGOMERY, D. R., GREENBERG, H. M., GLEIN, C. R. & MOORE, J. M. 2006. Light-toned layered deposits in Juventae Chasma, Mars. *Icarus*, **181**, 26–51.

CHAPMAN, M. G. & TANAKA, K. L. 2001. Interior trough deposits on Mars: Subice volcanoes? *Journal of Geophysical Research – Planets*, **106**, 10 087–10 100.

CHIPERA, S. J. & VANIMAN, D. T. 2007. Experimental stability of magnesium sulfate hydrates that may be present on Mars. *Geochimica et Cosmochimica Acta*, **71**, 241–250.

CHRISTENSEN, P. R., BANDFIELD, J. L. *ET AL.* 2001*a*. Mars global surveyor Thermal Emission Spectrometer experiment: investigation description and surface science results. *Journal of Geophysical Research – Planets*, **106**, 23 823–23 872.

CHRISTENSEN, P. R., MORRIS, R. V., LANE, M. D., BANDFIELD, J. L. & MALIN, M. C. 2001*b*. Global mapping of the martian hematite mineral deposits: remnants of water-driven processes on early Mars. *Journal of Geophysical Research*, **106**, 23 873–23 886.

CHRISTENSEN, P. R., JAKOSKY, B. M. *ET AL.* 2004. The Thermal Emission Imaging System (THEMIS) for the Mars 2001 Odyssey Mission. *Space Science Reviews*, **110**, 85–130.

CLARK, B. C., MORRIS, R. V. *ET AL.* 2005. Chemistry and mineralogy of outcrops at Meridiani Planum. *Earth and Planetary Science Letters*, **240**, 73–94.

EDGETT, K. S. & CHRISTENSEN, P. R. 1991. The particle size of Martian aeolian dunes. *Journal of Geophysical Research*, **96**, 22 765–22 776.

FUETEN, F., STESKY, R. M. & MACKINNON, P. 2005. Structural attitudes of large-scale layering in Valles Marineris, Mars, calculated from Mars Orbiter Laser Altimeter data and Mars Orbiter Camera imagery. *Icarus*, **175**, 68–77.

GENDRIN, A., MANGOLD, N. *ET AL.* 2005. Sulfates in Martian layered terrains: the OMEGA/Mars express view. *Science*, **307**, 1587–1591.

GLOTCH, T. D. & CHRISTENSEN, P. R. 2005. Geologic and mineralogic mapping of Aram Chaos: Evidence for a water-rich history. *Journal of Geophysical Research – Planets*, **110**, E09006, doi: 10.1029/2004JE002389.

GLOTCH, T. D. & ROGERS, A. D. 2007. Evidence for aqueous deposition of hematite- and sulfate-rich light-toned layered deposits in Aureum and Iani Chaos, Mars. *Journal of Geophysical Research – Planets*, **112**, E06001, doi: 10.1029/2006JE002863.

GOLOMBEK, M. P. & BRIDGES, N. T. 2000. Erosion rates on Mars and implications for climate change: constraints from the Pathfinder landing site. *Journal of Geophysical Research – Planets*, **105**, 1841–1854.

GREELEY, R., LANCASTER, N., LEE, S. & THOMAS, P. 1992. *Martian aeolian processes, sediments, and features*. *In*: KIEFFER, H. H., JACOSKY, B., SNYDER, C. & MATTHEWS, M. (eds) *Mars*. University of Arizona Press, Tucson, AZ, 730–766.

GWINNER, K., SCHOLTEN, F. *ET AL.* 2010. Topography of Mars from global mapping by HRSC high-resolution digital terrain models and orthoimages: characteristics and performance. *Earth and Planetary Science Letters*, **294**, 506–519, doi: 10.1016/j.epsl.2009.11.007.

HARTMANN, W. K. & NEUKUM, G. 2001. Cratering chronology and the evolution of Mars. *Space Science Reviews*, **96**, 165–194.

HEAD, J. W., GREELEY, R. *ET AL.* 2001. Geological processes and evolution. *Space Science Reviews*, **96**, 263–292.

JAKOSKY, B. M. 1986. On the thermal properties of Martian fines. *Icarus*, **66**, 117–124.

JAKOSKY, B. M., MELLON, M. T., KIEFFER, H. H., CHRISTENSEN, P. R., VARNES, E. S. & LEE, S. W. 2000. The thermal inertia of Mars from the Mars Global Surveyor Thermal Emission Spectrometer. *Journal of Geophysical Research (Planets)*, **105**, 9643–9652.

JAUMANN, R., NEUKUM, G. *ET AL.* & THE HRSC CO-INVESTIGATOR TEAM 2007. The high-resolution stereo camera (HRSC) experiment on Mars express: instrument aspects and experiment conduct from interplanetary cruise through the nominal mission. *Planetary and Space Science*, **55**, 928–952.

KARR, C., JR. , 1975. *Infrared and Raman Spectroscopy of Lunar and Terrestrial Minerals*. Academic Press, New York.

KNEISSL, T., VAN GASSELT, S. & NEUKUM, G. 2010. Measurement of strike and dip of geologic layers from remote sensing data – New software tool for ArcGIS. *In*: *Proceedings of the 41st Annual Lunar and Planetary Science Conference, March 1–5, 2010, The Woodlands, Texas*. Lunar and Planetary Institute, Houston, TX, sAbstract 1640.

LE DEIT, L., LE MOUÉLIC, S. *ET AL.* 2008. Ferric oxides in East Candor Chasma, Valles Marineris (Mars) inferred from analysis of OMEGA/Mars Express data: Identification and geological interpretation. *Journal of Geophysical Research – Planets*, **113**, E07001, doi: 10.1029/2007JE002950.

LICHTENBERG, K. A., ARVIDSON, R. E. *ET AL.* & THE CRISM SCIENCE TEAM. 2009. Stratigraphy and relationship of hydrated minerals in the Layered Deposits of Aram Chaos, Mars. *In*: *Proceedings of the 40th Annual Lunar and Planetary Science Conference, March 23–27, 2009, The Woodlands, Texas*.

Lunar and Planetary Institute, Houston, TX, Abstract 2326.

Lucchitta, B. K., McEwen, A. S. et al. 1992. *The Canyon System on Mars. In*: Kieffer, H. H., Jacosky, B., Snyder, C. & Matthews, M. (eds) *Mars*. University of Arizona Press, Tucson, AZ, 453–492.

Lucchitta, B. K., Isbell, N. K. & Howington-Kraus, A. 1994. Topography of Valles Marineris: Implications for erosional and structural history. *Journal of Geophysical Research – Planets*, **99**, 3783–3798.

Malin, M. C. & Edgett, K. S. 2000. Sedimentary rocks of Early Mars. *Science*, **290**, 1927–1937.

Malin, M. C., Danielson, G. E., Ingersoll, A. P., Masursky, H., Veverka, J., Ravine, M. A. & Soulanille, T. A. 1992. Mars observer camera. *Journal of Geophysical Research – Planets*, **97**, 7699–7718.

Malin, M. C., Bell, J. F., III. et al. 2007. Context Camera Investigation on board the Mars Reconnaissance Orbiter. *Journal of Geophysical Research*, **112**, E05S04, doi: 10.1029/2006JE002808.

Mangold, N., Gendrin, A. et al. 2008. Spectral and geological study of the sulfate-rich region of West Candor Chasma, Mars. *Icarus*, **194**, 519–543.

Marion, G. M. & Kargel, J. S. 2005. Stability of magnesium sulfate minerals in Martian environments. *In*: *Proceedings of the 36th Annual Lunar and Planetary Science Conference, March 14–18, 2005, League City, Texas*. Lunar and Planetary Institute, Houston, TX, Abstract 2290.

Matthes, S. 2001. Die sedimentäre Abfolge, Sedimente und Sedimentgesteine. *In*: Matthes, S. (ed.) *Mineralogie*. Springer-Verlag, Berlin, 301–349.

McCauley, J. F. 1978. Geologic Map of the coprates quadrangle of Mars. United States Geological Survey Miscellaneous Field Studies Map, **I-896**.

McEwen, A. S., Eliason, E. M. et al. 2007. Mars Reconnaissance Orbiter's High Resolution Imaging Science Experiment (HiRISE). *Journal of Geophysical Research – Planets*, **112**, E05S02, doi: 10.1029/2005JE002605.

Milliken, R. E., Grotzinger, J. P., Beyer, R. A., Murchie, S., McEwen, A. & The CRISM Science Team 2007. Evidence for salt tectonics in Valles Marineris, Mars. *In*: *Seventh International Conference on Mars, July 9–13, 2007, Pasadena, California*. Lunar and Planetary Institute, Houston, TX, Abstract 3383.

Murchie, S., Roach, L. et al. 2009. Evidence for the origin of layered deposits in Candor Chasma, Mars, from mineral composition and hydrologic modeling. *Journal of Geophysical Research – Planets*, **114**, E00D05, doi: 10.1029/2009JE003343.

Murchie, S. L., Arvidson, R. et al. 2007. Compact Reconnaissance Imaging Spectrometer for Mars (CRISM) on Mars Reconnaissance Orbiter (MRO). *Journal of Geophysical Research (Planets)*, **112**, E05S03, doi: 10.1029/2006JE002682.

Pelkey, S. M., Jakosky, B. M. & Mellon, M. T. 2001. Thermal inertia of crater-related wind streaks on Mars. *Journal of Geophysical Research – Planets*, **106**, 23 909–23 920.

Peterson, C. 1981. A secondary origin for the central plateau of Hebes Chasma. *In*: *Proceedings of the 12th Annual Lunar and Planetary Science Conference, March 16–20, 1981, Houston, Texas*. Lunar and Planetary Institute, Houston, TX, Abstract 1459.

Putzig, D. E., Mellon, M. T., Kretke, K. A. & Arvidson, R. E. 2005. Global thermal inertia and surface properties of Mars from the MGS mapping mission. *Icarus*, **173**, 325–341.

Roach, L. H., Mustard, J. F., Swayze, G. A., Milliken, R. E., Bishop, J. L., Murchie, S. L. & Lichtenberg, K. 2009. Hydrated mineral stratigraphy of Ius Chasma, Valles Marineris. *Icarus*, **206**, 253–268.

Roach, L. H., Mustard, J. F., Lane, M. D., Bishop, J. L. & Murchie, S. L. 2010. Diagenetic haematite and sulfate assemblages in Valles Marineris. *Icarus*, **207**, 659–674.

Rossi, A. P., Neukum, G. et al. 2008. Large-scale spring deposits on Mars? *Journal of Geophysical Research (Planets)*, **113**, E08016, doi: 10.1029/2007JE003062.

Schultz, R. A. 1998. Multiple-process origin of Valles Marineris basins and troughs, Mars. *Planetary and Space Science*, **46**, 827–834.

Scott, D. H. & Tanaka, K. L. 1986. *Geologic Map of the Western Equatorial Region of Mars*. United States Geological Survey Miscellaneous Field Studies Map, **I-1802-A**.

Sharp, R. P. 1973. Mars: fretted and chaotic terrains. *Journal of Geophysical Research – Planets*, **78**, 4073–4083.

Smith, D. E., Zuber, M. T. et al. 2001. Mars Orbiter Laser Altimeter: experiment summary after the first year of global mapping of Mars. *Journal of Geophysical Research – Planets*, **106**, 23 689–23 722.

Sowe, M. 2009. *Interior layered deposits in chaotic terrains on Mars*. PhD thesis, Freie Universitaet Berlin.

Sowe, M., Roach, L. H., Hauber, E., Jaumann, R., Mustard, J. L. & Neukum, G. 2008. Interior Layered Deposits on Mars: insights from elevation, image- and spectral data of Ganges Mensa. *In*: *Third European Planetary Science Congress (2008), Westfälische Wilhelms Universität, Schloßplatz, Münster, Germany, 21–26 September 2008*. Europlanet, Toulouse, Abstract 510.

Squyres, S. W., Arvidson, R. E. et al. 2004. The opportunity Rover's Athena Science Investigation at Meridiani Planum, Mars. *Science*, **306**, 1698–1703.

Vaniman, D. T., Bish, D. L., Chipera, S. J., Fialips, C. I., Carey, J. W. & Feldman, W. C. 2004. Magnesium sulfate salts and the history of water on Mars. *Nature*, **431** 663–665.

Warren, J. K. 2006. *Evaporites – Sediments, Resources and Hydrocarbons*. Springer, Berlin.

Wendt, L., Gross, C., McGuire, P. C., Combe, J. P. & Neukum, G. 2009. Analysis of Juventae Chasma sulfate Mound B using the Multiple-Endmember linear spectral unmixing Model (MELSUM) on CRISM data. *In*: *Proceedings of the 40th Annual Lunar and Planetary Science Conference, March 23–27, 2009, The Woodlands, Texas*. Lunar and Planetary Institute, Houston, TX, Abstract 1531.

Index

Page numbers in *italic* refer to Figures. Page numbers in **bold** refer to Tables.